T0331790

Nonparametric System Identification

Presenting a thorough overview of the theoretical foundations of nonparametric systems identification for nonlinear block-oriented systems, Włodzimierz Greblicki and Mirosław Pawlak show that nonparametric regression can be successfully applied to system identification, and they highlight what you can achieve in doing so.

Starting with the basic ideas behind nonparametric methods, various algorithms for nonlinear block-oriented systems of cascade and parallel forms are discussed in detail. Emphasis is placed on the most popular systems, Hammerstein and Wiener, which have applications in engineering, biology, and financial modeling.

Algorithms using trigonometric, Legendre, Laguerre, and Hermite series are investigated, and the kernel algorithm, its semirecursive versions, and fully recursive modifications are covered. The theories of modern nonparametric regression, approximation, and orthogonal expansions are also provided, as are new approaches to system identification. The authors show how to identify nonlinear subsystems so that their characteristics can be obtained even when little information exists, which is of particular significance for practical application. Detailed information about all the tools used is provided in the appendices.

This book is aimed at researchers and practitioners in systems theory, signal processing, and communications. It will also appeal to researchers in fields such as mechanics, economics, and biology, where experimental data are used to obtain models of systems.

Włodzimierz Greblicki is a professor at the Institute of Computer Engineering, Control, and Robotics at the Wrocław University of Technology, Poland.

Mirosław Pawlak is a professor in the Department of Electrical and Computer Engineering at the University of Manitoba, Canada. He was awarded his Ph.D. from the Wrocław University of Technology, Poland.

Both authors have published extensively over the years in the area of nonparametric theory and applications.

Nonparametric System Identification

WŁODZIMIERZ GREBLICKI
Wrocław University of Technology

MIROSŁAW PAWLAK
University of Manitoba, Canada

CAMBRIDGE
UNIVERSITY PRESS

Shaftesbury Road, Cambridge CB2 8EA, United Kingdom

One Liberty Plaza, 20th Floor, New York, NY 10006, USA

477 Williamstown Road, Port Melbourne, VIC 3207, Australia

314–321, 3rd Floor, Plot 3, Splendor Forum, Jasola District Centre, New Delhi – 110025, India

103 Penang Road, #05–06/07, Visioncrest Commercial, Singapore 238467

Cambridge University Press is part of Cambridge University Press & Assessment, a department of the University of Cambridge.

We share the University's mission to contribute to society through the pursuit of education, learning and research at the highest international levels of excellence.

www.cambridge.org
Information on this title: www.cambridge.org/9780521868044

First published 2008
First paperback edition 2012

A catalogue record for this publication is available from the British Library

Library of Congress Cataloging-in-Publication data
Greblicki, Włodzimierz.
Nonparametric system identification / W. Greblicki, M. Pawlak.
   p. cm.
Includes bibliographical references and index.
ISBN 978-0-521-86804-4 (hardcover)
1. System identification. 2. Nonparametric signal detection. 3. Signal processing – Mathematics.
4. Mathematical optimization. I. Pawlak, M. (Mirosław), 1954– II. Title.
QA402.G7315 2008
003´.1–dc22   2008014570

ISBN   978-0-521-86804-4   Hardback
ISBN   978-1-107-41062-6   Paperback

Contents

Preface   *page* ix

1   Introduction   1

2   Discrete-time Hammerstein systems   3
2.1   The system   3
2.2   Nonlinear subsystem   4
2.3   Dynamic subsystem identification   8
2.4   Bibliographic notes   9

3   Kernel algorithms   11
3.1   Motivation   11
3.2   Consistency   13
3.3   Applicable kernels   14
3.4   Convergence rate   16
3.5   The mean-squared error   21
3.6   Simulation example   21
3.7   Lemmas and proofs   24
3.8   Bibliographic notes   29

4   Semirecursive kernel algorithms   30
4.1   Introduction   30
4.2   Consistency and convergence rate   31
4.3   Simulation example   34
4.4   Proofs and lemmas   35
4.5   Bibliographic notes   43

5   Recursive kernel algorithms   44
5.1   Introduction   44
5.2   Relation to stochastic approximation   44
5.3   Consistency and convergence rate   46
5.4   Simulation example   49
5.5   Auxiliary results, lemmas, and proofs   51
5.6   Bibliographic notes   58

6   Orthogonal series algorithms   59
   6.1   Introduction   59
   6.2   Fourier series estimate   61
   6.3   Legendre series estimate   64
   6.4   Laguerre series estimate   66
   6.5   Hermite series estimate   68
   6.6   Wavelet estimate   69
   6.7   Local and global errors   70
   6.8   Simulation example   71
   6.9   Lemmas and proofs   72
   6.10   Bibliographic notes   78

7   Algorithms with ordered observations   80
   7.1   Introduction   80
   7.2   Kernel estimates   81
   7.3   Orthogonal series estimates   85
   7.4   Lemmas and proofs   89
   7.5   Bibliographic notes   99

8   Continuous-time Hammerstein systems   101
   8.1   Identification problem   101
   8.2   Kernel algorithm   103
   8.3   Orthogonal series algorithms   106
   8.4   Lemmas and proofs   108
   8.5   Bibliographic notes   112

9   Discrete-time Wiener systems   113
   9.1   The system   113
   9.2   Nonlinear subsystem   114
   9.3   Dynamic subsystem identification   119
   9.4   Lemmas   121
   9.5   Bibliographic notes   122

10   Kernel and orthogonal series algorithms   123
   10.1   Kernel algorithms   123
   10.2   Orthogonal series algorithms   126
   10.3   Simulation example   129
   10.4   Lemmas and proofs   130
   10.5   Bibliographic notes   142

11   Continuous-time Wiener system   143
   11.1   Identification problem   143
   11.2   Nonlinear subsystem   144
   11.3   Dynamic subsystem   146
   11.4   Lemmas   146
   11.5   Bibliographic notes   148

12   Other block-oriented nonlinear systems   149
   12.1   Series-parallel, block-oriented systems   149
   12.2   Block-oriented systems with nonlinear dynamics   173
   12.3   Concluding remarks   218
   12.4   Bibliographical notes   220

13   Multivariate nonlinear block-oriented systems   222
   13.1   Multivariate nonparametric regression   222
   13.2   Additive modeling and regression analysis   228
   13.3   Multivariate systems   242
   13.4   Concluding remarks   248
   13.5   Bibliographic notes   248

14   Semiparametric identification   250
   14.1   Introduction   250
   14.2   Semiparametric models   252
   14.3   Statistical inference for semiparametric models   255
   14.4   Statistical inference for semiparametric Wiener models   264
   14.5   Statistical inference for semiparametric Hammerstein models   286
   14.6   Statistical inference for semiparametric parallel models   287
   14.7   Direct estimators for semiparametric systems   290
   14.8   Concluding remarks   309
   14.9   Auxiliary results, lemmas, and proofs   310
   14.10   Bibliographical notes   316

A   Convolution and kernel functions   319
   A.1   Introduction   319
   A.2   Convergence   320
   A.3   Applications to probability   328
   A.4   Lemmas   329

B   Orthogonal functions   331
   B.1   Introduction   331
   B.2   Fourier series   333
   B.3   Legendre series   340
   B.4   Laguerre series   345
   B.5   Hermite series   351
   B.6   Wavelets   355

C   Probability and statistics   359
   C.1   White noise   359
   C.2   Convergence of random variables   361
   C.3   Stochastic approximation   364
   C.4   Order statistics   365
   References   371
   Index   387

To my wife, Helena, and my children, Jerzy, Maria, and Magdalena – WG
To my parents and family and those whom I love – MP

Preface

The aim of this book is to show that the nonparametric regression can be applied successfully to nonlinear system identification. It gathers what has been done in the area so far and presents main ideas, results, and some new recent developments.

The study of nonparametric regression estimation began with works published by Cencov, Watson, and Nadaraya in the 1960s. The history of nonparametric regression in system identification began about ten years later. Such methods have been applied to the identification of composite systems consisting of nonlinear memoryless systems and linear dynamic ones. Therefore, the approach is strictly connected with so-called block-oriented methods developed since Narendra and Gallman's work published in 1966. Hammerstein and Wiener structures are most popular and have received the greatest attention in numerous applications. Fundamental for nonparametric methods is the observation that the unknown characteristic of the nonlinear subsystem or its inverse can be represented as regression functions.

In terms of the a priori information, standard identification methods and algorithms work when it is parametric, that is, when our knowledge about the system is rather large; for example, when we know that the nonlinear subsystem has a polynomial characteristic. In this book, the information is much smaller, nonparametric. The mentioned characteristic can be, for example, any integrable or bounded or, even, any Borel function.

It can thus be said that this book associates block-oriented system identification with nonparametric regression estimation and shows how to identify nonlinear subsystems, that is, to recover their characteristics when the a priori information is small. Because of this, the approach should be of interest not only to researchers but also to people interested in applications.

Chapters 2–7 are devoted to discrete-time Hammerstein systems. Chapter 2 presents basic discussion of the Hammerstein system and its relationship with the concept of the nonparametric regression. The nonparametric kernel algorithm is presented in Chapter 3, its semirecursive versions are examined in Chapter 4, and Chapter 5 deals with fully recursive modifications derived from the idea of stochastic approximation. Next, Chapter 6 is concerned with the nonparametric orthogonal series method. Algorithms using trigonometric, Legendre, Laguerre, and Hermite series are investigated. Some space is devoted to estimation methods based on wavelets. Nonparametric algorithms based on ordered observations are presented and examined in Chapter 7. Chapter 8 discusses the nonparametric algorithms when applied to continuous-time Hammerstein systems.

The Wiener system is identified in Chapters 9–11. Chapter 9 presents the motivation for nonparametric algorithms that are studied in the next two chapters devoted to the discrete and continuous-time Wiener systems, respectively. Chapter 12 is concerned with the generalization of our theory to other block-oriented nonlinear systems. This includes, among others, parallel models, cascade-parallel models, sandwich models, and generalized Hammerstein systems possessing local memory. In Chapter 13, the multivariate versions of block-oriented systems are examined. The common problem of multivariate systems, that is, the curse of dimensionality, is cured by using low-dimensional approximations. With respect to this issue, models of the additive form are introduced and examined. In Chapter 14, we develop identification algorithms for a semiparametric class of block-oriented systems. Such systems are characterized by a mixture of finite dimensional parameters and nonparametric functions being typically a set of univariate functions.

The reader is encouraged to look into the appendices, in which fundamental information about tools used in the book is presented in detail. Appendix A is strictly related to kernel algorithms, and Appendix B is tied with the orthogonal series nonparametric curve estimates. Appendix C recalls some facts from probability theory and presents results from the theory of order statistics used extensively in Chapter 7.

Over the years, our work has benefited greatly from the advice and support of a number of friends and colleagues with interest in ideas of nonparametric estimation, pattern recognition, and nonlinear system modeling. There are too many names to list here, but special mention is due to Adam Krzyżak, as well as Danuta Rutkowska, Leszek Rutkowski, Alexander Georgiev, Simon Liao, Pradeepa Yahampath, and Yongqing Xin – our past Ph.D. students, now professors at universities in Canada, the United States, and Poland. Cooperation with them has been a great pleasure and given us a lot of satisfaction. We are deeply indebted to Zygmunt Hasiewicz, Ewaryst Rafajłowicz, Uli Stadtmüller, Ewa Rafajłowicz, Hajo Holzmann, and Andrzej Kozek, who have contributed greatly to our research in the area of nonlinear system identification, pattern recognition, and nonparametric inference.

Last, but by no means least, we would like to thank Mount-first Ng for helping us with a number of typesetting problems. Ed Shwedyk and January Gnitecki have provided support for correcting English grammar.

We also thank Anna Littlewood, from Cambridge University Press, for being a very supportive and patient editor. Research presented in this monograph was partially supported by research grants from Wrocław University of Technology, Wrocław, Poland, and NSERC of Canada.

Wrocław, Winnipeg   *Włodzimierz Greblicki, Mirosław Pawlak*
February 2008

1 Introduction

System identification, as a particular process of statistical inference, exploits two types of information. The first is experiment; the other, called a priori, is known before making any measurements. In a wide sense, the a priori information concerns the system itself and signals entering the system. Elements of the information are, for example:

- the nature of the signals, which may be random or nonrandom, white or correlated, stationary or not, their distributions can be known in full or partially (up to some parameters) or completely unknown,
- general information about the system, which can be, for example, continuous or discrete in the time domain, stationary or not,
- the structure of the system, which can be of the Hammerstein or Wiener type, or other,
- the knowledge about subsystems, that is, about nonlinear characteristics and linear dynamics.

In other words, the a priori information is related to the theory of the phenomena taking place in the system (a real physical process) or can be interpreted as a hypothesis (if so, results of the identification should be necessarily validated) or can be abstract in nature.

This book deals with systems consisting of nonlinear memoryless and linear dynamic subsystems, for example, Hammerstein and Wiener systems and other related structures. With respect to them, the a priori information is understood in a narrow sense because it relates to the subsystems only and concerns the a priori knowledge about their descriptions. We refer to such systems as block-oriented.

The characteristic of the nonlinear subsystem is recovered with the help of nonparametric regression estimates. The kernel and orthogonal series methods are used. Ordered statistics are also applied. Both offline and online algorithms are investigated. We examine only these estimation methods and nonlinear models for which we are able to deliver fundamental results in terms of consistency and convergence rates. There are other techniques, for example, neural networks, which may exhibit a promising performance but their statistical accuracy is mostly unknown.

For the theory of nonparametric regression, see Efromovich [78], Györfi, Kohler, Krzyżak, and Walk [140], Härdle [150], Prakasa Rao [241], Simonoff [278], or Wand and Jones [310]. Nonparametric wavelet estimates are discussed in Antoniadis and Oppenheim [6], Härdle, Kerkyacharian, Picard, and Tsybakov [151], Ogden [223], and Walter and Shen [308].

Parametric methods are beyond the scope of this book; nevertheless, we mention Brockwell and Davies [33], Ljung [198], Norton [221], Zhu [332], and Söderström and Stoica [280].

Nonlinear system identification within the parametric framework is studied by Nells [218], Westwick and Kearney [316], Marmarelis and Marmarelis [207], Bendat [16], and Mathews and Sicuranza [208]. These books present identification algorithms based mostly on the theory of Wiener and Volterra expansions of nonlinear systems. A comprehensive list of references concerning nonlinear system identification and applications has been given by Giannakis and Serpendin [102], see also the 2005 special issue on system identification of the IEEE Trans. on Automatic Control [199]. A nonparametric statistical inference for time series is presented in Bosq [26], Fan and Yao [89], and Györfi, Härdle, Sarda, and Vieu [139].

It should be stressed that nonparametric and parametric methods are supposed to be applied in different situations. The first are used when the a priori information is nonparametric, that is, when we wish to recover an infinite-dimensional object with underlying assumptions as weak as possible. Clearly, in such a case, parametric methods can only approximate, but not estimate, the unknown characteristics. When the information is parametric, parametric methods are the natural choice. If, however, the unknown characteristic is a complicated function of parameters convergence analysis becomes difficult. Moreover, serious computational problems can occur. In such circumstances, one can resort to nonparametric algorithms because, from the computational viewpoint, they are not discouraging. On the contrary, they are simple but consume computer memory, because, for example, kernel estimates require all data to be stored. Nevertheless it can be said that the two approaches do not compete with each other since they are designed to be applied in quite different situations. The situations differ from each other by the amount of the a priori information about the identified system. However, a compromise between these two separate worlds can be made by restricting a class of nonparametric models to those that consist of a finite dimensional parameter and nonlinear characteristics, which run through a nonparametric class of univariate functions. Such semiparametric models can be efficiently identified, and the theory of semiparametric identification is examined in this book. The methodology of semiparametric statistical inference is examined in Härdle, Müller, Sperlich, and Werwatz [152], Ruppert, Wand, and Carroll [259], and Yatchev [329].

For two number sequences a_n and b_n, $a_n = O(b_n)$ means that a_n/b_n is bounded in absolute value as $n \to \infty$. In particular, $a_n = O(1)$ denotes that a_n is bounded, that is, that $\sup_n |a_n| < \infty$. Writing $a_n \sim b_n$, we mean that a_n/b_n has a nonzero limit as $n \to \infty$.

Throughout the book, "almost everywhere" means "almost everywhere with respect to the Lebesgue measure," whereas "almost everywhere (μ)" means "almost everywhere with respect to the measure μ."

2 Discrete-time Hammerstein systems

In this chapter, we discuss some preliminary aspects of the discrete-time Hammerstein system. In Section 2.1 we form the input–output equations of the system. A fundamental relationship between the system nonlinearity and the nonparametric regression is established in Section 2.2. The use of the correlation theory for recovering the linear subsystem is discussed in Section 2.3.

2.1 The system

A Hammerstein system, shown in Figure 2.1, consists of a nonlinear memoryless subsystem with a characteristic $m(\bullet)$ followed by a linear dynamic one with an impulse response $\{\lambda_n\}$. The output signal W_n of the linear part is disturbed by Z_n and $Y_n = W_n + Z_n$ is the output of the whole system. Neither V_n nor W_n is available to measurement. Our goal is to identify the system, that is, to recover both $m(\bullet)$ and $\{\lambda_n\}$, from observations

$$(U_1, Y_1), (U_2, Y_2), \ldots, (U_n, Y_n), \ldots \tag{2.1}$$

taken at the input and output of the whole system.

Signals coming to the system, that is, the input $\{\ldots, U_{-1}, U_0, U_1, \ldots\}$ and disturbance $\{\ldots, Z_{-1}, Z_0, Z_1, \ldots\}$ are mutually independent stationary white random signals. The disturbance has zero mean and finite variance, that is, $E Z_n = 0$ and $\mathrm{var}\,[Z_n] = \sigma_Z^2 < \infty$.

Regarding the nonlinear subsystem, we assume that $m(\bullet)$ is a Borel measurable function. Therefore, V_n is a random variable. The dynamic subsystem is described by the state equation

$$\begin{cases} X_{n+1} = AX_n + bV_n \\ \quad W_n = c^T X_n, \end{cases} \tag{2.2}$$

where X_n is a state vector at time n, A is a matrix, b and c are vectors. Thus,

$$\lambda_n = \begin{cases} 0, & \text{for } n = 0, -1, -2, \ldots \\ c^T A^{n-1} b, & \text{for } n = 1, 2, 3, \ldots, \end{cases}$$

and

$$W_n = \sum_{i=-\infty}^{n} \lambda_{n-i} m(U_i). \tag{2.3}$$

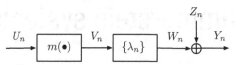

Figure 2.1 The discrete-time Hammerstein system.

Neither b nor c is known. The matrix A and its dimension are also unknown. Nevertheless, the matrix A is stable, all its eigenvalues lie in the unit circle. Therefore, assuming that

$$E m^2(U) < \infty, \qquad (2.4)$$

the time index at U is dropped, we conclude that both X_n as well as W_n are random variables. Clearly random processes $\{\ldots, X_{-1}, X_0, X_1, \ldots\}$ and $\{\ldots, W_{-1}, W_0, W_1, \ldots\}$ are stationary. Consequently, the output process $\{\ldots, Y_{-1}, Y_0, Y_1, \ldots\}$ is also a stationary stochastic process. Therefore, the problem is well posed in the sense that all signals are random variables. In the light of this, we estimate both $m(\bullet)$ and $\{\lambda_n\}$ from random observations (2.1).

The restrictions imposed on the signals entering the system and both subsystems apply whenever the Hammerstein system is concerned. They will not be repeated in further considerations, neither lemmas nor theorems.

Input random variables U_ns may have a probability density denoted by $f(\bullet)$ or may be distributed quite arbitrarily. Nevertheless (2.4) holds. It should be emphasized that, apart from few cases, (2.4) is the only restriction in which the nonlinearity is involved.

Assumption (2.4) is irrelevant to identification algorithms and has been imposed for only one reason: to guarantee that both W_n and Y_n are random variables. Nevertheless it certainly has an influence on the restrictions imposed on both $m(\bullet)$ and the distribution of U to meet (2.4). If, for example, U is bounded, (2.4) is satisfied for any $m(\bullet)$. The restriction also holds, if $E U^2 < \infty$ and $|m(u)| \le \alpha + \beta|u|$ with any α, β. In yet another example, $E U^4 < \infty$ and $|m(u)| \le \alpha + \beta u^2$. For Gaussian U and $|m(u)| \le W(u)$, where W is an arbitrary polynomial, (2.4) is also met. Anyway, the a priori information about the characteristic is nonparametric because $m(\bullet)$ cannot be represented in a parametric form. This is because the class of all possible characteristics is very wide.

The family of all stable dynamic subsystems also cannot be parameterized, because its order is unknown. Therefore, the a priori information about the impulse response is nonparametric, too. To form a conclusion we infer about both subsystems under nonparametric a priori information.

In the following chapters, for simplicity, U, W, Y, and Z stand for U_n, W_n, Y_n, and Z_n, respectively.

2.2   Nonlinear subsystem

2.2.1   The problem and the motivation for algorithms

Fix $p \ge 1$ and observe that, since $Y_p = Z_p + \sum_{i=-\infty}^{p} \lambda_{p-i} m(U_i)$ and $\{U_n\}$ is a white process,

$$E\{Y_p | U_0 = u\} = \mu(u),$$

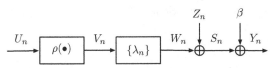

Figure 2.2 The equivalent Hammerstein system.

where

$$\mu(u) = \lambda_p m(u) + \alpha_p$$

with $\alpha_p = Em(U)\sum_{i=1,i\neq p}^{\infty}\lambda_i$. Estimating the regression $E\{Y_p|U_0 = u\}$, we thus recover $m(\bullet)$ up to some unknown constants λ_p and α_p. If $Em(U) = 0$, which is the case, for example, when the distribution of U is symmetrical with respect to zero and $m(\bullet)$ is an even function then $\alpha_p = 0$ and we estimate $m(\bullet)$ only up to the multiplicative constant λ_p.

Since $Y_{p+n} = \mu(U_n) + \xi_{p+n} + Z_{p+n}$ with $\xi_{p+n} = \sum_{i=-\infty,i\neq n}^{p+n}\lambda_{p+n-i}m(U_i)$, it can be said that we estimate $\mu(u)$ from pairs

$$(U_0, Y_p), (U_1, Y_{p+1}), \ldots, (U_n, Y_{p+n}), \ldots,$$

and that the regression $\mu(u)$ is corrupted by the noise $Z_{p+n} + \xi_{p+n}$. The first component of noise is white with zero mean. Because of dynamics the other noise component is correlated. Its mean $E\xi_n = \alpha_p$ is usually nonzero and the variance is equal to $\text{var}[m(U)]\sum_{i=1,i\neq p}^{\infty}\lambda_i^2$. Thus, main difficulties in the analysis of any estimate of $\mu(\bullet)$ are caused by the correlation of $\{\xi_n\}$, that is, the system itself but not by the white disturbance Z_n coming from outside.

Every algorithm estimating the nonlinearity in Hammerstein systems studied in this book, the estimate is denoted here as $\hat{\mu}(U_0, \ldots, U_n; Y_p, \ldots, Y_{p+n})$, is linear with respect to output observations, which means that

$$\hat{\mu}(U_0, \ldots, U_n; \theta_p + \eta_p, \ldots, \theta_{p+n} + \eta_{p+n})$$
$$= \hat{\mu}(U_0, \ldots, U_n; \theta_p, \ldots, \theta_{p+n}) + \hat{\mu}(U_0, \ldots, U_n; \eta_p, \ldots, \eta_{p+n}) \quad (2.5)$$

and has a natural property that, for any number θ,

$$\hat{\mu}(U_0, \ldots, U_n; \theta, \ldots, \theta) \to \theta \text{ as } n \to \infty \quad (2.6)$$

in an appropriate stochastic sense. This property, or rather its consequence, is exploited when proving consistency. To explain this, observe that with respect to U_n and Y_n, the identified system shown in Figure 2.1 is equivalent to that in Figure 2.2 with nonlinearity $\rho(u) = m(u) - Em(U)$ and an additional disturbance $\beta = Em(U)\sum_{i=1}^{\infty}\lambda_i$. In the equivalent system, $E\rho(U) = 0$ and $E\{Y_p|U_0 = u\} = \mu(u)$. From (2.5) and (2.6), it follows that

$$\hat{\mu}(U_0, \ldots, U_n; Y_p, \ldots, Y_{p+n}) = \hat{\mu}(U_0, \ldots, U_n; S_p + \beta, \ldots, S_{p+n} + \beta)$$
$$= \hat{\mu}(U_0, \ldots, U_n; S_p, \ldots, S_{p+n})$$
$$+ \hat{\mu}(U_0, \ldots, U_n; \beta, \ldots, \beta)$$

with $\hat{\mu}(U_0, \ldots, U_n; \beta, \ldots, \beta) \to \beta$ as $n \to \infty$. Hence, if

$$\hat{\mu}(U_0, \ldots, U_n; S_p, \ldots, S_{p+n}) \to E\{S_p | U_0 = u\}, \text{ as } n \to \infty,$$

we have

$$\hat{\mu}(U_0, \ldots, U_n; Y_p, \ldots, Y_{p+n}) \to E\{Y_p | U_0 = u\}, \text{ as } n \to \infty,$$

where convergence is understood in the same sense as that in (2.6).

Thus, if the estimate recovers the regression $E\{S_p | U_0 = u\}$ from observations

$$(U_0, S_p), (U_1, S_{1+p}), (U_2, S_{2+p}), \ldots,$$

it also recovers $E\{Y_p | U_0 = u\}$ from

$$(U_0, Y_p), (U_1, Y_{1+p}), (U_2, Y_{2+p}), \ldots.$$

We can say that if the estimate works properly when applied to the system with input U_n and output S_n (in which $E\rho(U) = 0$), it behaves properly also when applied to the system with input U_n and output Y_n (in which $Em(U)$ may be nonzero).

The result of the reasoning is given in the following remark:

REMARK 2.1 *Let an estimate have properties (2.5) and (2.6). If the estimate is consistent for $Em(U) = 0$, then it is consistent for $Em(U) \neq 0$, too.*

Owing to the remark, with no loss of generality, in all proofs of consistency of algorithms recovering the nonlinearity, we assume that $Em(U) = 0$.

In parametric problems the nonlinearity is usually a polynomial $m(u) = \alpha_0 + \alpha_1 u + \cdots + \alpha_q u^q$ of a fixed degree with unknown true values of parameters $\alpha_0, \ldots, \alpha_q$. Therefore, to apply parametric methods, we must have a great deal more a priori information about the subsystem. It seems that in many applications, it is impossible to represent $m(\bullet)$ in a parametric form.

Since the system with the following $ARMA$ type difference equation:

$$w_n + a_{k-1} w_{n-1} + \cdots + a_0 w_{n-k} = b_{k-1} m(u_{n-1}) + \cdots + b_0 m(u_{n-k})$$

can be described by (2.2), all presented methods can be used to recover the nonlinearity $m(\bullet)$ in the previous $ARMA$ system.

It will be convenient to denote

$$\phi(u) = E\left\{ W_p^2 | U_0 = u \right\}. \tag{2.7}$$

Since $W_p = \sum_{i=-\infty}^{p-1} \lambda_i m(U_i)$, denoting $c_0 = Em^2(U) \sum_{i=1, i \neq p}^{\infty} \lambda_i^2 + E^2 m(U) (\sum_{i=1, i \neq p}^{\infty} \lambda_i)^2$, $c_1 = 2\lambda_p Em(U) \sum_{i=1, i \neq p}^{\infty} \lambda_i$, and $c_2 = \lambda_p^2$, we find

$$\phi(u) = c_0 + c_1 m(u) + c_2 m^2(u).$$

To avoid complicated notation, we do not denote explicitly the dependence of the estimated regression and other functions on p and simply write $\mu(\bullet)$ and $\phi(\bullet)$.

Results presented in further chapters can be easily generalized on the system shown in Figure 2.3, where $\{\ldots, \xi_0, \xi_1, \xi_2, \ldots\}$ is another zero mean noise. Moreover, $\{Z_n\}$ can

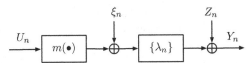

Figure 2.3 Possible generalization of the system shown in Figure 2.1.

be correlated, that is, it can be the output of a stable linear dynamic system stimulated by white random noise. So can $\{\xi_n\}$.

It is worth noting that a class of stochastic processes generated by the output process $\{Y_n\}$ of the Hammerstein system is different from the class of strong mixing processes considered extensively in the statistical literature concerning the nonparametric inference from dependent data, see, for example, [26] and [89]. Indeed, the ARMA process $\{X_n\}$ in which $X_{n+1} = aX_n + V_n$, where $0 < a \le 1/2$, and where V_ns are Bernoulli random variables is not strong mixing, see [4] and [5]. Such a process can be easily generated by the Hammerstein system if the input of the whole system has a normal density and a nonlinear characteristic takes two different values. In the light of that, the strong mixing approach developed in the statistical literature does not apply in general to the identification problem of nonlinear systems.

2.2.2 Simulation example

In the chapters devoted to the Hammerstein system, the behavior of the identification algorithms presented in this book is illustrated with results of simulation examples. In all examples, the system is described by the following scalar equation:

$$X_{n+1} = aX_n + m(U_n),$$

where

$$m(u) = (1 - e^{-|u|})\,\mathrm{sign}(u),$$

(see Figure 2.4). The input signal has a normal density with zero mean and variance 1. In all algorithms, $p = 1$, which means that $\mu(u) = m(u)$. For $a = 0.5$, an example of a

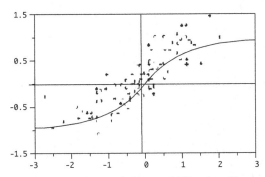

Figure 2.4 The characteristic m and 200 pairs of input–output observations; $a = 0.5$.

cloud of 200 input–output observations, we infer from is presented in Figure 2.4. The quality of each estimate, denoted here by $\hat{m}(u)$, is measured with

$$\text{MISE} = \int_{-3}^{3} (\hat{m}(u) - m(u))^2 du.$$

2.3   Dynamic subsystem identification

Passing to the dynamic subsystem, we use (2.3) and recall $EZ_n = 0$ to notice $E\{Y_iU_0\} = \sum_{j=-\infty}^{i} \lambda_{i-j} E\{m(U_i)U_0\} = \lambda_i E\{m(U)U\}$. Denoting $\kappa_i = \lambda_i E\{Um(U)\}$, we obtain

$$\kappa_i = E\{Y_iU_0\},$$

which can be estimated in the following way:

$$\hat{\kappa}_i = \frac{1}{n} \sum_{j=1}^{n-i} Y_{i+j}U_j.$$

THEOREM 2.1 *For any i,*

$$\lim_{n \to \infty} E(\hat{\kappa}_i - \kappa_i)^2 = 0.$$

Proof. The estimate is unbiased, that is, $E\hat{\kappa}_i = E\{Y_iU_0\} = \kappa_i$. Moreover, $\text{var}[\hat{\kappa}_i] = P_n + Q_n + R_n$ with

$$P_n = \frac{1}{n^2} \text{var}\left[\sum_{j=1}^{n} Z_{i+j}U_j\right] = \frac{1}{n^2} \sum_{j=1}^{n} \text{var}\left[Z_{i+j}U_j\right] = \frac{1}{n}\sigma_Z^2 EU^2,$$

$$Q_n = \frac{1}{n} \text{var}\left[W_iU_0\right],$$

and

$$R_n = \frac{1}{n^2} \sum_{j=1}^{n} \sum_{j=1, j\neq i}^{n} \text{cov}\left[W_{i+j}U_j, W_{i+m}U_m\right]$$

$$= \frac{1}{n^2} \sum_{j=1}^{n} (n-j) \text{cov}\left[W_{i+j}U_j, W_iU_0\right].$$

Since $W_i = \sum_{j=-\infty}^{i} \lambda_{i-j}m(U_j)$, $Q_n = n^{-1}\lambda_i^2 \text{var}\left[m(U)U\right]$. For the same reason, for $j > 0$,

$$\text{cov}\left[W_{i+j}U_j, W_iU_0\right] = \sum_{p=-\infty}^{i+j} \sum_{q=-\infty}^{i} \lambda_{i+j-p}\lambda_{i-q} \text{cov}\left[m(U_p)U_j, m(U_q)U_0\right]$$

$$= E^2\{Um(U)\}\lambda_{i+j}\lambda_{i-j}$$

See Lemma C.3 in Appendix C, which leads to

$$|R_n| \leq \frac{1}{n^2} E^2\{Um(U)\} \sum_{j=1}^{n} (n-j)|\lambda_{i+j}\lambda_{i-j}| \leq \frac{1}{n} E^2\{Um(U)\} \max_s |\lambda_s| \sum_{j=1}^{\infty} |\lambda_j|.$$

Thus,

$$E(\hat{\kappa}_i - \kappa_i)^2 = \text{var}\,[\hat{\kappa}_i] = O\left(\frac{1}{n}\right) \tag{2.8}$$

which completes the proof. ∎

The theorem establishes convergence of the local error $E(\hat{\kappa}_i - \kappa_i)^2$ to zero as $n \to \infty$. As an estimate of the whole impulse response $\{\kappa_1, \kappa_2, \kappa_3, \ldots\}$, we take a sequence $\{\hat{\kappa}_1, \hat{\kappa}_2, \hat{\kappa}_3, \ldots, \hat{\kappa}_{N(n)}, 0, 0, \ldots\}$ and find the mean summed square error (MSSE) is equal to

$$\text{MSSE}(\hat{\kappa}) = \sum_{i=1}^{N(n)} E(\hat{\kappa}_i - \kappa_i)^2 + \sum_{i=N(n)+1}^{\infty} \kappa_i^2.$$

From (2.8), it follows that the error is not greater than

$$O\left(\frac{N(n)}{n}\right) + \sum_{i=N(n)+1}^{\infty} \kappa_i^2.$$

Therefore, if $N(n) \to \infty$ as $n \to \infty$ and $N(n)/n \to 0$ as $n \to \infty$,

$$\lim_{n \to \infty} \text{MSSE}(\hat{\kappa}) = 0.$$

The identity $\lambda_s \tau = E\{Y_s U_0\}$, where $\tau = E\{Um(U)\}$, allows us to form a nonparametric estimate of the linear subsystem in the frequency domain. Indeed, formation of the Fourier transform of the identity yields

$$\Lambda(\omega)\tau = S_{YU}(\omega), \quad |\omega| \leq \pi, \tag{2.9}$$

where $S_{YU}(\omega) = \sum_{s=-\infty}^{\infty} \kappa_s e^{-is\omega}$ is the cross-spectral density function of the processes $\{Y_n\}$ and $\{U_n\}$. Moreover,

$$\Lambda(\omega) = \sum_{s=0}^{\infty} \lambda_s e^{-is\omega}$$

is the transfer function of the linear subsystem. Note also that if $\lambda_0 = 1$, then $\tau = \kappa_0$. See Chapter 12 for further discussion on the frequency domain identification of linear systems.

2.4 Bibliographic notes

Various aspects of parametric identification algorithms of discrete-time Hammerstein systems have been studied by Narendra and Gallman [216]; Haist, Chang, and Luus

[142], Thatchachar and Ramaswamy [289], Kaminskas [175], Gallman [92], Billings [19], Billings and Fakhouri [20,24], Shih and Kung [276], Kung and Shih [190], Liao and Sethares [195], Verhaegen and Westwick [301], Giri, Chaoui, and Rochidi [103], Ninness and Gibson [220], Bai [11,12], and Vörös [305]. The analysis of block–oriented systems and, in particular, Hammerstein ones, useful for various aspects of identification and its applications can be found in Bendat [16], Chen [45], Marmarelis and Marmarelis [207], Mathews and Sicuranza [208], Nells [218], and Westwick and Kearney [316].

Sometimes results concerning Hammerstein systems are given, however not explicitly, in works devoted to more complicated Hammerstein–Wiener or Wiener–Hammerstein structures, see, for example, Gardiner [94], Billings and Fakhouri [22, 23], Fakhouri, Billlings, and Wormald [86], Hunter and Korenberg [168], Korenberg and Hunter [177], Emara-ShaBaik, Moustafa, and Talaq [79], Boutayeb and Darouach [27], Vandersteen, Rolain, and Schoukens [296], Bai [10], Bershad, Celka, and McLaughlin [18], and Zhu [333].

The nonparametric approach offers a number of algorithms to recover the characteristics of the nonlinear subsystem. The most popular kernel estimate can be used in the offline version, see Chapter 3. For semirecursive and fully recursive forms, see Chapter 4 and Chapter 5, respectively. Nonparametric orthogonal series identification algorithms, see Chapter 6, utilize trigonometric, Legendre, Laguerre, Hermite functions or wavelets. Both classes of estimates can be modified to use ordered input observations (see Chapter 7), which makes them insensitive to the roughness of the input density.

The Hammerstein model has been used in various and diverse areas. Eskinat, Johnson, and Luyben [82] applied it to describe processes in distillation columns and heat exchangers. The hysteresis phenomenon in ferrites was analyzed by Hsu and Ngo [166], pH processes were analyzed by Patwardhan, Lakshminarayanan, and Shah [227], biological systems were studied by Hunter and Korenberg [168], and Emerson, Korenberg, and Citron [80] described some neuronal processes. The use of the Hammerstein model for modeling aspects of financial volatility processes is presented in Capobianco [38]. In Giannakis and Serpendin [102] a comprehensive bibliography on nonlinear system identification is given, see also the 2005 special issue on system identification of the IEEE Trans. on Automatic Control [199].

It is also worth noting that the concept of the Hammerstein model originates from the theory of nonlinear integral equations developed by Hammerstein in 1930 [148], see also Tricomi [292].

3   Kernel algorithms

The kernel algorithm is just the kernel estimate of a regression function. This is the most popular nonparametric estimation method and is very convenient from the computational viewpoint. In Section 3.1, an intuitive motivation for the algorithm is presented and in Section 3.2, its pointwise consistency is shown. Some results hold for any input signal density, that is, are density-free; some are even distribution-free, that is, they hold for any distribution of the input signal. In Section 3.3, the attention is focused on a class of applicable kernel functions. The convergence rate is studied in Section 3.4.

3.1   Motivation

It is obvious that

$$\lim_{h \to 0} \frac{1}{2h} \int_{u-h}^{u+h} \mu(v) f(v) dv = \mu(u) f(u)$$

at every continuity point $u \in R$ of both $m(\bullet)$ and $f(\bullet)$, since $\mu(u) = \lambda_p m(u) + \alpha_p$. Because the formula can be rewritten in the following form:

$$\lim_{h \to 0} \int \mu(v) f(v) \frac{1}{h} K \left(\frac{u - v}{h} \right) dv = \mu(u) f(u), \tag{3.1}$$

where

$$K(u) = \begin{cases} \dfrac{1}{2}, & \text{for } |u| < 1 \\ 0, & \text{otherwise,} \end{cases} \tag{3.2}$$

is the rectangular, sometimes called the window kernel (see Figure 3.1), one can expect that the convergence holds also for other kernel functions. This expectation is justified by the fact that for a suitably selected $K(\bullet)$,

$$\frac{1}{h} K \left(\frac{u - v}{h} \right)$$

gets close to the Dirac impulse $\delta(\bullet)$ located at the point u and that

$$\int \mu(v) f(v) \frac{1}{h} K \left(\frac{u - v}{h} \right) dv$$

Figure 3.1 Rectangular kernel (3.2).

converges to $\int \mu(v)f(v)\delta(u - v)dv = \mu(u)f(u)$ as $h \to 0$.

Because $\mu(u) = E\{Y_p|U_0 = u\}$, we get

$$\int \mu(u)f(v)\frac{1}{h}K\left(\frac{u-v}{h}\right)dv = \frac{1}{h}\int E\{Y_p|U_0 = v\}K\left(\frac{u-v}{h}\right)f(v)dv$$
$$= \frac{1}{h}E\left\{Y_pK\left(\frac{u - U_0}{h}\right)\right\},$$

which suggests the following estimate of $\mu(u)f(u)$:

$$\frac{1}{nh}\sum_{i=1}^{n}Y_{p+i}K\left(\frac{u - U_i}{h}\right).$$

For similar reasons,

$$\frac{1}{nh}\sum_{i=1}^{n}K\left(\frac{u - U_i}{h}\right)$$

is a good candidate for an estimate of

$$\int f(v)\frac{1}{h}K\left(\frac{u-v}{h}\right)dv,$$

which converges to $f(u)$ as $h \to 0$. Thus,

$$\hat{\mu}(u) = \frac{\displaystyle\sum_{i=1}^{n}Y_{p+i}K\left(\frac{u - U_i}{h_n}\right)}{\displaystyle\sum_{i=1}^{n}K\left(\frac{u - U_i}{h_n}\right)} \tag{3.3}$$

with h_n tending to zero, is a kernel estimate of $\mu(u)$. The parameter h_n is called a bandwidth. Note that the above formula is of the ratio form and we always treat the case $0/0$ as 0.

In light of this, crucial problems are the choice of the kernel $K(\bullet)$ and the number sequence $\{h_n\}$. From now on, we denote $g(u) = \mu(u)f(u)$.

It is worth mentioning that there is a wide range of kernel estimates [88, 140, 172] available for finding a curve in data. The most prominent are: the classical Nadaraya–Watson estimator, defined in (3.3), local linear and polynomial kernel estimates, convolution type kernel estimates, and various recursive kernel methods. Some of these techniques are thoroughly examined in this book.

3.2   Consistency

On the kernel function, the following restrictions are imposed:

$$\sup_{-\infty < u < \infty} |K(u)| < \infty, \tag{3.4}$$

$$\int |K(u)| du < \infty, \tag{3.5}$$

$$|u|^{1+\varepsilon} K(u) \to 0 \quad \text{as } |u| \to \infty, \tag{3.6}$$

where the parameter $\varepsilon \geq 0$ controls the tail decay of the kernel function. The sequence $\{h_n\}$ of positive numbers is such that

$$h_n \to 0 \quad \text{as } n \to \infty, \tag{3.7}$$

$$n h_n \to \infty \quad \text{as } n \to \infty. \tag{3.8}$$

Convergence results presented in Theorems 3.1 and 3.2 are density-free because the density $f(\bullet)$ of the input signal can be of any shape. The proof is given is Section 3.7.1.

THEOREM 3.1 *Let U have a density $f(\bullet)$ and let $E m^2(U) < \infty$. Let the Borel measurable kernel $K(\bullet)$ satisfy (3.4), (3.5), and (3.6) with $\varepsilon = 0$. Let the sequence $\{h_n\}$ satisfy (3.7) and (3.8). Then,*

$$\hat{\mu}(u) \to \mu(u) \text{ as } n \to \infty \text{ in probability} \tag{3.9}$$

at every $u \in R$ where both $m(\bullet)$ and $f(\bullet)$ are continuous and $f(u) > 0$.

Taking Lemma A.2 into account and arguing as in the proof of Theorem 3.1, we easily obtain the result given in the following remark:

REMARK 3.1 *Let U have a probability density $f(\bullet)$ such that $\sup_u |f(u)| < \infty$. Let $\sup_u |m(u)| < \infty$. Let the Borel measurable kernel satisfy (3.4) and (3.5). If, moreover, (3.7) and (3.8) hold, then convergence (3.9) takes place at every u where both $m(\bullet)$ and $f(\bullet)$ are continuous and $f(u) > 0$.*

The next theorem is the "almost everywhere" version of Theorem 3.1. The restriction imposed on the kernel and number sequence are the same as in Theorem 3.1 with the only exception that (3.6) holds with some $\varepsilon > 0$ but not with $\varepsilon = 0$.

THEOREM 3.2 *Let U have a probability density $f(\bullet)$ and let $E m^2(U) < \infty$. Let the Borel measurable satisfy (3.4), (3.5), and (3.6) with some $\varepsilon > 0$. Let the sequence $\{h_n\}$ of positive numbers satisfy (3.7) and (3.8). Then, convergence (3.9) takes place at every Lebesgue point $u \in R$ of both $m(\bullet)$ and $f(\bullet)$, where $f(u) > 0$, and, a fortiori, at almost every u where $f(u) > 0$, that is, at almost every u belonging to support of $f(\bullet)$.*

Proof. The proof is very much like that of Theorem 3.1. The difference is that we apply Lemma A.9 rather than Lemma A.8. ∎

The algorithm converges also when the input signal has not a density, when the distribution of U is of any shape. The proof of the theorem is in Section 3.7.1.

THEOREM 3.3 *Let $Em^2(U) < \infty$. Let $H(\bullet)$ be a nonnegative nonincreasing Borel function defined on $[0, \infty)$, continuous and positive at $t = 0$ and such that*

$$tH(t) \to 0 \ as \ t \to \infty.$$

Let, for some c_1 and c_2,

$$c_1 H(|u|) \leq K(u) \leq c_2 H(|u|).$$

Let the sequence $\{h_n\}$ of positive numbers satisfy (3.7) and (3.8). Then convergence (3.9) takes place at almost every (ζ) $u \in R$, where ζ is the probability measure of U.

Restrictions (3.7) and (3.8) are satisfied by a wide class of number sequences. If $h_n = cn^{-\delta}$ with $c > 0$, they are satisfied for $0 < \delta < 1$. The problem of kernel selection is discussed in Section 3.3.

3.3   Applicable kernels

In Theorems 3.1 and 3.2, U has a probability density denoted by $f(\bullet)$. The first theorem establishes convergence at every u where $m(\bullet)$ and $f(\bullet)$ are continuous and, moreover, $f(u) > 0$. The other does it for every Lebesgue point of both $m(\bullet)$ and $f(\bullet)$, that is, for almost every (with respect to the Lebesgue measure) u where $f(u) > 0$, that is, at almost every (ζ) point. In Theorem 3.3 the kernel satisfies restrictions (3.4), (3.5), and (3.6) with $\varepsilon = 0$. In Theorems 3.1 and 3.2, (3.6) holds with $\varepsilon > 0$.

If both $m(\bullet)$ and $f(\bullet)$ are bounded and continuous, we can apply kernels satisfying only (3.4) and (3.5), see Remark 3.1. In Theorem 3.3, U has an arbitrary distribution, which means that it may not have a density.

In the light of this to achieve convergence at Lebesgue points and, a fortiori, continuity points, we can apply the following kernel functions:

- the rectangular kernel (3.2),
- the triangle kernel

$$K(u) = \begin{cases} 1 - |u|, & \text{for } |u| < 1 \\ 0, & \text{otherwise,} \end{cases}$$

- the parabolic kernel

$$K(u) = \begin{cases} \dfrac{3}{4} \left(1 - u^2\right), & \text{for } |u| < 1 \\ 0, & \text{otherwise,} \end{cases}$$

- the Gauss–Weierstrass kernel (see Figure 3.2)

$$K(u) = \frac{1}{\sqrt{2\pi}} e^{-u^2/2}, \tag{3.10}$$

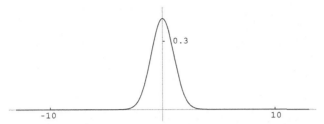

Figure 3.2 Gauss–Weierstrass kernel (3.10).

- the Poisson kernel

$$K(u) = \frac{1}{\pi} \frac{1}{1 + u^2},$$

- the Fejér kernel (see Figure 3.3)

$$K(u) = \frac{1}{\pi} \frac{\sin^2 u}{u^2}, \tag{3.11}$$

- the Lebesgue kernel

$$K(u) = \frac{1}{2} e^{-|u|}.$$

All these kernels satisfy (3.4), (3.5), and (3.6) for some $\varepsilon > 0$. The kernel

$$K(u) = \begin{cases} \dfrac{1}{4e}, & \text{for } |u| \le e \\[2mm] \dfrac{1}{4|u| \ln^2 |u|}, & \text{otherwise,} \end{cases} \tag{3.12}$$

satisfies (3.4), (3.5), and (3.6) with $\varepsilon = 0$ only. In turn, kernels

$$K(u) = \frac{1}{\pi} \frac{\sin u}{u}, \tag{3.13}$$

(see Figure 3.4) and

$$K(u) = \sqrt{\frac{2}{\pi}} \cos u^2, \tag{3.14}$$

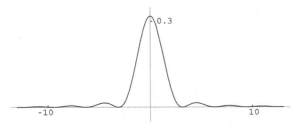

Figure 3.3 Fejér kernel (3.11).

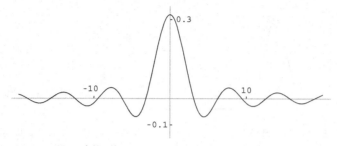

Figure 3.4 Kernel (3.13).

(see Figure 3.5), satisfy (3.4) and (3.5), but not (3.6), even with $\varepsilon = 0$. For all presented kernels, $\int K(u)du = 1$. Observe that they can be continuous or not and can have compact or unbounded support.

Notice that Theorem 3.3 admits the following one:

$$K(u) = \begin{cases} \dfrac{1}{e}, & \text{for } |u| \le e \\ \dfrac{1}{|u| \ln |u|}, & \text{otherwise}, \end{cases}$$

for which $\int K(u)du = \infty$. Restrictions imposed by the theorem are illustrated in Figure 3.6.

3.4   Convergence rate

In this section, both the characteristic $m(\bullet)$ and an input density $f(\bullet)$ are smooth functions and have q derivatives. Proper selection of the kernel and number sequence increases the speed where the estimate converges. We now find the convergence rate.

In our analysis, the kernel satisfies the following additional restrictions:

$$\int v^i K(v)dv = 0, \quad \text{for } i = 1, 2, \ldots, q - 1, \tag{3.15}$$

and

$$\int |v^{q-1/2} K(v)|dv < \infty, \tag{3.16}$$

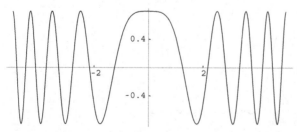

Figure 3.5 Kernel (3.14).

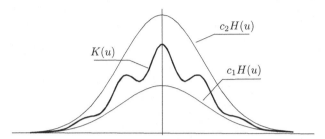

Figure 3.6 A kernel satisfying restrictions of Theorem 3.3.

see the analysis in Section A.2.2. For simplicity of notation, $\int K(v)dv = 1$. For a fixed u, we get

$$E\hat{f}(u) = \frac{1}{h_n} \int f(v)K\left(\frac{u-v}{h_n}\right) dv = \int f(u+vh_n)K(-v)dv,$$

which yields

$$\text{bias}[\hat{f}(u)] = E\hat{f}(u) - f(u) = \int (f(u+vh_n) - f(u))K(-v)dv.$$

Assuming that $f^{(q)}(\bullet)$ is square integrable and applying (A.17), we find $\text{bias}[\hat{f}(u)] = O(h_n^{q-1/2})$. We next recall (3.27) and write $\text{var}[\hat{f}(u)] = O(1/nh_n)$, which leads to

$$E(\hat{f}(u) - f(u))^2 = O(h_n^{2q-1}) + O\left(\frac{1}{nh_n}\right).$$

Thus, selecting

$$h_n \sim n^{-1/2q}, \tag{3.17}$$

we finally obtain

$$E(\hat{f}(u) - f(u))^2 = O(n^{-1+1/2q}).$$

Needless to say that if the qth derivative of $g(u)$ is square integrable, for the same reasons, $E(\hat{g}(u) - g(u))^2$ is of the same order. Hence, applying Lemma C.9, we finally obtain the following convergence rate:

$$P\{|\hat{\mu}(u) - \mu(u)| > \varepsilon|\mu(u)|\} = O(n^{-1+1/2q})$$

for any $\varepsilon > 0$, and

$$|\hat{\mu}(u) - \mu(u)| = O(n^{-1/2+1/4q}) \text{ as } n \to \infty \text{ in probability.}$$

If $f^{(q)}(u)$ is bounded, $\text{bias}[\hat{f}(u)] = O(h_n^q)$, see (A.18); and, for

$$h_n \sim n^{-1/(2q+1)},$$

$$E(\hat{f}(u) - f(u))^2 = O(n^{-1+1/(2q+1)}).$$

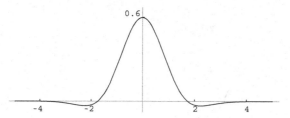

Figure 3.7 Kernel G_4.

If, in addition, the qth derivative of $g(u)$ is bounded, $E(\hat{g}(u) - g(u))^2$ is of the same order and, as a consequence,

$$P\{|\hat{\mu}(u) - \mu(u)| > \varepsilon|\mu(u)|\} = O(n^{-1+1/(2q+1)})$$

for any $\varepsilon > 0$, and

$$|\hat{\mu}(u) - \mu(u)| = O(n^{-1/2+1/(4q+2)}) \text{ as } n \to \infty \text{ in probability,}$$

which means that the rate is slightly better.

The rate $O(n^{-q/(2q+1)})$ in probability, obtained above is known to be optimal within the class of q differentiable input densities and nonlinear characteristics, see [285].

It is not difficult to construct kernels satisfying (3.15) such that $\int K(v)dv = 1$. For example, starting from the Gauss–Weierstrass kernel (3.10) denoted now as $G(\bullet)$ we observe that $\int u^i G(u)du = 0$ for odd i, and $\int u^i G(u)du = 1 \times 3 \times \cdots \times (i-1)$ for even i. Thus, for

$$G_2(u) = G(u) = \frac{1}{\sqrt{2\pi}}e^{-u^2/2},$$

(3.15) is satisfied for $q = 2$. For the same reasons, for

$$G_4(u) = \frac{1}{2}(3 - u^2)G(u) = \frac{1}{2\sqrt{2\pi}}(3 - u^2)e^{-u^2/2}, \tag{3.18}$$

(see Figure 3.7), and

$$G_6(u) = \frac{1}{8}(15 - 10u^2 + u^4)G(u) = \frac{1}{8\sqrt{2\pi}}(15 - 10u^2 + u^4)e^{-u^2/2}$$

(3.15) hold for $q = 4$ and $q = 6$, respectively.

In turn, for rectangle kernel (3.2) denoted now as $W(\bullet)$, $\int u^i W(u)du$ equals zero for odd i and $1/(i+1)$ for even i. Thus for $W_2(u) = W(u)$, (3.15) holds with $q = 2$, while for

$$W_4(u) = \frac{1}{4}(9 - 15u^2)W(u) = \begin{cases} \frac{1}{8}(9 - 15u^2), & \text{for } |u| \le 1 \\ 0, & \text{otherwise,} \end{cases} \tag{3.19}$$

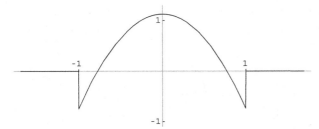

Figure 3.8 Kernel W_4.

with $q = 4$. For $q = 6$, we find

$$W_6(u) = \frac{5}{64}(45 - 210u^2 + 189u^4)W(u) \tag{3.20}$$

$$= \begin{cases} \dfrac{5}{128}\left(45 - 210u^2 + 189u^4\right), & \text{for } |u| \leq 1 \\ 0, & \text{otherwise.} \end{cases}$$

Kernels $W_4(u)$ and $W_6(u)$ are shown in Figures 3.8 and 3.9, respectively.

There is a formal way of generating kernel functions satisfying Conditions (3.15) and (3.16) for an arbitrary value of q. This technique relies on the theory of orthogonal polynomials that is examined in Chapter 6. In particular, if one wishes to obtain kernels defined on a compact interval then we can use a class of Legendre orthogonal polynomials, see Section 6.3 for various properties of this class. Hence, let $\{p_\ell(u); 0 \leq \ell \leq \infty\}$ be a set of the orthonormal Legendre polynomials defined on $[-1, 1]$, that is, $\int_{-1}^{1} p_\ell(u)p_j(u)du = \delta_{\ell j}$, $\delta_{\ell j}$ being the Kronecker delta function and $p_\ell(u) = \sqrt{\frac{2\ell+1}{2}} P_\ell(u)$, where $P_\ell(u)$ is the ℓth order Legendre polynomial.

The following lemma describes the procedure for generation of a kernel function of order q with a support defined on $[-1, 1]$.

LEMMA 3.1 *A kernel function*

$$K(u) = \sum_{j=0}^{q-1} p_j(0)p_j(u), \quad |u| \leq 1 \tag{3.21}$$

satisfies Condition (3.15).

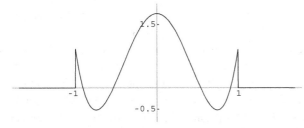

Figure 3.9 Kernel W_6.

Proof. For $i \leq q - 1$ consider $\int_{-1}^{1} u^i K(u) du$. Since u^i can be expanded into the Legendre series, that is, $u^i = \sum_{\ell=0}^{i} a_\ell p_\ell(u)$, where $a_\ell = \int_{-1}^{1} u^i p_\ell(u) du$ then for $K(u)$ defined in (3.21), we have

$$\int_{-1}^{1} u^i K(u) dv = \sum_{\ell=0}^{i} \sum_{j=0}^{q-1} a_\ell p_j(0) \int_{-1}^{1} p_\ell(u) p_j(u) du$$

$$= \sum_{\ell=0}^{i} a_\ell p_\ell(0) = 0^i = \begin{cases} 1 & \text{if } i = 0 \\ 0 & \text{if } i = 1, 2, \dots, q - 1. \end{cases}$$

The proof of Lemma 3.1 has been completed. ∎

It is worth noting that $P_\ell(0) = 0$ for $\ell = 1, 3, 5, \dots$ and $P_\ell(-u) = P_\ell(u)$ for $\ell = 0, 2, 4, \dots$. Consequently, the kernel in (3.21) is symmetric and all terms in (3.21) with odd values of j are equal zero.

Since $p_0(u) = \sqrt{\frac{1}{2}}$ and $p_2(u) = \sqrt{\frac{5}{2}} \left(\frac{3}{2} u^2 - \frac{1}{2} \right)$, it is easy to verify that the kernel in (3.21) with $q = 4$ is given by

$$K(u) = \left(\frac{9}{8} - \frac{15}{8} u^2 \right), \quad |u| \leq 1.$$

This confirms the form of the kernel $W_4(v)$ given in (3.19).

The result of Lemma 3.1 can be extended to a larger class of orthogonal polynomials defined on the set S, that is, when we have the system of functions $\{p_\ell(u); 0 \leq \ell \leq \infty\}$ defined on S, which satisfies

$$\int_{S} p_\ell(u) p_j(u) w(u) du = \delta_{\ell j},$$

where $w(u)$ is the weight function being positive on S and such that $w(0) = 1$. Then formula (3.21) takes the following modified form:

$$K(u) = \sum_{j=0}^{q-1} p_j(0) p_j(u) w(u). \tag{3.22}$$

In particular, if $w(u) = e^{-u^2}$, $-\infty < u < \infty$ and $\{p_\ell(u)\}$ are the orthonormal Hermite polynomials (see Section 6.5) then for $q = 4$, we can obtain the kernel in (3.18).

The rate depends on the smoothness of both $m(\bullet)$ and $f(\bullet)$, the bandwidth h_n, and the kernel. It is not surprising that the smoother curves, that is, the more derivatives of $m(\bullet)$ and $f(\bullet)$ exist, the greater speed can be achieved. If the number q increases to infinity, the derived rate becomes close to $n^{-1/2}$, that is, the rate typical for parametric inference.

As far as the bandwidth and kernel, the rate depends heavier on h_n. Deeper analysis shows that, for $h_n = cn^{-\delta}$, the choice of δ is much more important than c.

3.5 The mean-squared error

In this section, we assume that both the nonlinearity $m(\bullet)$ and the noise Z are bounded, that is, that

$$\sup_u |m(u)| + |Z| \le C$$

for some C. Applying (C.2), we get the following bound for the mean-squared error:

$$E(\hat{\mu}(u) - \mu(u))^2 \le 2\frac{1}{f(u)^2} E(\hat{g}(u) - \mu(u)g(u))^2 + 2C^2 \frac{1}{f^2(u)} E(\hat{f}(u) - f(u))^2.$$

Going through the proof of Theorem 3.1 we recall that $E(\hat{f}_n(u) - f(u))^2$ converges to zero as $n \to \infty$. Arguing as in the proof we verify that so does $E(\hat{g}_n(u) - g(u))^2$. Finally,

$$\lim_{n \to \infty} E(\hat{\mu}(u) - \mu(u))^2 = 0.$$

For the kernel satisfying (3.4), (3.5), and (3.6) with $\varepsilon = 0$, the convergence holds at every u where both $m(\bullet)$ and $f(\bullet)$ are continuous and $f(u) > 0$. If $\varepsilon > 0$, it holds the error vanishes at almost every u.

Moreover, for $m(\bullet)$ and $f(\bullet)$ having q-bounded derivatives, the kernel satisfying (3.15), and h_n, as in (3.17), we have

$$E(\hat{\mu}(u) - \mu(u))^2 = O(n^{-1+1/(2q+1)}),$$

see Lemma C.10. Hence, the mean-squared error tends to zero as fast as $O(n^{-2q/(2q+1)})$. This is known to be the optimal rate of convergence, see [285].

3.6 Simulation example

In the simulation example the system is as that in Section 2.2.2. The rectangular kernel is applied. To show the contribution of the dynamics to the identification error we set $Z_n \equiv 0$. For $a = 0.5$ realizations of the estimate are shown in Figure 3.10. The MISE versus n, for $a = 0.0, 0.25, 0.5$, and $a = 0.75$, is shown in Figure 3.11. It is clear that the greater a is the greater the error becomes. The influence of the variance of output noise is shown in Figure 3.12.

For $a = 0.0, 0.25, 0.5$, and $a = 0.75$, the MISE versus $h(n)$ is presented in Figures 3.13, 3.14, 3.15, and 3.16, respectively. Results suggest that a too small $h(n)$ should be avoided. For $h_n = n^{-\delta}$ with δ in the interval $[-0.25, 1.2]$, the MISE is shown in Figure 3.17.

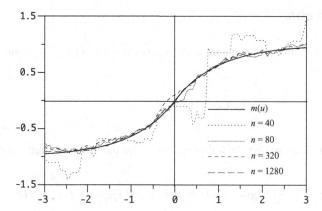

Figure 3.10 Realizations of the estimate; $a = 0.5$, $h_n = n^{-2/5}$ (example in Section 3.6).

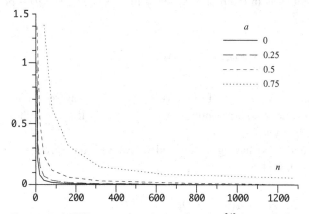

Figure 3.11 MISE versus n, various a; $h_n = n^{-2/5}$ (example in Section 3.6).

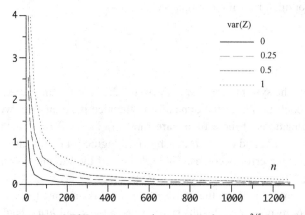

Figure 3.12 MISE versus n, various var(Z); $h_n = n^{-2/5}$ (example in Section 3.6).

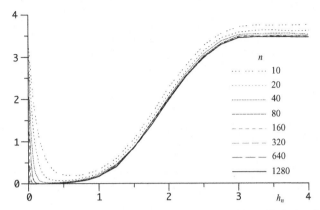

Figure 3.13 MISE versus h_n, various n; $a = 0.0$ (example in Section 3.6).

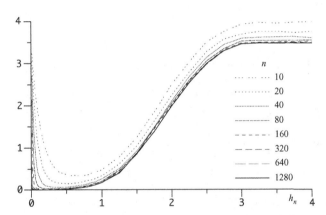

Figure 3.14 MISE versus h_n, various n; $a = 0.25$ (example in Section 3.6).

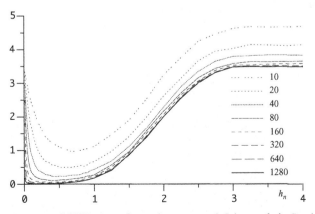

Figure 3.15 MISE versus h_n, various n; $a = 0.5$ (example in Section 3.6).

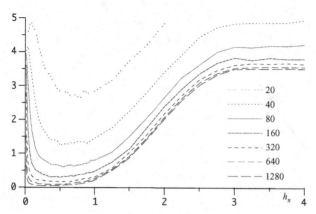

Figure 3.16 MISE versus h_n, various n; $a = 0.75$ (example in Section 3.6).

3.7   Lemmas and proofs

3.7.1   Lemmas

In Lemma 3.2, U has a density, in Lemma 3.3, the distribution of U is arbitrary.

LEMMA 3.2 *Let U have a probability density. Let $Em(U) = 0$, $\mathrm{var}[m(U)] < \infty$. Let the kernel $K(\bullet)$ satisfy (3.4), (3.5). If (3.6) holds with $\varepsilon = 0$, then, for $i \neq 0$,*

$$\sup_{h>0} \left| \mathrm{cov}\left[W_{p+i} \frac{1}{h} K\left(\frac{u - U_i}{h} \right), W_p \frac{1}{h} K\left(\frac{u - U_0}{h} \right) \right] \right|$$
$$\leq (|\lambda_p \lambda_{p+i}| + |\lambda_p \lambda_{p-i}| + |\lambda_{p+i} \lambda_{p-i}|)\omega(u),$$

where $\omega(u)$ is finite at every continuity point u of both $m(\bullet)$ and $f(\bullet)$. If $\varepsilon > 0$, the property holds at almost every $u \in R$.

Proof. We prove the continuous version of the lemma. The "almost everywhere" version can be verified in a similar way.

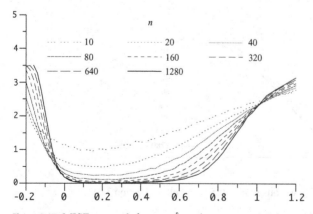

Figure 3.17 MISE versus δ, $h_n = n^{-\delta}$, various n; $a = 0.5$ (example in Section 3.6).

Since $W_{p+i} = \sum_{q=-\infty}^{p+i} \lambda_{p+i-q} m(U_q)$ and $W_p = \sum_{r=-\infty}^{p} \lambda_{p-r} m(U_r)$, the covariance in the assertion equals

$$\sum_{q=-\infty}^{p+i} \sum_{r=-\infty}^{p} \lambda_{p+i-q} \lambda_{p-r} \operatorname{cov}\left[m(U_q)\frac{1}{h}K\left(\frac{u-U_i}{h}\right), m(U_r)\frac{1}{h}K\left(\frac{u-U_0}{h}\right)\right].$$

Applying Lemma C.2, we find that the above formula is equal to

$$(\lambda_p \lambda_{p+i} + \lambda_p \lambda_{p-i})\frac{1}{h}E\left\{K\left(\frac{u-U}{h_n}\right)\right\}\frac{1}{h}E\left\{m^2(U)K\left(\frac{u-U}{h}\right)\right\}$$

$$+ \lambda_{p+i}\lambda_{p-i}\frac{1}{h^2}E^2\left\{m(U)K\left(\frac{u-U}{h}\right)\right\}.$$

Let u be a point where both $m(\bullet)$ and $f(\bullet)$ are continuous. It suffices to apply Lemmas A.8 and A.9 to find that the following formulas

$$\sup_{h>0}\left|\frac{1}{h}EK\left(\frac{u-U}{h}\right)\right|, \quad \sup_{h>0}E\left|m(U)\frac{1}{h}K\left(\frac{u-U}{h}\right)\right|,$$

$$\sup_{h>0}E\left|m^2(U)\frac{1}{h}K\left(\frac{u-U}{h}\right)\right|,$$

are finite. ∎

In the next lemma, U has an arbitrary distribution.

LEMMA 3.3 *Let $Em(U) = 0$ and $\operatorname{var}[m(U)] < \infty$. If the kernel satisfies the restrictions of Theorem 3.3, then*

$$\limsup_{h\to 0}\left|\frac{\operatorname{cov}\left[W_{p+i}K\left(\frac{u-U_i}{h}\right), W_p K\left(\frac{u-U_0}{h}\right)\right]}{E^2 K\left(\frac{u-U}{h}\right)}\right|$$

$$\leq (|\lambda_p \lambda_{p+i}| + |\lambda_p \lambda_{p-i}| + |\lambda_{p+i}\lambda_{p-i}|)\theta(u),$$

where some $\theta(u)$ is finite at almost every (ς) $u \in R$, where ς is the distribution of U.

Proof. The proof is similar to that of Lemma 3.2. Lemma A.10, rather than Lemmas A.8 and A.9, should be employed. ∎

3.7.2   Proofs

Proof of Theorem 3.1
For the sake of the proof $Em(U) = 0$, see Remark 2.1. Observe that $\hat{\mu}(u) = \hat{g}(u)/\hat{f}(u)$ with

$$\hat{g}(u) = \frac{1}{nh_n}\sum_{i=1}^{n} Y_{p+i}K\left(\frac{u-U_i}{h_n}\right) \qquad (3.23)$$

and

$$\hat{f}(u) = \frac{1}{nh_n} \sum_{i=1}^{n} K\left(\frac{u - U_i}{h_n}\right). \tag{3.24}$$

Fix $u \in R$ and suppose that both $m(\bullet)$ and $f(\bullet)$ are continuous at the point. We will now show that

$$\hat{g}(u) \to g(u) \int K(v)dv \to 0 \quad \text{as } n \to \infty \text{ in probability,} \tag{3.25}$$

where, we recall, $g(u) = \mu(u)f(u)$. Since

$$E\hat{g}(u) = \frac{1}{h_n} E\left\{E\left\{Y_p \mid U_0\right\} K\left(\frac{u - U_0}{h_n}\right)\right\} = \frac{1}{h_n} E\left\{\mu(U)K\left(\frac{u - U}{h_n}\right)\right\},$$

applying Lemma A.8, we conclude that

$$E\hat{g}(u) \to g(u) \int K(v)dv \text{ as } n \to \infty.$$

In turn, since $Y_n = W_n + Z_n$,

$$\text{var}[\hat{g}(u)] = P_n(u) + Q_n(u) + R_n(u),$$

where

$$P_n(u) = \frac{1}{nh_n} \sigma_Z^2 \frac{1}{h_n} EK^2\left(\frac{u - U}{h_n}\right),$$

$$Q_n(u) = \frac{1}{nh_n} \frac{1}{h_n} \text{var}\left[W_p K\left(\frac{u - U_0}{h_n}\right)\right],$$

and

$$R_n(u) = \frac{1}{n^2 h_n^2} \sum_{i=1}^{n} \sum_{\substack{j=1 \\ j \neq i}}^{n} \text{cov}\left[W_{p+i} K\left(\frac{u - U_i}{h_n}\right), W_{p+j} K\left(\frac{u - U_j}{h_n}\right)\right]$$

$$= \frac{2}{n^2 h_n^2} \sum_{i=1}^{n} (n - i) \text{cov}\left[W_{p+i} K\left(\frac{u - U_i}{h_n}\right), W_p K\left(\frac{u - U_0}{h_n}\right)\right].$$

In view of Lemma A.8,

$$nh_n P_n(u) \to \sigma_Z^2 f(u) \int K^2(v)dv \quad \text{as } n \to \infty.$$

Since

$$\text{var}\left[W_p K\left(\frac{u - U_0}{h_n}\right)\right]$$

$$= E\left\{W_p^2 K^2\left(\frac{u - U_0}{h_n}\right)\right\} - E^2\left\{W_p K\left(\frac{u - U_0}{h_n}\right)\right\}$$

$$= E\left\{\phi(U) K^2\left(\frac{u - U}{h}\right)\right\} - E^2\left\{\mu(U)K\left(\frac{u - U}{h}\right)\right\}, \tag{3.26}$$

where $\phi(\bullet)$ is as in (2.7), by Lemma A.8,

$$nh_n Q_n(u) \rightarrow \phi(u) f(u) \int K^2(v) dv \quad \text{as } n \rightarrow \infty.$$

Passing to $R_n(u)$, we apply Lemma 3.2 to obtain

$$|R_n(u)| \leq 2\omega(u) \frac{1}{n^2} \sum_{i=1}^{n} (n-i)(|\lambda_p \lambda_{p+i}| + |\lambda_p \lambda_{p-i}| + |\lambda_{p+i} \lambda_{p-i}|)$$

$$\leq 6\omega(u)(\max_n |\lambda_n|) \frac{1}{n} \sum_{i=1}^{\infty} |\lambda_i| = O\left(\frac{1}{n}\right).$$

Finally,

$$nh_n \operatorname{var}[\hat{g}(u)] \rightarrow \left(\sigma_Z^2 + \phi(u)\right) f(u) \int K^2(v) dv \text{ as } n \rightarrow \infty.$$

In this way, we have verified (3.25).

Using similar arguments, we show that $E \hat{f}(u) \rightarrow f(u) \int K(v) dv$ as $n \rightarrow \infty$ and

$$nh_n \operatorname{var}[\hat{f}(u)] \rightarrow f(u) \int K^2(v) dv \text{ as } n \rightarrow \infty, \quad (3.27)$$

and then we conclude that $\hat{f}(u) \rightarrow f(u) \int K(v) dv \rightarrow 0$ as $n \rightarrow \infty$ in probability. The proof has been completed. ∎

Proof of Theorem 3.3

In general, the idea of the proof is similar to that of Theorem 3.1. Some modifications, however, are necessary.

Recalling Remark 2.1, with no loss of generality, we assume that $Em(U) = 0$ and begin with the observation that $\hat{\mu}(u) = \hat{\xi}(u)/\hat{\eta}(u)$, where

$$\hat{\xi}(u) = \frac{1}{n E K\left(\frac{u-U}{h_n}\right)} \sum_{i=1}^{n} Y_{p+i} K\left(\frac{u - U_i}{h_n}\right)$$

and

$$\hat{\eta}(u) = \frac{1}{n E K\left(\frac{u-U}{h_n}\right)} \sum_{i=1}^{n} K\left(\frac{u - U_i}{h_n}\right).$$

Obviously,

$$E\hat{\xi}(u) = \frac{E\left\{Y_1 K\left(\frac{u - U_0}{h_n}\right)\right\}}{E K\left(\frac{u - U}{h_n}\right)} = \frac{E\left\{\mu(U) K\left(\frac{u - U}{h_n}\right)\right\}}{E K\left(\frac{u - U}{h_n}\right)},$$

which, by Lemma A.10, converges to $\mu(u)$ as $n \rightarrow \infty$ for almost every (ζ) $u \in R$.

Since $Y_n = W_n + Z_n$, $\text{var}[\hat{\xi}(u)] = P_n(u) + Q_n(u) + R_n(u)$, where

$$P_n(u) = \sigma_Z^2 \frac{EK^2\left(\dfrac{u-U}{h_n}\right)}{nE^2K\left(\dfrac{u-U}{h_n}\right)},$$

$$Q_n(u) = \frac{\text{var}\left[W_p K\left(\dfrac{u-U_0}{h_n}\right)\right]}{nE^2K\left(\dfrac{u-U}{h_n}\right)},$$

and

$$R_n(u) = 2\sum_{i=1}^{n}(n-i)\frac{\text{cov}\left[W_{p+i}K\left(\dfrac{u-U_i}{h_n}\right), W_p K\left(\dfrac{u-U_0}{h_n}\right)\right]}{E^2K\left(\dfrac{u-U}{h_n}\right)}.$$

Since,

$$P_n(u) \le \frac{1}{nh_n}\sigma_Z^2 \frac{h_n}{EK\left(\dfrac{u-U}{h_n}\right)}\sup_v K(v),$$

applying Lemma A.11 we find $P_n(u) = O_u(1/nh_n)$ as $n \to \infty$ for almost every $(\zeta)\, u$.
Lemmas A.10 and A.11, together with

$$Q_n(u) = \frac{1}{nh_n}\frac{h_n}{EK\left(\dfrac{u-U}{h_n}\right)} \times \left[\frac{E\left\{\phi(U)K^2\left(\dfrac{u-U}{h_n}\right)\right\}}{EK\left(\dfrac{u-U}{h_n}\right)}\right.$$

$$\left. - EK\left(\dfrac{u-U}{h_n}\right)\frac{E^2\left\{\mu(u)K\left(\dfrac{u-U}{h_n}\right)\right\}}{E^2K\left(\dfrac{u-U}{h_n}\right)}\right],$$

entail $Q_n(u) = O_u(1/nh_n)$ as $n \to \infty$ for almost every $(\zeta)\, u$.

Application of Lemma 3.3 leads to the conclusion that, at almost every $(\zeta)\, u \in R$,

$$|R_n(u)| \le 2\theta(u)\frac{1}{n^2}\sum_{i=1}^{n}(n-i)(|\lambda_p\lambda_{p+i}| + |\lambda_p\lambda_{p-i}| + |\lambda_{p+i}\lambda_{p-i}|)$$

$$\le 6\theta(u)(\max_n|\lambda_n|)\frac{1}{n}\sum_{i=1}^{\infty}|\lambda_i| = O\left(\frac{1}{n}\right).$$

Finally, $\text{var}[\hat{\xi}(u)] = O(1/nh_n)$ at almost every $(\zeta)\, u$.

In this way, we have shown that $E(\hat{\xi}(u) - \mu(u))^2 \to 0$ as $n \to \infty$ at almost every (ζ) u. Since, for the same reasons $E(\hat{\xi}(u) - 1)^2 \to 0$ at almost every $(\zeta)\, u$, the theorem follows. ∎

3.8   Bibliographic notes

The kernel regression estimate has been proposed independently by Nadaraya [215] and Watson [312] and was the subject of studies performed by Rosenblatt [257], Collomb [55], Greblicki [105], Greblicki and Krzyżak [121], Chu and Marron [51], Fan [87], Müller and Song [212], Jones, Davies, and Parkand [172], and many others. A comprehensive overview of various kernel methods is presented in Wand and Jones [310]. At first, the density of U was assumed to exist. Since Stone [284], consistency for any distribution has been examined. Later, distribution-free properties were studied by Spiegelman and Sacks [282], Devroye and Wagner [73, 74], Devroye [71], Krzyżak and Pawlak [187, 188], Greblicki, Krzyżak, and Pawlak [122], Kozek and Pawlak [179], among others. In particular, the monograph by Györfi, Kohler, Krzyżak, and Walk [140] examines the problem of a distribution-free theory of nonparametric regression.

The kernel regression estimate has been derived in a natural way from the kernel estimate (3.24) of a probability density function introduced by Parzen [226], generalized to multivariate cases by Cacoullos [37] and examined by a number of authors, see, for example, Rosenblatt [256], Van Ryzin [297, 298], Deheuvels [65], Wahba [306], Devroye and Wagner [72], Devroye and Györfi [68], and Csörgo and Mielniczuk [58]. See also Härdle [150], Prakasa Rao [241], or Silverman [277] and papers cited therein.

In all mentioned works, however, the kernel estimate is of form (3.3) with $p = 0$, while independent observations (U_i, Y_i)s come from a model $Y_n = m(U_n) + Z_n$. In the context of the Hammerstein system, it means that dynamics is just missing because the linear subsystem is reduced to a simple delay.

The nonparametric kernel regression estimate has been applied to recover the nonlinear characteristic in a Hammerstein system by Greblicki and Pawlak [126]. In Greblicki and Pawlak [129], the input signal has an arbitrary distribution. Not a state equation, but a convolution to describe the dynamic subsystem, has been applied in Greblicki and Pawlak [127]. The kernel estimate has also been discussed in Krzyżak [182, 183], as well as Krzyżak and Partyka [185]. For very specific distributions of the input signal, the nonparametric kernel regression estimate has been studied by Lang [193].

4 Semirecursive kernel algorithms

This chapter is devoted to semirecursive kernel algorithms, modifications of those examined in Chapter 3. Their numerators and denominators can be calculated online. We show consistency and examine convergence rate. The results for all input densities and all input distributions are established.

4.1 Introduction

We examine the following semirecursive kernel estimates:

$$\tilde{\mu}_n(u) = \frac{\sum_{i=1}^{n} \frac{1}{h_i} Y_{p+i} K\left(\frac{u - U_i}{h_i}\right)}{\sum_{i=1}^{n} \frac{1}{h_i} K\left(\frac{u - U_i}{h_i}\right)} \tag{4.1}$$

and

$$\bar{\mu}_n(u) = \frac{\sum_{i=1}^{n} Y_{p+i} K\left(\frac{u - U_i}{h_i}\right)}{\sum_{i=1}^{n} K\left(\frac{u - U_i}{h_i}\right)}, \tag{4.2}$$

modifications of (3.3). To demonstrate recursiveness, we observe that $\tilde{\mu}_n(u) = \tilde{g}_n(u)/\tilde{f}_n(u)$, where

$$\tilde{g}_n(u) = \frac{1}{n} \sum_{i=1}^{n} Y_{p+i} \frac{1}{h_i} K\left(\frac{u - U_i}{h_i}\right)$$

and

$$\tilde{f}_n(u) = \frac{1}{n} \sum_{i=1}^{n} \frac{1}{h_i} K\left(\frac{u - U_i}{h_i}\right).$$

Therefore,

$$\tilde{g}_n(u) = \tilde{g}_{n-1}(u) - \frac{1}{n}\left(\tilde{g}_{n-1}(u) - Y_{p+n} \frac{1}{h_n} K\left(\frac{u - U_n}{h_n}\right)\right)$$

and

$$\tilde{f}_n(u) = \tilde{f}_{n-1}(u) - \frac{1}{n}\left(\tilde{f}_{n-1}(u) - \frac{1}{h_n}K\left(\frac{u - U_n}{h_n}\right)\right).$$

For the other estimate, $\bar{\mu}_n(u) = \bar{g}_n(u)/\bar{f}_n(u)$ with

$$\bar{g}_n(u) = \frac{1}{\sum_{i=1}^{n} h_i}\sum_{i=1}^{n} Y_{p+i}K\left(\frac{u - U_i}{h_i}\right)$$

and

$$\bar{f}_n(u) = \frac{1}{\sum_{i=1}^{n} h_i}\sum_{i=1}^{n} K\left(\frac{u - U_i}{h_i}\right).$$

Both $\bar{g}_n(u)$ and $\bar{f}_n(u)$ can be calculated with the following recurrence formulas:

$$\bar{g}_n(u) = \bar{g}_{n-1}(u) - \frac{h_n}{\sum_{i=1}^{n} h_i}\left(\bar{g}_{n-1}(u) - \frac{1}{h_n}Y_{p+n}K\left(\frac{u - U_n}{h_n}\right)\right)$$

and

$$\bar{f}_n(u) = \bar{g}_{n-1}(u) - \frac{h_n}{\sum_{i=1}^{n} h_i}\left(\bar{f}_{n-1}(u) - \frac{1}{h_n}K\left(\frac{u - U_n}{h_n}\right)\right).$$

In both estimates, the starting points

$$\tilde{g}_1(u) = \bar{g}_1(u) = \frac{1}{h_1}Y_{p+1}K\left(\frac{u - U_1}{h_1}\right)$$

and

$$\tilde{f}_1(u) = \bar{f}_1(u) = \frac{1}{h_1}K\left(\frac{u - U_1}{h_1}\right)$$

are the same.

Thus, both estimates are semirecursive because their numerators and denominators can be calculated recursively, but not they themselves.

4.2   Consistency and convergence rate

In Theorems 4.1 and 4.2, the input signal has a density; in Theorem 4.3, its distribution is arbitrary.

THEOREM 4.1 *Let U have a density $f(\bullet)$ and let $Em^2(U) < \infty$. Let the Borel measurable kernel $K(\bullet)$ satisfy (3.4), (3.5), and (3.6) with $\varepsilon = 0$. Let the sequence $\{h_n\}$ satisfy the following restrictions:*

$$h_n \to 0 \text{ as } n \to \infty, \tag{4.3}$$

$$\frac{1}{n^2}\sum_{i=1}^{n} \frac{1}{h_i} \to 0 \text{ as } n \to \infty. \tag{4.4}$$

Then,

$$\tilde{\mu}_n(u) \to \mu(u) \text{ as } n \to \infty \text{ in probability.} \qquad (4.5)$$

at every $u \in R$ where both $m(\bullet)$ and $f(\bullet)$ are continuous and $f(u) > 0$. If, (3.6) holds for some $\varepsilon > 0$, then the convergence takes place at every Lebesgue point $u \in R$ of both $m(\bullet)$ and $f(\bullet)$, such that $f(u) > 0$; a fortiori, at almost every u belonging to support of $f(\bullet)$.

THEOREM 4.2 *Let U have a density $f(\bullet)$ and let $Em^2(U) < \infty$. Let the Borel measurable kernel $K(\bullet)$ satisfy (3.4), (3.5), and (3.6) with $\varepsilon = 0$. Let the sequence $\{h_n\}$ satisfy (4.3) and*

$$\sum_{n=1}^{\infty} h_i = \infty. \qquad (4.6)$$

Then,

$$\tilde{\mu}_n(u) \to \mu(u) \text{ as } n \to \infty \text{ in probability.} \qquad (4.7)$$

at every $u \in R$ where both $m(\bullet)$ and $f(\bullet)$ are continuous and $f(u) > 0$. If, (3.6) holds for some $\varepsilon > 0$, then the convergence takes place at every Lebesgue point $u \in R$ of both $m(\bullet)$ and $f(\bullet)$, such that $f(u) > 0$; a fortiori, at almost every u belonging to support of $f(\bullet)$.

Estimate (4.2) is consistent not only for U having a density but also for any distribution. In the next theorem, the kernel is the same as in Theorem 3.3.

THEOREM 4.3 *Let $Em^2(U) < \infty$. Let the kernel $K(\bullet)$ satisfy the restrictions of Theorem 3.3. Let the sequence $\{h_n\}$ of positive numbers satisfy (4.3) and (4.6). Then, convergence (4.7) takes place at almost every (ζ) point $u \in R$, where ζ is the probability measure of U.*

Estimate (4.1) converges if the number sequence satisfies (4.3) and (4.4), while (4.2) if (4.3) and (4.6) hold. Thus, for $h_n = cn^{-\delta}$ with $c > 0$, both converge if $0 < \delta < 1$. By selecting δ, we decide on the speed where the estimates converge. As in Section 3.4, in addition to the standard restrictions, the kernel satisfies restrictions (3.15) and (3.16). Both $f(\bullet)$ and $m(\bullet)$ have q derivatives. Moreover, $f^{(q)}(\bullet)$ and $g^{(q)}(\bullet)$ are square integrable. To study the convergence rate, we write

$$\frac{1}{h}EK\left(\frac{u-U}{h}\right) = \frac{1}{h}\int f(v)K\left(\frac{u-v}{h}\right)dv = \int f(u+vh)K(-v)dv.$$

Beginning with (4.1), we have

$$E\tilde{f}_n(u) = \frac{1}{n}\sum_{i=1}^{n}\frac{1}{h_i}EK\left(\frac{u-U}{h_i}\right) = \frac{1}{n}\sum_{i=1}^{n}\frac{1}{h_i}\int f(v)K\left(\frac{u-v}{h_i}\right)dv$$

$$= \frac{1}{n}\sum_{i=1}^{n}\int f(u+vh_i)K(-v)dv$$

and find

$$\text{bias}[\tilde{f}_n(u)] = E\tilde{f}_n(u) - f(u) \int K(v)dv = \frac{1}{n}\sum_{i=1}^{n}\int (f(u+vh_i) - f(u))K(-v)dv.$$

Applying (A.17), we obtain

$$\text{bias}[\tilde{f}_n(u)] = \frac{1}{n}\sum_{i=1}^{n}O\left(h_i^{q-1/2}\right) = O\left(\frac{1}{n}\sum_{i=1}^{n}h_i^{q-1/2}\right).$$

Recalling (4.11), we find

$$E(\tilde{f}_n(u) - f(u))^2 = O\left(\frac{1}{n^2}\left(\sum_{i=1}^{n}h_i^{q-1/2}\right)^2\right) + O\left(\frac{1}{n^2}\sum_{i=1}^{n}\frac{1}{h_i}\right)$$

with the first term incurred by squared bias and the other by variance. Hence, for

$$h_n \sim n^{-1/2q}, \tag{4.8}$$

that is, the same as in (3.17) applied in the offline estimate,

$$E(\hat{f}(u) - f(u))^2 = O(n^{-1+1/2q}),$$

Since the same rate holds for $\tilde{g}_n(u)$, that is, $E(\hat{g}(u) - g(u))^2 = O(n^{-1+1/2q})$, we finally obtain

$$P\{|\tilde{\mu}(u) - \mu(u)| > \varepsilon|\mu(u)|\} = O(n^{-1+1/2q})$$

for any $\varepsilon > 0$, and

$$|\tilde{\mu}(u) - \mu(u)| = O(n^{-1/2+1/4q}) \text{ as } n \to \infty \text{ in probability.}$$

Considering estimate (4.2) next, for obvious reasons, we write

$$\text{bias}[\bar{f}_n(u)] = \frac{1}{\sum_{i=1}^{n}h_i}\sum_{i=1}^{n}O(h_i^{q+1/2}) = O\left(\frac{\sum_{i=1}^{n}h_i^{q+1/2}}{\sum_{i=1}^{n}h_i}\right)$$

and, due to (4.12),

$$E(\bar{f}_n(u) - f(u))^2 = O\left(\frac{\left(\sum_{i=1}^{n}h_i^{q+1/2}\right)^2}{\left(\sum_{i=1}^{n}h_i\right)^2}\right) + O\left(\frac{1}{\sum_{i=1}^{n}h_i}\right),$$

which, for h_n selected as in (4.8), becomes

$$E(\bar{f}_n(u) - f(u))^2 = O(n^{-1+1/2q}).$$

Since $E(\bar{g}_n(u) - g(u))^2 = O(n^{-1+1/2q})$, we come to the conclusion that

$$P\{|\bar{\mu}_n(u) - \mu(u)| > \varepsilon|\mu(u)|\} = O(n^{-1+1/2q})$$

for any $\varepsilon > 0$, and

$$|\bar{\mu}_n(u) - \mu(u)| = O(n^{-1/2+1/4q}) \text{ as } n \to \infty \text{ in probability.}$$

If the qth derivatives of both $f(u)$ and $g(u)$ are bounded, using (A.18), we obtain

$$P\{|\tilde{\mu}(u) - \mu(u)| > \varepsilon|\mu(u)|\} = O(n^{-1+1/(2q+1)})$$

for any $\varepsilon > 0$, and

$$|\tilde{\mu}(u) - \mu(u)| = O(n^{-1/2+1/(4q+2)}) \text{ as } n \to \infty \text{ in probability,}$$

that is, somewhat faster convergence. The same rate holds also for $\bar{\mu}_n(u)$.

4.3   Simulation example

In the system as in Section 2.2.2, $a = 0.5$ and $Z_n = 0$. Since $\mu(u) = m(u)$, we just estimate $m(u)$ and rewrite them in the following forms:

$$\tilde{m}_n(u) = \frac{\sum_{i=1}^{n} \frac{1}{h_i} Y_{1+i} K\left(\frac{u - U_i}{h_i}\right)}{\sum_{i=1}^{n} \frac{1}{h_i} K\left(\frac{u - U_i}{h_i}\right)}, \qquad (4.9)$$

and

$$\tilde{m}_n(u) = \frac{\sum_{i=1}^{n} Y_{1+i} K\left(\frac{u - U_i}{h_i}\right)}{\sum_{i=1}^{n} K\left(\frac{u - U_i}{h_i}\right)}. \qquad (4.10)$$

For the rectangular kernel and $h_n = n^{-1/5}$, the MISE for both estimates is shown in Figure 5.5 in Section 5.4. For $h_n = n^{-\delta}$ with δ varying in the interval $[-0.25, 1.5]$, the error is shown in Figures 4.1 and 4.2.

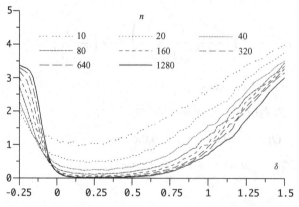

Figure 4.1 Estimate (4.9); MISE versus δ, various n; $h_n = n^{-\delta}$ (Section 4.3).

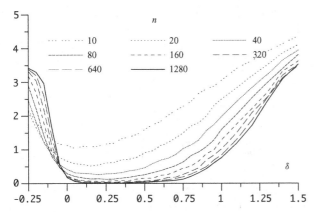

Figure 4.2 Estimate (4.10); MISE versus δ, various n; $h_n = n^{-\delta}$ (Section 4.3).

4.4   Proofs and lemmas

4.4.1   Lemmas

The system

LEMMA 4.1 *Let U have a probability density $f(\bullet)$. Let $Em(U) = 0$ and $\mathrm{var}[m(U)]$ $< \infty$. Let $n \neq 0$. Let kernel satisfy (3.4), (3.5). If (3.6) holds with $\varepsilon = 0$, then,*

$$\sup_{h>0,H>0} \left| \mathrm{cov}\left[W_{p+i}\frac{1}{h}K\left(\frac{u-U_i}{h}\right), W_p\frac{1}{H}K\left(\frac{u-U_0}{H}\right)\right]\right|$$
$$\leq (|\lambda_p\lambda_{p+i-j}| + |\lambda_p\lambda_{p-i+j}| + |\lambda_{p+i-j}\lambda_{p-i+j}|)\rho(u),$$

where $\rho(u)$ is finite at every continuity point u of both $m(\bullet)$ and $f(\bullet)$. If $\varepsilon > 0$, the property holds at almost every $u \in R$.

Proof. As $W_{p+i} = \sum_{q=-\infty}^{p+i} \lambda_{p+i-q}m(U_q)$ and $W_p = \sum_{r=-\infty}^{p} \lambda_{p-r}m(U_r)$, the covariance in the assertion equals (see Lemma C.2)

$$\sum_{q=-\infty}^{p+i} \sum_{r=-\infty}^{p} \lambda_{p+i-q}\lambda_{p-r}\, \mathrm{cov}\left[m(U_q)\frac{1}{h}K\left(\frac{u-U_i}{h}\right), m(U_r)\frac{1}{H}K\left(\frac{u-U_0}{H}\right)\right],$$

which is equal to

$$= \lambda_p\lambda_{p+i-j}\frac{1}{h}E\left\{K\left(\frac{u-U}{h}\right)\right\}\frac{1}{H}E\left\{m^2(U)K\left(\frac{u-U}{H}\right)\right\}$$
$$+ \lambda_p\lambda_{p-i+j}\frac{1}{h}E\left\{K\left(\frac{u-U}{h}\right)\right\}\frac{1}{h}E\left\{m^2(U)K\left(\frac{u-U}{h}\right)\right\}$$
$$+ \lambda_{p+i-j}\lambda_{p-i+j}\frac{1}{h}E\left\{m(U)K\left(\frac{u-U}{h}\right)\right\}\frac{1}{H}E\left\{m(U)K\left(\frac{u-U}{H}\right)\right\}.$$

Let u be a point where both $m(\bullet)$ and $f(\bullet)$ are continuous. It suffices to apply Lemma A.8 to find that the following formulas

$$\sup_{h \neq 0} \left| \frac{1}{h} E K \left(\frac{u - U}{h} \right) \right|,$$

$$\sup_{h \neq 0} E \left| m(U) \frac{1}{h} K \left(\frac{u - U}{h} \right) \right|,$$

$$\sup_{h \neq 0} E \left| m^2(U) \frac{1}{h} K \left(\frac{u - U}{h} \right) \right|,$$

are finite at every continuity point of both $m(\bullet)$ and $f(\bullet)$. The "almost everywhere" version of the lemma can be verified in a similar way. ∎

In the next lemma, U has an arbitrary distribution.

LEMMA 4.2 *Let $Em(U) = 0$ and $\mathrm{var}[m(U)] < \infty$. If the kernel satisfies the restrictions of Theorem 3.3, then*

$$\sup_{h>0, H>0} \left| \frac{\mathrm{cov} \left[W_{n+p} K \left(\frac{u - U_n}{h} \right), W_p K \left(\frac{u - U_0}{H} \right) \right]}{E K \left(\frac{u - U}{h} \right) E K \left(\frac{u - U}{H} \right)} \right|$$

$$\leq (|\lambda_p \lambda_{p+i-j}| + |\lambda_p \lambda_{p-i+j}| + |\lambda_{p+i-j} \lambda_{p-i+j}|) \eta(u),$$

where $\eta(u)$ is finite at almost every (ζ) $u \in R$, where ζ is the distribution of U.

Proof. It suffices to apply arguments used in the proof of Lemma 3.3. ∎

Number sequences

LEMMA 4.3 *If (4.3) and (4.4) hold, then*

$$\lim_{n \to \infty} \frac{\frac{1}{n}}{\frac{1}{n^2} \sum_{i=1}^{n} \frac{1}{h_i}} = 0.$$

Proof. From

$$n^2 = \left(\sum_{i=1}^{n} h_i^{1/2} \frac{1}{h_i^{1/2}} \right)^2 \leq \sum_{i=1}^{n} h_i \sum_{i=1}^{n} \frac{1}{h_i}$$

it follows that

$$\frac{\frac{1}{n}}{\frac{1}{n^2} \sum_{i=1}^{n} \frac{1}{h_i}} \leq \frac{1}{n} \sum_{i=1}^{n} h_i,$$

which converges to zero as $n \to \infty$. ∎

LEMMA 4.4 (TOEPLITZ) *If $\sum_{i=1}^{n} a_n \to \infty$ and $x_n \to x$ as $n \to \infty$, then*

$$\frac{\displaystyle\sum_{i=1}^{n} a_n x_n}{\displaystyle\sum_{i=1}^{n} a_n} \to x \text{ as } n \to \infty.$$

Proof. The proof is immediate. For any $\varepsilon > 0$, there exists N such that $|x_n| < \varepsilon$ for $n > N$. Hence,

$$\frac{\sum_{i=1}^{n} a_n x_n}{\sum_{i=1}^{n} a_n} - x = \frac{\sum_{i=1}^{N} a_n (x_n - x)}{\sum_{i=1}^{n} a_n} + \frac{\sum_{i=N+1}^{n} a_n (x_n - x)}{\sum_{i=1}^{n} a_n},$$

where the first term is bounded in absolute value by $c / \sum_{i=1}^{n} a_n$ for some c, and the other by ε. ∎

4.4.2   Proofs

Proof of Theorem 4.1

We give the continuous version of the proof. To verify the "almost everywhere" version, it suffices to apply Lemma A.9 rather than Lemma A.8.

Suppose that both $m(\bullet)$ and $f(\bullet)$ are continuous at $u \in R$. We start from the observation that

$$E\tilde{g}_n(u) = \frac{1}{n} \sum_{i=1}^{n} \frac{1}{h_i} E\left\{ E\left\{ Y_p \mid U_0 \right\} K\left(\frac{u - U_0}{h_i} \right) \right\}$$

$$= \frac{1}{n} \sum_{i=1}^{n} \frac{1}{h_i} E\left\{ \mu(U) K\left(\frac{u - U}{h_i} \right) \right\}.$$

Since

$$\frac{1}{h_i} E\left\{ \mu(U) K\left(\frac{u - U}{h_i} \right) \right\} \to g(u) \int K(v) dv \text{ as } i \to \infty,$$

(see Lemma A.8) we conclude that $E\tilde{g}_n(u) \to g(u) \int K(v) dv$ as $n \to \infty$, where, according to our notation, $g(u) = \mu(u) f(u)$.

To examine variance, we write $\text{var}[\tilde{g}_n(u)] = P_n(u) + Q_n(u) + R_n(u)$ with

$$P_n(u) = \sigma_Z^2 \frac{1}{n^2} \sum_{i=1}^{n} \frac{1}{h_i^2} \text{var}\left[K\left(\frac{u - U}{h_i} \right) \right],$$

$$Q_n(u) = \frac{1}{n^2} \sum_{i=1}^{n} \text{var}\left[W_p \frac{1}{h_i} K\left(\frac{u - U_0}{h_i} \right) \right],$$

and

$$R_n(u) = \frac{1}{n^2} \sum_{i=1}^{n} \sum_{\substack{j=1 \\ j \neq i}}^{n} \text{cov} \left[W_{p+i} \frac{1}{h_i} K \left(\frac{u - U_i}{h_i} \right), W_{p+j} \frac{1}{h_j} K \left(\frac{u - U_j}{h_j} \right) \right]$$

$$= \frac{1}{n^2} \sum_{i=1}^{n} \sum_{\substack{j=1 \\ j \neq i}}^{n} \text{cov} \left[W_{p+i-j} \frac{1}{h_i} K \left(\frac{u - U_{i-j}}{h_i} \right), W_{p} \frac{1}{h_j} K \left(\frac{u - U_0}{h_j} \right) \right].$$

Since

$$P_n(u) = \sigma_Z^2 \frac{1}{n^2} \sum_{i=1}^{n} \frac{1}{h_i} \left[\frac{1}{h_i} E K^2 \left(\frac{u - U}{h_i} \right) - h_i \frac{1}{h_i^2} E^2 K \left(\frac{u - U}{h_i} \right) \right],$$

using Lemma A.8, we find the quantity in square brackets converges to $f(u) \int K^2(v) dv$ as $i \to \infty$. Noticing that $\sum_{n=1}^{\infty} h_n^{-1} = \infty$ and applying Toeplitz Lemma 4.4, we conclude that

$$\frac{1}{\frac{1}{n^2} \sum_{i=1}^{n} \frac{1}{h_i}} P_n(u) \to \sigma_Z^2 f(u) \int K^2(v) dv \text{ as } n \to \infty.$$

For the same reasons, observing

$$Q_n(u) = \frac{1}{n^2} \sum_{i=1}^{n} \frac{1}{h_i} \left(\frac{1}{h_i} E \left\{ \phi(U) K^2 \left(\frac{u - U}{h_i} \right) \right\} - h_i \frac{1}{h_i^2} E^2 K \left(\frac{u - U}{h_i} \right) \right),$$

where $\phi(\bullet)$ is as in (2.7), we obtain

$$\frac{1}{\frac{1}{n^2} \sum_{i=1}^{n} \frac{1}{h_i}} Q_n(u) \to \phi(u) f(u) \int K^2(v) dv \text{ as } n \to \infty.$$

Moreover, using Lemma 4.1,

$$|R_n(u)| \leq \frac{1}{n^2} \rho(u) \sum_{i=1}^{n} \sum_{j=1}^{n} (|\lambda_p \lambda_{p+i-j}| + |\lambda_p \lambda_{p-i+j}| + |\lambda_{p+i-j} \lambda_{p-i+j}|)$$

$$\leq \frac{1}{n} \rho(u) (\max_n |\lambda_n|) \sum_{n=1}^{\infty} |\lambda_n| = O \left(\frac{1}{n} \right).$$

Using Lemma 4.3, we conclude that $R_n(u)$ vanishes faster than both $P_n(u)$ and $Q_n(u)$ and then we obtain

$$\frac{1}{\frac{1}{n^2} \sum_{i=1}^{n} \frac{1}{h_i}} \text{var}[\tilde{g}_n(u)] \to (\sigma_Z^2 + \phi(u) f(u)) \int K^2(v) dv \text{ as } n \to \infty. \qquad (4.11)$$

For similar reasons, $E \tilde{f}_n(u) \to f(u) \int K(v) dv$ as $n \to \infty$, and

$$\frac{1}{\frac{1}{n^2} \sum_{i=1}^{n} \frac{1}{h_i}} \text{var}[\tilde{f}_n(u)] \to f(u) \int K^2(v) dv \text{ as } n \to \infty,$$

which completes the proof. ■

Proof of Theorem 4.2

Suppose that both $m(\bullet)$ and $f(\bullet)$ are continuous at a point $u \in R$. Evidently,

$$E\bar{g}_n(u) = \frac{1}{\sum_{i=1}^{n} h_i} \sum_{i=1}^{n} h_i \frac{1}{h_i} E\left\{ E\{Y_p \mid U_0\} K\left(\frac{u - U_0}{h_i}\right)\right\}$$

$$= \frac{1}{\sum_{i=1}^{n} h_i} \sum_{i=1}^{n} h_i \frac{1}{h_i} E\left\{ \mu(U) K\left(\frac{u - U}{h_i}\right)\right\}.$$

Since (4.6) holds and

$$\frac{1}{h_i} E\left\{ \mu(U) K\left(\frac{u - U}{h_i}\right)\right\} \to g(u) \int K(v)dv \text{ as } n \to \infty,$$

(see Lemma A.8) an application of Toeplitz lemma 4.4 gives

$$E\bar{g}_n(u) \to g(u) \int K(v)dv \text{ as } n \to \infty.$$

To examine variance, we write $\mathrm{var}[\bar{g}_n(u)] = P_n(u) + Q_n(u) + R_n(u)$, where

$$P_n(u) = \sigma_Z^2 \frac{1}{\left(\sum_{i=1}^{n} h_i\right)^2} \sum_{i=1}^{n} \mathrm{var}\left[K\left(\frac{u - U}{h_i}\right)\right],$$

$$Q_n(u) = \frac{1}{\left(\sum_{i=1}^{n} h_i\right)^2} \sum_{i=1}^{n} \mathrm{var}\left[W_p K\left(\frac{u - U_0}{h_i}\right)\right],$$

and

$$R_n(u) = \frac{1}{\left(\sum_{i=1}^{n} h_i\right)^2} \sum_{i=1}^{n}\sum_{\substack{j=1 \\ j \neq i}}^{n} \mathrm{cov}\left[W_{p+i} K\left(\frac{u - U_i}{h_i}\right), W_{p+j} K\left(\frac{u - U_j}{h_j}\right)\right]$$

$$= \frac{1}{\left(\sum_{i=1}^{n} h_i\right)^2} \sum_{i=1}^{n}\sum_{\substack{j=1 \\ j \neq i}}^{n} \mathrm{cov}\left[W_{p+i-j} K\left(\frac{u - U_{i-j}}{h_i}\right), W_p K\left(\frac{u - U_0}{h_j}\right)\right].$$

Since

$$P_n(u) = \sigma_Z^2 \frac{1}{\sum_{i=1}^{n} h_i} P_{1n}(u)$$

with

$$P_{1n}(u) = \frac{1}{\sum_{i=1}^{n} h_i} \sum_{i=1}^{n} h_i \left(\frac{1}{h_i} EK^2\left(\frac{u - U}{h_i}\right) - h_i \frac{1}{h_i^2} E^2 K\left(\frac{u - U}{h_i}\right)\right)$$

converging, due to (4.6) and Toeplitz lemma 4.4, to the same limit as

$$\frac{1}{h_n} EK^2\left(\frac{u - U}{h_n}\right) - h_n \frac{1}{h_n^2} E^2 K\left(\frac{u - U}{h_n}\right),$$

we get $P_n(u) \sum_{i=1}^{n} h_i \to \sigma_Z^2 f(u) \int K^2(v)dv$ as $n \to \infty$.

For the same reasons, observing

$$Q_n(u) = \frac{1}{\sum_{i=1}^{n} h_i} \sum_{i=1}^{n} h_i \left(\frac{1}{h_i} E \left\{ \phi(U) K^2 \left(\frac{u-U}{h_i} \right) \right\} - h_i \frac{1}{h_i^2} E^2 K \left(\frac{u-U}{h_i} \right) \right),$$

where $\phi(\bullet)$ is as in (2.7), we obtain $Q_n(u) \sum_{i=1}^{n} h_i \to \phi(u) f(u) \int K^2(v) dv$ as $n \to \infty$.

Applying Lemma 4.1, we get

$$R_n(u) \sum_{i=1}^{n} h_i \leq \rho(u) \frac{1}{\sum_{i=1}^{n} h_i} \sum_{i=1}^{n} h_i \sum_{j=1}^{n} h_j (|\lambda_p \lambda_{p+i-j}| + |\lambda_p \lambda_{p-i+j}|$$

$$+ |\lambda_{p+i-j} \lambda_{p-i+j}|) \leq 3\rho(u) (\max_n h_n)(\max_n |\lambda_n|) \frac{1}{\sum_{i=1}^{n} h_i} \sum_{i=1}^{n} h_i \alpha_i,$$

where $\alpha_i = \sum_{j=i-p}^{\infty} |\lambda_i|$. Since $\lim_{i\to\infty} \alpha_i = 0$, applying Toeplitz lemma 4.4, we get $\lim_{n\to\infty} R_n(u) \sum_{i=1}^{n} h_i = 0$, which means that $R_n(u)$ vanishes faster than both $P_n(u)$ and $Q_n(u)$. Finally,

$$\mathrm{var}[\bar{g}_n(u)] \sum_{i=1}^{n} h_i \to \left(\sigma_Z^2 + \phi(u) f(u) \right) \int K^2(v) dv \text{ as } n \to \infty. \tag{4.12}$$

Since, for the same reasons, $E \bar{f}_n(u) \to f(u) \int K(v) dv$ as $n \to \infty$ and $\mathrm{var}[\bar{f}_n(u)] \sum_{i=1}^{n} h_i \to f(u) \int K^2(v) dv$ as $n \to \infty$, the proof has been completed. ∎

Proof of Theorem 4.3

Each convergence in the proof holds for almost every (ζ) $u \in R$. In a preparatory step, we show that

$$\sum_{n=1}^{\infty} E K \left(\frac{u-U}{h_n} \right) = \infty. \tag{4.13}$$

Since, by Lemma A.5,

$$\frac{1}{h_i} E K \left(\frac{u-U}{h_i} \right)$$

converges to a nonzero limit or increases to infinity, as $i \to \infty$, in view of (4.6),

$$\sum_{n=1}^{\infty} E K \left(\frac{u-U}{h_n} \right) = \sum_{n=1}^{\infty} h_n \frac{1}{h_n} E K \left(\frac{u-U}{h_n} \right) = \infty. \tag{4.14}$$

Thus (4.13) holds.

In general, the construction of the proof is the same as those in the previous section. We begin with the observation that $\bar{\mu}_n(u) = \hat{\xi}_n(u)/\hat{\eta}_n(u)$, where

$$\hat{\xi}_n(u) = \frac{\sum_{i=1}^{n} Y_{p+i} K\left(\frac{u - U_i}{h_i}\right)}{\sum_{i=1}^{n} E K\left(\frac{u - U}{h_i}\right)}$$

and

$$\hat{\eta}_n(u) = \frac{\sum_{i=1}^{n} K\left(\frac{u - U_i}{h_i}\right)}{\sum_{i=1}^{n} E K\left(\frac{u - U}{h_i}\right)}.$$

To examine the bias of $\hat{\xi}_n(u)$, observe that

$$E\hat{\xi}_n(u) = \frac{\sum_{i=1}^{n} E\left\{\mu(U) K\left(\frac{u - U}{h_i}\right)\right\}}{\sum_{i=1}^{n} E K\left(\frac{u - U}{h_i}\right)} = \frac{\sum_{i=1}^{n} a_i(u) E K\left(\frac{u - U}{h_i}\right)}{\sum_{i=1}^{n} E K\left(\frac{u - U}{h_i}\right)}$$

with

$$a_i(u) = \frac{E\left\{\mu(U) K\left(\frac{u - U}{h_i}\right)\right\}}{E K\left(\frac{u - U}{h_i}\right)}$$

converging to $\mu(u)$ as $i \to \infty$. Since (4.13) holds, an application of Toeplitz lemma 4.4 yields $E\hat{\xi}_n(u) \to \mu(u)$ as $n \to \infty$.

Passing to variance, we obtain

$$\text{var}[\hat{\xi}_n(u)] = P_n(u) + Q_n(u) + R_n(u)$$

with

$$P_n(u) = \sigma_Z^2 \frac{1}{\left(\sum_{i=1}^{n} E K\left(\frac{u-U}{h_i}\right)\right)^2} \sum_{i=1}^{n} \text{var}\left[K\left(\frac{u - U}{h_i}\right)\right],$$

$$Q_n(u) = \frac{1}{\left(\sum_{i=1}^{n} E K\left(\frac{u-U}{h_i}\right)\right)^2} \sum_{i=1}^{n} \text{var}\left[W_p K\left(\frac{u - U_0}{h_i}\right)\right],$$

and

$R_n(u)$

$$= \frac{1}{\left(\sum_{i=1}^{n} EK\left(\frac{u-U}{h_i}\right)\right)^2} \sum_{i=1}^{n}\sum_{\substack{j=1 \\ j\neq i}}^{n} \text{cov}\left[W_{p+i}K\left(\frac{u-U_i}{h_i}\right), W_{p+j}K\left(\frac{u-U_j}{h_j}\right)\right]$$

$$= \frac{1}{\left(\sum_{i=1}^{n} EK\left(\frac{u-U}{h_i}\right)\right)^2} \sum_{i=1}^{n}\sum_{\substack{j=1 \\ j\neq i}}^{n} \text{cov}\left[W_{p+i-j}K\left(\frac{u-U_{i-j}}{h_i}\right), W_p K\left(\frac{u-U_0}{h_j}\right)\right].$$

Rewriting

$$P_n(u) \stackrel{!}{=} \sigma_Z^2 \frac{1}{\sum_{i=1}^{n} EK\left(\frac{u-U}{h_i}\right)} \frac{\sum_{i=1}^{n} b_i(u)EK\left(\frac{u-U}{h_i}\right)}{\sum_{i=1}^{n} EK\left(\frac{u-U}{h_i}\right)}$$

with

$$b_i(u) = \frac{\text{var}\left[K\left(\frac{u-U}{h_i}\right)\right]}{EK\left(\frac{u-U}{h_i}\right)} \leq \frac{EK^2\left(\frac{u-U}{h_i}\right)}{EK\left(\frac{u-U}{h_i}\right)} \leq \kappa,$$

where $\kappa = \sup_v K(v)$, and applying (4.13), we find $P_n(u)$ converging to zero as $n \to \infty$. For similar reasons, $Q_n(u)$ converges to zero, too.

Clearly,

$$R_n(u)\sum_{i=1}^{n} EK\left(\frac{u-U}{h_i}\right) = \frac{1}{\sum_{i=1}^{n} EK\left(\frac{u-U}{h_i}\right)} \sum_{i=1}^{n}\sum_{\substack{j=1 \\ j\neq i}}^{n} E^2 K\left(\frac{u-U}{h_i}\right)$$

$$\times \frac{\text{cov}\left[W_{p+i-j}K\left(\frac{u-U_{i-j}}{h_i}\right), W_p K\left(\frac{u-U_0}{h_j}\right)\right]}{E^2 K\left(\frac{u-U}{h_i}\right)},$$

which is bounded by

$$\kappa\eta(u)\frac{1}{\sum_{i=1}^{n} EK\left(\frac{u-U}{h_i}\right)}$$

$$\times \sum_{i=1}^{n}\sum_{j=1}^{n} EK\left(\frac{u-U}{h_i}\right)\left(|\lambda_p\lambda_{p+i-j}| + |\lambda_p\lambda_{p-i+j}| + |\lambda_{p+i-j}\lambda_{p-i+j}|\right)$$

$$\leq 3\kappa\eta(u)(\max_n h_n)(\max_n |\lambda_n|)\frac{1}{\sum_{i=1}^{n} EK\left(\frac{u-U}{h_i}\right)} \sum_{i=1}^{n} \gamma_i EK\left(\frac{u-U}{h_i}\right),$$

where $\gamma_i = \sum_{j=i-p}^{\infty} |\lambda_i|$. Since (4.14) holds and $\lim_{i \to \infty} \gamma_i = 0$, using Toeplitz lemma 4.4, we find $R_n(u)$ converging to zero as $n \to \infty$.

Finally,

$$E(\hat{\xi}_n(u) - \mu(u))^2 \to 0 \text{ as } n \to \infty.$$

This and the fact that $E(\hat{\eta}_n(u) - 1)^2 \to 0$ as $n \to \infty$ complete the proof. ∎

4.5 Bibliographic notes

Recursive kernel estimates were applied to recover a probability density by Wolverton and Wagner [325, 326], Yamato [328], Davies [62], Reytö and Révész [251], Davies and Wegman [63], Wegman and Davies [314], and Hosni and Gado [164]. They have led to semirecursive regression estimates, see Greblicki [105], Ahmad and Lin [1], Devroye and Wagner [74], Krzyżak and Pawlak [186, 187], and Greblicki and Pawlak [128]. Greblicki and Pawlak [130] applied them to recover the nonlinear characteristic in Hammerstein systems. A semirecursive algorithm of the stochastic approximation type was examined by Greblicki [118].

5 Recursive kernel algorithms

This chapter is devoted to fully recursive kernel algorithms. The algorithms are motivated by stochastic approximation methods and this is explained in Section 5.2. In Section 5.3 the consistency and convergence rate of the algorithms are examined.

5.1 Introduction

In this chapter, to recover $\mu(u)$, we use the following two recursive estimates:

$$\hat{\mu}_n(u) = \hat{\mu}_{n-1}(u) - \gamma_n \frac{1}{h_n} K\left(\frac{u - U_n}{h_n}\right) [\hat{\mu}_{n-1}(u) - Y_{n+p}] \qquad (5.1)$$

with

$$\hat{\mu}_0(u) = Y_p \frac{1}{h_0} K\left(\frac{u - U_0}{h_0}\right)$$

and

$$\tilde{\mu}_n(u) = \tilde{\mu}_{n-1}(u) - \gamma_n K\left(\frac{u - U_n}{h_n}\right) [\tilde{\mu}_{n-1}(u) - Y_{n+p}] \qquad (5.2)$$

with

$$\tilde{\mu}_0(u) = Y_p K\left(\frac{u - U_0}{h_0}\right).$$

Both are closely related to the kernel estimate and its semirecursive versions studied in Chapters 3 and 4, respectively. They are, however, fully recursive. In addition, contrary to the kernel estimate and its semirecursive counterpart, the algorithms in (5.1) and (5.2) are not of a quotient form.

5.2 Relation to stochastic approximation

We will now present the relationship between the examined algorithms and the classical stochastic approximation framework, see Section C.3 and, for example, Wasan [311].

Rewriting (5.1) as

$$\hat{\mu}_n(u) = \hat{\mu}_{n-1}(u) - \gamma_n \left[\hat{\mu}_{n-1}(u) \frac{1}{h_n} K \left(\frac{u - U_n}{h_n} \right) - Y_{n+p} \frac{1}{h_n} K \left(\frac{u - U_n}{h_n} \right) \right], \quad (5.3)$$

we notice that the algorithm is founded on the obvious expectation that

$$|\hat{\mu}_n(u) - \mu_n(u)| \to 0 \text{ as } n \to \infty, \quad (5.4)$$

where

$$\mu_n(u) = \frac{\frac{1}{h_n} E \left\{ Y_{n+p} K \left(\frac{u - U_n}{h_n} \right) \right\}}{\frac{1}{h_n} E K \left(\frac{u - U_n}{h_n} \right)}$$

is the solution of the following time-varying equation:

$$E \left\{ \mu_n(u) \frac{1}{h_n} K \left(\frac{u - U_n}{h_n} \right) - Y_{n+p} \frac{1}{h_n} K \left(\frac{u - U_n}{h_n} \right) \right\} = 0. \quad (5.5)$$

The quantity under the sign of expectation on the left-hand side is just the expression in square brackets in (5.3) with $\hat{\mu}_{n-1}(u)$ replaced by $\mu_n(u)$. To examine $\mu_n(u)$, observe that

$$\mu_n(u) = \frac{\frac{1}{h_n} E \left\{ E\{Y_p \mid U_0\} K \left(\frac{u - U}{h_n} \right) \right\}}{\frac{1}{h_n} E K \left(\frac{u - U}{h_n} \right)}$$

$$= \frac{\frac{1}{h_n} E \left\{ \mu(U) K \left(\frac{u - U}{h_n} \right) \right\}}{\frac{1}{h_n} E K \left(\frac{u - U}{h_n} \right)}.$$

Therefore, if the kernel is well chosen and $h_n \to 0$ as $n \to \infty$,

$$\frac{1}{h_n} K \left(\frac{u - v}{h_n} \right)$$

gets close to the Dirac impulse located at $u = v$ as $n \to \infty$, that is, to $\delta (u - v) \int K(v) dv$, and thus we can expect that

$$\frac{1}{h_n} E \left\{ m(U) K \left(\frac{u - U}{h_n} \right) \right\} \to g(u) \int K(v) dv \text{ as } n \to \infty,$$

where, as usual, $g(u) = \mu(u) f(u)$, and

$$\frac{1}{h_n} E K \left(\frac{u - U}{h_n} \right) \to f(u) \int K(v) dv \text{ as } n \to \infty.$$

It justifies our expectation that $\mu_n(u) \to \mu(u)$ as $n \to \infty$. Combining this with (5.4), we finally obtain

$$\hat{\mu}_n(u) \to \mu(u) \text{ as } n \to \infty,$$

that is, the desired convergence of procedure (5.1).

In light of this, we can say that (5.1), and, a fortiori (5.3), solves the nonstationary nonparametric regression equation (5.5). The estimate is in the form of a stochastic approximation procedure generalized to the nonparametric case. Observe, moreover, that the equation is nonstationary despite the fact that the system is time invariant. It is caused by the fact that

$$\frac{1}{h}K\left(\frac{u-U_0}{h}\right) \quad \text{and} \quad \frac{1}{h}Y_p K\left(\frac{u-U_0}{h}\right)$$

are only asymptotically unbiased estimates of

$$f(u)\int K(v)dv \quad \text{and} \quad g(u)\int K(v)dv,$$

respectively.

The procedure has yet another specific feature that leads to analytical difficulties and requires special attention. Its gain, that is,

$$\gamma_n K\left(\frac{u-U_n}{h_n}\right),$$

is random. In the classical stochastic approximation, the gain is deterministic.

Clearly, all of the above remarks also refer to (5.2).

5.3   Consistency and convergence rate

Depending on the estimate nonnegative number sequences, $\{\gamma_n\}$ and $\{h_n\}$ satisfy some of the following restrictions:

$$h_n \to 0 \text{ as } n \to \infty, \tag{5.6}$$

$$\gamma_n \to 0 \text{ as } n \to \infty, \tag{5.7}$$

$$\sum_{n=1}^{\infty} \gamma_n = \infty, \tag{5.8}$$

$$\frac{\gamma_n}{h_n} \to 0 \text{ as } n \to \infty, \tag{5.9}$$

and

$$\sum_{n=1}^{\infty} \gamma_n h_n = \infty. \tag{5.10}$$

As such sequences one can apply, for example, $h_n \sim n^{-\delta}$ and $\gamma_n \sim n^{-\gamma}$. Algorithm (5.1) converges if restrictions (5.6)–(5.9) are satisfied, that is, if $0 < \delta < \gamma \leq 1$, while (5.2) does if (5.6), (5.7), and (5.10) hold, that is, if $0 < \delta, 0 < \gamma, \gamma + \delta \leq 1$.

We will now examine estimate (5.1). Its bias and variance are studied in Lemmas 5.5 and 5.7, respectively. Using Lemmas 5.5 and 5.7, we are in a position to give the following theorems.

THEOREM 5.1 *Let U have a probability density $f(\bullet)$ and let $Em^2(U) < \infty$. Let the kernel K satisfy (3.4), (3.5), and (3.6) with $\varepsilon = 0$. Let number sequences $\{h_n\}$ and $\{\gamma_n\}$ satisfy (5.6)–(5.9). Then*

$$E(\hat{\mu}_n(u) - \mu(u))^2 \to 0 \text{ as } n \to \infty, \tag{5.11}$$

at every $u \in R$ where both $m(\bullet)$ and $f(\bullet)$ are continuous and $f(u) > 0$.

Our next theorem deals with algorithm (5.2). Its proof is in Section 5.5.

THEOREM 5.2 *Let U have a probability density $f(\bullet)$ and let $Em^2(U) < \infty$. Let the kernel $K(\bullet)$ satisfy (3.4), (3.5), and (3.6) with $\varepsilon = 0$. Let number sequences $\{h_n\}$ and $\{\gamma_n\}$ satisfy (3.7), (5.7), and (5.10). Then,*

$$E(\tilde{\mu}_n(u) - \mu(u))^2 \to 0 \text{ as } n \to \infty, \tag{5.12}$$

at every point $u \in R$ where both $m(\bullet)$ and $f(\bullet)$ are continuous and $f(u) > 0$.

It is obvious that if $\varepsilon = 0$ is replaced by $\varepsilon > 0$, both algorithms converge at every Lebesgue point $u \in R$ of both $m(\bullet)$ and $f(\bullet)$ at which $f(u) > 0$, and, a fortiori, at every $u \in R$ where both $m(\bullet)$ and $f(\bullet)$ are continuous and $f(u) > 0$.

Imposing some smoothness restrictions on $m(\bullet)$ and $f(\bullet)$ we give the convergence rate for both algorithms. We assume that both $m(\bullet)$ and $f(\bullet)$ have q derivatives and that, in addition, $f^{(q)}(\bullet)$ and the qth derivative of $f(u)m(u)$ are square integrable. Moreover,

$$h_n \sim n^{-\delta} \quad \text{and} \quad \gamma_n \sim n^{-\gamma}.$$

In addition to those in Theorem 5.1, the kernel satisfies the assumptions in Section 3.4. Denoting $r_n(u) = K_n(u) - f(u) \int K(v)dv$ and $R_n(u) = L_n(u) - \mu(u)f(u) \int K(v)dv$, and then using (5.18), we get

$$\text{bias}[\hat{\mu}_n(u)] = (1 - \gamma_n K_n(u))b_{n-1}(u) + \gamma_n[R_n(u) - \mu(u)r_n(u)],$$

where $K_n(u)$ and $L_n(u)$ are as in (5.19) and (5.20), respectively. Applying (A.17) we find $r_n(u) = O(h_n^{q-1/2}) = O(n^{-\delta q + \delta/2})$ and, for the same reasons, $R_n(u) = O(n^{-\delta q + \delta/2})$. Hence,

$$|\text{bias}[\hat{\mu}_n(u)]| = |1 - cn^{-\gamma}K_n(u)| \, |\text{bias}[\hat{\mu}_{n-1}(u)]| + n^{-\gamma}n^{-\delta q + \delta/2}O(1).$$

Assuming that $f(u) > 0$ we find $K_n(u)$ converging to a positive limit as $n \to \infty$ and get, for n large enough,

$$|\text{bias}[\hat{\mu}_n(u)]| \leq (1 - cn^{-\gamma}K_n(u))| \, \text{bias}[\hat{\mu}_{n-1}(u)]| + n^{-\gamma}n^{-\delta q + \delta/2}O(1).$$

Application of Lemma 5.2 yields

$$\text{bias}[\hat{\mu}_n(u)] = O(n^{-\delta q + \delta/2}).$$

Since, by Lemma (5.4), $\sum_{i=1}^{n} \gamma_i \| A^{n-i} \| = O(n^{-\gamma})$, from (5.25) it follows that

$$\text{var}[\hat{\mu}_n(u)] \leq (1 - ca_n(u)n^{-\gamma}) \text{var}[\hat{\mu}_{n-1}(u)] + n^{-2\gamma+\delta} O(1).$$

Applying Lemma 5.2 again, we find

$$\text{var}[\hat{\mu}_n(u)] = O(n^{\delta-\gamma}).$$

Combining results concerning bias and variance, we get

$$E(\hat{\mu}_n(u) - \mu(u))^2 = O(n^{-2\delta q+\delta}) + O(n^{\delta-\gamma}).$$

Setting $\gamma = 1 - \varepsilon$, that is,

$$\gamma_n \sim n^{-1+\varepsilon}$$

and selecting $\delta = 1/2q$, that is,

$$h_n \sim n^{-1/2q},$$

we find

$$E(\hat{\mu}_n(u) - \mu(u))^2 = O(n^{-1+1/2q+\varepsilon}).$$

Therefore, to obtain fast convergence a small ε should be chosen. For the same γ_n and h_n, the rate holds also for the other estimate, that is,

$$E(\tilde{\mu}_n(u) - \mu(u))^2 = O(n^{-1+1/2q+\varepsilon}).$$

If the qth derivatives of $f(u)$ and $f(u)m(u)$ are bounded, we use (A.18) and obtain a better rate for bias, see also Sections 3.4 and 4.2. Therefore, we obtain

$$E(\hat{\mu}_n(u) - \mu(u))^2 = O(n^{-1+1/(2q+1)+\varepsilon})$$

and

$$E(\tilde{\mu}_n(u) - \mu(u))^2 = O(n^{-1+1/(2q+1)+\varepsilon}).$$

It is interesting to compare this rate by comparison with semirecursive and offline algorithms. The comparison is made under the assumption that the appropriate qth derivatives are bounded, that is, for the case when the better rate holds. First observe that from the result concerning the mean-squared error it follows that both algorithms converge in probability as fast as $O(n^{-1/2+1/(4q+2)+\varepsilon/2})$, where $\varepsilon > 0$ can be selected arbitrarily small, see Lemma C.9. The off-line algorithm (3.3) as well as semirecursive (4.1) and (4.1) converge in probability at the rate $O(n^{-1/2+1/(4q+2)})$, (see Sections 3.4 and 4.2), that is, somewhat faster. The $n^{\varepsilon/2}$ is just the price paid for full recursiveness.

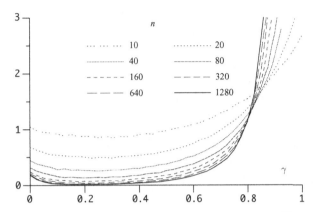

Figure 5.1 MISE versus γ for estimate (5.1) with $\gamma_n = n^{-0.8}$ and $h_n = n^{-\gamma}$, various n (example in Section 5.4).

We want, however, to recall and stress here that, contrary to fully recursive, semi-recursive and offline algorithms are of a quotient form. Besides, in the semirecursive case only numerators and denominators are calculated in a recursive way.

5.4   Simulation example

In the simulation example, the system is the same as in Section 2.2.2 but $Z_n = 0$ and $a = 0.5$. In estimates (5.1) and (5.2), the kernel is rectangular, $p = 1$, $h_n = n^{-\delta}$, and $\gamma_n = n^{-\gamma}$. Clearly, $\mu(u) = m(u)$. For $\gamma = 0.8$ and various δ, the MISE is shown in Figures 5.1 and 5.2. For $\delta = 0.2$ and various γ, the error is in Figures 5.3 and 5.4. The fact that (5.1) converges if $\gamma + \delta < 1$ is well apparent. The MISE for both estimates as well as for kernel ones studied earlier, that is, for (3.3), (4.1), (4.2), is shown in Figure 5.5. All behave similarly, nevertheless, some price is paid for semirecursiveness and full recursiveness.

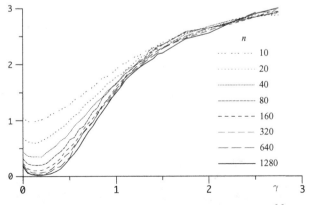

Figure 5.2 MISE versus γ for estimate (5.2) with $\gamma_n = n^{-0.8}$ and $h_n = n^{-\gamma}$, various n (example in Section 5.4).

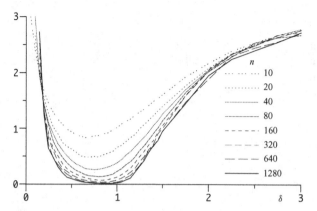

Figure 5.3 MISE versus δ for estimate (5.1) with $h_n = n^{-0.2}$ and $\gamma_n = n^{-\gamma}$, various n (example in Section 5.4).

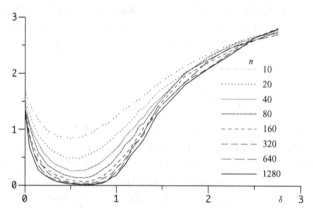

Figure 5.4 MISE versus δ for estimate (5.2) with $h_n = n^{-0.2}$ and $\gamma_n = n^{-\gamma}$, various n (example in Section 5.4).

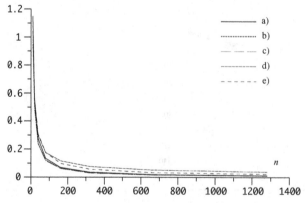

Figure 5.5 MISE versus n for kernels estimates, a) – (3.3), b) – (4.1), c) – (4.2), d) – (5.1), e) – (5.2); $h_n = n^{-2/5}$, $\gamma_n = n^{-4/5}$ for (5.1) and $\gamma_n = n^{-1/2}$ for (5.2) (example in Section 5.4).

5.5   Auxiliary results, lemmas, and proofs

5.5.1   Auxiliary results

LEMMA 5.1 *Let (5.7) and (5.8) hold. Let for $n = 1, 2, \ldots,$*

$$\xi_n = (1 - \gamma_n A_n)\xi_{n-1} + \gamma_n B_n$$

with $\xi_0 = B_0/A_0$, where $A_n \to A$, $A \neq 0$, $B_n \to B$, as $n \to \infty$. Then,

$$\xi_n \to \frac{B}{A} \text{ as } n \to \infty.$$

Proof. Denoting $c_n = \gamma_n A_n$, $n = 1, 2, \ldots$, we get

$$\xi_n = (1 - c_n)\xi_{n-1} + c_n D_n$$

with $D_n = B_n/A_n$. Therefore, $\xi_n = \sum_{i=0}^{n} d_i(n)D_i$, where

$$d_i(n) = \begin{cases} c_i \prod_{j=i+1}^{n}(1 - c_j), & \text{for } i = 0, 1, \ldots, n - 1 \\ c_n, & \text{for } i = n. \end{cases}$$

with $c_0 = 1$. Denoting $s_n = \sum_{i=0}^{n} d_i(n)$ we get $s_n = (1 - c_n)s_{n-1} + c_n$. Observing $s_1 = 1$, we obtain

$$\sum_{i=0}^{n} d_i(n) = 1. \tag{5.13}$$

Now let

$$\rho_i = \begin{cases} c_0, & \text{for } i = 0 \\ \dfrac{c_i}{\prod_{j=1}^{i}(1 - c_j)}, & \text{for } i = 1, 2, \ldots. \end{cases} \tag{5.14}$$

Using (5.13), we obtain $\sum_{i=1}^{n} \rho_i \prod_{j=1}^{n}(1 - c_j) = 1$. Thus, $d_i(n) = \rho_i / \sum_{j=1}^{n} \rho_j$, which leads to

$$\xi_n = \frac{\sum_{i=0}^{n} \rho_i D_i}{\sum_{j=0}^{n} \rho_j}. \tag{5.15}$$

Observing $\sum_{n=0}^{\infty} c_n = \sum_{n=0}^{\infty} \gamma_n A_n$, recalling that A_n converges to a positive limit as $n \to \infty$ and using (5.8), we find $\sum_{n=0}^{\infty} c_n = \infty$. On the other hand, from (5.14) we conclude that, since $c_n \to 0$ as $n \to \infty$, we have $c_n \leq \rho_n$ for n large enough. Finally $\sum_{n=0}^{\infty} \rho_n = \infty$. Owing to that, (5.15) and Toeplitz lemma 4.4, ξ_n and D_n have the same limit as $n \to \infty$. Now because $D_n \to B/A$ as $n \to \infty$, we obtain $\xi_n \to B/A$ as $n \to \infty$ and complete the proof. ∎

The following lemma given can be found in Chung [53] or Fabian [85]. The final part of our proof is somewhat simpler since we apply our Lemma 5.1.

LEMMA 5.2 *Let*

$$\xi_n = \left(1 - \frac{A_n}{n^\alpha}\right)\xi_{n-1} + \frac{B_n}{n^{\alpha+\beta}},$$
(5.16)

where $0 < \alpha \le 1, 0 < \beta$, *and where* $A_n \to A$, $A \ne 0$, $B_n \to B$, *as* $n \to \infty$. *Let C equal A for* $0 < \alpha < 1$, *and* $A - \beta$ *for* $\alpha = 1$, *respectively. If* $C > 0$, *then, for any* ξ_0,

$$n^\beta \xi_n \to \frac{B}{C} \text{ as } n \to \infty.$$

Proof. Observe that $(1 + n^{-1})^\beta = 1 + \beta_n n^{-1}$ with $\beta_n \to \beta$ as $n \to \infty$. Thus,

$$\left(1 + \frac{1}{n}\right)^\beta \left(1 - \frac{A_n}{n^\alpha}\right) = 1 - \frac{A_n}{n^\alpha} + \frac{\beta_n}{n} + \frac{A_n \beta_n}{n^{1+\alpha}} = 1 - \frac{C_n}{n^\alpha}$$

with $C_n \to C$ as $n \to \infty$. Multiplying left- and right-hand sides of (5.16) by $(n + 1)^\beta$ and $n^\beta(1 + n^{-1})^\beta$, respectively, and denoting $\lambda_n = (n + 1)^\beta \xi_n$, we obtain

$$\lambda_n = (1 - \gamma_n)\lambda_{n-1} + \gamma_n \frac{b_n}{C_n}$$

with $b_n = B_n(1 + n^{-1})^\beta$ and $\gamma_n = C_n n^{-\alpha}$. Since $b_n/C_n \to B/C$ as $n \to \infty$, an application of Lemma 5.1 yields $\lambda_n \to B/C$ as $n \to \infty$, and completes the proof. ∎

LEMMA 5.3 *Let* $\lim_{n\to\infty} \gamma_n = 0$ *and let* $\sum_{n=1}^\infty |\beta_n| < \infty$. *Then,*

$$\lim_{n\to\infty} \sum_{i=1}^n \gamma_i \beta_{n-i} = 0.$$

Proof. Fix a positive ε. Since γ_n converges to zero, there exists N such that $|\gamma_n| < \varepsilon$ for $n > N$. Therefore, the examined quantity equals $\sum_{i=1}^N \gamma_i \beta_{n-i} + \sum_{i=N+1}^n \gamma_i \beta_{n-1}$. The first term is bounded in absolute value by $(\sup_n |\beta_n|) \sum_{i=1}^N |\gamma_{n-i}|$ and converges to zero as n increases to infinity since $\gamma_{n-i} \to 0$ as $n \to \infty$. Observing that the absolute value of the other term is not greater than $\varepsilon \sum_{i=1}^\infty |\beta_i|$ and noticing that ε can be arbitrarily small, we complete the proof. ∎

LEMMA 5.4 *Let* $0 < q < 1$. *Then,*

$$\sum_{i=1}^n \frac{1}{i^\alpha} q^{n-i} = O\left(\frac{1}{n^\alpha}\right).$$

Proof. It suffices to notice that

$$\sum_{i=1}^n \frac{1}{i^\alpha} q^{n-i} = \frac{1}{n^\alpha} \sum_{i=0}^{n-1} \frac{n^\alpha}{(n-i)^\alpha} q^i \le \frac{1}{n^\alpha} \sum_{i=0}^{n-1} i^\alpha q^i \le \frac{c}{n^\alpha},$$

where $c = \sum_{i=0}^\infty i^\alpha q^i$. ∎

5.5.2   Lemmas

LEMMA 5.5 (BIAS OF (5.3)) *Let U have a probability density and let $Em^2(U) < \infty$. Let the kernel $K(\bullet)$ satisfy (3.4), (3.5), and (3.6) with $\varepsilon = 0$. Let number sequences $\{h_n\}$ and $\{\gamma_n\}$ satisfy (5.6)–(5.8). Then,*

$$E \hat{\mu}_n(u) \to \mu(u) \text{ as } n \to \infty$$

at every $u \in R$ where both $m(\bullet)$ and $f(\bullet)$ are continuous, and $f(u) > 0$.

Proof. As usual, for the sake of the proof, $Em(U) = 0$. Let $u \in R$ be a point where both $m(u)$ and $f(u)$ are continuous, and $f(u) > 0$. We begin by rewriting (5.1) in the following form:

$$\hat{\mu}_n(u) = \left[1 - \frac{\gamma_n}{h_n} K\left(\frac{u - U_n}{h_n}\right)\right] \hat{\mu}_{n-1}(u) + \frac{\gamma_n}{h_n} Y_{n+p} K\left(\frac{u - U_n}{h_n}\right). \tag{5.17}$$

Since $\hat{\mu}_{n-1}(u)$ and U_n are independent,

$$E \hat{\mu}_n(u) = (1 - \gamma_n K_n(u)) E \hat{\mu}_{n-1}(u) + \gamma_n L_n(u) \tag{5.18}$$

with

$$K_n(u) = \frac{1}{h_n} E \left\{ K\left(\frac{u - U}{h_n}\right)\right\}, \tag{5.19}$$

and

$$L_n(u) = \frac{1}{h_n} E \left\{ \mu(U) K\left(\frac{u - U}{h_n}\right)\right\}. \tag{5.20}$$

Now, applying (5.6) and Lemma A.8, we find that $L_n(u)$ and $K_n(u)$ converge to $f(u)\mu(u) \int K(v)dv$ and $f(u) \int K(v)dv$ as $n \to \infty$, respectively. Using then (5.7), (5.8), and Lemma 5.1, we find $E \hat{\mu}_n(u)$ converging to the same limit as $L_n(u)/K_n(u)$ as $n \to \infty$. Since the limit equals $\mu(u)$, we have completed the proof. ∎

LEMMA 5.6 *Let U have a probability density and let $Em^2(U) < \infty$. Let the kernel $K(\bullet)$ satisfy (3.4), (3.5), and (3.6) with $\varepsilon = 0$. Let number sequences $\{h_n\}$ and $\{\gamma_n\}$ satisfy (5.6)–(5.9). Then,*

$$\text{cov}[\hat{\mu}_n(u), c^T A X_{n+1}] = O(1) \sum_{i=1}^{n} \gamma_i \left\| A^{n-i} \right\|$$

at every $u \in R$ where both $m(\bullet)$ and $f(\bullet)$ are continuous. If (3.6) holds with some $\varepsilon > 0$, the convergence takes place at every Lebesgue $u \in R$ point of both $m(\bullet)$ and $f(\bullet)$.

Proof. In the proof, $u \in R$ is a point where both $m(u)$ and $f(u)$ are continuous and $f(u) > 0$. Since (5.6) holds $K_n(u) \to f(u) \int K(v)dv$ and $L_n(u) \to f(u)\mu(u) \int K(v)dv$ as $n \to \infty$, see Lemma A.8, where $f(u) \int K(v)dv > 0$. We recall that $K_n(u)$ and $L_n(u)$ are defined in (5.19) and (5.20), respectively. Moreover, since (5.7) is satisfied, for the sake of simplicity, we assume that

$$\varepsilon < 1 - \gamma_n K_n(u) < 1, \tag{5.21}$$

for all n, for some $\varepsilon > 0$. We define

$$\beta_i(u; n) = \begin{cases} \prod_{j=i+1}^{n} \left(1 - \frac{\gamma_j}{h_j} K\left(\frac{u - U_j}{h_j}\right)\right), & \text{for } i = 0, \ldots, n-1 \\ 1, & \text{for } i = n \end{cases}$$

with $\gamma_0 = 1$, and observe that, for $i < j$,

$$\text{cov}[m(U_j), \beta_i(u; n)] = -\gamma_j \, \text{cov}\left[m(U), \frac{1}{h_j} K\left(\frac{u - U}{h_j}\right)\right] \prod_{\substack{p=i+1 \\ p \neq j}}^{n} (1 - \gamma_p K_p(u)).$$

By (5.21), and the fact that the covariance in the expression converges as $j \to \infty$, we get

$$\text{cov}[m(U_j), \beta_i(u; n)] = \gamma_j O_j(1) \prod_{p=i+1}^{n} (1 - \gamma_p K_p(u)). \tag{5.22}$$

We now pass to the main part of the proof. Iterating (5.17), we get

$$\hat{\mu}_n(u) = \sum_{i=0}^{n} \gamma_i \beta_i(u; n) Y_{i+1} \frac{1}{h_i} K\left(\frac{u - U_i}{h_i}\right),$$

which leads to

$$\text{cov}[X_{n+1}, \hat{\mu}_n(u)] = c^T \sum_{i=0}^{n} \gamma_i \, \text{cov}\left[X_{n+1}, \beta_i(u; n) Y_{i+1} \frac{1}{h_i} K\left(\frac{u - U_i}{h_i}\right)\right]$$

$$= S_1(u) + S_2(u)$$

with

$$S_1(u) = \sum_{i=0}^{n} \gamma_i c^T A^{n-i} \, \text{cov}\left[X_{i+1}, \beta_i(u; n) Y_{i+1} \frac{1}{h_i} K\left(\frac{u - U_i}{h_i}\right)\right]$$

and

$$S_2(u) = \sum_{i=0}^{n} \gamma_i \sum_{j=i+1}^{n} c^T A^{n-j} b \, \text{cov}\left[m(U_j), \beta_i(u; n) Y_{i+1} \frac{1}{h_i} K\left(\frac{u - U_i}{h_i}\right)\right].$$

We have used here the fact that $X_{n+1} = A^{n-i} X_{i+1} + \sum_{j=i+1}^{n} A^{n-j} bm(U_j)$.
Because $\beta_i(u; n)$ is independent of both U_i and X_{i+1}, we obtain

$$S_1(u) = \sum_{i=1}^{n} \gamma_i c^T A^{n-i} C_i(u) E \beta_i(u; n)$$

with

$$C_i(u) = \text{cov}\left[c^T X_{i+1}, Y_{i+1} \frac{1}{h_i} K\left(\frac{u - U_i}{h_i}\right)\right].$$

Having $Y_{i+1} = c^T X_{i+1} + Z_{i+1}$ we find $C_i(u) = K_i(u) \operatorname{cov}[X, X] c$ converging as $i \to \infty$. Therefore, since $|E\beta_i(n; u)| \le 1$, we get

$$\|S_1(u)\| = O(1) \sum_{i=1}^{n} \gamma_i \|A^{n-i}\|.$$

Since, for $i < j$, Y_{i+1} and U_i are independent of both $\beta_i(u; n)$ and U_j, we have

$$S_2(u) = \sum_{i=1}^{n} \gamma_i L_i(u) \sum_{j=i+1}^{n} c^T A^{n-j} b \operatorname{cov}[m(U_j), \beta_i(u; n)].$$

Using (5.22), we find the quantity equal to

$$O(1) \sum_{i=1}^{n} \frac{L_i(u)}{K_i(u)} \left[\gamma_i K_i(u) \prod_{p=i+1}^{n} (1 - \gamma_p K_p(u)) \right] \sum_{j=i+1}^{n} c^T A^{n-j} b \gamma_j$$

$$= O(1) \sum_{i=1}^{n} \frac{L_i(u)}{K_i(u)} \delta_i(n; u) \sum_{j=i+1}^{n} c^T A^{n-j} b \gamma_j,$$

where $\delta_i(n; u) = \gamma_i K_i(u) E\beta_i(u; n)\}$. Since $L_n(u)/K_n(u) \to \mu(u)$ as $n \to \infty$, we obtain

$$S_2(u) = O(1) \sum_{i=1}^{n} \delta_i(n; u) \sum_{j=i+1}^{n} \gamma_j \|A^{n-j}\| = O(u) \sum_{i=1}^{n} \gamma_i \|A^{n-i}\|$$

since, by (5.13), $\sum_{i=1}^{n} \delta_i(n; u) = 1$. The proof has thus been completed. ∎

LEMMA 5.7 (VARIANCE OF (5.3)) *Let U have a probability density and let $Em(U) = 0$ and $Em^2(U) < \infty$. Let the kernel $K(\bullet)$ satisfy (3.4), (3.5), and (3.6) with $\varepsilon = 0$. Let number sequences $\{h_n\}$ and $\{\gamma_n\}$ satisfy (5.6)–(5.9). Then*

$$\operatorname{var}[\hat{\mu}_n(u)] \to 0 \text{ as } n \to \infty$$

at every $u \in R$ where both $m(\bullet)$ and $f(\bullet)$ are continuous, and $f(u) > 0$.

Proof. Fix $u \in R$ and assume that both $m(u)$ and $f(u)$ are continuous at the point u and that $f(u) > 0$. From (5.17) we get

$$\operatorname{var}[\hat{\mu}_n(u)] = V_1(u) + V_2(u) + 2V_3(u),$$

where

$$V_1(u) = \operatorname{var}\left[\left(1 - \frac{\gamma_n}{h_n} K\left(\frac{u - U_n}{h_n}\right) \right) \hat{\mu}_{n-1}(u) \right],$$

$$V_2(u) = \frac{\gamma_n^2}{h_n^2} \operatorname{var}\left[Y_{n+1} K\left(\frac{u - U_n}{h_n}\right) \right],$$

and

$$V_3(u) = \operatorname{cov}\left[\left(1 - \frac{\gamma_n}{h_n} K\left(\frac{u - U_n}{h_n}\right) \right) \hat{\mu}_{n-1}(u), \frac{\gamma_n}{h_n} Y_{n+p} K\left(\frac{u - U_n}{h_n}\right) \right].$$

For independent X, Y, we have $\mathrm{var}[XY] = EX^2 \, \mathrm{var}[Y] + E^2 Y \, \mathrm{var}[X]$. Thus, since U_n is independent of $\hat{\mu}_{n-1}(u)$, we obtain

$$V_1(u) = (1 - \gamma_n a_n(u)) \, \mathrm{var}[\hat{\mu}_{n-1}(u)] + \frac{\gamma_n^2}{h_n}(M_n(u) - h_n K_n^2(u)) E^2 \hat{\mu}_{n-1}(u),$$

where

$$M_n(u) = \frac{1}{h_n} E K^2 \left(\frac{u - U}{h_n} \right),$$

and

$$a_n(u) = 2 K_n(u) - \frac{\gamma_n}{h_n} M_n(u). \tag{5.23}$$

In view of Lemma A.8 and (5.9), we find $M_n(u)$ convergent as $n \to \infty$ and get

$$a_n(u) \to a(u) \text{ as } n \to \infty \tag{5.24}$$

with $a(u) = 2f(u) \int K(v) dv > 0$. Since, moreover, by Lemma 5.5, $E\hat{\mu}_{n-1}(u)$ also converges as $n \to \infty$, we finally obtain

$$V_1(u) = (1 - \gamma_n a_n(u)) \, \mathrm{var}[\hat{\mu}_{n-1}(u)] + \frac{\gamma_n^2}{h_n} O(1).$$

In turn, $V_2(u) = (\gamma_n^2 / h_n) P_n(u)$ with

$$P_n(u) = \frac{1}{h_n} \, \mathrm{var} \left[Y_1 K \left(\frac{u - U_0}{h_n} \right) \right]$$

converging as $n \to \infty$, see Lemma A.8 again. Thus,

$$V_2(u) = \frac{\gamma_n^2}{h_n} O(1).$$

Passing to V_3 we observe that $Y_{n+p} = c^T A^p X_n + c^T A^{p-1} bm(U_n) + \xi_{n+1} + Z_{n+p}$ with

$$\xi_{n+1} = \sum_{i=n+1}^{n+p-1} c^T A^{n+p-1-i} bm(U_i).$$

Since Z_{n+p} is independent of both U_n and $\hat{\mu}_{n-1}(u)$, we have $V_3(u) = V_{31}(u) + V_{32}(u) + V_{33}(u)$ with

$$V_{31}(u) = \gamma_n \, \mathrm{cov} \left[\left(1 - \frac{\gamma_n}{h_n} K \left(\frac{u - U_n}{h_n} \right) \right) \hat{\mu}_{n-1}(u), \frac{1}{h_n} K \left(\frac{u - U_n}{h_n} \right) c^T A^p X_n \right],$$

$$V_{32}(u) = c^T A^{p-1} b \gamma_n \, \mathrm{cov} \left[\left(1 - \frac{\gamma_n}{h_n} K \left(\frac{u - U_n}{h_n} \right) \right) \hat{\mu}_{n-1}(u), \frac{1}{h_n} K \left(\frac{u - U_n}{h_n} \right) m(U_n) \right],$$

and

$$V_{33}(u) = \gamma_n \, \mathrm{cov} \left[\left(1 - \frac{\gamma_n}{h_n} K \left(\frac{u - U_n}{h_n} \right) \right) \hat{\mu}_{n-1}(u), \frac{1}{h_n} K \left(\frac{u - U_n}{h_n} \right) \xi_n \right].$$

For any random variables X, Y, V, W, such that pairs (X, Y) and (V, W) are independent, we have $\text{cov}[XV, YW] = E\{XY\}\text{cov}[V, W] + \text{cov}[X, Y]\,EVEW$. Owing to this and the fact that $EX = 0$, we obtain

$$V_{31}(u) = \gamma_n \left(K_n(u) - \frac{\gamma_n}{h_n} M_n(u) \right) \text{cov}[\hat\mu_{n-1}(u), c^T A^p X_n].$$

Since both $K_n(u)$ and $M_n(u)$ converge as $n \to \infty$, recalling (5.9) and applying Lemma 5.6, we find

$$V_{31}(u) = \gamma_n O(1) \sum_{i=1}^{n} \gamma_i \left\| A^{n-i} \right\| + \frac{\gamma_n^2}{h_n} O(1).$$

Using again the fact that $\hat\mu_{n-1}(u)$ and U_n are independent, we get

$$V_{32}(u) = -c^T A^{p-1} b \gamma_n E\hat\mu_{n-1}(u)$$
$$\times \text{cov}\left[\frac{\gamma_n}{h_n} K\left(\frac{u - U_n}{h_n} \right), \frac{1}{h_n} K\left(\frac{u - U_n}{h_n} \right) \mu(U_n) \right]$$
$$= -c^T A^{p-1} b \frac{\gamma_n^2}{h_n} (S_n(u) - h_n K_n(u) L_n(u)) E\hat\mu_{n-1}(u)$$

with

$$S_n(u) = \frac{1}{h_n} E\left\{ \mu(U) K^2 \left(\frac{u - U}{h_n} \right) \right\}$$

converging as $n \to \infty$, see Lemma A.8. Hence,

$$V_{32}(u) = O(1) \frac{\gamma_n^2}{h_n}.$$

Since ξ_{n+1} is independent of both U_n and $\hat\mu_{n-1}(u)$, we obtain $V_{33}(u) = 0$. Thus,

$$V_3(u) = O(1)\gamma_n \sum_{i=1}^{n} \gamma_i \left\| A^{n-i} \right\| + \frac{\gamma_n^2}{h_n} O(1).$$

In this way, we have finally shown that

$$\text{var}[\hat\mu_n(u)] \leq (1 - \gamma_n a_n(u))\,\text{var}[\hat\mu_{n-1}(u)] + \gamma_n d_n(u), \tag{5.25}$$

where

$$d_n(u) = O(1) \left(\frac{\gamma_n}{h_n} + \sum_{i=1}^{n} \gamma_i \left\| A^{n-i} \right\| \right).$$

We now apply (5.9) and Lemma 5.3 to find that $d_n(u) \to 0$ as $n \to \infty$. Thus, in view of (5.7), (5.8), and (5.24), an application of Lemma 5.1 completes the proof. ∎

5.5.3   Proofs

Proof of Theorem 5.2

Since the proof is very similar to that of Theorem 5.1, we focus our attention on main points only. As far as bias is concerned, we have

$$E\tilde{\mu}_n(u) = (1 - \gamma_n h_n K_n(u))E\tilde{\mu}_{n-1}(u) + \gamma_n h_n L_n(u)$$

and easily verify that $E\tilde{\mu}_n(u) \to \mu(u)$ as $n \to \infty$. Examining variance, we employ the "almost everywhere" version of Lemma A.8 and obtain

$$\text{var}[\tilde{\mu}_n(u)] \le (1 - \gamma_n h_n A_n(u))\, \text{var}[\tilde{\mu}_{n-1}(u)] + \gamma_n h_n B_n(u), \qquad (5.26)$$

where

$$B_n(u) = O(1)\left(\gamma_n + \sum_{i=1}^{n} \gamma_i h_i \left\| A^{n-i} \right\|\right)$$

with some $A_n(u) \to a(u)$ as $n \to \infty$, where $a(u)$ is as in (5.23). Application of Lemma 5.1 completes the proof. ■

5.6   Bibliographic notes

Estimate (5.1) has been studied by Györfi [138], Révész [250], and Györfi and Walk [141]. All of those authors have, however, assumed that Y_i and Y_j are independent for $i \ne j$. In other words, in their works $A = 0$, that is, the dynamic subsystem is a simple delay, which means that the whole system is just memoryless. Dynamics is present in Greblicki [117], where both (5.1) and (5.2) were examined.

6   Orthogonal series algorithms

This chapter is devoted to another class of nonparametric regression estimators making use of ideas from orthogonal series. Various orthogonal representations are used starting with trigonometric, or Fourier, series in Section 6.2. Nonparametric estimates stemming from the class of classical orthogonal polynomials are studied in Section 6.3 (Legendre polynomials), Section 6.4 (Laguerre polynomials), and Section 6.5 (Hermite polynomials). Wavelet type orthogonal expansions are examined in Section 6.6. Here, the detailed studies are confined to the Haar wavelet orthogonal basis.

6.1   Introduction

We begin with the presentation of the idea behind the class of orthogonal series algorithms. Suppose that the values of the input signal are in a set D, which can be an interval or a half real line $[0, \infty)$ or the whole real line R.

Let $\varphi_0, \varphi_1, \ldots$ be a system of orthonormal functions in D, that is, let

$$\int_D \varphi_i(u)\varphi_j(u)du = \begin{cases} 1, & \text{for } i = j \\ 0, & \text{otherwise.} \end{cases}$$

Assuming that $\int_D |f(u)|du < \infty$, we expand the density $f(u)$ of U in the series, that is, write

$$f(u) \sim \sum_{k=0}^{\infty} \beta_k \varphi_k(u) \tag{6.1}$$

with

$$\beta_k = \int_D f(u)\varphi_k(u)du = E\varphi_k(U).$$

The expression $S_N(u) = \sum_{k=0}^{N} \beta_k \varphi_k(u)$ is a partial sum of the expansion. Estimating coefficients β_ks in a natural way, that is, defining its estimate as

$$\hat{\beta}_k = \frac{1}{n} \sum_{i=1}^{n} \varphi_k(U_i),$$

we take

$$\hat{f}(u) = \sum_{k=0}^{N(n)} \hat{\beta}_k \varphi_k(u)$$

as an estimate of $f(u)$, where $N(n)$ is a sequence of integers increasing to infinity with n.

Denoting $g(u) = \mu(u)f(u)$ and assuming that $\int_D |g(u)| < \infty$, that is, that $\int_D |m(u)f(u)|du = E|m(U)| < \infty$, for similar reasons, we write

$$g(u) \sim \sum_{k=0}^{\infty} \alpha_k \varphi_k(u),$$

where

$$\alpha_k = \int_D \varphi_k(u)\mu(u)f(u)du = E\{\mu(U)\varphi_k(U)\} = E\{Y_p\varphi_k(U_0)\}.$$

Therefore, we define

$$\hat{g}(u) = \sum_{k=0}^{N(n)} \hat{\alpha}_k \varphi_k(u),$$

where

$$\hat{\alpha}_k = \frac{1}{n} \sum_{i=1}^{n} Y_{p+i} \varphi_k(U_i),$$

as an estimate of $g(u) = \mu(u)f(u)$.

In light of this,

$$\hat{\mu}(u) = \frac{\displaystyle\sum_{i=1}^{n} Y_{p+i}\varphi_k(U_i)}{\displaystyle\sum_{i=1}^{n}\varphi_k(U_i)}$$

is an orthogonal series estimate of the regression function $\mu(u) = E\{Y_p|U_0 = u\}$. The function

$$K_N(u, v) = \sum_{k=0}^{N} \varphi_k(u)\varphi_k(v)$$

is called a kernel function of the orthonormal system $\{\varphi_k(u)\}$, while the number N is called the order of the kernel. Using the kernel, we can rewrite the estimate in the following form:

$$\hat{\mu}(u) = \frac{\displaystyle\sum_{i=1}^{n} Y_{p+i} K_{N(n)}(u, (U_i))}{\displaystyle\sum_{i=1}^{n} K_{N(n)}(u, (U_i))}$$

close to the kernel estimate examined in Chapter 3.

Different orthogonal series should be applied for different D, for $D = [-\pi, \pi]$ and $D = [-1, 1]$, the Fourier and Legendre series are obvious choices. For $D = [0, \infty)$

Table 6.1. Convergence rate for numerators and denominators of various estimates.

Estimate	Bias	Variance	MSE	$f^{(q)}$ and $g^{(q)}$
Fourier	$O(N^{-q+\frac{1}{2}}(n))$	$O\left(\dfrac{N(n)}{n}\right)$		square
Legendre	$O(N^{-q+\frac{1}{2}}(n))$	$O\left(\dfrac{N(n)}{n}\right)$		integrable
Laguerre	$O(N^{-\frac{q}{2}+\frac{1}{4}}(n))$	$O\left(\dfrac{N^{\frac{1}{2}}(n)}{n}\right)$	$O(n^{-1+\frac{1}{2q}})$	or
Hermite	$O(N^{-\frac{q}{2}+\frac{1}{4}}(n))$	$O\left(\dfrac{N^{\frac{1}{2}}(n)}{n}\right)$		similar
kernel	$O(h_n^{q-\frac{1}{2}})$	$O\left(\dfrac{1}{nh_n}\right)$		property
	$O(h_n^{q})$	$O\left(\dfrac{1}{nh_n}\right)$	$O(n^{-1+\frac{1}{2q+1}})$	bounded

the Laguerre system can be applied, for $D = (-\infty, \infty)$, we can use the Hermite functions. We can also employ wavelets, for example, for $D = [-1/2, 1/2]$, we use Haar ones. Fundamental properties of these orthogonal systems are presented in consecutive sections of Appendix B. Also, useful lemmas are derived therein.

Basic results concerning the rate of pointwise convergence of estimates with various orthogonal systems are given in Table 6.1. In the fourth column, the mean square error (MSE for brevity) defined as $E(\hat{f}(u) - f(u))^2$, where $\hat{f}$ is the estimate of the denominator in the appropriate estimate, is given. We recall that the rate holds also for $\hat{g}(u)$, that is, the estimate of the numerator $g(u) = \mu(u)f(u)$. In the last row, we give properties of the kernel estimate. It follows from the table that the overall quality of all estimates is very much the same.

6.2   Fourier series estimate

Here $D = [-\pi, \pi]$ and $|U| \leq \pi$. To estimate the characteristic, we apply the trigonometric orthonormal system

$$\frac{1}{\sqrt{2\pi}}, \frac{1}{\sqrt{\pi}}\cos u, \frac{1}{\sqrt{\pi}}\sin u, \frac{1}{\sqrt{\pi}}\cos 2u, \frac{1}{\sqrt{\pi}}\sin 2u, \ldots,$$

(see Section B.2). We assume that $\int_{-\pi}^{\pi} |f(u)|du < \infty$ and $\int_{-\pi}^{\pi} |g(u)|du < \infty$. Clearly,

$$g(u) \sim a_0 + \sum_{k=1}^{\infty} a_k \cos ku + \sum_{k=1}^{\infty} b_k \sin ku,$$

where

$$a_0 = \frac{1}{2\pi} \int_{-\pi}^{\pi} g(u)du,$$

and, for $k = 1, 2, \ldots,$

$$a_k = \frac{1}{\pi} \int_{-\pi}^{\pi} g(u) \cos(ku)du, \quad b_k = \frac{1}{\pi} \int_{-\pi}^{\pi} g(u) \sin(ku)du.$$

For the same reasons,

$$f(u) \sim \frac{1}{2\pi} + \sum_{k=1}^{\infty} \alpha_k \cos ku + \sum_{k=1}^{\infty} \beta_k \sin ku$$

where

$$\alpha_k = \frac{1}{\pi} \int_{-\pi}^{\pi} f(u) \cos(ku)du, \quad \beta_k = \frac{1}{\pi} \int_{-\pi}^{\pi} g(u) \sin(ku)du.$$

Therefore, the trigonometric series estimate of $\mu(u)$ has the following form:

$$\hat{\mu}(u) = \frac{\hat{a}_0 + \sum_{k=1}^{N(n)} \hat{a}_k \cos ku + \sum_{k=1}^{N(n)} \hat{b}_k \sin ku}{\frac{1}{2\pi} + \sum_{k=1}^{N(n)} \hat{\alpha}_k \cos ku + \sum_{k=1}^{N(n)} \hat{\beta}_k \sin ku}$$

with

$$\hat{a}_0 = \frac{1}{2\pi n} \sum_{i=1}^{n} Y_{p+i}, \quad \hat{a}_k = \frac{1}{\pi n} \sum_{i=1}^{n} Y_{p+i} \cos kU_i,$$

$$\hat{b}_k = \frac{1}{\pi n} \sum_{i=1}^{n} Y_{p+i} \sin kU_i,$$

and

$$\hat{\alpha}_k = \frac{1}{\pi n} \sum_{i=1}^{n} \cos kU_i, \quad \hat{\beta}_k = \frac{1}{\pi n} \sum_{i=1}^{n} \sin kU_i.$$

Applying the complex version of the Fourier system (see Section B.2), we get

$$g(u) \sim \sum_{k=-\infty}^{\infty} c_k e^{iku} \quad \text{and} \quad f(u) \sim \sum_{k=-\infty}^{\infty} d_k e^{iku},$$

where

$$c_k = \frac{1}{2\pi} \int_{-\pi}^{\pi} e^{-ikv} g(v)dv \quad \text{and} \quad d_k = \frac{1}{2\pi} \int_{-\pi}^{\pi} e^{-ikv} f(v)dv,$$

and rewrite the estimate as

$$\hat{\mu}(u) = \frac{\sum\limits_{k=-N(n)}^{N(n)} \hat{c}_k e^{iku}}{\sum\limits_{k=-N(n)}^{N(n)} \hat{d}_k e^{iku}},$$

where

$$\hat{c}_k = \frac{1}{2\pi n} \sum_{j=1}^{n} Y_{p+j} e^{-ikU_j} \quad \text{and} \quad \hat{d}_k = \frac{1}{2\pi n} \sum_{j=1}^{n} e^{-ikU_j}.$$

Using the Dirichlet kernel, see Section B.2,

$$D_n(u) = \frac{\sin\left(n + \frac{1}{2}\right) u}{2\pi \sin \frac{1}{2} u},$$

we can also write

$$\hat{\mu}(u) = \frac{\sum\limits_{i=1}^{n} Y_{p+i} D_{N(n)}(U_i - u)}{\sum\limits_{i=1}^{n} D_{N(n)}(U_i - u)}.$$

THEOREM 6.1 *Let $Em^2(U) < \infty$. Let $\int_{-\pi}^{\pi} |f(u)| du < \infty$. If*

$$N(n) \to \infty \text{ as } n \to \infty \tag{6.2}$$

and

$$\frac{N(n)}{n} \to 0 \text{ as } n \to \infty, \tag{6.3}$$

then

$$\hat{\mu}(u) \to \mu(u) \text{ as } n \to \infty \text{ in probability}$$

at every $u \in (-\pi, \pi)$ where both $m(\bullet)$ and $f(\bullet)$ are differentiable and $f(u) > 0$.

The proof of the theorem, in which Corollary B.2 has been applied, is given in Section 6.9. Using Corollary B.2 rather than B.1, we easily verify the next result in which $f(\bullet)$ is a Lipschitz function.

THEOREM 6.2 *Under the restrictions of Theorem 6.1,*

$$\hat{\mu}(u) \to \mu(u) \text{ as } n \to \infty \text{ in probability}$$

at every $u \in (-\pi, \pi)$ where both $f(\bullet)$ and $m(\bullet)$ satisfy a Lipschitz condition, and $f(u) > 0$.

In turn, making use of Carleson and Hunt result, (see Theorem B.2), we get a result on convergence at almost every point.

THEOREM 6.3 *Let $Em^2(U) < \infty$. Let, moreover, $\int_{-\pi}^{\pi} |f(u)|^s du < \infty$ and $\int_{-\pi}^{\pi} |g(u)|^s$ $du < \infty$ with some $s > 1$. If (6.2) and (6.3) hold, then*

$$\hat{\mu}(u) \to \mu(u) \text{ as } n \to \infty \text{ in probability}$$

at almost every $u \in (-\pi, \pi)$, where $f(u) > 0$.

To establish the speed where the estimate converges we apply (B.14). Assuming that $m^{(i)}(\pm\pi) = f^{(i)}(\pm\pi) = 0$, for $i = 0, 1, \ldots, (q-1)$, and that $\int_{-\pi}^{\pi}(f^{(q)}(v))^2 dv < \infty$ and $\int_{-\pi}^{\pi}(g^{(q)}(v))^2 dv < \infty$, we find

$$\text{bias}[\hat{g}(u)] = E\hat{g}(u) - g(u) = O(N^{-q+1/2}(n)).$$

Recalling (6.9), we write

$$E(\hat{g}(u) - g(u))^2 = \text{bias}^2[\hat{g}(u)] + \text{var}[\hat{g}(u)]$$

$$= O(N^{-2q+1}(n)) + O\left(\frac{N(n)}{n}\right)$$

and by selecting

$$N(n) \sim n^{1/2q}, \tag{6.4}$$

we obtain $E(\hat{g}(u) - g(u))^2 = O(n^{-1+1/2q})$. Since the same rate holds for $\hat{f}(u)$, applying Lemma C.7, we conclude that

$$|\hat{\mu}(u) - \mu(u)| = O(n^{-1/2+1/4q}) \text{ in probability}$$

and

$$P\{|\hat{\mu}(u) - \mu(u)| > \varepsilon|\mu(u)|\} = O(n^{-1+1/2q})$$

for every $\varepsilon > 0$.

6.3   Legendre series estimate

In this section, $|U| \leq 1$ and $D = [-1, 1]$. Legendre orthogonal polynomials $P_0(u)$, $P_1(u), \ldots$ satisfy Rodrigues' formula

$$P_k(u) = \frac{1}{2^k k!} \frac{d^k}{du^k}(u^2 - 1)^k,$$

(see Section B.3). The system $p_0(u), p_1(u), \ldots$, where

$$p_k(u) = \sqrt{\frac{2k+1}{2}} P_k(u),$$

is orthonormal. Making use of the system, we obtain the following estimate:

$$\tilde{\mu}(u) = \frac{\displaystyle\sum_{k=0}^{N(n)} \tilde{a}_k p_k(u)}{\displaystyle\sum_{k=0}^{N(n)} \tilde{b}_k p_k(u)},$$

where

$$\tilde{a}_k = \frac{1}{n}\sum_{i=1}^{n} Y_{p+i}\, p_k(U_i) \quad \text{and} \quad \tilde{b}_k = \frac{1}{n}\sum_{i=1}^{n} p_k(U_i)$$

are estimates of

$$a_k = \int_{-1}^{1} \mu(v) f(v) p_k(v)\,dv = E\{Y_p p_k(U_0)\}$$

and

$$b_k = \int_{-1}^{1} f(v) p_k(v)\,dv = E\{p_k(U)\},$$

respectively.

The estimate can be rewritten in the so-called kernel form:

$$\tilde{\mu}(u) = \frac{\displaystyle\sum_{i=1}^{n} Y_{p+i} k_{N(n)}(u, U_i)}{\displaystyle\sum_{i=1}^{n} k_{N(n)}(u, U_i)},$$

where

$$k_n(u, v) = \sum_{k=0}^{n} p_k(u) p_k(v) = \frac{n+1}{2}\frac{P_n(u)P_{n+1}(v) - P_{n+1}(u)P_n(v)}{v - u}$$

is the kernel of the Legendre system.

THEOREM 6.4 *Let* $Em^2(U) < \infty$. *Let* $\int_{-1}^{1} |f(u)|(1 - u^2)^{-1/4}du < \infty$ *and* $\int_{-1}^{1} |f(u)$ $m(u)|(1 - u^2)^{-1/4}du < \infty$. *If (6.2) and (6.3) hold, then*

$$\hat{\mu}(u) \to \mu(u) \text{ as } n \to \infty \text{ in probability}$$

at every $u \in (-1, 1)$ *where both* $m(\bullet)$ *and* $f(\bullet)$ *are differentiable and* $f(u) > 0$.

The next two theorems are obvious if we refer to Corollaries B.4 and B.5. Applying Corollary B.2 rather than Corollary B.1, we easily verify the next result.

THEOREM 6.5 *Under the restrictions of Theorem 6.4,*

$$\tilde{\mu}(u) \to \mu(u) \text{ as } n \to \infty \text{ in probability}$$

at every $u \in (-1, 1)$, where both $f(\bullet)$ and $m(\bullet)$ satisfy a Lipschitz condition, and $f(u) > 0$.

THEOREM 6.6 Let $\int_{-1}^{1} |f(u)|^s du < \infty$ with any $s > 1$. Under the restrictions of Theorem 6.4,

$$\tilde{\mu}(u) \to \mu(u) \text{ as } n \to \infty \text{ in probability}$$

at almost every $u \in (-1, 1)$, where $f(u) > 0$.

Assuming that both $f(\bullet)$ and $m(\bullet)$ have q derivatives, $\int_{-1}^{1}(f^{(q)}(v))^2 dx < \infty$, $\int_{-1}^{1}(g^{(q)}(v))^2 dv < \infty$ and $f^{(i)}(\pm 1) = 0 = g^{(i)}(\pm 1) = 0$, $i = 0, 1, \ldots, q - 1$, we observe that bias$[\tilde{g}(u)] = O(N^{-q+1/2})$, see (B.26). Applying (6.10) we thus find $E(\tilde{g}(u) - g(u))^2 = O(N^{-2q+1}) + O(N(n)/n)$. Therefore, for $N(n)$ as in (6.4), $E(\tilde{g}(u) - g(u))^2$ and, for similar reasons, $E(\tilde{f}(u) - f(u))^2$ are of order $O(n^{-1+1/2q})$. Finally, by applying Lemma C.7, we conclude that

$$|\tilde{\mu}(u) - \mu(u)| = O(n^{-1/2+1/4q}) \text{ in probability},$$

and

$$P\{|\tilde{\mu}(u) - \mu(u)| > \varepsilon |\mu(u)|\} = O(n^{-1+1/2q})$$

for every $\varepsilon > 0$. The convergence rate is the same as that obtained for the Fourier series estimate.

6.4   Laguerre series estimate

Now $U \geq 0$ and $D = [0, \infty)$. We apply Laguerre orthonormal functions $l_0(u), l_1(u), \ldots$, where

$$l_k(u) = e^{-u/2} L_k(u)$$

with Laguerre polynomials $L_0(u), L_1(u), \ldots$ defined by Rodrigues' formula

$$L_k(u) = e^u \frac{1}{k!} \frac{d^k}{du^k}(u^k e^{-u}),$$

(see Section B.4), to obtain the following estimate:

$$\bar{\mu}(u) = \frac{\sum\limits_{k=0}^{N(n)} \bar{a}_k l_k(u)}{\sum\limits_{k=0}^{N(n)} \bar{b}_k l_k(u)},$$

where

$$\bar{a}_k = \frac{1}{n} \sum_{i=1}^{n} Y_{p+i} l_k(U_i) \quad \text{and} \quad \bar{b}_k = \frac{1}{n} \sum_{i=1}^{n} l_k(U_i).$$

The kernel representation has the following form:

$$\bar{\mu}(u) = \frac{\displaystyle\sum_{i=1}^{n} Y_{p+i} k_{N(n)}(u, U_i)}{\displaystyle\sum_{i=1}^{n} k_{N(n)}(u, U_i)},$$

where

$$k_n(u, v) = \sum_{k=0}^{n} l_k(u) l_k(v) = (n+1)\frac{l_n(u)l_{n+1}(v) - l_{n+1}(u)l_n(v)}{v - u}$$

is the kernel of the Laguerre system.

THEOREM 6.7 *Let $Em^2(U) < \infty$. If (6.2) holds and*

$$\frac{\sqrt{N(n)}}{n} \to 0 \text{ as } n \to \infty, \tag{6.5}$$

then

$$\bar{\mu}(u) \to \mu(u) \text{ as } n \to \infty \text{ in probability}$$

at every $u \in (0, \infty)$, where both $m(\bullet)$ and $f(\bullet)$ are differentiable and $f(u) > 0$.

THEOREM 6.8 *Under the restrictions of Theorem 6.7,*

$$\bar{\mu}(u) \to \mu(u) \text{ as } n \to \infty \text{ in probability}$$

at every $u \in (0, \infty)$, where both $m(\bullet)$ and $f(\bullet)$ satisfy a Lipschitz condition and $f(u) > 0$.

THEOREM 6.9 *Let $\int_{-1}^{1} |f(u)|^s du < \infty$ with any $s > 1$. Under the restrictions of Theorem 6.7,*

$$\bar{\mu}(u) \to \mu(u) \text{ as } n \to \infty \text{ in probability}$$

at almost every $u \in (0, \infty)$, where $f(u) > 0$.

Suppose that some integral $g^{(q)}(\bullet)$ is involved in is finite, (see Section B.4), and further that $\lim_{u \to \infty} g(u)e^{-u/2+\delta} = 0$ and $\lim_{u \to 0} u^i g^{(i)}(u) = 0$, for $i = 0, 1, 2, \ldots, q - 1$. Then, see (B.33), bias$[\bar{g}(u)] = O(N^{-q/2+1/4})$. This, combined with (6.11), yields $E(\bar{g}(u) - g(u))^2 = O(N^{-q+1/2}) + O(N^{1/2}(n)/n)$. Thus, for

$$N \sim n^{1/q}, \tag{6.6}$$

we get $E(\bar{g}(u) - g(u))^2 = O(n^{-1+1/2q})$. For similar reasons, the rate holds for $\bar{f}(u)$, which leads to

$$|\bar{\mu}(u) - \mu(u)| = O(n^{-1/2+1/4q}) \text{ in probability,}$$

and

$$P\{|\bar{\mu}(u) - \mu(u)| > \varepsilon|\mu(u)|\} = O(n^{-1+1/2q})$$

for every $\varepsilon > 0$, that is, the same rate as that derived for the Fourier and Legendre series estimates.

6.5   Hermite series estimate

For any U, we can use the Hermite orthonormal system $h_0(u), h_1(u), \ldots$ where

$$h_k(u) = \frac{1}{\sqrt{2^k k! \sqrt{\pi}}} H_k(u) e^{-u^2/2},$$

with the Hermite polynomials satisfying Rodrigues' formula

$$H_k(u) = (-1)^k e^{u^2} \frac{d^k}{du^k} e^{-u^2},$$

(see Section B.5). The system is orthonormal over the whole real line $(-\infty, \infty)$.
     The estimate is of the following form:

$$\check{\mu}(u) = \frac{\displaystyle\sum_{k=0}^{N(n)} \check{a}_k h_k(u)}{\displaystyle\sum_{k=0}^{N(n)} \check{b}_k h_k(u)},$$

where

$$\check{a}_k = \frac{1}{n} \sum_{i=1}^{n} Y_{p+i} h_k(U_i) \quad \text{and} \quad \check{b}_k = \frac{1}{n} \sum_{i=1}^{n} h_k(U_i).$$

Using the kernel function

$$k_n(u, v) = \sum_{k=0}^{n} h_k(u) h_k(v) = (n+1) \frac{h_n(u) h_{n+1}(v) - h_{n+1}(u) h_n(v)}{v - u},$$

we write

$$\check{\mu}(u) = \frac{\displaystyle\sum_{i=1}^{n} Y_{p+i} k_{N(n)}(u, U_i)}{\displaystyle\sum_{i=1}^{n} k_{N(n)}(u, U_i)}.$$

THEOREM 6.10 *Let $Em^2(U) < \infty$. If (6.2) and (6.5) hold, then*

$$\check{\mu}(u) \to \mu(u) \text{ as } n \to \infty \text{ in probability}$$

at every $u \in (-\infty, \infty)$, where both $m(\bullet)$ and $f(\bullet)$ are differentiable and $f(u) > 0$.

If, for $i = 0, 1, \ldots, q - 1$, $\lim_{|u| \to \infty} g^{(i)}(u) e^{-u^2/2+\delta} = 0$, with some $\delta > 0$, and some integral $g^{(q)}$ is involved in is finite, bias$[\check{g}(u)] = O(n^{-q/2+1/4})$, see (B.46). As var$[\check{g}(u)] = O(N^{1/2}(n)/n)$, for $N(n)$ selected as in (6.6), $E(\check{g}(u) - g(u))^2 = O(n^{-1+1/2q})$. For similar reasons, under appropriate restrictions imposed on $f(\bullet)$, $E(\check{f}(u) - f(u))^2 = O(n^{-1+1/2q})$. Therefore,

$$|\check{\mu}(u) - \mu(u)| = O(n^{-1/2+1/4q}) \text{ in probability}$$

and

$$P\{|\check{\mu}(u) - \mu(u)| > \varepsilon|\mu(u)|\} = O(n^{-1+1/2q})$$

for every $\varepsilon > 0$. The rate is the same as that derived for the Fourier, Legendre, and Hermite series estimates.

6.6 Wavelet estimate

In this section, we apply Haar wavelets, see Section B.6. Now $-1/2 \leq U \leq 1/2$ and $D = [-1/2, 1/2]$. For each m, wavelets

$$\phi_{m,0}, \phi_{m,1}, \phi_{m2}, \ldots, \phi_{m,2^m-1}$$

are defined as

$$\phi_{mn}(x) = 2^{m/2}\phi(2^m x - n),$$

where $\phi_{mn}(x) = I_{A_{mn}}(x)$ and, where

$$A_{mn} = \left[-\frac{1}{2} + \frac{n}{2^m}, -\frac{1}{2} + \frac{n+1}{2^m} \right).$$

We define the following wavelet expansions:

$$S_m(g) = 2^m \sum_{k=0}^{2^m-1} a_{mk}\phi_{mk}(u)$$

with $a_{mk} = \int_{-1//2}^{1/2} g(u)\phi_{mk}(u)du = E\{Y_p\phi_{mk}(U_0)\}$ and

$$S_m(f) = 2^m \sum_{k=0}^{2^m-1} b_{mk}\phi_{mk}(u)$$

with $b_{mk} = \int_{-1/2}^{1/2} f(u)\phi_{mk}(u)du = E\phi_{mk}(U)$ of $g(\bullet)$ and $f(\bullet)$, respectively. Therefore, the estimate of $\mu(u)$ has the following form:

$$\check{\mu}(u) = \frac{\sum_{k=0}^{2^{M(n)}-1} \check{a}_{M(n),k}\phi_{M(n),k}(u)}{\sum_{k=0}^{2^{M(n)}-1} \check{b}_{M(n),k}\phi_{M(n),k}(u)}$$

with

$$\breve{a}_{M(n),k} = \frac{1}{n} \sum_{i=1}^{n} Y_{p+i} \phi_{M(n),k}(U_i)$$

and

$$\breve{b}_{M(n),k} = \frac{1}{n} \sum_{i=1}^{n} \phi_{M(n),k}(U_i).$$

THEOREM 6.11 *Let $\int_0^1 |m(u)| du < \infty$. If*

$$M(n) \to \infty, \ as \ n \to \infty,$$

$$\frac{2^{M(n)}}{n} \to 0 \ as \ n \to \infty,$$

then

$$\breve{\mu}(u) \to \mu(u) \ as \ n \to \infty \ in \ probability$$

at every $u \in (-1, 1)$, where both $f(\bullet)$ and $m(\bullet)$ are continuous and $f(u) > 0$.

Assuming that both $m(\bullet)$ and $f(\bullet)$ satisfy a Lipschitz inequality, that is, that $|m(u) - m(v)| \le c_1 |u - v|$ and $|f(u) - f(v)| \le c_2 |u - v|$ for all $u, v \in [0, 1]$, where c_1 and c_2 are some constants, we notice that bias$[\breve{f}(u)] = O(1/2^{M(n)})$, see (B.47). As var$[\breve{f}(u)] = O(2^{M(n)}/n)$, we thus get

$$E(\breve{f}(u) - f(u))^2 = O\left(\frac{1}{2^{2M(n)}}\right) + O\left(\frac{2^{M(n)}}{n}\right).$$

Thus, selecting $2^{M(n)} \sim n^{1/3}$, for example,

$$M(n) = \frac{1 + \varepsilon_n}{3 \log 2} \log n,$$

where $\lim_{n \to \infty} \varepsilon_n = 0$, and we obtain $E(\breve{f}(u) - f(u))^2 = O(n^{-2/3})$. Since $E(\breve{g}(u) - g(u))^2$ diminishes at the same speed,

$$|\breve{\mu}(u) - \mu(u)| = O(n^{-1/3}) \ in \ probability$$

and

$$P\{|\breve{\mu}(u) - \mu(u)| > \varepsilon |\mu(u)|\} = O(n^{-2/3})$$

for every $\varepsilon > 0$. The rate is the same as that of the kernel estimate obtained for both $m(\bullet)$ and $f(\bullet)$ with bounded derivatives, that is, under very similar restrictions (see Table 6.1).

6.7   Local and global errors

In previous sections we examined local properties of estimates employing Fourier, Legendre, Laguerre, and Hermite series estimates. Their properties are very much alike. Roughly speaking, for $f(\bullet)$ and $m(\bullet)$ having q derivatives,

$$|\mu_n(u) - \mu(u)| = O(n^{-1/2+1/4q}) \ in \ probability,$$

and

$$P\{|\mu_n(u) - \mu(u)| > \varepsilon|\mu(u)|\} = O(n^{-1+1/2q})$$

for every $\varepsilon > 0$, where $\mu_n(u)$ is any of those four estimates. The rate holds, however, if $f(\bullet)$ and $m(\bullet)$ behave appropriately not only in a neighborhood of the point u but also in the whole set D, and, in particular, at its ends, where D is understood as in Section 6.1.

Furthermore, let us assume that both $m(\bullet)$ and Z are bounded, and that the input density $f(u)$ is well separated from zero, that is, $f(u) > \epsilon$ for all $u \in D$, some positive ϵ. Then, we can apply Lemma C.10 and, for the Fourier estimate $\hat{\mu}(u)$, we obtain the following bound for the global mean integrated squared error (MISE)

$$E(\hat{\mu}(U) - \mu(U))^2$$
$$\leq 2 \int_D \frac{1}{f(u)} E(\hat{g}(u) - g(u))^2 du + 2C^2 \int_D \frac{1}{f(u)} E(\hat{f}(u) - f(u))^2 du. \quad (6.7)$$

Apart from the conclusion that

$$\lim_{n\to\infty} E(\hat{\mu}(U) - \mu(U))^2 = 0$$

we can also conclude that

$$E(\hat{\mu}(U) - \mu(U))^2 = O(n^{-1+1/2q}),$$

provided that $N(n)$ satisfies (6.4) and both $f(\bullet)$ and $m(\bullet)$ have q-derivatives and satisfy some additional restrictions mentioned in Section 6.2. The rate holds also for the other estimates if $N(n)$ is selected in the way suggested in consecutive sections. It is worth mentioning that the rate $O(n^{-2q/(2q+1)})$ obtained for the MISE is optimal for the class of $f(\bullet)$ and $m(\bullet)$ possessing q-bounded derivatives, see [286].

6.8 Simulation example

In the simulation example, as usual, the system is the same as that in Section 2.2.2 with $a = 0.5$. Since we estimate not on the interval $[-\pi, \pi]$, we apply the system

$$\frac{1}{\sqrt{6}}, \frac{1}{\sqrt{6}} \cos\frac{\pi}{3}u, \frac{1}{\sqrt{3}} \sin\frac{\pi}{3}u, \frac{1}{\sqrt{3}} \cos\frac{2\pi}{3}u, \frac{1}{\sqrt{3}} \sin\frac{2\pi}{3}u, \ldots$$

orthonormal over the interval $[-3, 3]$. This adaptation and the fact that $p = 1$ lead to the following estimate:

$$\hat{m}(u) = \frac{\sum_{i=1}^{n} Y_{1+i} D_{N(n)}\left(\frac{\pi}{3}(U_i - u)\right)}{\sum_{i=1}^{n} D_{N(n)}\left(\frac{\pi}{3}(U_i - u)\right)}.$$

Results are shown in Figures 6.1, 6.2, and 6.3. The MISE versus N, that is, for $N(n)$ fixed and equal to N, are shown in Figure 6.1. It is clear that large N is not recommended.

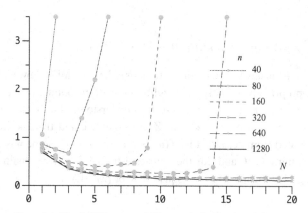

Figure 6.1 The MISE versus N (example in Section 6.8).

For $N(n) = n^{1/4}$, the MISE varying with n is presented in Figure 6.2. Examples of the estimate are in Figure 6.3.

6.9   Lemmas and proofs

6.9.1   Lemmas

LEMMA 6.1 *Let* $Em(U) = 0$ *and* $\text{var}[m(U)] < \infty$ *and let* $i \neq 0$. *If* $\int_{-\pi}^{\pi} |f(v)|dv < \infty$, *then*

$$\sup_N \left| \text{cov} \left[W_{p+i} D_N(U_i - u), W_p D_N(U_0 - u) \right] \right|$$

$$\leq (|\lambda_p \lambda_{p+i}| + |\lambda_p \lambda_{p-i}| + |\lambda_{p+i} \lambda_{p-i}|)\rho(u),$$

where $\rho(u)$ *is finite at every* $u \in (-\pi, \pi)$, *where both* $m(\bullet)$ *and* $f(\bullet)$ *are differentiable, or satisfy a Lipschitz condition. If* $\int_{-\pi}^{\pi} |f(v)|^p dv < \infty$ *with* $p > 0$, *then the inequality holds at almost every* $u \in (-\pi, \pi)$.

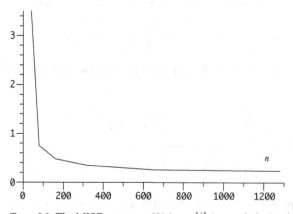

Figure 6.2 The MISE versus n, $N(n) = n^{1/4}$ (example in Section 6.8).

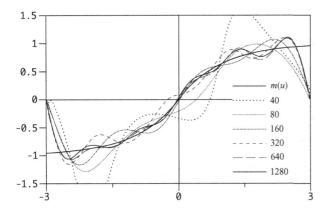

Figure 6.3 Realizations of the estimate (example in Section 6.8).

Proof. The proof is similar to that of Lemma 3.2. Bearing in mind that $W_{p+i} = \sum_{q=-\infty}^{p+i} \lambda_{p+i-q} m(U_q)$ and $W_p = \sum_{r=-\infty}^{p} \lambda_{p-r} m(U_r)$, we find the covariance in the assertion equal to

$$\sum_{q=-\infty}^{p+i} \sum_{r=-\infty}^{p} \lambda_{p+i-q} \lambda_{p-r} \, \mathrm{cov} \left[m(U_q) D_N(U_i - u), m(U_r) D_N(U_0 - u) \right]$$

$$= (\lambda_p \lambda_{p+i} + \lambda_p \lambda_{p-i}) E \left\{ D_N(U - u) \right\} E \left\{ m^2(U) D_N(U - u) \right\}$$

$$+ \lambda_{p+i} \lambda_{p-i} E^2 D_N(U - u),$$

(see Lemma C.2). An application of Corollaries B.1 and B.2 as well as Theorem B.2 completes the proof. ∎

In a similar way, for the Legendre, Laguerre, and Hermite series, we can show that, under appropriate restrictions,

$$\sup_N | \mathrm{cov}[W_{p+i} k_N(u, U_i), W_p k_N(u, U_0)]|$$

$$\leq (|\lambda_p \lambda_{p+i}| + |\lambda_p \lambda_{p-i}| + |\lambda_{p+i} \lambda_{p-i}|) \psi(u), \tag{6.8}$$

where $k_N(\bullet, \bullet)$ is the kernel of the series, and where $\psi(u)$ is finite at specific points, that is, at every u where both $m(\bullet)$ and $f(\bullet)$ are differentiable, or satisfy a Lipschitz condition or at almost every u.

6.9.2 Proofs

Proof of Theorem 6.1

Fix $u \in (-\pi, \pi)$ and suppose that both $m(\bullet)$ and $f(\bullet)$ are differentiable at the point u. Denoting $g(u) = \mu(u) f(u)$ and

$$\hat{g}(u) = \frac{1}{n} \sum_{i=1}^{n} Y_{p+i} D_{N(n)}(U_i - u),$$

we get

$$E\hat{g}(u) = E\{Y_p D_{N(n)}(U_0 - u)\} = E\{E\{Y_p|U_0\}D_{N(n)}(U_0 - u)\}$$

$$= E\{\mu(U)D_{N(n)}(U - u)\} = \int_{-\pi}^{\pi} D_{N(n)}(v - u)\mu(v)f(v)dv$$

$$= \int_{-\pi}^{\pi} D_{N(n)}(v - u)g(v)dv.$$

Thus, by Corollary B.1 and (6.2),

$$\lim_{n\to\infty} \int_{-\pi}^{\pi} D_{N(n)}(v - u)g(v)dv = g(u).$$

In turn,

$$\mathrm{var}[\hat{g}(u)] = P_n(u) + Q_n(u)$$

with

$$P_n(u) = \frac{1}{n}\, \mathrm{var}\left[Y_p D_{N(n)}(U_0 - u)\right],$$

and, since disturbance Z_n is independent of the input signal,

$$Q_n(u) = \frac{1}{n^2}\sum_{i=1}^{n}\sum_{j=1,j\neq i}^{n} \mathrm{cov}\left[Y_{p+i}D_{N(n)}(U_i - u), Y_{p+j}D_{N(n)}(U_j - u)\right]$$

$$= \frac{1}{n^2}\sum_{i=1}^{p}(n - i)\,\mathrm{cov}\left[Y_{p+i}D_{N(n)}(U_i - u), Y_p D_{N(n)}(U_0 - u)\right]$$

$$+ \frac{1}{n^2}\sum_{i=p+1}^{n}(n - i)\,\mathrm{cov}\left[Y_{p+i}D_{N(n)}(U_i - u), Y_p D_{N(n)}(U_0 - u)\right].$$

As $\pi D_n^2(u) = (n + 1/2)F_{2n+1}(u)$, where $F_n(u)$ is the Fejèr kernel, and $E\{Y_p^2|U_0 = u\} = \sigma_Z^2 + \phi(u)$, see (2.7) and (B.7), we have

$$nP_n(u) = E\left\{Y_p^2 D_{N(n)}^2(U_0 - u)\right\} - E^2\left\{Y_p D_{N(n)}(U_0 - u)\right\}$$

$$= \frac{2N(n) + 1}{2\pi} E\left\{(\sigma_Z^2 + \phi(U))F_{2N(n)+1}(U - u)\right\} - E^2\left\{\mu(U)D_{N(n)}(U - u)\right\}.$$

By Corollary B.1 and Lebesgue's Theorem B.3,

$$E\left\{(\sigma_Z^2 + \phi(U))F_N(U - u)\right\} = \int_{-\pi}^{\pi}(\sigma_Z^2 + \phi(v))f(v)F_N(v - u)dv$$

and

$$E\left\{\mu(U)D_N(U - u)\right\} = \int_{-\pi}^{\pi}\mu(v)f(v)D_N(v - u)dv$$

converge, respectively, to $(\sigma_Z^2 + \phi(u))f(u)$ and $\mu(u)f(u)$ as $N \to \infty$. Hence,

$$\frac{n}{N(n)}P_n(u) \to (\sigma_Z^2 + \phi(u))f(u) \text{ as } n \to \infty.$$

To examine $Q_n(u)$, we apply Corollary B.1 once again and find that $E D_N (U - u)$, $E\{m(U)D_N(U - u)\}$, and $E\{m^2(U)D_N(U - u)\}$ all are bounded as $N \to \infty$. From this and Lemma 6.1, it follows that

$$|R_n(u)| \leq 2\rho(u)\frac{1}{n^2}\sum_{i=1}^{n}(n - i)(|\lambda_p\lambda_{p+i}| + |\lambda_p\lambda_{p-i}| + |\lambda_{p+i}\lambda_{p-i}|)$$

$$\leq 6\rho(u)(\max_n |\lambda_n|)\frac{1}{n}\sum_{i=1}^{\infty}|\lambda_i| = O\left(\frac{1}{n}\right).$$

Thus,

$$\lim_{n\to\infty} \frac{n}{N(n)} \text{var}[\hat{g}(u)] = (\sigma_Z^2 + \phi(u))f(u), \tag{6.9}$$

and, finally,

$$E(\hat{g}(u) - g(u))^2 \to 0, \quad \text{as } n \to \infty.$$

Since, for similar reasons, $E(\hat{f}(u) - f(u))^2 \to 0$, the theorem follows. ■

Proof of Theorem 6.4

The proof is similar to the proof of Theorem 6.1. Fix $u \in (-1, 1)$ and suppose that both $m(\bullet)$ and $f(\bullet)$ are differentiable at the point u. Denoting $g(u) = \mu(u)f(u)$ and

$$\tilde{g}(u) = \frac{1}{n}\sum_{i=1}^{n}Y_{p+i}k_{N(n)}(u, U_i),$$

we get

$$E\tilde{g}(u) = E\{Y_p k_{N(n)}(u, U_0)\} = E\{\mu(U)k_{N(n)}(u, U)\}$$

$$= \int_{-1}^{1} k_{N(n)}(u, v)\mu(v)f(v)dv = \int_{-1}^{1} k_{N(n)}(u, v)g(v)dv$$

which, by Corollary B.3, converges to $g(v) = \mu(u)f(u)$.
    In turn,

$$\text{var}[\tilde{g}(u)] = P_n(u) + Q_n(u)$$

with

$$P_n(u) = \frac{1}{n}\text{var}\left[Y_p k_{N(n)}(u, U_0)\right],$$

and, since disturbance Z_n is independent of the input signal,

$$Q_n(u) = \frac{1}{n^2} \sum_{i=1}^{n} \sum_{j=1,j\neq i}^{n} \text{cov}\left[Y_{p+i}k_{N(n)}(u, U_i), Y_{p+j}k_{N(n)}(u, U_j)\right]$$

$$= \frac{1}{n^2} \sum_{i=1}^{p} (n-i) \text{cov}\left[Y_{p+i}k_{N(n)}(u, U_i), Y_p k_{N(n)}(u, U_0)\right]$$

$$+ \frac{1}{n^2} \sum_{i=p+1}^{n} (n-i) \text{cov}\left[Y_{p+i}k_{N(n)}(u, U_i), Y_p k_{N(n)}(u, U_0)\right].$$

Clearly,

$$n P_n(u) = E\left\{Y_p^2 k_{N(n)}^2(u, U_0)\right\} - E^2\left\{Y_p k_{N(n)}(u, U_0)\right\}$$

$$= E\left\{(\sigma_Z^2 + \phi(U))k_{N(n)}^2(u, U)\right\} - E^2\left\{\mu(U)k_{N(n)}(u, U)\right\}.$$

By Corollary B.3 and Lemma B.3,

$$E\left\{\mu(U)k_N(u, U)\right\} = \int_{-1}^{1} k_N(u, v)g(v)dv$$

converges, and

$$\frac{1}{N}E\left\{(\sigma_Z^2 + \phi(U))k_N^2(u, U)\right\} = \frac{1}{N}\int_{-1}^{1}(\sigma_Z^2 + \phi(v))f(v)k_N^2(u, v)dv$$

converges to $\pi^{-1}(1-u^2)^{1/2}(\sigma_Z^2 + \phi(u))f(u)$ as $N \to \infty$. Thus,

$$\frac{n}{N(n)}P_n(u) \to \frac{1}{\pi\sqrt{1-u^2}}(\sigma_Z^2 + \phi(u))f(u) \text{ as } n \to \infty.$$

Proceeding as in the proof of Theorem 6.1 and using (6.8), we find

$$Q_n(u) = O\left(\frac{1}{n}\right)$$

and conclude that

$$\lim_{n\to\infty} \frac{n}{N(n)} \text{var}[\hat{g}(u)] = \frac{1}{\pi\sqrt{1-u^2}}(\sigma_Z^2 + \phi(u))f(u). \tag{6.10}$$

Hence,

$$E(\tilde{g}(u) - g(u))^2 \to 0 \text{ as } n \to \infty.$$

Since $E(\tilde{f}(u) - f(u))^2 \to 0$ as $n \to \infty$, the proof is complete. ∎

Proof of Theorem 6.7
The proof is similar to that of Theorems 6.1 and 6.4. We only focus our attention on differences. Now,

$$\frac{1}{\sqrt{N}}E\left\{(\sigma_Z^2 + \phi(U))k_N^2(u, U)\right\} = \frac{1}{N}\int_{-1}^{1}(\sigma_Z^2 + \phi(v))f(v)k_N^2(u, v)dv$$

converges to $\pi^{-1}u^{-1/2}f(u)$ as $n \to \infty$, see (B.4). Thus

$$\frac{n}{\sqrt{N(n)}}P_n(u) \to \frac{1}{\pi\sqrt{u}}(\sigma_Z^2 + \phi(u))f(u) \text{ as } n \to \infty,$$

and, as a consequence,

$$\frac{n}{\sqrt{N(n)}}\text{var}[\bar{g}(u)] \to \frac{1}{\pi\sqrt{u}}(\sigma_Z^2 + \phi(u))f(u) \text{ as } n \to \infty. \tag{6.11}$$

Hence,

$$E(\bar{g}(u) - g(u))^2 \to 0 \text{ as } n \to \infty.$$

Since $E(\bar{f}(u) - f(u))^2 \to 0$ as $n \to \infty$, the proof is complete. ∎

Proof of Theorem 6.10

The proof is similar to that of Theorems 6.1 and 6.4. The difference is that now

$$\frac{1}{\sqrt{N}}E\left\{(\sigma_Z^2 + \phi(U))k_N^2(u, U)\right\} = \frac{1}{N}\int_{-1}^{1}(\sigma_Z^2 + \phi(v))f(v)k_N^2(u, v)dv$$

converges to $\sqrt{2}\pi^{-1}f(u)$ as $N \to \infty$, see (B.5). Thus

$$\frac{n}{\sqrt{N(n)}}P_n(u) \to \frac{1}{\sqrt{2\pi}}(\sigma_Z^2 + \phi(u))f(u) \text{ as } n \to \infty,$$

and, consequently,

$$\frac{n}{\sqrt{N(n)}}\text{var}[\check{g}(u)] \to \frac{1}{\sqrt{2\pi}}(\sigma_Z^2 + \phi(u))f(u) \text{ as } n \to \infty.$$

Hence,

$$E(\check{g}(u) - g(u))^2 \to 0 \text{ as } n \to \infty.$$

Since $E(\check{f}(u) - f(u))^2 \to 0$ as $n \to \infty$, the proof is complete. ∎

Proof of Theorem 6.11

Denoting

$$\check{g}(u) = 2^{M(n)}\sum_{k=0}^{2^{M(n)}-1}\check{a}_{M(n),k}\phi_{M(n),k}(u)$$

and

$$\check{f}(u) = 2^{M(n)}\sum_{k=0}^{2^{M(n)}-1}\check{b}_{M(n),k}\phi_{M(n),k}(u),$$

we observe $\check{\mu}(u) = \check{g}(u)/\check{f}(u)$. Clearly, $E\check{b}_{M(n),k} = b_{M(n),k}$, and thus

$$E\check{f}(u) = 2^{M(n)}\sum_{k=0}^{2^{M(n)}-1}b_{M(n),k}\phi_{M(n),k}(u)$$

is the $M(n)$th partial sum of the expansion of $f(\bullet)$. Therefore, by Theorem B.15,

$$E \check{f}(u) \to f(u) \text{ as } n \to \infty$$

at every continuity point u of $f(\bullet)$.

Furthermore,

$$\check{f}(u) = \frac{1}{n} \sum_{n=1}^{n} k_{M(n)}(u, U_i),$$

where

$$k_n(u, v) = 2^m \sum_{n=0}^{2^m-1} \phi_n(v)\phi_n(v)$$

is the kernel function. Thus,

$$\text{var}[\check{f}(u)] = \frac{1}{n^2} \text{var}\left[\sum_{n=1}^{n} k_{M(n)}(u, U_i)\right] = \frac{1}{n} \text{var}\left[k_{M(n)}(u, U)\right]$$

$$\leq \frac{1}{n} E k_{M(n)}^2(u, U) = \frac{2^{M(n)}}{n} \frac{1}{2^{M(n)}} \int_{-1/2}^{1/2} k_{M(n)}^2(u, v) f(v) dv.$$

Applying Lemma B.6, we come to a conclusion that $E(\check{f}(u) - f(u))^2 \to 0$ as $n \to \infty$ at every point where $f(\bullet)$ is continuous.

Leaving the verification of the fact that $E(\check{g}(u) - g(u))^2 \to 0$ as $n \to \infty$ at every point where $g(\bullet)$ is continuous, to the reader, we complete the proof. ■

6.10   Bibliographic notes

The orthogonal density estimate was proposed by Cencov [42] and then studied by Van Ryzin [298], Schwartz [275], Crain [57], Kronmal and Tarter [180], Watson [313], Specht [281], Mirzahmedov and Hasimov [210], Greblicki [105], Wahba [306, 307], Bleuez and Bosq [25], Walter [309], Viollaz [303], Anderson and de Figueiredo [3], Hall [143–146], Greblicki and Rutkowski [137], Greblicki and Pawlak [123, 125], and Liebscher [196].

The regression function nonparametric orthogonal series estimates have been examined by Greblicki [105, 106], Greblicki, Rutkowska, and Rutkowski [136], Greblicki and Pawlak [124], and Chen and Tin [49].

Regarding the Hammerstein system identification, the nonparametric estimate with the trigonometric and Hermite series was applied by Greblicki [107]. In [131, 135], Greblicki and Pawlak employed the Legendre and Laguerre series. The approach has been also studied in Greblicki and Pawlak [132, 133], Krzyżak [181, 184], Krzyżak, Sąsiadek, and Kégl [189], and Pawlak [229]. In Pawlak and Greblicki [233] the nonparametric orthogonal series algorithm for identification of the Hammerstein system with a correlated input process is proposed. The use of Laguerre series in the identification problem of linear systems is summarized in Heuberger, Van den Hof, and Wahlberg [163]. The use of wavelets in the statistical inference is examined in Härdle, Kerkyacharian, Picard, and Tsybakov [151], Percival and Walden [238], and Vidakovic [302]. Wavelet expansions have been applied in system identification by Pawlak and Hasiewicz [234], Hasiewicz [154–156], Hasiewicz and Śliwiński [159], and Staszewski [283].

7 Algorithms with ordered observations

This chapter is concerned with nonparametric identification algorithms based on ordered input observations. Both kernel (Section 7.2) and orthogonal series (Section 7.3) estimates are examined.

7.1   Introduction

In this chapter, we assume that the input signal is bounded, that is, that support of the input density $f(\bullet)$ is an interval, say $[a, b]$. Therefore, it can be said that input observations $U_1, U_2, \ldots, U_n$, that is, points randomly scattered over the interval $[a, b]$, split the interval into subintervals. We rearrange the sequence $U_1, U_2, \ldots, U_n$ of input observations into a new one $U_{(1)}, U_{(2)}, \ldots, U_{(n)}$, in which $U_{(1)} < U_{(2)} < \cdots < U_{(n)}$. Ties, that is, events that $U_{(i)} = U_{(j)}$, for $i \neq j$, have zero probability, since U_ns have a density. Moreover, we define $U_{(0)} = a$ and $U_{(n+1)} = b$. The sequence $U_{(1)}, U_{(2)}, \ldots, U_{(n)}$ is called the order statistics of $U_1, U_2, \ldots, U_n$. We then rearrange the sequence

$$(U_1, Y_{p+1}), (U_2, Y_{p+2}), \ldots, (U_n, Y_{p+n}) \tag{7.1}$$

of input–output observations into the following one:

$$(U_{(1)}, Y_{[p+1]}), (U_{(2)}, Y_{[p+2]}), \ldots, (U_{(n)}, Y_{[p+n]}). \tag{7.2}$$

Observe that $Y_{[p+i]}$s are not ordered, but just paired with $U_{(i)}$s. Clearly

$$\mu(u) = E\left\{Y_{[p+i]}|U_{(i)} = u\right\} = E\left\{Y_{p+j}|U_j = u\right\}. \tag{7.3}$$

Algorithms examined in this chapter use ordered sequence (7.2) rather than (7.1). It makes possible for us to approximate some integrals by referring to the Riemann definition. As a result, we can obtain nonparametric estimates that do not suffer instability due to the ratio form of the previously introduced methods.

We assume that a density $f(\bullet)$ of the input signal exists and that

$$0 < \delta \leq f(u), \tag{7.4}$$

for all $u \in [a, b]$ and some $\delta > 0$. The characteristic $m(\bullet)$ is a Lipschitz function, that is,

$$|m(u) - m(v)| \leq c_m |u - v| \tag{7.5}$$

for some c_m and all $u, v \in [a, b]$, and, a fortiori, bounded. For convenience, we denote

$$\max_{a \leq u \leq b} |m(u)| = M. \tag{7.6}$$

Unless otherwise stated, all integrals are taken over the whole real line R.

7.2   Kernel estimates

In this section, $-1 \leq U \leq 1$, that is, $[a, b] = [-1, 1]$.

7.2.1   Motivation and estimates

The motivation for the kernel estimate studied in Chapter 3 is based on the observation that, for the kernel such that $\int K(v)dv = 1$,

$$\lim_{h \to 0} \int_{-1}^{1} \mu(v) \frac{1}{h} K\left(\frac{u-v}{h}\right) dv = \mu(u) \tag{7.7}$$

since

$$\frac{1}{h} K\left(\frac{u-v}{h}\right)$$

gets close to a Dirac delta located at the point u as $h \to 0$. The idea of the kernel estimates examined in this chapter is based on the numerical evaluation of the integral in (7.7) from the rearranged version of the training sequence. Hence, according to the Riemann definition,

$$S_n(u; h) = \sum_{i=1}^{n} \mu(U_{(i)}) \frac{1}{h} K\left(\frac{u - U_{(i)}}{h}\right) (U_{(i)} - U_{(i-1)})$$

is a natural approximation of the integral in (7.7). Thus, since

$$\max_{1 \leq i \leq n+1} (U_{(i)} - U_{(i-1)}) \to 0 \text{ as } n \to \infty \text{ almost surely,}$$

(see Darling [59] and the Slud Lemma C.11), then

$$S_n(u; h) \to \int_{-1}^{1} \mu(v) \frac{1}{h} K\left(\frac{u-v}{h}\right) dv \text{ as } h \to 0$$

almost surely, as well. Thus, (7.3) and (7.7) suggest the following estimate of $\mu(u)$:

$$\hat{\mu}(u) = \sum_{i=1}^{n} Y_{[p+i]} \frac{1}{h_n} K\left(\frac{u - U_{(i)}}{h_n}\right) (U_{(i)} - U_{(i-1)})$$

$$= \sum_{i=1}^{n} Y_{[p+i]} \frac{1}{h_n} \int_{U_{(i-1)}}^{U_{(i)}} K\left(\frac{u - U_{(i)}}{h_n}\right) dv \tag{7.8}$$

and its modification

$$\tilde{\mu}(u) = \sum_{i=1}^{n} Y_{[p+i]} \frac{1}{h_n} \int_{U_{(i-1)}}^{U_{(i)}} K\left(\frac{u-v}{h_n}\right) dv, \tag{7.9}$$

which is expected to behave even better. Some improvement is also possible by application of symmetrical spacings, that is, $(U_{(i+1)} - U_{(i-1)})/2$ instead of $(U_{(i)} - U_{(i-1)})$.

7.2.2   Consistency and convergence rate

The kernel satisfies the standard restriction

$$\sup_{-\infty < u < \infty} |K(u)| < \infty \tag{7.10}$$

and is such that

$$\int K(u) du = 1. \tag{7.11}$$

Since

$$Y_{p+n} = \mu(U_n) + \xi_{p+n} + Z_{p+n} \tag{7.12}$$

with $\xi_{p+n} = \sum_{i=-\infty, i \neq n}^{p+n} c^T A^{p+n-i-1} bm(U_i)$, observations of $\mu(U_n)$ are disturbed by external white noise Z_{p+n} and by the correlated ξ_{p+n} noise, which is incurred by the system. We recall that, after ordering,

$$(U_i, \xi_{p+i}, Z_{p+i}, Y_{p+i})$$

becomes

$$(U_{(j)}, \xi_{[p+j]}, Z_{[p+j]}, Y_{[p+j]})$$

with some j, that is, that $U_i = U_{(j)}, \xi_{p+i} = \xi_{[p+j]}, Z_{p+i} = Z_{[p+j]}, Y_{p+i} = Y_{[p+j]}$. Indices at $\xi_{[p+j]}, Z_{[p+j]}, Y_{[p+j]}$ are induced by that of $U_{(j)}$. Consistency of the algorithm is established in the following theorem whose proof, like all the others, is in Section 7.4.

THEOREM 7.1 *Let $f(\bullet)$ satisfy (7.4). Let $m(\bullet)$ satisfy (7.5). Let the kernel $K(\bullet)$ satisfy (7.10) and (7.11). Let the number sequence be such that*

$$h_n \to 0, \text{ as } n \to \infty, \tag{7.13}$$

$$nh_n \to \infty \text{ as } n \to \infty. \tag{7.14}$$

Then,

$$\lim_{n \to \infty} E \int (\tilde{\mu}(u) - \mu(u))^2 du = 0.$$

Estimate (7.8) behaves slightly worse because we do not take full advantage of the fact that $K(\bullet)$ is known. In (7.9), the integral

$$\int_{U_{(i-1)}}^{U_{(i)}} K\left(\frac{u-v}{h_n}\right) dv$$

appears while, in (7.8), only its approximation

$$K\left(\frac{u-U_{(i)}}{h_n}\right)(U_{(i)} - U_{(i-1)})$$

is used.

THEOREM 7.2 *Let $f(\bullet)$ satisfy (7.4). Let $m(\bullet)$ satisfy (7.5). Let the kernel $K(\bullet)$ satisfy (7.10), (7.11), and a Lipschitz inequality*

$$|K(u) - K(v)| \leq c_K |u - v| \qquad (7.15)$$

for some c_k, for all $u, v \in R$. Let the number sequence satisfy (7.13) and

$$nh_n^2 \to 0 \text{ as } n \to \infty.$$

Then,

$$E \int (\hat{\mu}(u) - \mu(u))^2 du \to 0 \text{ as } n \to \infty.$$

We now assume that $m(\bullet)$ has q derivatives. The kernel $K(\bullet)$ has bounded support, that is, that $K(u) > 0$ on a bounded set, and satisfies restrictions (3.15) and (3.16), that is, the same as those in Section 3.4.

Let $\epsilon > 0$ and let $u \in (-1 + \epsilon, 1 - \epsilon)$. Since, for h small enough, support of $K(\bullet)$ is a subset of

$$\left[\frac{-1-u}{h}, \frac{1-u}{h}\right],$$

we can write

$$\mu(u) = \mu(u) \int_{-1}^{1} K(v)dv = \mu(u) \int_{-(1+u)/h}^{(1-u)/h} K(-v)dv.$$

Since

$$\mu_h(u) = \int_{-1}^{1} \mu(v)\frac{1}{h}K\left(\frac{u-v}{h}\right) dv = \int_{-(1+u)/h}^{(1-u)/h} \mu(u+hv)K(-v)dv,$$

we get

$$\mu_h(u) - \mu(u) = \int_{-(1+u)/h}^{(1-u)/h} [\mu(u+hv) - \mu(u)] K(-v)dv.$$

Expanding $\mu(u)$ in a Taylor series, assuming that $\int (m^{(q)}(u))^2 du < \infty$, and making use of (A.17), we find $|\mu_h(u) - \mu(u)| \leq ch^{q-1/2}$ for some c, which yields

$$\int_{-1+\epsilon}^{1-\epsilon} (\mu_h(u) - \mu(u))^2 du = O\left(h^{2q-1}\right).$$

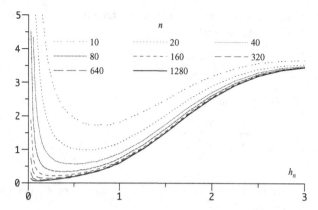

Figure 7.1 MISE versus h_n; $a = 0.5$; various n.

Recalling (7.25), we conclude that

$$E \int_{-1+\epsilon}^{1-\epsilon} (\tilde{\mu}(u) - \mu(u))^2 du = O(h_n^{2q-1}) + O\left(\frac{1}{nh_n}\right).$$

Therefore, for $h_n \sim n^{1/2q}$,

$$E \int_{-1+\epsilon}^{1-\epsilon} (\tilde{\mu}(u) - \mu(u))^2 du = O(n^{-1+1/2q}).$$

If $m^{(q)}(\bullet)$ is bounded, we apply (A.18) to get $|\mu_h(u) - \mu(u)| \leq ch^q$, select $h_n \sim n^{-1/(2q+1)}$ and obtain

$$E \int_{-1+\epsilon}^{1-\epsilon} (\tilde{\mu}(u) - \mu(u))^2 du = O(n^{-1+1/(2q+1)}). \tag{7.16}$$

Contrary to other kernel algorithms, the rate is independent of the shape of the probability density $f(\bullet)$ of the input signal, provided that the density is bounded from zero. Irregularities of $f(\bullet)$ do not worsen the rate. It depends however on the number q of existing derivatives of $m(\bullet)$. The larger q, that is, the smoother characteristic, the greater speed of convergence. For large q, the rate gets very close to n^{-1}, that is, the rate typical for the parametric inference.

7.2.3   Simulation example

The system is the same as that in Section 2.2.2. We assume that $a = 0.5$ and $Z_n = 0$. The estimate of $m(u)$ is of the following form:

$$\hat{m}(u) = \sum_{i=1}^{n} Y_{[1+i]} \frac{1}{h_n} K\left(\frac{u - U_{(i)}}{h_n}\right) (U_{(i)} - U_{(i-1)}). \tag{7.17}$$

The MISE versus h_n is shown in Figure 7.1. Realizations of the estimate are in Figure 7.2.

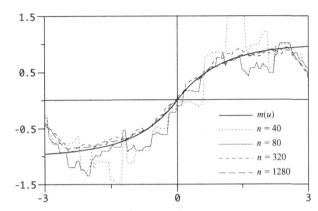

Figure 7.2 Realizations of estimate (7.17) in which $h_n = 0.25n^{-2/5}$.

7.3 Orthogonal series estimates

7.3.1 Motivation

Let $\{\varphi_0, \varphi_1, \varphi_2, \ldots\}$ be an orthonormal complete system of functions over the interval $[a, b]$, see Appendix B. Since $EY^2 < \infty$, $\int_a^b |\mu(v)|dv < \infty$, we write

$$\mu(u) \sim \sum_{k=0}^{\infty} a_k\varphi_k(u),$$

where

$$a_k = \int_a^b \mu(u)\varphi_k(u)du,$$

$k = 0, 1, 2, \ldots$. Because $\mu(u) = E\{Y_{[p+i]}|U_{(i)} = u\}$, the coefficient a_k, that is, the integral can be estimated in the following way:

$$\hat{a}_k = \sum_{i=1}^{n} Y_{[p+i]}\varphi_k(U_{(i)})(U_{(i)} - U_{(i-1)}).$$

Therefore,

$$\hat{\mu}(u) = \sum_{k=0}^{N(n)} \hat{a}_k\varphi_k(u),$$

is the estimate of $\mu(u)$. An alternative form of the estimate $\hat{\mu}(u)$ can be obtained by using the kernel of the orthogonal series. Hence, let

$$k_N(u, v) = \sum_{k=0}^{N} \varphi_k(u)\varphi_k(v),$$

be the kernel of $\{\varphi_k(u)\}$. Then, we have

$$\hat{\mu}(u) = \sum_{i=1}^{n} Y_{[p+i]}k(u, U_{(i)})(U_{(i)} - U_{(i-1)}).$$

Another, and better estimate, is of the following form:

$$\tilde{\mu}(u) = \sum_{k=0}^{N(n)} \tilde{a}_k \varphi_k(u) = \sum_{i=1}^{n} Y_{[p+i]} \int_{U_{(i-1)}}^{U_{(i)}} k(u, v) dv,$$

where

$$\tilde{a}_k = \sum_{i=1}^{n} Y_{[p+i]} \int_{U_{(i-1)}}^{U_{(i)}} \varphi_k(v) dv.$$

7.3.2   Fourier series estimate

In this subsection, $|U| \leq \pi$, $[a, b] = [-\pi, \pi]$, and we apply the trigonometric series, (see Section B.2). Clearly,

$$\mu(u) \sim \frac{1}{2} a_0 + \sum_{k=1}^{\infty} (a_k \cos ku + b_k \sin ku),$$

where

$$a_0 = \frac{1}{\pi} \int_{-\pi}^{\pi} \mu(v) dv,$$

and

$$a_k = \frac{1}{\pi} \int_{-\pi}^{\pi} \mu(v) \cos(kv) dv, \quad b_k = \frac{1}{\pi} \int_{-\pi}^{\pi} \mu(v) \sin(kv) dv,$$

$k = 1, 2, \ldots$. In the estimate

$$\bar{\mu}(u) = \frac{1}{2} \bar{a}_0 + \sum_{k=1}^{N(n)} (\bar{a}_k \cos ku + \bar{b}_k \sin ku),$$

$$\bar{a}_0 = \sum_{i=1}^{n} Y_{[p+i]} \int_{U_{(i-1)}}^{U_{(i)}} dv$$

and

$$\bar{a}_k = \sum_{i=1}^{n} Y_{[p+i]} \int_{U_{(i-1)}}^{U_{(i)}} \cos(kv) dv, \quad \bar{b}_k = \sum_{i=1}^{n} Y_{[p+i]} \int_{U_{(i-1)}}^{U_{(i)}} \sin(kv) dv.$$

Using the Dirichlet kernel, we rewrite the estimate as follows:

$$\bar{\mu}(u) = \sum_{i=1}^{n} Y_{[p+i]} \int_{U_{(i-1)}}^{U_{(i)}} D_{N(n)}(v - u) dv.$$

Referring to the complex version of the series, we get

$$\mu(u) \sim \sum_{k=-\infty}^{\infty} c_k e^{iku}$$

with

$$c_k = \frac{1}{2\pi} \int_{-\pi}^{\pi} \mu(v) e^{-ikv} dv.$$

Therefore, the estimate can be represented in yet another form:

$$\bar{\mu}(u) = \sum_{k=-N(n)}^{N(n)} \bar{c}_k e^{iku},$$

where

$$\bar{c}_k = \frac{1}{2\pi} Y_{[p+i]} \int_{U_{(i-1)}}^{U_{(i)}} e^{-ikv} dv.$$

THEOREM 7.3 *Let $f(\bullet)$ satisfy (7.4). Let $m(\bullet)$ satisfy (7.5). If*

$$N(n) \to \infty \text{ as } n \to \infty, \tag{7.18}$$

$$\frac{N(n)}{n} \to 0 \text{ as } n \to \infty, \tag{7.19}$$

then

$$\lim_{n \to \infty} E \int_{-\pi}^{\pi} (\bar{\mu}(u) - \mu(u))^2 du = 0.$$

Assuming that $m^{(p)}(\pm\pi) = 0$, for $p = 1, \ldots, q-1$ and $\int_{-\pi}^{\pi} (m^{(q)}(u))^2 du < \infty$, we find

$$\int_{-\pi}^{\pi} (\mu_N(u) - \mu(u))^2 du = O(N^{-2q}),$$

See the result in (B.13) of Appendix B. Note first that

$$E \int_{-\pi}^{\pi} (\bar{\mu}(u) - \mu(u))^2 du$$
$$\leq 2E \int_{-\pi}^{\pi} (\bar{\mu}(u) - \mu_N(u))^2 du + 2 \int_{-\pi}^{\pi} (\mu_N(u) - \mu(u))^2 du.$$

Then, recalling the result established in (7.28) (see the proof of Theorem 7.3), we have

$$E \int_{-\pi}^{\pi} (\bar{\mu}(u) - \mu(u))^2 du = O(N^{-2q}) + O\left(\frac{N(n)}{n}\right).$$

Therefore, selecting

$$N(n) \sim n^{1/(1+2q)} \tag{7.20}$$

we obtain

$$E \int_{-\pi}^{\pi} (\bar{\mu}(u) - \mu(u))^2 du = O(n^{-1+1/(2q+1)}),$$

that is, the same rate as that of the kernel estimate, see (7.16).

7.3.3   Legendre series estimate

For $|U| \leq 1$, that is, for $[a, b] = [-1, 1]$, we apply the Legendre series (see Section B.3). In such a case,

$$\mu(u) \sim \sum_{k=0}^{\infty} a_k p_k(u),$$

where

$$a_k = \int_{-1}^{1} \mu(v) p_k(v) dv,$$

$k = 0, 1, 2, \ldots$. The estimate is of the following form:

$$\check{\mu}(u) = \sum_{k=0}^{N(n)} \check{a}_k p_k(u),$$

where

$$\check{a}_k = \sum_{i=1}^{n} Y_{[p+i]} \int_{U_{(i-1)}}^{U_{(i)}} p_k(v) dv.$$

Using the kernel we get

$$\check{\mu}(u) = \sum_{i=1}^{n} Y_{[p+i]} \int_{U_{(i-1)}}^{U_{(i)}} k(u, v) dv.$$

The next theorem and a result concerning convergence rate are given without proofs, since it suffices to repeat arguments used in the proof of Theorem 7.3 and apply facts presented in Section B.3.

THEOREM 7.4 *Let $f(\bullet)$ satisfy (7.4). Let $m(\bullet)$ satisfy (7.5). If $N(n)$ satisfies (7.18) and (7.19), then*

$$\lim_{n \to \infty} E \int_{-1}^{1} (\check{\mu}(u) - \mu(u))^2 du = 0.$$

Assuming that $m^{(p)}(\pm 1) = 0$, for $p = 1, \ldots, q - 1$ and $\int_{-1}^{1} (m^{(q)}(u))^2 du < \infty$, we obtain

$$E \int_{-1}^{1} (\check{\mu}(u) - \mu(u))^2 du = O(n^{-2q/(2q+1)}),$$

provided that (7.20) holds.

7.4   Lemmas and proofs

7.4.1   Lemmas

LEMMA 7.1 *Let $m(\bullet)$ be a Borel function satisfying (7.6) and let $Em(U) = 0$. Then, for any nonnegative Borel function $g(\bullet)$ and for any i, j, k, m, all different,*

$$E\left\{\xi^2_{[p+i]}\left|g(U_{(i)}, U_{(j)})\right\} \le \frac{1}{n}\lambda M^2 E\left\{g(U_{(i)}, U_{(j)})\right\},$$

and

$$E\left\{\left|\xi_{[p+i]}\xi_{[p+j]}\left|g(U_{(i)}, U_{(j)}, U_{(k)}, U_{(m)})\right\}\right.\right.$$
$$\le \frac{1}{n}\rho M^2 E\left\{g(U_{(i)}, U_{(j)}, U_{(k)}, U_{(m)})\right\}$$

with some λ and ρ dependent on A, b, and c only.

The lemma is an immediate consequence of the next one in which, for any matrix $P = [p_{ij}]$, $|P|$ denotes a matrix $[|p_{ij}|]$, for two matrices P and Q, $P \le Q$ means that the inequality holds for each pair of corresponding elements of the matrices, respectively.

LEMMA 7.2 *Let $m(\bullet)$ be a Borel function satisfying restrictions of Lemma 7.1. Let $0 \le p$, and let*

$$\eta_{p+n} = \sum_{\substack{i=-\infty \\ i \ne n}}^{p+n-1} A^{p+n-i-1}m(U_i).$$

Then, for i, j, k, m, all different,

$$E\left\{\left|\eta_{[p+i]}\eta^T_{[p+i]}\right|\left|U_{(i)}, U_{(j)}, U_{(k)}, U_{(m)}\right\} \le \frac{1}{n}M^2 P,\right.$$

$$E\left\{\left|\eta_{[p+i]}\eta^T_{[p+j]}\right|\left|U_{(i)}, U_{(j)}, U_{(k)}, U_{(m)}\right\} \le \frac{1}{n}M^2 Q,\right.$$

where matrices P and Q depend on A only.

Proof. We verify the second part of the assertion. Let $r \ne s$. Let i, j, k, and m be all different. In the first part of the proof, we show that

$$E\left\{\left|\eta_{[p+r]}\eta^T_{[p+s]}\right|\left|U_{(i)}, U_{(j)}, U_{(k)}, U_{(m)}\right\} \le \Gamma_{rstv}\right. \tag{7.21}$$

for some matrix Γ_{rstv}, such that

$$\sum_{\substack{r=1 \\ r,s,t,v \text{ all different}}}^{n}\sum_{s=1}^{n}\sum_{t=1}^{n}\sum_{v=1}^{n}\Gamma_{rstv} = O(n^3)M^2 DD^T,$$

where $D = \sum_{n=0}^{\infty}|A^n|$.

Suppose that $s < r$. Since $\eta_{p+s} = \sum_{k=-\infty, k\neq s}^{p+s-1} A^{p+s-k-1} m(U_k)$ and

$$
\eta_{p+r} = \sum_{\substack{m=-\infty \\ m\neq r}}^{p+r-1} A^{p+r-m-1} m(U_m) + \sum_{\substack{m=p+s \\ m\neq r}}^{p+r-1} A^{p+r-m-1} m(U_m)
$$

$$
= A^{r-s} \sum_{\substack{m=-\infty \\ m\neq r}}^{p+s-1} A^{p+s-m-1} m(U_m) + \sum_{\substack{m=p+s \\ m\neq r}}^{p+r-1} A^{p+r-m-1} m(U_m),
$$

we get

$$
E\{\eta_{p+r}\eta_{p+s}^T | U_r, U_s, U_t, U_v\} = P_{rstv} + Q_{srtv}
$$

with

$$
P_{rstv} = A^{r-s} \sum_{\substack{m=-\infty \\ m\neq r}}^{p+s-1} \sum_{\substack{k=-\infty \\ k\neq s}}^{p+s-1} A^{p+s-m-1} (A^T)^{p+s-k-1}
$$

$$
\times E\{m(U_m)m(U_k) | U_r, U_s, U_t, U_v\}
$$

and

$$
Q_{rstv} = \sum_{\substack{m=p+r \\ m\neq r}}^{p+r-1} \sum_{\substack{k=-\infty \\ k\neq s}}^{p+s-1} A^{p+r-m-1} (A^T)^{p+s-k-1}
$$

$$
\times E\{m(U_m)m(U_k) | U_r, U_s, U_t, U_v\}.
$$

For any r and s, $|P_{rstv}| \leq M^2 A^{|s-r|} D D^T$, where $D = \sum_{n=0}^{\infty} |A^n|$. For $p \leq r - s$, $|Q_{rstv}| \leq M^2 D D^T$. Let $r - s < p$. Because, for $m \neq k$,

$$
E\{m(U_m)m(U_k) | U_r, U_s, U_t, U_v\}
$$

$$
= E\{m(U_m) | U_r, U_s, U_t, U_v\} E\{m(U_k) | U_r, U_s, U_t, U_v\},
$$

we obtain

$$
Q_{rstv} = \sum_{\substack{m=p+r \\ m\neq r}}^{p+r-1} A^{p+r-m-1} E\{m(U_k) | U_r, U_s, U_t, U_v\}
$$

$$
\times \sum_{\substack{k=-\infty \\ k\neq s}}^{p+s-1} (A^T)^{p+s-k-1} E\{m(U_m) | U_r, U_s, U_t, U_v\}
$$

$$
= \sum_{\substack{m=p+r \\ m\neq r}}^{p+r-1} A^{p+r-m-1} E\{m(U_k) | U_s, U_t, U_v\}
$$

$$
\times \sum_{\substack{k=-\infty \\ k\neq s}}^{p+s-1} (A^T)^{p+s-k-1} E\{m(U_m) | U_t, U_v\}.
$$

Thus,

$$|Q_{rstv}| \leq (\delta_{m-t} + \delta_{m-v}) \sum_{\substack{m=p+r \\ m \neq r}}^{p+r-1} |A^{p+r-m-1}| \, |E\{m(U_k)|U_s, U_t, U_v\}|$$

$$\leq (\delta_{m-t} + \delta_{m-v}) M^2 D D^T.$$

Finally,

$$\sum_{\substack{r=1 \\ r,s,t,v \text{ all different}}}^{n} \sum_{s=1}^{n} \sum_{t=1}^{n} \sum_{v=1}^{n} (P_{rstv} + Q_{rstv}) = O(n^3) M^2 D D^T$$

and (7.21) follows.

Define the following event:

$$A_{rstv} = \{U_r = U_{(i)}, U_s = U_{(j)}, U_t = U_{(k)}, U_v = U_{(m)}\}$$
$$= \{U_r \text{ is } i\text{th}, U_s \text{ is } j\text{th}, U_t \text{ is } k\text{th}, \text{ and } U_v \text{ is } m\text{th in order}\}$$

and observe that (7.21) implies

$$E\left\{|\eta_{[p+i]}\eta_{[p+j]}^T| \, \big| A_{rstv}, U_{(i)}, U_{(j)}, U_{(k)}, U_{(m)}\right\} \leq \Gamma_{rstv},$$

for $i \neq j$. Therefore,

$$E\left\{|\eta_{[p+i]}\eta_{[p+j]}^T| \, \big| U_{(i)}, U_{(j)}, U_{(k)}, U_{(m)}\right\}$$

$$= \sum_{\substack{r=1 \\ r,s,t,v \text{ all different}}}^{n} \sum_{s=1}^{n} \sum_{t=1}^{n} \sum_{v=1}^{n} E\left\{|\eta_{[p+i]}\eta_{[p+j]}^T| \, \big| A_{rstv}, U_{(i)}, U_{(j)}, U_{(k)}, U_{(m)}\right\} P\{A_{rstv}\}$$

$$\leq \sum_{\substack{r=1 \\ r,s,t,v \text{ all different}}}^{n} \sum_{s=1}^{n} \sum_{t=1}^{n} \sum_{v=1}^{n} \Gamma_{rstv} P\{A_{rstv}\},$$

and the given fact that

$$P\{A_{rstv}\} = \frac{1}{n(n-1)(n-2)(n-3)}$$

yields the assertion. To verify the last equality, observe that

$$P\left\{U_r = U_{(i)} \, \big| U_s = U_{(j)}, U_t = U_{(k)}, U_v = U_{(m)}\right\} = \frac{1}{n-3},$$

implies

$$P\{A_{rstv}\} = \frac{1}{n-3} P\left\{U_s = U_{(j)}, U_t = U_{(k)}, U_v = U_{(m)}\right\}.$$

In turn,

$$P\left\{U_s = U_{(j)} \, \big| U_t = U_{(k)}, U_v = U_{(m)}\right\} = \frac{1}{n-2}$$

implies

$$P\{A_{rstv}\} = \frac{1}{(n-3)(n-2)} P\{U_s = U_{(k)}, U_t = U_{(m)}\}$$

and so on. The proof is completed. ∎

For a density $f(\bullet)$ satisfying (7.4) observe

$$U_{(i)} - U_{(i-1)} = \int_{U_{(i-1)}}^{U_{(i)}} dv \leq \frac{1}{\delta} \int_{U_{(i-1)}}^{U_{(i)}} f(v)dv.$$

Therefore, by applying results presented in Section C.4.2, we obtain

LEMMA 7.3 *For any $f(\bullet)$ satisfying (7.4),*

$$E(U_{(i)} - U_{(i-1)})^p = \frac{p!}{(n+1)\cdots(n+p)},$$

and, a fortiori,

$$E(U_{(i)} - U_{(i-1)}) = \frac{2}{n+1},$$

$$E(U_{(i)} - U_{(i-1)})^2 = \frac{8}{(n+1)(n+2)},$$

and

$$E(U_{(i)} - U_{(i-1)})^3 = \frac{48}{(n+1)(n+2)(n+3)}.$$

Moreover,

$$E(U_{(i)} - U_{(i-1)})(U_{(j)} - U_{(j-1)}) = \frac{4}{(n+1)(n+2)}.$$

7.4.2   Proofs

Proof of Theorem 7.1
To prove consistency, we show that

$$\lim_{n\to\infty} E \int (\tilde{\mu}(u) - \mu_{h_n}(u))^2 du = 0,$$

where

$$\mu_{h_n}(u) = \int_{-1}^{1} \mu(v) \frac{1}{h_n} K\left(\frac{u-v}{h_n}\right) dv,$$

and that also

$$\lim_{n\to\infty} \int (\mu_{h_n}(u) - \mu(u))^2 du = 0. \tag{7.22}$$

From (7.12) it follows that

$$(\tilde{\mu}(u) - \mu_{h_n}(u))^2 \le 3 P_n^2(u) + 3 Q_n^2(u) + 3 R_n^2(u),$$

where

$$P_n(u) = \sum_{i=1}^{n} Z_{[p+i]} \frac{1}{h_n} \int_{U_{(i-1)}}^{U_{(i)}} K\left(\frac{u-v}{h_n}\right) dv,$$

$$Q_n(u) = \sum_{i=1}^{n} \xi_{[p+i]} \frac{1}{h_n} \int_{U_{(i-1)}}^{U_{(i)}} K\left(\frac{u-v}{h_n}\right) dv,$$

and

$$R_n(u) = \sum_{i=1}^{n} \frac{1}{h_n} \int_{U_{(i-1)}}^{U_{(i)}} \mu(U_{(i)}) K\left(\frac{u-v}{h_n}\right) dv - \mu_{h_n}(u).$$

Since

$$E P_n^2(u) = \sigma_Z^2 \frac{1}{h_n^2} \sum_{i=1}^{n} E\left(\int_{U_{(i-1)}}^{U_{(i)}} K\left(\frac{u-v}{h_n}\right) dv\right)^2$$

and

$$\int \left| \int_{U_{(i-1)}}^{U_{(i)}} \frac{1}{h} K\left(\frac{u-v}{h}\right) dv \int_{U_{(i-1)}}^{U_{(i)}} \frac{1}{h} K\left(\frac{u-w}{h}\right) dw \right| du$$

$$\le \frac{\kappa}{h} \int_{U_{(i-1)}}^{U_{(i)}} dv \int_{U_{(i-1)}}^{U_{(i)}} \left[\int \left| \frac{1}{h} K\left(\frac{u-w}{h}\right) \right| du \right] dw = \gamma \kappa \frac{1}{h} (U_{(i)} - U_{(i-1)})^2, \quad (7.23)$$

where $\kappa = \sup_v |K(v)|$ and $\gamma = \int |K(v)| dv$. Applying Lemma 7.3 we get

$$E \int P_n^2(u) du \le \gamma \kappa \sigma_Z^2 \frac{1}{h_n} \sum_{i=1}^{n} E(U_{(i)} - U_{(i-1)})^2 = O\left(\frac{1}{nh_n}\right).$$

In turn, $E Q_n^2(u) = E Q_{1n}(u) + E Q_{2n}(u)$ with

$$Q_{1n}(u) = \frac{1}{h_n^2} \sum_{i=1}^{n} \xi_{[p+i]}^2 \left(\int_{U_{(i-1)}}^{U_{(i)}} K\left(\frac{u-v}{h_n}\right) dv\right)^2$$

and

$$Q_{2n}(u) = \frac{1}{h_n^2} \sum_{i=1}^{n} \sum_{j=1, j\neq i}^{n} \xi_{[i]} \xi_{[j]} \int_{U_{(i-1)}}^{U_{(i)}} K\left(\frac{u-v}{h_n}\right) dv \int_{U_{(j-1)}}^{U_{(j)}} K\left(\frac{u-v}{h_n}\right) dv.$$

Applying Lemma 7.1 and using (7.23) again, we find

$$E \int Q_{1n}(u) du \le \lambda \kappa M^2 \frac{1}{nh_n} \sum_{i=1}^{n} E(U_{(i)} - U_{(i-1)})^2 = O\left(\frac{1}{nh_n}\right).$$

Another application of Lemma 7.1 gives

$$E\,Q_{2n}(u) \leq \rho M^2 \frac{1}{nh_n^2} \sum_{i=1}^{n} \sum_{\substack{j=1 \\ j\neq i}}^{n} E\left\{ \int_{U_{(i-1)}}^{U_{(i)}} K\left(\frac{u-v}{h_n}\right) dv \int_{U_{(j-1)}}^{U_{(j)}} K\left(\frac{u-w}{h_n}\right) dw \right\}$$

which, after yet another application of (7.23) and Lemma 7.3, yields

$$E \int Q_{2n}(u)du \leq \rho \kappa M^2 \frac{1}{nh_n} \sum_{i=1}^{n} \sum_{\substack{j=1 \\ j\neq i}}^{n} E(U_{(i)} - U_{(i-1)})E(U_{(j)} - U_{(j-1)})$$

$$= O\left(\frac{1}{nh_n}\right).$$

Hence,

$$E \int Q_n(u)du = O\left(\frac{1}{nh_n}\right).$$

Passing to $R_n(u)$, we observe

$$\mu_h(u) = \sum_{i=1}^{n} \int_{U_{(i-1)}}^{U_{(i)}} \mu(v)\frac{1}{h}K\left(\frac{u-v}{h}\right) dv + r_h(u)$$

with

$$r_h(u) = \int_{U_{(n)}}^{1} \mu(v)\frac{1}{h}K\left(\frac{u-v}{h}\right) dv.$$

Thus $R_n(u) = R_{1n}(u) + r_{h_n}(u)$ with

$$R_{1n}(u) = \sum_{i=1}^{n} \int_{U_{(i-1)}}^{U_{(i)}} (\mu(U_{(i)}) - \mu(v))\frac{1}{h_n}K\left(\frac{u-v}{h_n}\right) dv.$$

Applying (7.23) again and (7.5), we get

$$\int R_{1n}^2(u)du \leq \gamma\kappa \frac{1}{h_n} \sum_{i=1}^{n} \sum_{j=1}^{n} \int_{U_{(i-1)}}^{U_{(i)}} |\mu(U_{(i)}) - \mu(v)|dv \int_{U_{(j-1)}}^{U_{(j)}} |\mu(U_{(j)}) - \mu(v)|dv$$

$$\leq \gamma\kappa c_m^2 \frac{1}{h_n} \sum_{i=1}^{n} \sum_{j=1}^{n} \int_{U_{(i-1)}}^{U_{(i)}} (U_{(i)} - v)dv \int_{U_{(j-1)}}^{U_{(j)}} (U_{(j)} - v)dv$$

$$= \gamma\kappa c_m^2 \frac{1}{4h_n} \sum_{i=1}^{n} (U_{(i)} - U_{(i-1)})^2 \sum_{j=1}^{n} (U_{(j)} - U_{(j-1)})^2,$$

which equals

$$\gamma \kappa c_m^2 \frac{1}{4h_n} \left(\sum_{i=1}^{n} (U_{(i)} - U_{(i-1)})^2 \right)^2$$

$$= \gamma \kappa c_m^2 \frac{1}{4h_n} \left(\sum_{i=1}^{n} (U_{(i)} - U_{(i-1)})^{1/2} (U_{(i)} - U_{(i-1)})^{3/2} \right)^2$$

$$\leq \gamma \kappa c_m^2 \frac{1}{4h_n} \sum_{i=1}^{n} (U_{(i)} - U_{(i-1)})^3. \qquad (7.24)$$

Hence,

$$E \int R_{1n}^2(u) du = O\left(\frac{1}{n^2 h_n} \right).$$

Since the verification that $E \int r_{h_n}^2(u) du = O(1/n^2)$ is easy, we can write

$$E \int R_n^2(u) du = O\left(\frac{1}{n^2 h_n} \right).$$

Thus

$$E \int (\tilde{\mu}(u) - \mu_{h_n}(u))^2 du = O\left(\frac{1}{nh_n} \right). \qquad (7.25)$$

To complete the proof it suffices to verify (7.22). It holds, thanks to Lemma A.7. ∎

Proof of Theorem 7.2

Since the proof is very similar to that of Theorem 7.1, we point out the main differences. Clearly,

$$E \int (\hat{\mu}(u) - \mu_{h_n}(u))^2 du = O\left(\frac{1}{n^2 h_n} \right) \text{ as } n \to \infty.$$

Moreover, $(\hat{\mu}(u) - \mu(u))^2 \leq 3S_n^2(u) + 3T_n^2(u) + 3V_n^2(u)$, where

$$S_n(u) = \sum_{i=1}^{n} Z_{[p+i]} \frac{1}{h_n} K\left(\frac{u - U_{(i)}}{h_n} \right) (U_{(i)} - U_{(i-1)}),$$

$$T_n(u) = \sum_{i=1}^{n} \xi_{[p+i]} \frac{1}{h_n} K\left(\frac{u - U_{(i)}}{h_n} \right) (U_{(i)} - U_{(i-1)}),$$

and

$$V_n(u) = \sum_{i=1}^{n} \frac{1}{h_n} \int_{U_{(i-1)}}^{U_{(i)}} \mu(U_{(i)}) K\left(\frac{u - U_{(i)}}{h_n} \right) dv - \mu_{h_n}(u).$$

Examining $V_n(u)$, we get $V_n(u) = V_{1n}(u) + r_{h_n}(u)$ with

$$V_{1n}(u) = \frac{1}{h_n} \sum_{i=1}^{n} \int_{U_{(i-1)}}^{U_{(i)}} \left[\mu(U_{(i)})K\left(\frac{u - U_{(i)}}{h_n}\right) dv - \mu(v)K\left(\frac{u - v}{h_n}\right) \right] dv.$$

Since both μ and K are Lipschitz,

$$\left| \mu(U_{(i)})K\left(\frac{u - U_{(i)}}{h}\right) - \mu(v)K\left(\frac{u - v}{h}\right) \right|$$

$$\leq \left| \mu(U_{(i)})K\left(\frac{u - U_{(i)}}{h}\right) - \mu(v)K\left(\frac{u - U_{(i)}}{h}\right) \right|$$

$$\left| \mu(v)K\left(\frac{u - U_{(i)}}{h}\right) - \mu(v)K\left(\frac{u - v}{h}\right) \right|$$

$$\leq c_m |U_{(i)} - v| \left| K\left(\frac{u - U_{(i)}}{h}\right) \right| + M \left| K\left(\frac{u - U_{(i)}}{h}\right) - K\left(\frac{u - v}{h}\right) \right|.$$

Defining

$$W_n(u) = \frac{1}{h_n} \sum_{i=1}^{n} \int_{U_{(i-1)}}^{U_{(i)}} \left| K\left(\frac{u - U_{(i)}}{h_n}\right) - K\left(\frac{u - v}{h_n}\right) \right| dv$$

and using (7.23) and (7.15), we get

$$\int W_n^2(u)du = 2\frac{1}{h_n^2} \int |K(v)|dv \sum_{i=1}^{n} \sum_{i=1}^{n} (U_i - v) \int_{U_{(j-1)}}^{U_{(j)}} dv$$

$$= \frac{1}{h_n^2} \int |K(v)|dv \sum_{i=1}^{n} \sum_{j=1}^{n} (U_{(i)} - U_{(i-1)})^2 (U_{(j)} - U_{(j-1)})$$

$$= \frac{1}{h_n^2} \int |K(v)|dv \sum_{i=1}^{n} (U_{(i)} - U_{(i-1)})^2.$$

Hence,

$$E \int W_n^2(u)du = O\left(\frac{1}{n^2 h_n}\right),$$

which results in slower convergence rate. ∎

Proof of Theorem 7.3

From (7.12) it follows that

$$(\bar{\mu}(u) - \mu(u))^2 \leq 3P_n^2(u) + 3Q_n^2(u) + 3R_n^2(u),$$

where

$$P_n(u) = \sum_{i=1}^{n} Z_{[p+i]} \int_{U_{(i-1)}}^{U_{(i)}} D_{N(n)}(v-u)dv,$$

$$Q_n(u) = \sum_{i=1}^{n} \xi_{[p+i]} \int_{U_{(i-1)}}^{U_{(i)}} D_{N(n)}(v-u)dv,$$

and

$$R_n(u) = \sum_{i=1}^{n} \int_{U_{(i-1)}}^{U_{(i)}} \mu(U_{(i)}) \int_{U_{(i-1)}}^{U_{(i)}} D_{N(n)}(v-u)dv - \mu_{N(n)}(u),$$

where

$$\mu_N(u) = \int_{-\pi}^{\pi} \mu(v)D_N(v-u)dv.$$

By the Schwartz inequality, (B.7), the equality $\int_{-\pi}^{\pi} F_n(u)du = 1$ (see Section B.2), we get

$$\int_{-\pi}^{\pi} \left(\int_{U_{(i-1)}}^{U_{(i)}} D_{N(n)}(v-u)dv \right)^2 du$$

$$\leq (U_{(i)} - U_{(i-1)}) \int_{-\pi}^{\pi} \left(\int_{U_{(i-1)}}^{U_{(i)}} D_{N(n)}^2(v-u)dv \right) du$$

$$= \left(N(n) + \frac{1}{2} \right) (U_{(i)} - U_{(i-1)}) \int_{U_{(i-1)}}^{U_{(i)}} \left(\int_{-\pi}^{\pi} F_{N(n)}(v-u)du \right) dv$$

$$= N(n)(U_{(i)} - U_{(i-1)})^2. \tag{7.26}$$

Applying Lemma 7.3, we obtain

$$E \int_{-\pi}^{\pi} P_n^2(u)du = O(N(n)) \sum_{i=1}^{n} E(U_{(i)} - U_{(i-1)})^2 = O\left(\frac{N(n)}{n} \right).$$

In turn,

$$E Q_n^2(u) = E Q_{1n}(u) + E Q_{2n}(u)$$

with

$$Q_{1n}(u) = \sum_{i=1}^{n} \xi_{[p+i]}^2 \left(\int_{U_{(i-1)}}^{U_{(i)}} D_{N(n)}(v-u)dv \right)^2$$

and

$$Q_{2n}(u) = \sum_{i=1}^{n} \sum_{j=1, j\neq i}^{n} \xi_{[i]}\xi_{[j]} \int_{U_{(i-1)}}^{U_{(i)}} D_{N(n)}(v-u)dv \int_{U_{(j-1)}}^{U_{(j)}} D_{N(n)}(v-u)dv.$$

Applying Lemma 7.1 and using (7.26), we find

$$E \int_{-\pi}^{\pi} Q_{1n}(u)du \leq \lambda M^2 \sum_{i=1}^{n} E(U_{(i)} - U_{(i-1)})^2 = O\left(\frac{N(n)}{n}\right).$$

Since

$$\left| \int_{-\pi}^{\pi} D_N(v-u)D_N(w-u)du \right|$$

$$\leq \left(\int_{-\pi}^{\pi} D_N^2(v-u)du \right)^{1/2} \left(\int_{-\pi}^{\pi} D_N^2(w-u)du \right)^{1/2}$$

$$= \left(N + \frac{1}{2} \right) \left(\int_{-\pi}^{\pi} F_{2N+1}(v-u)du \right)^{1/2} \left(\int_{-\pi}^{\pi} F_{2N+1}(w-u)du \right)^{1/2}$$

$$= O(N), \tag{7.27}$$

an application of Lemma 7.1 gives

$$E \int_{-\pi}^{\pi} Q_{2n}(u)du = O(N(n)) \sum_{i=1}^{n} \sum_{j=1, j\neq i}^{n} E\left\{ \int_{U_{(i-1)}}^{U_{(i)}} dv \int_{U_{(j-1)}}^{U_{(j)}} dw \right\}$$

$$= O(N(n)) \sum_{i=1}^{n} E(U_{(i-1)} - U_{(i)})^2 = O\left(\frac{N(n)}{n}\right).$$

Turning to $R_n(u)$, we observe

$$\mu_N(u) = \sum_{i=1}^{n} \int_{U_{(i-1)}}^{U_{(i)}} \mu(v)D_N(v-u)dv + r_{N(n)}(u)$$

with

$$r_{N(n)}(u) = \int_{U_{(n)}}^{\pi} \mu(v)D_N(v-u)dv.$$

Thus,

$$R_n(u) = R_{1n}(u) + r_{N(n)}(u)$$

with

$$R_{1n}(u) = \sum_{i=1}^{n} \int_{U_{(i-1)}}^{U_{(i)}} (\mu(U_{(i)}) - \mu(v))D_N(v-u)dv.$$

Applying (7.27) and (7.5), we get

$$
\int_{-\pi}^{\pi} R_{1n}^2(u)\,du = O(N(n)) \sum_{i=1}^{n} \sum_{j=1}^{n} \int_{U_{(i-1)}}^{U_{(i)}} |\mu(U_{(i)}) - \mu(v)|\,dv \int_{U_{(j-1)}}^{U_{(j)}} |\mu(U_{(j)}) - \mu(v)|\,dv
$$

$$
\leq O(N(n)) c_m^2 \sum_{i=1}^{n} \sum_{j=1}^{n} \int_{U_{(i-1)}}^{U_{(i)}} (U_{(i)} - v)\,dv \int_{U_{(j-1)}}^{U_{(j)}} (U_{(j)} - v)\,dv
$$

$$
= O(N(n)) c_m^2 \sum_{i=1}^{n} (U_{(i)} - U_{(i-1)})^2 \sum_{j=1}^{n} (U_{(j)} - U_{(j-1)})^2
$$

$$
= O(N(n)) c_m^2 \sum_{i=1}^{n} (U_{(i)} - U_{(i-1)})^3,
$$

see (7.24). Hence,

$$
E \int_{-\pi}^{\pi} R_{1n}^2(u)\,du = O\left(\frac{N(n)}{n^2} \right).
$$

Since the verification that $E \int_{-\pi}^{\pi} r_{N(n)}^2(u)\,du = O(1/n^2)$ is easy, we get

$$
E \int_{-\pi}^{\pi} R_n^2(u)\,du = O\left(\frac{N(n)}{n^2} \right).
$$

We have thus come to the conclusion that

$$
E \int_{-\pi}^{\pi} (\bar{\mu}(u) - \mu_N(u))^2\,du = O\left(\frac{N(n)}{n} \right) \quad \text{as } n \to \infty. \tag{7.28}
$$

Since μ_N is the partial sum of the Fourier expansion of μ,

$$
\lim_{N \to \infty} \int_{-\pi}^{\pi} (\mu_N(u) - \mu(u))^2\,du = 0,
$$

(see Section B.2), which completes the proof. ∎

7.5   Bibliographic notes

The idea of estimating a regression function via the approximation of the integral in (7.7) stems from Priestley and Chao [242] and has been developed by Benedetti [17], Clark [54], Schuster and Yakowitz [274], Cheng and Lin [50], Georgiev [96–101], Georgiev and Greblicki [95], Müller and Stadtmüller [213], and Isogai [170]. Orthogonal series estimates have been studied by Rutkowski [260–266], Gałkowski and Rutkowski [90], Rafajłowicz [245, 246], Rutkowski and Rafajłowicz [267], Eubank, Hart, and Speckman [84], Rafajłowicz and Skubalska-Rafajłowicz [248], and Rafajłowicz and Schwabe [247], among others.

In all those works, the regression is estimated from pairs $(u_1, Y_1), \ldots, (u_n, Y_n)$ such that

$$Y_n = m(u_n) + Z_n,$$

where nonrandom points $u_1, u_2, \ldots, u_n$, such that $u_{i-1} < u_i$, are usually scattered uniformly. Mack and Müller [201] showed that convolution type kernel estimates work properly also when the points are random, see also Chu and Marron [51], and Jones, Davies, and Park [172] for further discussion of such kernel estimates. Since such inputs U_is are independent random variables, the operation of ordering is necessary.

The idea of applying ordered input observations to the identification of the nonlinear part of the Hammerstein system has been proposed in Greblicki and Pawlak [134] in the context of the orthogonal series estimate. The kernel estimate of this type has been studied in Greblicki [110].

8 Continuous-time Hammerstein systems

In this chapter, we extend the theory developed in all of the preceding chapters to the case of continuous-time Hammerstein systems. The consistency and convergence rates of kernel (Section 8.2) and orthogonal series (Section 8.3) estimates are established. The problem of identification of the linear subsystem is examined in Section 8.1.2.

8.1   Identification problem

A continuous-time Hammerstein system shown in Figure 8.1 consists of a nonlinear subsystem with a characteristic $m(\bullet)$ and a linear dynamic one with the impulse response $\lambda(t)$. Our goal is to recover both $m(\bullet)$ and $\lambda(t)$ from observations $(U(t), Y(t))$ taken on the real half line $[0, \infty)$.

The input signal $\{U(t)\}$ is a stationary zero mean white random process with autocovariance function $\sigma_U^2 \delta(t)$, where $\delta(t)$ is the delta function. The nonlinear characteristic is a Borel function, such that

$$Em^2(U) < \infty,$$

compare with (2.4). The dynamic subsystem is described by the following continuous-time state equation:

$$\begin{cases} \dot{X}(t) = AX(t) + bV(t) \\ W(t) = c^T X(t) \end{cases} \tag{8.1}$$

The corresponding impulse response $\lambda(t) = c^T e^{At} b$, and it is assumed that the dynamic subsystem is stable, that is, all eigenvalues of A have negative real parts. Clearly,

$$W(t) = \int_{-\infty}^{t} \lambda(t - \tau) m(U(\tau)) d\tau. \tag{8.2}$$

Since $\int_0^\infty \lambda^2(t) dt < \infty$, $W(t)$ is a random variable while $\{W(t)\}$ is a stationary random process. The disturbance $\{Z(t)\}$ is a stationary zero mean random process independent of $\{U(t)\}$.

Neither the functional form of $m(\bullet)$ nor the matrix A or vectors b, c, are known. Thus the a priori information about both the subsystems is nonparametric.

For simplicity, U, W, Y, and Z stand for $U(t), W(t), Y(t)$, and $Z(t)$, respectively. Proofs of all theorems are in Section 8.4.

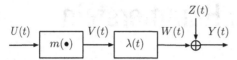

Figure 8.1 The continuous-time Hammerstein system.

8.1.1 Nonlinear subsystem identification

We now present the idea behind the identification methods used in this chapter to recover the nonlinear characteristic. From Lemma 8.1 it follows that for any fixed $s > 0$

$$E\{Y(t+s)|U(t) = u\} = \mu(u),$$

where $\mu(u) = \lambda(s)m(u) + Em(U) \int_0^\infty \lambda(\tau)d\tau$. Therefore, to identify the subsystem it suffices to estimate the regression.

Like those used to identify discrete systems, each algorithm estimating the nonlinearity in continuous-time ones, denoted here as $\hat{\mu}(U(t), Y(t))$, is linear with respect to output observations which means that

$$\hat{\mu}(U(t), \theta(t) + \eta(t)) = \hat{\mu}(U(t), \theta(t)) + \hat{\mu}(U(t), \eta(t)) \tag{8.3}$$

and has a natural property that, for any number θ,

$$\hat{\mu}(U(t), \theta) \to \theta \text{ as } t \to \infty \tag{8.4}$$

in an appropriate stochastic sense. Arguing as in Section 2.2 we come to a conclusion presented in the following remark:

REMARK 8.1 *Let an estimate have properties (8.3) and (8.4). If the estimate is consistent for $Em(U) = 0$, then it is consistent for $Em(U) \neq 0$, also.*

Thus, with no loss of generality, in all proofs of consistency of algorithms recovering the nonlinearity, we assume that $Em(U) = 0$.

We notice that a system with the following differential equation:

$$w^{(k)} + a_{k-1}w^{(k-1)} + \cdots + a_0 w = b_{k-1}m(u^{(k-1)}) + \cdots + b_0 m(u)$$

can be described by (8.1). Thus our methods can be used to recover its nonlinearity $m(\bullet)$.

8.1.2 Dynamic subsystem identification

Passing to the dynamic subsystem we use (8.2) and recall $EZ = 0$ to notice that

$$E\{Y(\tau + t)U(t)\} = \int_{-\infty}^{\tau+t} \lambda(\tau + t - \sigma)E\{m(U(\sigma))U(t)\} d\sigma$$

$$= \lambda(\tau)\{Em(U)U\}.$$

Denoting $\kappa(\tau) = \lambda(\tau)E\{Um(U)\}$, we obtain

$$\kappa(\tau) = E\{Y(\tau + t)U(t)\},$$

which can be estimated as follows:

$$\hat{\kappa}(\tau) = \frac{1}{t} \int_0^t Y(\tau + \xi)U(\xi)d\xi.$$

THEOREM 8.1 *For a fixed* τ,

$$\lim_{t \to \infty} E(\hat{\kappa}(\tau) - \kappa(\tau))^2 = 0.$$

As an estimate of the whole impulse response, we take

$$\hat{\kappa}_{T(t)}(\tau) = \begin{cases} \hat{\kappa}(\tau), & \text{for } 0 \leq \tau \leq T(t) \\ 0, & \text{otherwise.} \end{cases}$$

The mean integrated square error (MISE) between the functions $\kappa(\bullet)$ and $\hat{\kappa}_{T(t)}(\bullet)$ can be defined as:

$$\text{MISE} = \int_0^{T(t)} (\hat{\kappa}_{T(t)}(\tau) - \kappa(\tau))^2 d\tau.$$

From (8.11) it follows that

$$\text{MISE} = O\left(\frac{T(t)}{t}\right) + \int_{T(t)}^\infty \kappa(\tau)^2 d\tau.$$

Hence, if

$$T(t) \to \infty \text{ as } t \to \infty$$

and

$$\frac{T(t)}{t} \to \infty \text{ as } t \to \infty,$$

then

$$\lim_{t \to \infty} \text{MISE} = 0.$$

8.2   Kernel algorithm

A kernel estimate

$$\hat{\mu}(u) = \frac{\int_0^t Y(s + \tau)K\left(\frac{u - U(\tau)}{h(t)}\right)d\tau}{\int_0^t K\left(\frac{u - U(\tau)}{h(t)}\right)d\tau}$$

as well as its semirecursive modifications

$$\tilde{\mu}_t(u) = \frac{\int_0^t Y(s + \tau)\frac{1}{h(\tau)}K\left(\frac{u - U(\tau)}{h(\tau)}\right)d\tau}{\int_0^t \frac{1}{h(\tau)}K\left(\frac{u - U(\tau)}{h(\tau)}\right)d\tau}$$

and

$$\bar{\mu}_t(u) = \frac{\int_0^t Y(s+\tau)K\left(\frac{u-U(\tau)}{h(\tau)}\right)d\tau}{\int_0^t K\left(\frac{u-U(\tau)}{h(\tau)}\right)d\tau}$$

are continuous-time versions of those studied in Chapters 3 and 4.

Denoting

$$\hat{g}_t(u) = \frac{1}{t}\int_0^t Y(s+\tau)\frac{1}{h(\tau)}K\left(\frac{u-U(\tau)}{h(\tau)}\right)d\tau,$$

$$\hat{f}_t(u) = \frac{1}{t}\int_0^t \frac{1}{h(\tau)}K\left(\frac{u-U(\tau)}{h(\tau)}\right)d\tau$$

and noticing that $\hat{\mu}_t(u) = \hat{g}_t(u)/\hat{f}_t(u)$, we can write

$$\frac{d}{dt}\hat{g}_t(u) = -\frac{1}{t}\left[\hat{g}_t(u) - Y(s+t)\frac{1}{h(t)}K\left(\frac{u-U(t)}{h(t)}\right)\right],$$

$$\frac{d}{dt}\hat{f}_t(u) = -\frac{1}{t}\left[\hat{f}_t(u) - \frac{1}{h(t)}K\left(\frac{u-U(t)}{h(t)}\right)\right],$$

with $\hat{g}_0(u) = \hat{f}_0(u) = 0$. Similarly, we denote

$$\bar{g}_t(u) = \frac{1}{\int_0^t h(\tau)d\tau}\int_0^t Y(s+\tau)K\left(\frac{u-U(\tau)}{h(\tau)}\right)d\tau,$$

$$\bar{f}_t(u) = \frac{1}{\int_0^t h(\tau)d\tau}\int_0^t K\left(\frac{u-U(\tau)}{h(\tau)}\right)d\tau.$$

Then, observe that $\bar{\mu}_t(u) = \bar{g}_t(u)/\bar{f}_t(u)$ and write

$$\frac{d}{dt}\bar{g}_t(u) = -\gamma(t)\left[\bar{g}_t(u) - Y(s+t)\frac{1}{h(t)}K\left(\frac{u-U(t)}{h(t)}\right)\right],$$

$$\frac{d}{dt}\bar{f}_t(u) = -\gamma(t)\left[\bar{f}_t(u) - \frac{1}{h(t)}K\left(\frac{u-U(t)}{h(t)}\right)\right],$$

with $\gamma(t) = h(t)/\int_0^t h(\xi)d\xi$ and $\bar{g}_0(u) = \bar{f}_0(u) = 0$.

THEOREM 8.2 *Let U have a density $f(\bullet)$ and let $Em^2(U) < \infty$. Let*

$$\sup_{-\infty < u < \infty} |K(u)| < \infty, \tag{8.5}$$

$$\int |K(u)|du < \infty, \tag{8.6}$$

$$|u|^{1+\varepsilon}K(u) \to 0 \text{ as } |u| \to \infty \tag{8.7}$$

with $\varepsilon = 0$. Let

$$h(t) \to 0 \text{ as } t \to \infty \qquad (8.8)$$

$$th(t) \to \infty \text{ as } t \to \infty. \qquad (8.9)$$

Then,

$$\hat{\mu}(u) \to \mu(u) \text{ as } t \to \infty \text{ in probability} \qquad (8.10)$$

at every $u \in R$ where both $m(\bullet)$ and $f(\bullet)$ are continuous and $f(u) > 0$.

The next theorem is analogous with Theorem 3.2.

THEOREM 8.3 *Let U have a probability density $f(\bullet)$ and let $Em^2(U) < \infty$. Let the Borel measurable satisfy (8.5), (8.6), and (8.7) for some $\varepsilon > 0$. Let $h(t)$ satisfy (8.8) and (8.9). Then convergence (8.10) takes place at every Lebesgue point $u \in R$ of both $m(\bullet)$ and $f(\bullet)$ where $f(u) > 0$, and, a fortiori, at almost every u where $f(u) > 0$, that is, at almost every u belonging to the support of $f(\bullet)$.*

Next two theorems correspond with Theorems 4.1 and 4.2, respectively.

THEOREM 8.4 *Let U have a density $f(\bullet)$ and let $Em^2(U) < \infty$. Let the Borel measurable kernel $K(\bullet)$ satisfy (8.5), (8.6) and (8.7) with $\varepsilon = 0$. Let $h(t)$ satisfy (8.8), and let*

$$\frac{1}{t^2} \int_0^t \frac{1}{h(\tau)} d\tau \to 0 \text{ as } t \to \infty.$$

Then,

$$\tilde{\mu}_t(u) \to \mu(u) \text{ as } t \to \infty \text{ in probability.}$$

at every $u \in R$ where both $m(\bullet)$ and $f(\bullet)$ are continuous and $f(u) > 0$. If, (8.7) holds for some $\varepsilon > 0$, then the convergence takes place at every Lebesgue point $u \in R$ of both $m(\bullet)$ and $f(\bullet)$, such that $f(u) > 0$; a fortiori, at almost every u belonging to support of $f(\bullet)$.

THEOREM 8.5 *Let U have a density $f(\bullet)$ and let $Em^2(U) < \infty$. Let the Borel measurable kernel $K(\bullet)$ satisfy (8.5), (8.6) and (8.7) with $\varepsilon = 0$. Let $h(t)$ satisfy (8.8) and let*

$$\int_0^\infty h(t)dt = \infty.$$

Then,

$$\bar{\mu}_t(u) \to \mu(u) \text{ as } t \to \infty \text{ in probability.}$$

at every point $u \in R$ where both $m(\bullet)$ and $f(\bullet)$ are continuous and $f(u) > 0$. If, (8.7) holds with some $\varepsilon > 0$, then the convergence takes place at every Lebesgue point $u \in R$ of both $m(\bullet)$ and $f(\bullet)$, such that $f(u) > 0$; a fortiori, at almost every point belonging to support of $f(\bullet)$.

Using arguments used in Sections 3.4 and 4.2 we easily come to a conclusion that if both $f^{(q)}(u)$ and $g^{(q)}(u)$ are square integrable, $K(\bullet)$ is selected to satisfy restrictions imposed in Section 3.4, and $h(t) \sim t^{-1/2q}$, we get

$$P\{|\hat{\mu}(u) - \mu(u)| > \varepsilon \mu(u)\} = O(t^{-1/2+1/4q})$$

with any $\varepsilon > 0$, and

$$|\hat{\mu}(u) - \mu(u)| = O(t^{-1+1/2q}) \text{ as } t \to \infty \text{ in probability.}$$

The rate is true also for $\tilde{\mu}_t(u)$ and $\bar{\mu}_t(u)$. If the derivatives are bounded and $h(t) \sim t^{-1/(2q+1)}$, the rate is $O(t^{-1/2+1/(4q+2)})$ and $O(t^{-1+1/(2q+1)})$, respectively.

8.3   Orthogonal series algorithms

We now show how to apply the orthogonal series approach presented in Chapter 6. The trigonometric series estimate serves as an example. To estimate the characteristic, we apply the trigonometric orthonormal system

$$\frac{1}{\sqrt{2\pi}}, \frac{1}{\sqrt{\pi}}\cos u, \frac{1}{\sqrt{\pi}}\sin u, \frac{1}{\sqrt{\pi}}\cos 2u, \frac{1}{\sqrt{\pi}}\sin 2u, \ldots,$$

(see also Section B.2). As in Section 6.2, we assume that $\int_{-\pi}^{\pi}|f(u)|du < \infty$ and $\int_{-\pi}^{\pi}|g(u)|du < \infty$, where, as usual, $g(u) = \mu(u)f(u)$. Clearly,

$$g(u) \sim a_0 + \sum_{k=1}^{\infty} a_k \cos ku + \sum_{k=1}^{\infty} b_k \sin ku,$$

where

$$a_0 = \frac{1}{2\pi}\int_{-\pi}^{\pi} g(u)du,$$

and, for $k = 1, 2, \ldots,$

$$a_k = \frac{1}{\pi}\int_{-\pi}^{\pi} g(u)\cos(ku)du, \; b_k = \frac{1}{\pi}\int_{-\pi}^{\pi} g(u)\sin(ku)du.$$

For the same reasons,

$$f(u) \sim \frac{1}{2\pi} + \sum_{k=1}^{\infty} \alpha_k \cos ku + \sum_{k=1}^{\infty} \beta_k \sin ku,$$

where

$$\alpha_k = \frac{1}{\pi}\int_{-\pi}^{\pi} f(u)\cos(ku)du, \; \beta_k = \frac{1}{\pi}\int_{-\pi}^{\pi} g(u)\sin(ku)du.$$

The trigonometric series estimate of $\mu(u)$ has the following form:

$$\hat{\mu}(u) = \frac{\hat{a}_0 + \sum_{k=1}^{N(t)} \hat{a}_k \cos ku + \sum_{k=1}^{N(t)} \hat{b}_k \sin ku}{\frac{1}{2\pi} + \sum_{k=1}^{N(t)} \hat{\alpha}_k \cos ku + \sum_{k=1}^{N(t)} \hat{\beta}_k \sin ku}$$

with

$$\hat{a}_0 = \frac{1}{2\pi t} \int_0^t Y(s+\tau)d\tau, \quad \hat{a}_k = \frac{1}{\pi t} \int_0^t Y(s+\tau)\cos[kU(\tau)]d\tau,$$

$$\hat{b}_k = \frac{1}{\pi t} \int_0^t Y(s+\tau)\sin[kU(\tau)]d\tau,$$

and

$$\hat{\alpha}_k = \frac{1}{\pi t} \int_0^t \cos[kU(\tau)]d\tau, \quad \hat{\beta}_k = \frac{1}{\pi t} \int_0^t \sin[kU(\tau)]d\tau.$$

Using the Dirichlet kernel $D_n(u)$ (see Section B.2), we can also write

$$\hat{\mu}(u) = \frac{\int_0^t Y(s+\tau)D_{N(t)}(U(\tau)-u)d\tau}{\int_0^t D_{N(t)}(U(\tau)-u)d\tau}.$$

The proof of the next theorem is omitted.

THEOREM 8.6 *Let $Em^2(U) < \infty$. Let $\int_{-\pi}^{\pi} |f(u)|du < \infty$. If*

$$N(t) \to \infty \text{ as } t \to \infty,$$

$$\frac{N(t)}{t} \to 0 \text{ as } t \to \infty,$$

then,

$$\hat{\mu}(u) \to \mu(u) \text{ as } t \to \infty \text{ in probability}$$

at every $u \in (-\pi, \pi)$ where both $m(\bullet)$ and $f(\bullet)$ are differentiable and $f(u) > 0$.

Assume that $m^{(i)}(-\pi) = m^{(i)}(\pi) = f^{(i)}(-\pi) = f^{(i)}(\pi) = 0$, for $i = 0, 1, \ldots,$ $(q-1)$, that $\int_{-\pi}^{\pi} (f^{(q)}(v))^2 dv < \infty$ and $\int_{-\pi}^{\pi} (g^{(q)}(v))^2 dv < \infty$. Then, proceeding as in Section 6.2, we verify that

$$|\hat{\mu}(u) - \mu(u)| = O(t^{-1/2+1/4q}) \text{ in probability,}$$

and, for any $\varepsilon > 0$,

$$P\{|\hat{\mu}(u) - \mu(u)| > \varepsilon|\mu(u)|\} = O(t^{-1+1/2q}),$$

provided that $N(t) \sim t^{1/2q}$.

8.4   Lemmas and proofs

8.4.1   Lemmas

LEMMA 8.1 *In the system,*

$$E\{W(t+s)|U(t) = u\} = \mu(u),$$

where $\mu(u) = \lambda(s)m(u) + \int_0^\infty \lambda(\tau)d\tau\, Em(U)$.

Proof. Let $\mu(u) = \lambda(s)m(u) + \int_0^\infty \lambda(\tau)d\tau\, EV$, $\rho(U(t)) = E\{W(t+s)|U(t)\}$. For any Borel function $\phi(\bullet)$, we get

$$E\{\rho(U(t))\phi(U(t))\} = E\{E\{W(t+s)|U(t)\}\phi(U(t))\}$$
$$= E\{W(t+s)\phi(U(t))\}.$$

Moreover, since $W(t+s) = \int_{-\infty}^{t+s} \lambda(t+s-\tau)m(U(\tau))d\tau$ and $\{U(t)\}$ is white noise, we get

$$E\{W(t+s)\phi(U(t))\} = \int_{-\infty}^{t+s} \lambda(t+s-\tau)E\{U(\tau))\phi(U(t))\}d\tau$$
$$= \lambda(s)E\{m(U)\phi(U)\} + \int_0^\infty \lambda(\tau)d\tau\, Em(U)E\phi(U).$$

Thus, for any $\phi(\bullet)$,

$$E\{\rho(U)\phi(U)\} = \lambda(s)E\{m(U)\phi(U\} + \int_0^\infty \lambda(\tau)d\tau\, Em(U)E\phi(U).$$

From this and

$$E\{\mu(U)\phi(U)\} = \lambda(s)E\{m(U)\phi(U)\} + \int_0^\infty \lambda(\tau)d\tau\, Em(U)E\phi(U),$$

it follows that for any $\phi(\bullet)$, $E\{\rho(U)\phi(U)\} = E\{\mu(U)\phi(U)\}$. Hence, $\rho(u) = \mu(u)$ with probability 1. ∎

LEMMA 8.2 *Let $\{U(t)\}$ be a stationary white random process, $EU(t) = 0$, and $Em(U) = 0$. For any $\varphi(\bullet)$ and $\psi(\bullet)$,*

$$\begin{aligned}
\text{cov}\,[W(s+\xi)&\varphi(U(\xi)), W(s+\eta)\psi(U(\eta))] \\
&= \delta(\xi - \eta)\lambda^2(s)\,\text{cov}\,[m(U)\varphi(U), m(U)\psi(U)] \\
&\quad + \delta(\xi - \eta)\int_0^\infty \lambda^2(t)dt\,\text{var}\,[m(U)]\,E\,\{\varphi(U)\psi(U)\} \\
&\quad + \lambda(s)\lambda(s+\xi-\eta)\lambda(s)E\varphi(U)E\,\{m^2(U)\psi(U)\} \\
&\quad + \lambda(s)\lambda(s+\eta-\xi)E\psi(U)E\,\{m^2(U)\varphi(U)\} \\
&\quad + \lambda(s+\xi-\eta)\lambda(s+\eta-\xi)E\,\{m(U)\psi(U)\}\,E\,\{m(U)\varphi(U)\}
\end{aligned}$$

Proof. It suffices to observe that the covariance in the assertion is equal to

$$\int_{-\infty}^{s+\xi} \int_{-\infty}^{s+\eta} \lambda(s+\xi-t)\lambda(s+\eta-\tau)$$
$$\times \text{cov}\,[m(U(t))\varphi(U(\xi)), m(U(\tau))\psi(U(\eta))]\,dt d\tau$$

and apply Lemma C.5. ∎

LEMMA 8.3 *Let $EU = 0$. Then,*

$$\text{cov}\,[W(\xi+s)U(\xi), W(\eta+s)U(\eta)]$$
$$= \delta(\xi-\eta)\lambda^2(s)\,\text{var}\,[m(U)U] + \delta(\xi-\eta)\int_0^\infty \lambda^2(\xi)d\xi\,\text{var}\,[m(U)]\,\sigma_U^2$$
$$+ \lambda(\xi+s-\eta)\lambda(\eta+s-\xi)E^2\{m(U)U\}.$$

8.4.2   Proofs

Proof of Theorem 8.1
The estimate is unbiased, that is, $E\hat{\kappa}(\tau) = \kappa(\tau)$. Moreover,

$$\text{var}\,[\hat{\kappa}(\tau)] = \frac{1}{t}\sigma_Z^2 EU^2 + P_t,$$

where

$$P_t = \frac{1}{t^2}\,\text{var}\left[\int_0^t W(\tau+\xi)U(\xi)d\xi\right]$$
$$= \frac{1}{t^2}\int_0^t\int_0^t \text{cov}\,[W(\tau+\xi)U(\xi), W(\tau+\eta)U(\eta)]\,d\xi d\eta,$$

which, by Lemma 8.3, is equal to

$$\frac{1}{t}\lambda^2(\tau)\,\text{var}\,[m(U)U] + \frac{1}{t}\int_0^\infty \lambda^2(\xi)d\xi\,\text{var}\,[m(U)]\,\sigma_U^2$$
$$+ \frac{1}{t^2}E^2\{m(U)U\}\int_0^t\int_0^t \lambda(\tau+\xi-\eta)\lambda(\tau+\eta-\xi)d\xi d\eta.$$

Since

$$\int_0^\infty\int_0^\infty |\lambda(\tau+\xi-\eta)\lambda(\tau+\eta-\xi)|d\xi d\eta$$
$$= \int_0^\infty |\lambda(\tau+v)\lambda(\tau-v)|dv \le \max_t |\lambda(t)|\int_0^\infty |\lambda(v)|dv,$$

we get

$$|P_t| \le \frac{1}{t}\lambda^2(\tau)\,\text{var}\,[m(U)U] + \frac{1}{t}\int_0^\infty \lambda^2(\xi)d\xi\,\text{var}\,[m(U)]\,\sigma_U^2$$
$$+ \frac{1}{t^2}E^2\{m(U)U\}\max_t |\lambda(t)|\int_0^\infty |\lambda(v)|dv.$$

Therefore,

$$E(\hat{\kappa}(\tau) - \kappa(\tau))^2 = O\left(\frac{1}{t}\right),\qquad(8.11)$$

which completes the proof. ∎

Proof of Theorem 8.2

The proof is similar to that of Theorem 3.1. Here, we assume that $Em(U) = 0$ (see Remark 8.1). Observe that $\hat{\mu}(u) = \hat{g}(u)/\hat{f}(u)$ with

$$\hat{g}(u) = \frac{1}{th(t)} \int_0^t Y(s+\tau)K\left(\frac{u-U(\tau)}{h(\tau)}\right) d\tau$$

and

$$\hat{f}(u) = \frac{1}{th(t)} \int_0^t K\left(\frac{u-U(\tau)}{h(\tau)}\right) d\tau.$$

Fix a point $u \in R$ and suppose that both $m(u)$ and $f(u)$ are continuous at the point u. Since

$$E\hat{g}(u) = \frac{1}{h(t)} E\left\{ E\left\{Y(t+s)|U(t)\right\} K\left(\frac{u-U(t)}{h(t)}\right)\right\}$$

$$= \frac{1}{h(t)} E\left\{ \mu(U)K\left(\frac{u-U}{h(t)}\right)\right\},$$

applying Lemma A.8 and remembering that $g(u) = \mu(u)f(u)$, we conclude that

$$E\hat{g}(u) \to g(u) \int K(v)dv \text{ as } t \to \infty.$$

In turn, since $Y(t) = W(t) + Z(t)$,

$$\text{var}[\hat{g}(u)] = P_t(u) + R_t(u),$$

where

$$P_t(u) = \frac{1}{th(t)}\sigma_Z^2 \frac{1}{h(t)} EK^2\left(\frac{u-U}{h(t)}\right),$$

$R_t(u)$

$$= \frac{1}{t^2 h^2(t)}$$

$$\times \int_0^t \int_0^t \left[\text{cov}\, W(\xi+s)K\left(\frac{u-U(\xi)}{h(t)}\right), W(\eta+s)K\left(\frac{u-U(\eta)}{h(t)}\right)\right] d\xi\, d\eta.$$

In view of Lemma A.8,

$$th(t)P_t(u) \to \sigma_Z^2 f(u) \int K^2(v)dv \text{ as } t \to \infty.$$

Turning to $R_t(u)$, we apply Lemma 8.2 to obtain $R_t(u) = \sum_{i=1}^{5} R_{it}(u)$ with

$$R_{1t}(u) = \lambda^2(s) \, \text{var} \left[m(U)K \left(\frac{u - U}{h(t)} \right) \right],$$

$$R_{2t}(u) = \frac{1}{th(t)} \int_0^\infty \lambda^2(\tau)d\tau \, \text{var}\,[m(U)] \frac{1}{h(t)} EK^2 \left(\frac{u - U}{h(t)} \right),$$

$$R_{3t}(u) = R_{4t}(u)$$

$$= \frac{1}{t^2 h^2(t)} \lambda(s) EK \left(\frac{u - U}{h(t)} \right) E \left\{ m^2(U)K \left(\frac{u - U)}{h(t)} \right) \right\}$$

$$\times \int_0^t \int_0^t \lambda(s + \eta - \xi)d\xi d\eta$$

$$= |\lambda(s)| \int_0^\infty |\lambda(\tau)|d\tau \, O \left(\frac{1}{t} \right),$$

and

$$R_{5t}(u) = \frac{1}{t^2} \frac{1}{h^2(t)} E^2 \left\{ m(U)K \left(\frac{u - U}{h(t)} \right) \right\} \int_0^t \int_0^t \lambda(s + \xi - \eta)\lambda(s + \eta - \xi)d\xi d\eta$$

$$= \max_\tau |\lambda(\tau)| \int_0^\infty |\lambda(\tau)|d\tau \, O \left(\frac{1}{t} \right).$$

Since

$$th(t)R_{1t}(u) \to \lambda^2(s)m^2(u)f(u) \int K^2(v)dv \text{ as } t \to \infty$$

and

$$th(t)R_{2t}(u) \to \int_0^\infty \lambda^2(\tau)d\tau \, \text{var}\,[m(U)] f(u) \int K^2(v)dv \text{ as } t \to \infty,$$

(see Lemma A.1), we obtain

$$th(t)R_t(u) \to \varphi(u)f(u) \int K^2(v)dv \text{ as } t \to \infty,$$

where $\varphi(u) = \lambda^2(s)m^2(u) + \int_0^\infty \lambda^2(\tau)d\tau \, \text{var}[m(U)]$. Finally,

$$th(t) \, \text{var}[\hat{g}(u)] \to (\sigma_Z^2 + \varphi(u))f(u) \int K^2(v)dv.$$

In this way, we have verified that

$$E(\hat{g}(u) - \mu(u)f(u))^2 \to 0 \text{ as } t \to \infty.$$

Using similar arguments, we show that $E \hat{f}(u) \to f(u)$ as $t \to \infty$,

$$th(t) \, \text{var}[\hat{f}(u)] \to f(u) \int K^2(v)dv \text{ as } t \to \infty$$

and come to the conclusion that $E(\hat{f}(u) - f(u))^2 \to 0$ as $t \to \infty$. The proof is complete. ∎

8.5   Bibliographic notes

Continuous-time estimates recovering the nonlinearity have been examined in Greblicki [114,116]. The problem of nonparametric estimation of a continuous-time Hammerstein system from observation sampled in time has been studied in Greblicki [120]. In Bosq [26] the kernel regression estimate for continuous-time processes satisfying the strong mixing condition has been investigated.

9 Discrete-time Wiener systems

In this chapter, we examine the problem of nonparametric identification of a Wiener system, that is, a system in which a linear dynamic part is followed by a nonlinear and memoryless one. A fundamental relationship between the inverse of the system nonlinearity and the nonparametric regression is established in Section 9.2. The use of the correlation theory for recovering the linear subsystem is discussed in Section 9.3. Section 9.2.3 explores the issue of incorporating the monotonicity constraint into the estimation problem.

9.1 The system

A Wiener system shown in Figure 9.1 comprises a linear dynamic subsystem with an impulse response $\{\lambda_n\}$ and a nonlinear memoryless subsystem with a characteristic $m(\bullet)$ connected in a cascade. The output signal W_n of the linear part is disturbed by Z_n and $V_n = W_n + Z_n$ is the input of the nonlinear part, while

$$Y_n = m(V_n)$$

is the output of the whole system. Neither W_n nor V_n is available for measurement. The goal is to identify the system, that is, to recover both $\{\lambda_n\}$ and $m(\bullet)$, from observations

$$(U_1, Y_1), (U_2, Y_2), \ldots, (U_n, Y_n), \ldots \tag{9.1}$$

taken at the input and output of the whole system.

The input $\{\ldots, U_{-1}, U_0, U_1, \ldots\}$ and disturbance $\{\ldots, Z_{-1}, Z_0, Z_1, \ldots\}$ are mutually independent Gaussian stationary white random signals. The disturbance has zero mean and finite variance, that is, $E Z_n = 0$ and $\text{var}[Z_n] = \sigma_Z^2 < \infty$.

The dynamic subsystem is described by the state equation

$$\begin{cases} X_{n+1} = AX_n + bU_n \\ \quad W_n = c^T X_n, \end{cases} \tag{9.2}$$

where X_n is a state vector, A is a matrix, and b and c are vectors. Therefore,

$$\lambda_n = \begin{cases} 0, & \text{for } n = 0, -1, -2, \ldots \\ c^T A^{n-1} b, & \text{for } n = 1, 2, 3, \ldots \end{cases}$$

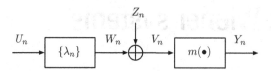

Figure 9.1 The discrete-time Wiener system.

and

$$W_n = \sum_{i=-\infty}^{n} \lambda_{n-i} U_i. \tag{9.3}$$

Neither the dimension of A, nor A itself, nor b, nor c are known. Nevertheless the matrix A is stable, all its eigenvalues lie in the unit circle. Therefore both X_n and W_n are random variables and $\{\ldots, X_{-1}, X_0, X_1, \ldots\}$ as well as $\{\ldots W_{-1}, W_0, W_1, \ldots\}$ are stationary correlated random processes. So is $\{\ldots Y_{-1}, Y_0, Y_1, \ldots\}$. Thus, all signals are random variables and we estimate both $\{\lambda_n\}$ and $m(\bullet)$ from random observations (9.1).

All the above restrictions concerning the input signal and disturbance as well as the system hold throughout all chapters entirely devoted to Wiener systems and will not repeated in relevant lemmas or theorems. In particular, we underline the assumptions concerning Gaussian and stationary properties of the signals coming to the system, and the stability of the dynamic subsystem.

Essential results on consistency of identification algorithms hold under an additional assumption that the nonlinear characteristic satisfies the Lipschitz inequality, that is, that

$$|m(v) - m(w)| \le c_m |v - w| \tag{9.4}$$

with some unknown c_m.

For simplicity of the notation, U, W, V, Y, and Z stand for U_n, W_n, V_n, Y_n, and Z_n, respectively.

9.2   Nonlinear subsystem

9.2.1   The problem and the motivation for algorithms

Let

$$\alpha = \frac{\sigma_U^2}{\sigma_U^2 \sum_{i=0}^{\infty} \lambda_i^2 + \sigma_Z^2}$$

and

$$\alpha_p = \lambda_p \alpha. \tag{9.5}$$

The basis for algorithms recovering the nonlinear characteristic is the following lemma:

LEMMA 9.1 *For any Borel measurable* $m(\bullet)$,

$$E\left\{U_n | V_{p+n} = v\right\} = \alpha_p v.$$

Proof. Since the pair (U_n, V_{p+n}) has a Gaussian distribution such that $EU_n = 0$ and $EV_n = 0$,

$$E\left\{U_n | V_{p+n} = v\right\} = \text{cov}\left[U_n, V_{p+n}\right] \frac{1}{\sigma_V^2} v.$$

As

$$\text{cov}\left[U_n, V_{p+n}\right] = \text{cov}\left[U_0, \sum_{i=-\infty}^{p} \lambda_{p-i} U_i\right] = \sum_{i=-\infty}^{p} \lambda_{p-i} \, \text{cov}\left[U_0, U_i\right]$$

$$= \lambda_p \sigma_U^2,$$

and $\sigma_V^2 = \sigma_U^2 \sum_{i=0}^{\infty} \lambda_i^2 + \sigma_Z^2$, we easily complete the proof. ■

By $m(R)$ we denote the image of the real line R under the mapping $m(\bullet)$ and assume that $m(\bullet)$ is invertible in the Cartesian product $R \times m(R)$. The inversion will be denoted by $m^{-1}(\bullet)$. Obviously, values of $m^{-1}(\bullet)$ lie in R, while $m(R)$ is its domain.

LEMMA 9.2 *Let* $m(\bullet)$ *be invertible in the Cartesian product* $R \times m(R)$. *For every* $y \in m(R)$,

$$E\left\{U_n | Y_{p+n} = y\right\} = \alpha_p m^{-1}(y).$$

Proof. Since

$$E\left\{U_n | Y_{p+n} = y\right\} = E\left\{U_n | V_{p+n} = m^{-1}(y)\right\},$$

an application of Lemma 9.1 completes the proof. ■

The lemma suggests that to recover $m^{-1}(y)$, we can just estimate the regression $E\left\{U_n | Y_{p+n} = y\right\}$. This goal can be, however, achieved only up to an unknown multiplicative constant α_p. For simplicity, we denote

$$v(y) = \alpha_p m^{-1}(y).$$

In Chapters 10, 11, we discuss a variety of estimates of the regression and show that they recover $v(y)$, that is, that

$$\hat{v}(y) \to v(y) \text{ as } n \to \infty$$

in an appropriate stochastic sense, where $\hat{v}(y)$ denotes such an estimate. Therefore, assuming that $\hat{v}(y)$ is invertible, we can expect that $\hat{v}^{-1}(v)$ converges to $v^{-1}(v) = m(v/\alpha_p)$, that is, that

$$\hat{v}(y) \to m(v/\alpha_p) \text{ as } n \to \infty$$

in the same manner. Thus we can recover $m(\bullet)$ up to some unknown dilation coefficient α_p^{-1}.

For $m(\bullet)$ satisfying Lipschitz inequality (9.4) Y_n may not have a density. Some results, however, require the density to exist. To assure that an additional restriction on $m(\bullet)$ should be imposed. In such cases, we assume that

$$m(\bullet) \text{ is strictly monotonous and has a bounded derivative.} \qquad (9.6)$$

Obviously such an $m(\bullet)$ satisfies (9.4). Owing to (9.6) the density denoted by $f(\bullet)$ exists and

$$f(y) = \begin{cases} f_V(m^{-1}(y)) \Big/ \left| \dfrac{d}{dy} m^{-1}(y) \right|, & \text{for } y \in m(R) \\ 0, & \text{otherwise,} \end{cases}$$

where $f_V(\bullet)$ is the density of V, that is, a normal density with zero mean and unknown variance σ_V^2.

Observe that (9.6) entails continuity of $m^{-1}(\bullet)$ in the set $m(R)$. Since $f_V(v) = (1/\sqrt{2\pi}\sigma_V)e^{-1/2\sigma_V^2}$ and $v(y) = \alpha_p m^{-1}(y)$, $f_V(v(y))$ is continuous at every point, where $m^{-1}(\bullet)$ is continuous, that is, in the whole set $m(R)$. Observing that

$$\frac{d}{dy}v(y) = \alpha_p \frac{d}{dy}m^{-1}(y) = \frac{\alpha_p}{\left. \dfrac{d}{dv}m(v) \right|_{v=m^{-1}(y)}} = \frac{\alpha_p}{m'(m^{-1}(y))}$$

we conclude that $v'(y)$ is also continuous in $m(R)$, provided that $m'(v) \neq 0$ for all $v \in (-\infty, \infty)$. Therefore, if (9.6) holds and $m'(\bullet)$ is nonzero in the whole real line, $f(\bullet)$ is continuous at every point $y \in m(R)$.

If $m'(v_0) = 0$, $v'(\bullet)$ is not continuous at a point y_0 such that $m^{-1}(y_0) = v_0$, that is, at the point y_0 where $y_0 = m(v_0)$. Also, $f(\bullet)$ is not continuous at this point.

9.2.2   Possible generalizations

In Section 9.2.1, the system is driven with zero mean random signal. If the mean is nonzero,

$$E\left\{ U_n | V_{n+p} = v \right\} = EU + \alpha_p \left(v - EU \sum_{i=0}^{\infty} \lambda_i \right),$$

which leads to

$$E\left\{ U_n | Y_{n+p} = y \right\} = \beta_p + \alpha_p m^{-1}(y) = \beta_p + v(v)$$

with $\beta_p = (1 - \alpha_p \sum_{i=0}^{\infty} \lambda_i)EU$. In such a case estimating the regression $E\{U_n | Y_{n+p} = y\}$, we recover $\beta_p + \alpha_p m^{-1}(y)$, where both β_p and α_p are unknown.

Moreover, results presented in further chapters can be extended to the system shown in Figure 9.2 in which $\{\xi_n\}$ is a white random Gaussian noise.

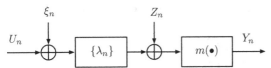

Figure 9.2 The Wiener system with an additional disturbance ξ_n.

9.2.3 Monotonicity-preserving algorithms

In many practical applications the nonlinear characteristics characterizing the Wiener and Hammerstein systems are known a priori to be nondecreasing. This assumption has been made in this chapter as the nonlinearity $m(\bullet)$ is assumed to be invertible. Unfortunately, common curve estimation methods are not very amenable to incorporate the monotonicity constraint. In fact, no estimation method can guarantee that the estimated characteristic is invertible. To include the monotone case in the particular identification algorithm one can simply add constraints forcing the solution to meet the monotonicity requirement. This framework can be easily incorporated in the classical parametric least squares approach to system identification by minimizing a penalized sum of squares. To illustrate our main ideas, let us consider the parametric version of the Hammerstein system:

$$X_{n+1} = AX_n + bV_n,$$
$$Y_n = c^T X_n + \varepsilon_n,$$

where ε_n is the output noise process and X_n is the d-dimensional state vector. Also let

$$V_n = m(U_n) = \sum_{i=1}^{r} \alpha_i g_i(U_n) \tag{9.7}$$

be the parametric representation of the nonlinear characteristic, that is, the basis functions (for example, polynomials) $g_i(\bullet)$ are selected in advance and the parameter r is assumed to be known. We can rewrite the input–output description of the Hammerstein system in the following equivalent form:

$$X_{n+1} = AX_n + \Theta G_n,$$
$$Y_n = c^T X_n + \varepsilon_n,$$

where $\Theta = b\alpha^T$ is the $d \times r$ matrix and G_n is the $r \times 1$ known vector of the transformed input signal U_n via the basis functions $g_i(\bullet)$, $i = 1, \ldots, r$.

An identification procedure starts with the initial estimates (obtained virtually by any linear system identification method) of A, c, Θ yielding the estimates $\hat{A}, \hat{c}, \hat{\Theta}$. Then the estimates of hidden parameters b, α are derived as the solution of the following least-squares problem:

$$(\hat{b}, \hat{\alpha}) = \arg \min_{b, \alpha} \|\hat{\Theta} - b\alpha^T\|^2.$$

All these derivations lead to the following estimate of the system nonlinearity

$$\hat{m}(u) = \sum_{i=1}^{r} \hat{\alpha}_i g_i(u).$$

The identical procedure applied to the Wiener system gives an estimate of the inverse function $m^{-1}(\bullet)$, that is, for the Wiener system, we have

$$\hat{m}^{-1}(u) = \sum_{i=1}^{r} \hat{\alpha}_i g_i(u).$$

It is clear that the monotonicity of $m(u)$ does not imply the monotonicity of $\hat{m}(u)$. Therefore, it is an important issue to construct the constrained solution $\bar{m}(u)$, which is a nondecreasing function of u. To do so, let us define the standardized form of the constrained solution

$$m_c(u) = \sum_{i=1}^{r} (\hat{\alpha}_i + \beta_i) g_i(u),$$

where $\{\beta_i\}$ is a new set of unknown parameters. Clearly, if the unconstrained solution $\hat{m}(u)$ meets the monotonicity requirement then we should have $\beta_i = 0, i = 1, \ldots, r$. The monotonicity of $m_c(u)$ gives the following condition for $\{\beta_i\}$

$$\sum_{i=1}^{r} \beta_i g_i^{(1)}(u) \geq -\sum_{i=1}^{r} \hat{\alpha}_i g_i^{(1)}(u),$$

where $g_i^{(1)}(u)$ is the derivative of $g_i(u)$. This formula can be written in the following vector notation form:

$$\beta^T h(u) \geq a(u), \tag{9.8}$$

where $\beta = (\beta_1, \ldots, \beta_r)^T$, $h(u) = (g_1^{(1)}(u), \ldots, g_r^{(1)}(u))^T$ and $a(u) = -\sum_{i=1}^{r} \hat{\alpha}_i g_i^{(1)}(u)$. Often one needs to do some normalization of the vector $h(u)$ such that $h^T(u)h(u) = 1$. Now, we are in a position to reformulate the aforementioned parametric identification procedure that takes into account the monotonicity constraint. Hence, we seek a solution of the following minimization problem

$$\hat{\beta} = \arg \min_{\beta} \|\hat{\Theta} - \hat{b}(\hat{\alpha} + \beta)^T\|^2$$

subject to the constraint in (9.8). Since $\hat{\Theta}, \hat{b}, \hat{\alpha}$ are already specified the problem of finding $\hat{\beta}$ is equivalent to the quadratic programming problem with linear constraints. It is worth noting that the constraint in (9.8) is required to hold for all u. The weaker requirement would ask for (9.8) to be satisfied only at the training input data points $U_i, i = 1, \ldots, n$.

All the aforementioned discussion would lead to the following algorithm for finding $\hat{\beta}$:

1. If $a(u) \leq 0$, for all u, that is, $\sum_{i=1}^{r} \hat{\alpha}_i g_i^{(1)}(u) \geq 0$ then set $\hat{\beta} = 0$. This is the case when the unconstrained solution $\hat{m}(u)$ meets the monotonicity constraint.

2. If $a(u) > 0$ for some u then $\hat{\beta} = 0$ is not the solution. In this case define

$$a(u^*) = max_u a(u)$$

and define

$$\beta^* = a(u^*)h(u^*).$$

Verify whether β^* satisfies (9.8). This can be done either for all u or for the input data points U_i, $i = 1, \ldots, n$. If (9.8) holds then set $\hat{\beta} = \beta^*$. The case if β^* does not meet (9.8) requires some further investigation and some suggestions can be found in [77], [252].

3. Exit the algorithm with the following monotonicity constrained estimate of $m(u)$

$$\bar{m}(u) = \sum_{i=1}^{r} (\hat{\alpha}_i + \hat{\beta}_i) g_i(u).$$

The consistency of this parametric estimate $\bar{m}(u)$ of $m(u)$ can be established using standard tools of the parametric statistical inference.

Thus far, we have assumed the parametric knowledge of the system nonlinearity defined by formula (9.7). Hence both the basis functions $g_i(u)$, $i = 1, \ldots, r$ and the parameter r are assumed to be known. The first step into the nonparametric extension of the proposed approach would be to allow the parameter r to be selected adaptively for known class of basis functions. A challenging issue would be to incorporate the monotonicity constraint into the nonparametric framework. A projection type approach for constrained nonparametric estimation [206] could be used here and this topic is left for future studies.

9.3 Dynamic subsystem identification

9.3.1 The motivation

Denote

$$\beta = \alpha E\{Vm(V)\} = \frac{\sigma_U^2}{\sigma_U^2 \sum_{i=0}^{\infty} \lambda_i^2 + \sigma_Z^2} E\{Vm(V)\}.$$

LEMMA 9.3 *If $E|Vm(V)| < \infty$, then*

$$E\{U_0 Y_i\} = \beta \lambda_i.$$

Proof. From Lemma 9.1, it follows that $E\{U_0|V_i\} = \alpha \lambda_i V_i$. Therefore, since

$$E\{U_0 Y_i|V_i\} = E\{U_0 m(V_i)|V_i\} = m(V_i)E\{U_0|V_i\},$$

we get $E\{U_0 Y_i|V_i\} = \alpha \lambda_i V_i m(V_i)$, which completes the proof. ∎

Since we can recover only $\beta \lambda_i$, it is convenient to denote $\kappa_i = \beta \lambda_i$.

9.3.2   The algorithm

Lemma 9.3 suggests the following estimate of $\kappa_i = \beta \lambda_i$:

$$\hat{\kappa}_i = \frac{1}{n} \sum_{j=1}^{n} U_j Y_{i+j}.$$

Observe that if (9.4) holds, there exist c_1 and c_2 such that $|m(v)| \le c_1 + c_2|v|$. Hence,

$$E|V m(V)| \le c_1 E|V| + c_2 E V^2 < \infty. \qquad (9.9)$$

For similar reasons, for any i,

$$E\{U_0^2 m^2(V_i)\} < \infty. \qquad (9.10)$$

THEOREM 9.1 *If $m(\bullet)$ satisfies (9.4), then, for every i,*

$$\lim_{n \to \infty} E\,(\hat{\kappa}_i - \kappa_i)^2 = 0.$$

Proof. Since the estimate is unbiased, we pass to the variance and get var$[\hat{\kappa}_i] = P_n + Q_n + R_n$, where $P_n = n^{-1}$ var$[U_0 Y_i]$,

$$Q_n = \frac{1}{n^2} \sum_{j=1}^{i} (n - j) \operatorname{cov} \left[U_0 m(V_i),\, U_j m(V_{i+j}) \right],$$

and

$$R_n = \frac{1}{n^2} \sum_{j=i+1}^{n} (n - j) \operatorname{cov} \left[U_0 m(V_i),\, U_j m(V_{i+j}) \right].$$

Clearly, $|P_n| \le n^{-1} E^{1/2} U^2 E^{1/2} Y^2 = O(n^{-1})$, and

$$|Q_n| \le \frac{1}{n} \sum_{j=1}^{i} \left| \operatorname{cov} \left[U_0 m(V_i),\, U_j m(V_{i+j}) \right] \right|$$

$$\le \frac{1}{n} \sum_{j=1}^{i} E \left| U_0 m(V_i) U_j m(V_{i+j}) \right| + \frac{1}{n} \sum_{j=1}^{i} E^2 \{U_0 m(V_i)\}$$

$$\le \frac{i}{n} E \{U_0^2 m^2(V_i)\} + \frac{i}{n} E^2 \{U_0 m(V_i)\} = O\left(\frac{1}{n}\right),$$

see (9.9) and (9.10). Thus, $|Q_n| = O(n^{-1})$. Moreover, by Lemma 9.4,

$$|R_n| \le \frac{1}{n} \sum_{j=i+1}^{n} \left| \operatorname{cov} \left[U_0 m(V_i),\, U_j m(V_{i+j}) \right] \right| \le \gamma \frac{1}{n} \sum_{j=i+1}^{n} \|A^j\|$$

$$= O\left(\frac{1}{n}\right),$$

since $\sum_{j=0}^{\infty} \left\| A^j \right\| < \infty$. Finally,

$$E\left(\hat{\kappa}_i - \kappa_i\right)^2 = \text{var}\,[\hat{\kappa}_i] = O\left(\frac{1}{n}\right), \tag{9.11}$$

which completes the proof. ∎

Sequence $\{\hat{\kappa}_1, \hat{\kappa}_2, \hat{\kappa}_3, \ldots, \hat{\kappa}_{N(n)}, 0, 0, \ldots\}$ is defined as an estimate of the whole impulse response $\{\kappa_1, \kappa_2, \kappa_3, \ldots\}$. Defining the mean-summed-square error (MSSE) as

$$\text{MSSE} = \sum_{i=1}^{N(n)} E(\hat{\kappa}_i - \kappa_i)^2 + \sum_{i=N(n)+1}^{\infty} \kappa_i^2,$$

we apply (9.11) to find that the error is not greater than

$$O\left(\frac{N(n)}{n}\right) + \sum_{i=N(n)+1}^{\infty} \kappa_i^2.$$

Therefore, if $N(n) \to \infty$ as $n \to \infty$ and $N(n)/n \to 0$ as $n \to \infty$, $\lim_{n\to\infty} \text{MSSE} = 0$.

9.4 Lemmas

LEMMA 9.4 *If (9.4) holds, then, for $0 < i < j$,*

$$\left| \text{cov}\left[U_0 m(V_i), U_j m(V_{i+j}) \right] \right| \le \gamma \left\| A^j \right\|$$

with a finite γ.

Proof. Suppose $0 < i < j$. Clearly, $V_{i+j} = c^T A^j X_i + \xi_{i+j}$, where $\xi_{i+j} = \sum_{q=i}^{i+j-1} c^T A^{j-1-q} b U_q + Z_{i+j}$. Since pairs (U_0, V_i) and (U_j, ξ_{i+j}) are independent, $\text{cov}[U_0 m(V_i), U_j m(\xi_{i+j})] = 0$. Hence,

$$\begin{aligned}
\text{cov}\left[U_0 m(V_i), U_j m(V_{i+j}) \right] &= \text{cov}\left[U_0 m(V_i), U_j [m(V_{i+j}) - m(\xi_{i+j})] \right] \\
&= E\left\{ U_0 U_j m(V_i)[m(V_{i+j}) - m(\xi_{i+j})] \right\} \\
&\quad - E\left\{ U_0 m(V_i) \right\} E\left\{ U_0 [m(V_{i+j}) - m(\xi_{i+j})] \right\}
\end{aligned}$$

Applying (9.4), we get

$$\left| m(V_{i+j}) - m(\xi_{i+j}) \right| \le c_m \left| V_{i+j} - \xi_{i+j} \right| \le c_m \left| c^T A^j X_i \right| \le c_m \left\| c \right\| \left\| A^j \right\| \left\| X_i \right\|.$$

Thus,

$$\begin{aligned}
&\left| \text{cov}\left[U_0 m(V_i), U_j m(V_{i+j}) \right] \right| \\
&\le c_m \left\| c \right\| \left\| A^j \right\| \left(E \left| U_0 U_j m(V_i) \right\| X_i \right\| + \left| E \left\{ U_0 m(V_i) \right\} \right| E \left| U_0 \left\| X_i \right\| \right| \right) \\
&= c_m \left\| c \right\| \left\| A^j \right\| \left(E \left| U_0 m(V_i) \right\| X_i \right\| E \left| U \right| + \left| E \left\{ U_0 m(V_i) \right\} \right| E \left| U_0 \left\| X_i \right\| \right| \right) \\
&= \gamma \left\| A^j \right\|
\end{aligned}$$

with a finite γ. ∎

We denote $\psi_n(y) = E\left\{ U_0^2 | Y_n = y \right\}$.

LEMMA 9.5 *In the system,*

$$\psi_n(y) = \sigma_U^2 \left(1 - \lambda_n^2 \frac{\sigma_U^2}{\sigma_V^2} \right) + \lambda_n^2 \frac{\sigma_U^4}{\sigma_V^4} v^2(y),$$

where $v(y) = m^{-1}(y)$. *Moreover,*

$$\sup_n |\psi_n(y)| \le \alpha + \beta v^2(y) = \psi(y),$$

where $\alpha = \sigma_U^2 (1 + (\max_n \lambda_n^2) \sigma_U^2 / \sigma_V^2)$ *and* $\beta = (\max_n \lambda_n^2) \sigma_U^4 / \sigma_V^4$.

Proof. Since the pair (U_0, V_n) has a Gaussian distribution such that $E U_0 = 0$, $E V_n = 0$ and $\text{cov}\,[U_0, V_n] = \lambda_n \sigma_U^2$, the conditional density of U_0 conditioned on $V_n = v$ is Gaussian with mean $\lambda_n (\sigma_U^2 / \sigma_V^2) v$ and variance $\sigma_U^2 [1 - \lambda_n^2 (\sigma_U^2 / \sigma_V^2)]$. Thus,

$$E \left\{ U_0^2 \mid V_n = v \right\} = \sigma_U^2 \left(1 - \lambda_n^2 \frac{\sigma_U^2}{\sigma_V^2} \right) + \lambda_n^2 \frac{\sigma_U^4}{\sigma_V^4} v^2,$$

which completes the proof. ■

9.5   Bibliographic notes

Possibilities of the identification of Wiener systems have been studied by Brillinger [31] and Bendat [16]. Parametric methods have been examined by Billings and Fakhouri [20], Westwick and Kearney [317], Wigren [319–321], Westwick and Verhaegen [315], Al-Duwaish, Karim, and Chandrasekar [2], Vandersteen, Rolain, and Schoukens [296], Ndorsjö [217], Korenberg and Hunter [178], Lacy, Erwin, and Bernstein [191], and Chen [44].

Larger classes of block systems including, in particular, Wiener systems, have attracted the attention of Billings and Fakhouri [21–24], Fakhouri, Billings, and Wormald [86], and Boutayeb and Darouach [27].

The Wiener system has been applied by den Brinker [66], Emerson, Korenberg, and Citron [80] to describe visual systems. Kalafatis, Arifin, Wang, and Cluett [173], Kalafatis, Wang, and Cluett [174], and Patwardhan, Lakshminarayanan, and Shah [227] used the Wiener model to identify a pH process. Another field of possible applications is biomedical engineering, see Korenberg and Hunter [177], Hunter and Korenberg [168], Celka and Colditz [41], and Westwick and Kearney [316].

A general approach for constrained nonparametric estimation in the context of the regression analysis was proposed in Mammen, Marron, Turlach, and Wand [206].

10 Kernel and orthogonal series algorithms

This chapter examines various nonparametric identification algorithms for recovering the nonlinearity in the Wiener system. Section 10.1 is devoted to kernel methods and Section 10.2 is concerned with methods using orthogonal expansions.

10.1 Kernel algorithms

10.1.1 Introduction

By Lemma 9.2 to recover $v(y) = \alpha_p m^{-1}(y)$, it suffices to estimate the regression $E\{U_n | Y_{p+n} = y\}$. This goal can be achieved with the following kernel estimate:

$$\hat{v}(y) = \frac{\sum_{i=1}^{n} U_i K\left(\dfrac{y - Y_{p+i}}{h_n}\right)}{\sum_{i=1}^{n} K\left(\dfrac{y - Y_{p+i}}{h_n}\right)} \tag{10.1}$$

and its semirecursive versions

$$\tilde{v}_n(y) = \frac{\sum_{i=1}^{n} \dfrac{1}{h_i} U_i K\left(\dfrac{y - Y_{p+i}}{h_i}\right)}{\sum_{i=1}^{n} \dfrac{1}{h_i} K\left(\dfrac{y - Y_{p+i}}{h_i}\right)} \tag{10.2}$$

and

$$\bar{v}_n(y) = \frac{\sum_{i=1}^{n} U_i K\left(\dfrac{y - Y_{p+i}}{h_i}\right)}{\sum_{i=1}^{n} K\left(\dfrac{y - Y_{p+i}}{h_i}\right)}. \tag{10.3}$$

Like those investigated in Chapter 4 their numerator and denominator can be calculated recursively, (see Section 4.1).

We study two cases. In one the nonlinear characteristic is so smooth, strictly monotonic and differentiable, that the output density $f(\bullet)$ of Y exists. In the other the characteristic is not differentiable but only a Lipschitz function. The density may not exist. All proofs are in Section 10.4.

10.1.2 Differentiable characteristic

In this section, we show that if the nonlinear characteristic is differentiable, that is, if the density $f(\bullet)$ of the output signal exists then all three kernel estimates are consistent. The presented theorems assume that $|y|^{1+\varepsilon}K(y) \to 0$ as $|y| \to \infty$, where $\varepsilon = 0$, and establish the convergence at continuity points. The verification of the consistency at almost every point holding for $\varepsilon > 0$ is left to the reader.

Offline estimate

We now show that offline estimate (10.1) is consistent.

THEOREM 10.1 *Let $m(\bullet)$ satisfy (9.6). Let the kernel satisfy the following restrictions:*

$$\sup_y |K(y)| < \infty, \tag{10.4}$$

$$\int |K(y)|dy < \infty, \tag{10.5}$$

$$yK(y) \to 0 \text{ as } |y| \to \infty, \tag{10.6}$$

and, moreover,

$$|K(y) - K(x)| \le c_K |y - x| \tag{10.7}$$

with some c_K, for all $x, y \in (-\infty, \infty)$. If

$$h_n \to 0 \text{ as } n \to \infty \tag{10.8}$$

and

$$nh_n^2 \to \infty \text{ as } n \to \infty, \tag{10.9}$$

then

$$\hat{v}(y) \to v(y) \text{ as } n \to \infty \text{ in probability} \tag{10.10}$$

at every $y \in m(R)$, where $f(\bullet)$ is continuous and $f(y) > 0$.

REMARK 10.1 *Notice that in Theorem 10.1, the kernel satisfies not only all the appropriate restrictions imposed by Theorem 3.1 and others dealing with the Hammerstein system but also (7.15). As a result, the rectangular kernel is not admitted. Concerning the number sequence, (10.8) and (10.9) are more restrictive than (3.7) and (3.8). A hypothesis is that (10.9) can be replaced by (3.8).*

Semirecursive estimates

In the two theorems presented in this section, the convergence of semirecursive estimates (10.2) and (10.3) is shown.

THEOREM 10.2 *Let the kernel satisfy (10.4), (10.5), (10.6), and (10.7). If the positive number sequence $\{h_n\}$ satisfies (10.8) and*

$$\frac{1}{n^2} \sum_{i=1}^{n} \frac{1}{h_i^2} \to 0 \text{ as } n \to \infty,$$

then

$$\tilde{v}(y) \to v(y) \text{ as } n \to \infty \text{ in probability}$$

at every point $y \in m(R)$ where $f(\bullet)$ is continuous and $f(y) > 0$.

THEOREM 10.3 *Let the kernel satisfy the appropriate restrictions of Theorem 10.2. If the positive monotonous number sequence $\{h_n\}$ satisfies (10.8) and*

$$h_n \sum_{i=1}^{n} h_i \to \infty \text{ as } n \to \infty,$$

then

$$\bar{v}(y) \to v(y) \text{ as } n \to \infty \text{ in probability}$$

at every $y \in m(R)$ where $f(\bullet)$ is continuous and $f(y) > 0$.

10.1.3   Lipschitz characteristic

In this section the nonlinear characteristic $m(\bullet)$ is invertible (see Section 9.2.1), and satisfies Lipschitz inequality (9.4) but may not be differentiable. Because of that, the output random variable Y may not have a density. In further considerations, the probability measure of Y is denoted by $\zeta(\bullet)$.

THEOREM 10.4 *Let $m(\bullet)$ be invertible in the Cartesian product $R \times m(R)$ and satisfy Lipschitz inequality (9.4). Let $H(\bullet)$ be a nonnegative nonincreasing Borel function defined on $[0, \infty)$, continuous and positive at $t = 0$ and such that*

$$tH(t) \to 0 \text{ as } t \to \infty.$$

Let, for some c_1 and c_2,

$$c_1 H(|y|) \le K(y) \le H(|y|)c_2.$$

Let, moreover, $K(\bullet)$ satisfy Lipschitz inequality (10.7). Let the sequence $\{h_n\}$ of positive numbers satisfy (10.8) and (10.9). Then convergence (10.10) takes place at almost every $(\zeta) y \in m(R)$.

10.1.4   Convergence rate

Like in Section 3.4, we assume that $f(\bullet)$ and $g(\bullet)$ have q derivatives bounded in a neighborhood of a point $y \in m(R)$. Selecting the kernel as described in Section 3.4, in view of (10.12), we obtain

$$E(\hat{f}(y) - f(y))^2 = O(h_n^{2q}) + O\left(\frac{1}{nh_n^2}\right).$$

If $m^{-1}(\bullet)$ has the same number of bounded derivatives, the same holds for $E(\hat{g}(y) - g(y))^2$. Finally, for $h_n \sim n^{-1/(2q+2)}$,

$$P\{|\hat{v}(y) - v(y)| > \varepsilon|v(y)|\} = O(n^{-1+1/(q+1)})$$

with any $\varepsilon > 0$, and

$$|\hat{v}(y) - v(y)| = O(n^{-1/2+1/(2q+2)})$$

in probability. Similar results can be obtained for semirecursive estimates.

10.2   Orthogonal series algorithms

10.2.1   Introduction

The nonlinear characteristic $m(\bullet)$ maps the whole real line R into the set $m(R)$. Assuming that the characteristic is smooth, that is, satisfies (9.6), we apply orthogonal series estimates.

10.2.2   Fourier series algorithm

For simplicity, we assume that $m(R) = [-\pi, \pi]$ and to recover $v(y) = \alpha_p m^{-1}(y)$ easily observe that $v(y) = g(y)/f(y)$, where $g(y) = v(y)f(y) = E\{U_0|Y_p = y\}f(y)$. Expanding $g(\bullet)$ in the Fourier series, we get

$$g(y) \sim a_0 + \sum_{k=0}^{\infty} a_k \cos ky + \sum_{k=0}^{\infty} b_k \sin ky$$

with

$$a_0 = \frac{1}{2\pi} \int_{-\pi}^{\pi} g(y)dy = \frac{1}{2\pi} EU,$$

and, for $k = 1, 2, \ldots,$

$$a_k = \frac{1}{\pi} \int_{-\pi}^{\pi} g(y) \cos(ky)dy = \frac{1}{\pi} E\{U_0 \cos(kY_p)\},$$

and

$$b_k = \frac{1}{\pi} \int_{-\pi}^{\pi} g(y) \sin(ky)dy = \frac{1}{\pi} E\{U_0 \sin(kY_p)\}.$$

For $f(\bullet)$, we obtain

$$f(y) \sim \frac{1}{2\pi} + \sum_{k=1}^{\infty} \alpha_k \cos ky + \sum_{k=1}^{\infty} \beta_k \sin ky$$

where

$$\alpha_k = \frac{1}{\pi} \int_{-\pi}^{\pi} f(y) \cos(ky) dy = \frac{1}{\pi} E \cos(kY),$$

and

$$\beta_k = \frac{1}{\pi} \int_{-\pi}^{\pi} f(y) \sin(ky) dy = \frac{1}{\pi} E \sin(kY),$$

$k = 1, 2, \ldots$. Therefore, the estimate of $v(y)$ has the following form:

$$\hat{v}(y) = \frac{\hat{a}_0 + \sum_{k=1}^{N(n)} \hat{a}_k \cos ky + \sum_{k=1}^{N(n)} \hat{b}_k \sin ky}{\frac{1}{2\pi} + \sum_{k=1}^{N(n)} \hat{\alpha}_k \cos ky + \sum_{k=1}^{N(n)} \hat{\beta}_k \sin ky},$$

where

$$\hat{a}_0 = \frac{1}{2\pi n} \sum_{i=1}^{n} U_i, \quad \hat{a}_k = \frac{1}{\pi n} \sum_{i=1}^{n} U_i \cos kY_{p+i},$$

$$\hat{b}_k = \frac{1}{\pi n} \sum_{i=1}^{n} U_i \sin kY_{p+i},$$

and

$$\hat{\alpha}_k = \frac{1}{\pi n} \sum_{i=1}^{n} \cos kY_i, \quad \hat{\beta}_k = \frac{1}{\pi n} \sum_{i=1}^{n} \sin kY_i.$$

In an alternative representation of the estimate, we employ the Dirichlet kernel, see formula (B.2) in Section B, and get

$$\hat{v}(y) = \frac{\sum_{i=1}^{n} U_i D_{N(n)}(Y_{p+i} - y)}{\sum_{i=1}^{n} D_{N(n)}(Y_{p+i} - y)}.$$

THEOREM 10.5 *Let (9.6) be satisfied. Let $\int_{-\pi}^{\pi} (m^{-1}(y)f(y))^2 dy < \infty$ and $\int_{-\pi}^{\pi} f(y)^2 dy < \infty$. If*

$$N(n) \rightarrow \infty \text{ as } n \rightarrow \infty$$

and

$$\frac{N^2(n)}{n} \to 0 \text{ as } n \to \infty.$$

then

$$\hat{v}(y) \to v(y) \text{ as } n \to \infty \text{ in probability}$$

at every $y \in (-\pi, \pi)$ where both $m^{-1}(\bullet)$ and $f(\bullet)$ are differentiable and $f(y) > 0$.

10.2.3   Legendre series algorithm

Assuming that $m(R) = [-1, 1]$ we can employ the Legendre series and get the following estimate:

$$\tilde{v}(y) = \frac{\displaystyle\sum_{i=1}^{n} U_i k_{N(n)}(y, Y_{p+i})}{\displaystyle\sum_{i=1}^{n} k_{N(n)}(y, Y_i)},$$

where

$$k_n(y, x) = \sum_{k=0}^{n} p_k(y)p_k(x) = \sum_{k=0}^{n} \frac{2k+1}{2} P_k(y)P_k(x)$$

$$= \frac{n+1}{2} \frac{P_n(y)P_{n+1}(x) - P_{n+1}(y)P_n(x)}{x - y} \tag{10.11}$$

is the kernel of the Legendre system, see Section B.3.

THEOREM 10.6 *Let (9.6) be satisfied. Let $\int_{-1}^{1} |f(y)|(1 - y^2)^{-1/4}dy < \infty$ and $\int_{-1}^{1} |f(y)m^{-1}(y)|(1 - y^2)^{-1/4}dy < \infty$. If (6.2) is satisfied and*

$$\frac{N^{7/2}(n)}{n} \to 0 \text{ as } n \to \infty,$$

then

$$\tilde{v}(y) \to v(y) \text{ as } n \to \infty \text{ in probability}$$

at every $y \in (-1, 1)$, where both $m^{-1}(\bullet)$ and $f(\bullet)$ are differentiable and $f(y) > 0$.

10.2.4   Hermite series algorithm

For $m(R) = (-\infty, \infty)$, we can employ the Hermite series and get the following estimate:

$$\check{v}(y) = \frac{\displaystyle\sum_{i=1}^{n} U_i k_{N(n)}(y, Y_{p+i})}{\displaystyle\sum_{i=1}^{n} k_{N(n)}(y, Y_i)},$$

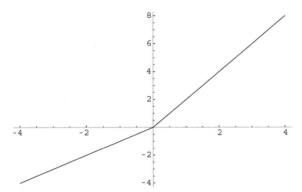

Figure 10.1 The nonlinear characteristic in the simulation example (Section 10.3).

where

$$k_n(x, y) = \sum_{k=0}^{n} h_k(x)h_k(y) = \sqrt{\frac{n+1}{2}} \frac{h_{n+1}(x)h_n(y) - h_n(x)h_{n+1}(y)}{y - x},$$

is the kernel of the Hermite system, see Section B.5.

THEOREM 10.7 *Let (9.6) be satisfied. If (6.2) holds and*

$$\frac{N^{13/6}(n)}{n} \to 0 \text{ as } n \to \infty,$$

then

$$\check{v}(y) \to v(y) \text{ as } n \to \infty \text{ in probability}$$

at every $y \in (-\infty, \infty)$ where both $m^{-1}(\bullet)$ and $f(\bullet)$ are differentiable and $f(y) > 0$.

10.3   Simulation example

In the simulation example the dynamic subsystem (see Figure 9.1), is described by the following equation: $X_{n+1} = aX_n + U_n$ with $a = 1/2$. The nonlinear characteristic is of the following form:

$$m(w) = \begin{cases} 2w, & \text{for } w \geq 0 \\ w, & \text{for } w < 0, \end{cases}$$

(see Figure 10.1). The input signal U_n has a Gaussian distribution with zero mean and variance 1. Therefore,

$$E\{U_n | Y_{n+1}\} = \alpha_1 m^{-1}(y)$$

with

$$\alpha_1 = \frac{1 - a^2}{1 + (1 - a^2)\sigma_Z^2}.$$

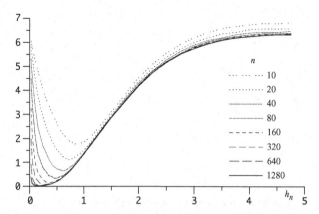

Figure 10.2 MISE versus h_n for estimate (10.1); various n.

Since the outer noise is absent, that is, $Z_n \equiv 0$, we have $\alpha_1 = 1 - a^2$. As $a = 1/2$,

$$E\{U_n | Y_{n+1}\} = \frac{3}{4} m^{-1}(y).$$

The accuracy of estimates is measured with

$$\text{MISE} = \int_{-3}^{3} \left(\hat{v}(y) - \frac{3}{4} m^{-1}(y) \right)^2 dy,$$

where $\hat{v}(y)$ stands for the involved estimate.

For offline estimate (10.1) the MISE is shown in Figure 10.2. The estimate behaves very much like its appropriate version applied to the identification of Hammerstein systems.

10.4   Lemmas and proofs

10.4.1   Lemmas

LEMMA 10.1 *Let the density $f(\bullet)$ of Y exist. If the kernel satisfies the appropriate restrictions of Theorem 10.2, then,*

$$\left| \text{cov}\left[U_i K\left(\frac{y - Y_{p+i}}{h_i} \right), U_j K\left(\frac{y - Y_{p+j}}{h_j} \right) \right] \right|$$

$$\leq \begin{cases} \sqrt{h_j h_j} \eta(y), & \text{for } 0 \leq |i - j| \leq p \\ \dfrac{h_j}{h_i} \|A^{i-j}\| \eta(y) & \text{for } p < i - j \\ \dfrac{h_i}{h_j} \|A^{j-i}\| \eta(y) & \text{for } p < j - i \end{cases}$$

with some $\eta(y)$ finite at every $y \in m(R)$ where both $m^{-1}(\bullet)$ and $f(\bullet)$ are continuous.

Proof. Fix $y \in m(R)$ and suppose that both $m^{-1}(\bullet)$ and $f(\bullet)$ are continuous at the point y.

For any i and j, the covariance in the assertion is equal to $Q_1(y) - Q_2(y)$, where

$$Q_1(y) = E\left\{U_i K\left(\frac{y - Y_{p+i}}{h_i}\right) U_j K\left(\frac{y - Y_{p+j}}{h_j}\right)\right\},$$

and

$$Q_2(y) = E\left\{U_i K\left(\frac{y - Y_{p+i}}{h_i}\right)\right\} E\left\{U_j K\left(\frac{y - Y_{p+j}}{h_j}\right)\right\}.$$

Since

$$|Q_1(y)| \leq E^{1/2}\left\{U_i^2 K^2\left(\frac{y - Y_{p+i}}{h_i}\right)\right\} E^{1/2}\left\{U_j^2 K^2\left(\frac{y - Y_{p+j}}{h_j}\right)\right\}$$

$$= \sqrt{h_i h_j}\, E^{1/2}\left\{\psi_p(Y)\frac{1}{h_i}K^2\left(\frac{y - Y}{h_i}\right)\right\} E^{1/2}\left\{\psi_p(Y)\frac{1}{h_j}K^2\left(\frac{y - Y}{h_j}\right)\right\}$$

with $\psi_p(\bullet)$ as in Lemma 9.5. Recalling Lemma A.8, we find that the quantity is not greater than $\sqrt{h_i h_j}\, \eta_1(y)$ for some finite $\eta_1(y)$. For the same reasons, $|Q_2(y)| \leq h_i h_j \eta_2(y)$ for some finite $\eta_2(y)$. For $0 \leq |i - j| \leq p$, the desired inequality follows.

Suppose that $p + j < i$. Obviously, $V_{p+i} = \sum_{n=-\infty}^{p+j-1} \lambda_{p+i-n} U_n + \xi_{p+i}$, where $\xi_{p+i} = \sum_{n=p+j}^{p+i} \lambda_{p+i-n} U_n + Z_{p+i}$. Because pairs (U_j, Y_{p+j}) and (U_i, ξ_{p+i}) are independent,

$$\mathrm{cov}\left[U_i K\left(\frac{y - m(\xi_{p+i})}{h_i}\right), U_j K\left(\frac{y - Y_{p+j}}{h_j}\right)\right] = 0.$$

Hence, denoting

$$\phi(y, Y_{p+i}, \xi_{p+i}) = K\left(\frac{y - Y_{p+i}}{h_i}\right) - K\left(\frac{y - m(\xi_{p+i})}{h_i}\right),$$

we find the examined covariance equal to

$$\mathrm{cov}\left[U_i \phi(y, Y_{p+i}, \xi_{p+i}), U_j K\left(\frac{y - Y_{p+j}}{h_j}\right)\right] = S_1(y) - S_2(y),$$

where

$$S_1(y) = E\left\{U_i U_j \phi(y, Y_{p+i}, \xi_{p+i}) K\left(\frac{y - Y_{p+j}}{h_j}\right)\right\}$$

and

$$S_2(y) = E\left\{U_j K\left(\frac{y - Y_{p+j}}{h_j}\right)\right\} E\left\{U_i \phi(y, Y_{p+i}, \xi_{p+i})\right\}$$

$$= h_j \frac{1}{h_j} E\left\{v(Y)K\left(\frac{y - Y}{h_j}\right)\right\} E\left\{U_i \phi(y, Y_{p+i}, \xi_{p+i})\right\}.$$

Since both $K(\bullet)$ and $m(\bullet)$ are Lipschitz functions, see (7.15) and (9.4),

$$|\phi(y, Y_{p+i}, \xi_{p+i})| \leq c_K c_m \frac{1}{h_i}|Y_{p+i} - m(\xi_{p+i})| = c_K c_m \frac{1}{h_i}\left|\sum_{n=-\infty}^{p+j-1} \lambda_{p+i-n} U_n\right|,$$

which yields

$|S_1(y)|$

$$\leq \gamma \frac{1}{h_i} \sum_{n=-\infty}^{p+j-1} |\lambda_{p+i-n}| E \left| U_i U_j U_n K \left(\frac{y - Y_{p+j}}{h_j} \right) \right|$$

$$\leq \gamma E|U| \frac{1}{h_i} \sum_{n=-\infty}^{p+j-1} |\lambda_{p+i-n}| E \left| U_j U_n K \left(\frac{y - Y_{p+j}}{h_j} \right) \right|$$

$$\leq \gamma \frac{1}{h_i} E^{1/2} \left\{ U_j^2 K^2 \left(\frac{y - Y_{p+j}}{h_j} \right) \right\} \sum_{n=-\infty}^{p+j-1} |\lambda_{p+i-n}| E^{1/2} \left\{ U_n^2 K^2 \left(\frac{y - Y_{p+j}}{h_j} \right) \right\},$$

which equals

$$\gamma \frac{1}{h_i} E^{1/2} \left\{ \psi_p(Y) K^2 \left(\frac{y - Y}{h_j} \right) \right\}$$

$$\times \sum_{n=-\infty}^{p+j-1} |\lambda_{p+i-n}| E^{1/2} \left\{ \psi_{p+j-n}(Y) K^2 \left(\frac{y - Y}{h_j} \right) \right\}$$

$$\leq \gamma \frac{h_j}{h_i} \frac{1}{h_j} E \left\{ \psi(Y) K^2 \left(\frac{y - Y}{h_j} \right) \right\} \sum_{n=-\infty}^{p+j-1} |\lambda_{p+i-n}| \leq \gamma \frac{h_j}{h_i} \eta_3(y) \sum_{n=-\infty}^{p+j-1} |\lambda_{p+i-n}|$$

with $\psi_p(\bullet)$ and $\psi(\bullet)$ as in Lemma 9.5. As

$$\sum_{n=-\infty}^{p+j-1} |\lambda_{p+i-n}| = \sum_{n=-\infty}^{p+j-1} \left\| c^T A^{p+i-n-1} b \right\| \leq \|c\| \|b\| \sum_{n=-\infty}^{p+j-1} \left\| A^{p+i-n-1} \right\|$$

$$\leq \|c\| \|b\| \left\| A^{i-j} \right\| \sum_{n=0}^{\infty} \|A^n\| = \delta \left\| A^{i-j} \right\|,$$

where $\delta = \|c\| \|b\| \sum_{n=0}^{\infty} \|A^n\|$, we get

$$|S_1(y)| \leq \frac{h_j}{h_i} \left\| A^{i-j} \right\| \eta_3(y),$$

with finite $\eta_3(y)$.

Passing to $S_2(y)$, we observe that

$$\left| E \left\{ U_i \phi(y, Y_{p+i}, \xi_{p+i}) \right\} \right| \leq c_K c_m \frac{1}{h_i} \sum_{n=-\infty}^{p+j-1} |\lambda_{p+i-n}| E|U_i U_n|$$

$$\leq c_K c_m E U^2 \frac{1}{h_i} \sum_{n=-\infty}^{p+j-1} |\lambda_{p+i-n}|$$

$$\leq c_K c_m \delta E U^2 \frac{1}{h_i} \left\| A^{i-j} \right\|,$$

which yields

$$|S_2(y)| \leq c_K c_m \delta E U^2 \frac{h_j}{h_i} \|A^{i-j}\| = \frac{h_j}{h_i} \|A^{i-j}\| \eta_4(y),$$

where $\eta_4(y)$ is finite. The proof is complete. ∎

COROLLARY 10.1 *Let the density $f(\bullet)$ of Y exist. If the kernel satisfies the appropriate restrictions of Theorem 10.1, then,*

$$\left| \text{cov} \left[U_i K \left(\frac{y - Y_{p+i}}{h} \right), U_j K \left(\frac{y - Y_{p+j}}{h} \right) \right] \right|$$

$$\leq \begin{cases} h\eta(y), & \text{for } 0 \leq |i - j| < p \\ \|A^{|i-j|}\| \eta(y) & \text{for } p < |i - j| \end{cases}$$

with some $\eta(y)$ finite at every $y \in m(R)$ where both $m^{-1}(\bullet)$ and $f(\bullet)$ are continuous.

LEMMA 10.2 *If the kernel satisfies the restrictions of Theorem 10.4, then*

$$\limsup_{h \to 0} \left| h \frac{\text{cov} \left[U_i K \left(\frac{y - Y_{p+i}}{h} \right), U_0 K \left(\frac{y - Y_p}{h} \right) \right]}{E K \left(\frac{y - Y}{h} \right)} \right|$$

$$\leq \begin{cases} \omega(y), & \text{for } i < p \\ \|A^{i-1}\| \omega(y), & \text{for } i \geq p, \end{cases}$$

where some $\omega(y)$ is finite at almost every (ζ) $y \in R$, where ζ is the distribution of Y.

Proof. Let $i < p$. Since

$$\left| \text{cov} \left[U_i K \left(\frac{y - Y_{p+i}}{h} \right), U_0 K \left(\frac{y - Y_p}{h} \right) \right] \right| \leq 2E \left\{ U_0^2 K^2 \left(\frac{y - Y_p}{h} \right) \right\},$$

we find the quantity in the assertion bounded in absolute value by

$$2 \frac{E \left\{ U_0^2 K^2 \left(\frac{y - Y_p}{h} \right) \right\}}{E K \left(\frac{y - Y}{h} \right)} = 2 \frac{E \left\{ \psi_p(y) K^2 \left(\frac{y - Y}{h} \right) \right\}}{E K \left(\frac{y - Y}{h} \right)},$$

where $\psi_p(\bullet)$ is as in Lemma 9.5. By virtue of Lemma A.10, the quantity is finite at almost every (ζ) $y \in m(R)$ and the first part of the lemma follows.

Let $p \leq i$. Arguing as in the proof of Lemma 10.1 and taking into account that now $h = h_j = h_i$, we get

$$h \frac{\mathrm{cov}\left[U_i K\left(\frac{y - Y_{p+i}}{h}\right), U_0 K\left(\frac{y - Y_p}{h}\right) \right]}{EK\left(\frac{y - Y}{h}\right)}$$

$$= h \frac{\mathrm{cov}\left[U_i \phi(y, Y_{p+i}, \xi_{p+i}), U_0 K\left(\frac{y - Y_p}{h}\right) \right]}{EK\left(\frac{y - Y}{h}\right)}$$

$$= \frac{h S_1(y)}{EK\left(\frac{y - Y}{h}\right)} - \frac{h S_2(y)}{EK\left(\frac{y - Y}{h}\right)},$$

where $S_1(y)$ and $S_2(y)$ are as in the proof of Lemma 10.1. Making use of the proof of Lemma 10.1, we obtain

$$\left| \frac{S_1(y)}{EK\left(\frac{y - Y}{h}\right)} \right| \leq \gamma \left\| A^{i-1} \right\| E|U|E\|X\| \frac{E\left| v(Y)K\left(\frac{y - Y}{h}\right) \right|}{EK\left(\frac{y - Y}{h}\right)}$$

and

$$\left| \frac{h S_2(y)}{EK\left(\frac{y - Y}{h}\right)} \right| \leq \delta \left\| A^{i-1} \right\| \frac{\left| E\left\{ v(Y)K\left(\frac{y - Y}{h}\right) \right\} \right|}{EK\left(\frac{y - Y}{h}\right)}.$$

By applying Lemma A.10, we complete the proof. ∎

10.4.2   Proofs

Proof of Theorem 10.1

Remembering that $v(\bullet)$ is continuous in $m(R)$, (see Section 9.2.1), we fix $y \in m(R)$ and suppose that $f(\bullet)$ is continuous at the point y. We notice that $\hat{v}(y) = \hat{g}(y)/\hat{f}(y)$, where

$$\hat{g}(y) = \frac{1}{nh_n} \sum_{i=1}^{n} U_i K\left(\frac{y - Y_{p+i}}{h_n}\right)$$

and

$$\hat{f}(y) = \frac{1}{nh_n} \sum_{i=1}^{n} K\left(\frac{y - Y_i}{h_n}\right).$$

Since

$$E\hat{g}(y) = \frac{1}{h_n}E\left\{U_0 K\left(\frac{y-Y_p}{h_n}\right)\right\} = \frac{1}{h_n}E\left\{v(Y)K\left(\frac{y-Y}{h_n}\right)\right\},$$

applying Lemma A.8, we get

$$E\hat{g}(y) \to v(y)f(y)\int K(x)dx \text{ as } n \to \infty.$$

In turn,

$$\text{var}\left[\hat{\xi}(y)\right] = P_n(y) + Q_n(y),$$

where

$$P_n(y) = \frac{1}{nh_n^2}\text{var}\left[U_0 K\left(\frac{y-Y_p}{h_n}\right)\right]$$

and

$$Q_n(y) = \frac{1}{n^2 h_n^2}\sum_{i=1}^{n}\sum_{j=1,j\neq i}^{n}\text{cov}\left[U_i K\left(\frac{y-Y_{p+i}}{h_n}\right), U_j K\left(\frac{y-Y_{p+j}}{h_n}\right)\right]$$

$$= \frac{1}{n^2 h_n^2}\sum_{i=1}^{n}(n-i)\text{cov}\left[U_0 K\left(\frac{y-Y_p}{h_n}\right), U_i K\left(\frac{y-Y_{p+i}}{h_n}\right)\right].$$

Clearly,

$$nh_n P_n(y) = \frac{1}{h_n}E\left\{U_0^2 K^2\left(\frac{y-Y_p}{h_n}\right)\right\} - \frac{1}{h_n}E^2\left\{U_0 K\left(\frac{y-Y_p}{h_n}\right)\right\}$$

$$= \frac{1}{h_n}E\left\{\psi_p(Y)K^2\left(\frac{y-Y}{h_n}\right)\right\} - \frac{1}{h_n}E^2\left\{v(Y)K\left(\frac{y-Y}{h_n}\right)\right\},$$

where $\psi_p(y)$ is as Lemma 9.5. Applying Lemma A.8, we observe that

$$\frac{1}{h_n}E\left\{\psi_p(Y)K^2\left(\frac{y-Y}{h_n}\right)\right\} \to \psi_p(y)f(y)\int K^2(x)dx \text{ as } n \to \infty$$

and

$$\frac{1}{h_n}E\left\{v(Y)K\left(\frac{y-Y}{h_n}\right)\right\} \to v(y)f(y)\int K(x)dx \text{ as } n \to \infty.$$

Therefore, $P_n(y) = O(1/nh_n)$.

Furthermore,

$$
|Q_n(y)| \leq \frac{1}{nh_n^2} \sum_{i=1}^n \left| \mathrm{cov}\left[U_0 K\left(\frac{y - Y_p}{h_n}\right), U_i K\left(\frac{y - Y_{p+i}}{h_n}\right) \right] \right|
$$

$$
= \frac{1}{nh_n^2} \sum_{i=1}^p \left| \mathrm{cov}\left[U_0 K\left(\frac{y - Y_p}{h_n}\right), U_i K\left(\frac{y - Y_{p+i}}{h_n}\right) \right] \right|
$$

$$
+ \frac{1}{nh_n^2} \sum_{i=p+1}^n \left| \mathrm{cov}\left[U_0 K\left(\frac{y - Y_p}{h_n}\right), U_i K\left(\frac{y - Y_{p+i}}{h_n}\right) \right] \right|.
$$

Due to (9.6), $m(\bullet)$ satisfies Lipschitz inequality, since it has a bounded derivative. Thus, applying Corollary 10.1, we find the quantity bounded by

$$
\frac{1}{nh_n^2} p\eta(y) + \frac{1}{nh_n^2}\eta(y) \sum_{i=p+1}^n \left\| A^{i-1} \right\| = \frac{1}{nh_n^2}\eta(y)\left(p + \sum_{i=1}^\infty \left\| A^i \right\| \right).
$$

Hence, $Q_n(y) = O(1/nh_n)$, which leads to the conclusion that

$$
\mathrm{var}\,[\hat{g}(y)] = O\left(\frac{1}{nh_n^2}\right). \tag{10.12}
$$

Finally,

$$
\hat{g}(y) \to v(y)f(y) \int K(x)dx \text{ as } n \to \infty \text{ in probability.}
$$

Since, for similar reasons, $\hat{f}(y) \to f(y) \int K(x)dx$ as $n \to \infty$ in probability, the proof is completed. ∎

Proof of Theorem 10.2

We begin our reasoning observing that $\tilde{v}_n(y) = \tilde{g}_n(y)/\tilde{f}_n(y)$, where

$$
\tilde{g}_n(y) = \frac{1}{n} \sum_{i=1}^n \frac{1}{h_i} U_i K\left(\frac{y - Y_{p+i}}{h_i}\right)
$$

and

$$
\tilde{f}_n(y) = \frac{1}{n} \sum_{i=1}^n \frac{1}{h_i} K\left(\frac{y - Y_i}{h_i}\right).
$$

Clearly,

$$
E\tilde{g}_n(y) = \frac{1}{n} \sum_{i=1}^n \frac{1}{h_i} E\left\{ U_0 K\left(\frac{y - Y_p}{h_i}\right) \right\} = \frac{1}{n} \sum_{i=1}^n \frac{1}{h_i} E\left\{ v(Y)K\left(\frac{y - Y}{h_i}\right) \right\}.
$$

Since

$$
\frac{1}{h_i} E\left\{ v(Y)K\left(\frac{y - Y}{h_i}\right) \right\} \to v(y)f(y) \int K(x)dx \text{ as } i \to \infty,
$$

see Lemma A.8,

$$E\tilde{g}_n(y) \to v(y)f(y) \int K(x)dx \text{ as } n \to \infty.$$

Furthermore, $\text{var}[\tilde{g}_n(y)] = R_n(y) + S_n(y)$ with

$$R_n(y) = \frac{1}{n^2} \sum_{i=1}^{n} \frac{1}{h_i^2} \text{var}\left[U_0 K\left(\frac{y - Y_p}{h_i}\right)\right],$$

and

$$S_n(y) = \frac{1}{n^2} \sum_{i=1}^{n} \sum_{\substack{j=1 \\ j\neq i}}^{n} \frac{1}{h_i h_j} \text{cov}\left[U_i K\left(\frac{y - Y_{p+i}}{h_i}\right), U_j K\left(\frac{y - Y_{p+j}}{h_j}\right)\right].$$

Since

$$R_n(y) = \frac{1}{n^2} \sum_{i=1}^{n} \frac{1}{h_i} \left[\frac{1}{h_i} E\left\{\psi_p(Y)K^2\left(\frac{y - Y}{h_i}\right)\right\} - h_i \frac{1}{h_i^2} E^2\left\{v(Y)K\left(\frac{y - Y}{h_i}\right)\right\}\right],$$

where $\psi_p(\bullet)$ is as in Lemma 9.5, we obtain

$$R_n(y) = O\left(\frac{1}{n^2} \sum_{i=1}^{n} \frac{1}{h_i}\right).$$

Moreover, by Corollary 10.1,

$$|S_n(y)| = \frac{1}{n^2} \eta(y) \left[\sum_{i=1}^{n} \sum_{j0\leq|i-j|\leq p}^{n} \frac{1}{h_i h_j} \sqrt{h_i h_j}\right.$$

$$\left. + \sum_{i=1}^{n} \sum_{jp<i-j}^{n} \frac{1}{h_i h_j} \frac{h_j}{h_i} \|A^{i-j}\| + \sum_{i=1}^{n} \sum_{jp<j-i}^{n} \frac{1}{h_i h_j} \frac{h_i}{h_j} \|A^{j-i}\|\right],$$

which is bounded by

$$\frac{1}{n^2} \eta(y) \left[\sum_{i=1}^{n} \frac{1}{\sqrt{h_i}} \sum_{j=1}^{n} \frac{1}{\sqrt{h_j}} + 2 \sum_{j=0}^{\infty} \|A^j\| \sum_{i=1}^{n} \frac{1}{h_i^2}\right]$$

$$\leq \frac{1}{n^2} \eta(y) \left[\sum_{i=1}^{n} \frac{1}{h_i} + 2 \sum_{j=0}^{\infty} \|A^j\| \sum_{i=1}^{n} \frac{1}{h_i^2}\right].$$

As the quantity converges to zero as $n \to \infty$, we obtain

$$\tilde{g}_n(y) \to v(y)f(y) \int K(x)dx \text{ as } n \to \infty \text{ in probability.}$$

Since $\tilde{f}_n(y) \to f(y) \int K(x)dx$ as $n \to \infty$ in probability, the proof is completed. ∎

Proof of Theorem 10.3

Clearly $\bar{v}_n(y) = \bar{g}_n(y)/\bar{f}_n(y)$, where

$$\bar{g}_n(y) = \frac{1}{\sum_{i=1}^n h_i} \sum_{i=1}^n U_i K\left(\frac{y - Y_{p+i}}{h_i}\right),$$

and

$$\bar{f}_n(Y) = \frac{1}{\sum_{i=1}^n h_i} \sum_{i=1}^n K\left(\frac{y - Y_i}{h_i}\right).$$

Since

$$E\bar{g}_n(y) = \frac{1}{\sum_{i=1}^n h_i} \sum_{i=1}^n h_i \frac{1}{h_i} E\left\{v(Y)K\left(\frac{y - Y}{h_i}\right)\right\},$$

an application of Lemma A.8 and that of Lemma 4.4 leads to the conclusion that

$$E\bar{g}_n(y) \to v(y)f(y) \int K(x)dx \text{ as } n \to \infty.$$

Passing to variance we get $\text{var}[\bar{g}_n(y)] = R_n(y) + S_n(y)$ with

$$R_n(y) = \frac{1}{\left(\sum_{i=1}^n h_i\right)^2} \sum_{i=1}^n \text{var}\left[U_0 K\left(\frac{y - Y_p}{h_i}\right)\right],$$

and

$$S_n(y) = \frac{1}{\left(\sum_{i=1}^n h_i\right)^2} \sum_{i=1}^n \sum_{\substack{j=1 \\ j \neq i}}^n \text{cov}\left[U_i K\left(\frac{y - Y_{p+i}}{h_i}\right), U_j K\left(\frac{y - Y_{p+j}}{h_j}\right)\right].$$

Since

$$R_n(y) \sum_{i=1}^n h_i = \frac{1}{\sum_{i=1}^n h_i} \sum_{i=1}^n h_i \left[\frac{1}{h_i} E\left\{\psi_p(Y)K^2\left(\frac{y - Y}{h_i}\right)\right\} \right.$$
$$\left. - h_i \frac{1}{h_i^2} E^2\left\{v(Y)K\left(\frac{y - Y}{h_i}\right)\right\}\right],$$

another application of Lemma A.8 and Lemma 4.4 yields

$$R_n(y) = O\left(\frac{1}{\sum_{i=1}^n h_i}\right).$$

Moreover,

$$|S_n(y)| = \frac{\eta(y)}{\left(\sum_{i=1}^n h_i\right)^2} \left[\sum_{i=1}^n \sum_{\substack{j \\ 0 \leq |i-j| < p}}^n \sqrt{h_i h_j} + \sum_{i=1}^n \sum_{\substack{j \\ p < i-j}}^n \frac{h_j}{h_i} \|A^{i-j}\| \right.$$
$$\left. + \sum_{i=1}^n \sum_{\substack{j \\ p < j-i}}^n \frac{h_i}{h_j} \|A^{j-i}\| \right],$$

which is bounded by

$$
\frac{\eta(y)}{\left(\sum_{i=1}^{n} h_i\right)^2} \left[\sum_{i=1}^{n} \sqrt{h_i} \sum_{j=1}^{n} \sqrt{h_j} + \sum_{i=1}^{n} \sum_{\substack{j \\ p<i-j}}^{n} \frac{h_j}{h_i} \|A^{i-j}\| + \sum_{i=1}^{n} h_i \sum_{\substack{j \\ p<j-i}}^{n} \frac{1}{h_j} \|A^{j-i}\| \right],
$$

which in turn is not greater than

$$
\frac{\eta(y)}{\left(\sum_{i=1}^{n} h_i\right)^2} \left[\sum_{i=1}^{n} h_i + \sum_{i=1}^{n} \sum_{\substack{j \\ p<i-j}}^{n} \frac{h_j}{h_i} \|A^{i-j}\| + \sum_{i=1}^{n} h_i \sum_{\substack{j \\ p<j-i}}^{n} \frac{1}{h_j} \|A^{j-i}\| \right].
$$

Since $\{h_n\}$ is monotonic, the quantity is bounded by

$$
\frac{\eta(y)}{\left(\sum_{i=1}^{n} h_i\right)^2} \left[\sum_{i=1}^{n} h_i + \frac{2}{h_n} \sum_{j=0}^{\infty} \|A^j\| \sum_{i=1}^{n} h_i \right] = \frac{\eta(y)}{\sum_{i=1}^{n} h_i} \left[1 + \frac{2}{h_n} \sum_{j=0}^{\infty} \|A^j\| \right],
$$

which converges to zero as $n \to \infty$. Thus,

$$
\bar{g}_n(y) \to v(y)f(y) \int K(x)dx \text{ as } n \to \infty \text{ in probability.}
$$

Moreover, since $\bar{f}_n(y) \to f(y) \int K(x)dx$ as $n \to \infty$ in probability, the proof is complete. ∎

Proof of Theorem 10.4

We have $\hat{v}(y) = \hat{\xi}(y)/\hat{\eta}(y)$, where

$$
\hat{\xi}(y) = \frac{1}{nEK\left(\frac{y-Y}{h_n}\right)} \sum_{i=1}^{n} U_i K\left(\frac{y - Y_{p+i}}{h_n}\right)
$$

and

$$
\hat{\eta}(y) = \frac{1}{nEK\left(\frac{y-Y}{h_n}\right)} \sum_{i=1}^{n} K\left(\frac{y - Y}{h_n}\right).
$$

Since

$$E\hat{\xi}(y) = \frac{E\left\{U_0 K\left(\frac{y-Y_p}{h_n}\right)\right\}}{EK\left(\frac{y-Y}{h_n}\right)} = \frac{E\left\{v(Y)K\left(\frac{y-Y}{h_n}\right)\right\}}{EK\left(\frac{y-Y}{h_n}\right)},$$

applying Lemma A.10, we get

$$E\hat{\xi}(y) \to v(y) \text{ as } n \to \infty$$

for almost every $(\zeta)\, y \in R$.

For the variance, we have

$$\text{var}[\hat{\xi}(y)] = P_n(y) + Q_n(y),$$

where

$$P_n(y) = \frac{1}{n E^2 K\left(\frac{y-Y}{h_n}\right)} \text{var}\left[U_0 K\left(\frac{y-Y_p}{h_n}\right)\right]$$

and

$$Q_n(y) = \frac{1}{n^2 E^2 K\left(\frac{y-Y}{h_n}\right)} \sum_{i=1}^{n}(n-i)\,\text{cov}\left[U_0 K\left(\frac{y-Y_p}{h_n}\right), U_i K\left(\frac{y-Y_{p+i}}{h_n}\right)\right].$$

Furhermore,

$$nh_n|P_n(y)| \leq \frac{h_n}{E^2 K\left(\frac{y-Y}{h_n}\right)} E\left\{U_0^2 K^2\left(\frac{y-Y_p}{h_n}\right)\right\}$$

$$\leq \frac{h_n}{EK\left(\frac{y-Y}{h_n}\right)} \frac{E\left\{\psi_p(Y)K^2\left(\frac{y-Y}{h_n}\right)\right\}}{EK\left(\frac{y-Y}{h_n}\right)},$$

where $\psi_p(\bullet)$ is as in Lemma 9.5. Due to Lemmas A.10 and A.11, the quantity converges to a finite limit as $n \to \infty$ at almost every $(\zeta)\, y \in R$. Thus, $P_n(y) = O(1/nh_n^2)$ at almost every $(\zeta)\, y \in R$.

Clearly,

$$
nh_n^2|Q_n(y)| \leq \frac{h_n}{EK\left(\frac{y-Y}{h_n}\right)} \sum_{i=1}^{n} h_n \frac{\left|\mathrm{cov}\left[U_0 K\left(\frac{y-Y_p}{h_n}\right), U_i K\left(\frac{y-Y_{p+i}}{h_n}\right)\right]\right|}{EK\left(\frac{y-Y}{h_n}\right)}
$$

$$
\leq \frac{h_n}{EK\left(\frac{y-Y}{h_n}\right)} \sum_{i=1}^{p} h_n \frac{\left|\mathrm{cov}\left[U_0 K\left(\frac{y-Y_p}{h_n}\right), U_i K\left(\frac{y-Y_{p+i}}{h_n}\right)\right]\right|}{EK\left(\frac{y-Y}{h_n}\right)}
$$

$$
+ \frac{h_n}{EK\left(\frac{y-Y}{h_n}\right)} \sum_{i=p+1}^{n} h_n \frac{\left|\mathrm{cov}\left[U_0 K\left(\frac{y-Y_p}{h_n}\right), U_i K\left(\frac{y-Y_{p+i}}{h_n}\right)\right]\right|}{EK\left(\frac{y-Y}{h_n}\right)}.
$$

Application of Lemma A.11 and then Lemma 10.2 leads to the conclusion that $Q_n(y) = O(1/nh_n^2)$ at almost every (ζ) $y \in R$. Thus,

$$
\mathrm{var}[\hat{\xi}(y)] = O\left(\frac{1}{nh_n^2}\right)
$$

at almost every (ζ) $y \in R$. Finally,

$$
\hat{\xi}(y) \rightarrow \nu(y) \text{ as } n \rightarrow \infty \text{ in probability,}
$$

at almost every (ζ) $y \in R$.

Since, for similar reasons, $\hat{\xi}(y) \rightarrow 1$ as $n \rightarrow \infty$ in probability, the proof has been completed. ∎

Proof of Theorem 10.5
In the proof of consistency, we apply the same arguments as in the proof of Theorem 10.1 dealing with the kernel estimate. The only essential difference is that we use

$$
|D_N(y) - D_N(x)| \leq \frac{1}{\pi} N|y - x|
$$

rather than (10.7). To verify this it suffices to notice that

$$
\frac{d}{dx} D_N(x) = \frac{d}{dx} \frac{1}{\pi}\left(\frac{1}{2} + \sum_{k=1}^{N} \cos kx\right) = -\frac{1}{\pi} \sum_{k=1}^{N} \sin kx,
$$

which is bounded in absolute value by N/π. ∎

Proof of Theorem 10.6
The proof is similar to that of Theorem 10.1. Examining the appropriate covariance, we take into account that the kernel is a Lipschitz function in the sense that, for any $x \in (-1, 1)$ and all $x, \xi \in [-1, 1]$, it satisfies the following inequality:

$$
|k_N(y, x) - k_N(y, \xi)| = O(N^{5/2})|x - \xi|.
$$

To verify the inequality we begin with (10.11) to get

$$\frac{\partial}{\partial x} k_N(y, x) = \sum_{k=0}^{N} \frac{2k+1}{2} P_k(y) P_k'(x)$$

and use the inequality (B.19), to obtain

$$\left| \frac{\partial}{\partial x} k_N(y, x) \right| \le \sum_{k=0}^{N} \frac{2k+1}{2} k |P_k(y)| = \sum_{k=0}^{N} O(k^2) |P_k(y)|.$$

Moreover, since

$$P_k(\cos \xi) = \sqrt{\frac{2}{\pi k \sin \xi}} \cos \left[\left(k + \frac{1}{2} \right) \xi + \frac{\pi}{4} \right] + O(k^{-3/2}),$$

for any $\xi \in (-\pi + \varepsilon, \pi - \varepsilon)$, any $\varepsilon > 0$, see Szegö [288, Theorem 8.21.8], we conclude that $P_k(y) = O(k^{-1/2})$ for any $y \in (-1, 1)$. Hence, for any such y,

$$\left| \frac{\partial}{\partial x} k_N(y, x) \right| \le \sum_{k=0}^{N} O(k^{3/2}) = O(N^{5/2}),$$

which leads to the desired inequality. ∎

Proof of Theorem 10.7
Since

$$\frac{\partial}{\partial x} k_N(y, x) = \sum_{k=0}^{n} h_k(y) h_k'(x),$$

using (B.42) and (B.43) we obtain, for any fixed y,

$$\max_{-\infty < x < \infty} \frac{\partial}{\partial x} |k_N(y, x)| \le \sum_{k=0}^{N} O(k^{-1/4}) O(k^{5/12}) = O(N^{7/6}).$$

Hence, for any x and all $x, \xi \in (-\infty, \infty)$,

$$|k_N(y, x) - k_N(y, \xi)| = O(N^{7/6}) |x - \xi|,$$

which completes the proof. ∎

10.5   Bibliographic notes

The offline kernel estimate has been examined in Greblicki [108, 111], and Krzyżak [184], while semirecursive versions, all with $p = 1$, have been studied by Greblicki [115]. Orthogonal series estimates have been studied by Greblicki [109], Krzyżak [184], and Krzyżak, Sąsiadek, and Kégl [189]. Algorithms are consistent also when the Wiener system is driven by a Gaussian random process which may not be white, see Greblicki [119].

11 Continuous-time Wiener system

In this chapter, we extend the theory developed in Chapters 9 and 10 to the case of a continuous-time Wiener system. The consistency of kernel estimates is examined in Section 11.2. The problem of identification of the linear subsystem is studied in Section 11.3.

11.1 Identification problem

In this chapter we identify a continuous-time Wiener system shown in Figure 11.1. The input random process $\{U(t); t \in (-\infty, \infty)\}$ is stationary, white, Gaussian, has zero mean and the autocovariance function $\sigma_U^2 \delta(t)$, where $\delta(t)$ is the Dirac impulse. The system consists of two subsystems. The first is a linear dynamic system and the second a memoryless nonlinearity. The dynamic subsystem is described by the following state-space equation:

$$\begin{cases} \dot{X}(t) = AX(t) + bU(t) \\ W(t) = c^T X(t) \end{cases}$$

in which A, b, and c are all unknown. Therefore, $\lambda(t) = c^T e^{At} b$, where $\lambda(t)$ is the impulse response of the subsystem. The dynamic subsystem is asymptotically stable, which means that all eigenvalues of A have negative real parts. Thus $\int_0^\infty \lambda^2(t)dt < \infty$ and, consequently, both $\{X(t); t \in (-\infty, \infty)\}$ and $\{W(t); t \in (-\infty, \infty)\}$ are stationary Gaussian processes.

The output of the subsystem is disturbed by additive noise $\{Z(t); t \in (-\infty, \infty)\}$. Therefore, $V(t) = W(t) + Z(t)$. The noise $\{Z(t)\}$ is a zero-mean stationary white Gaussian random process with autocovariance function $\sigma_Z^2 \delta(t)$. The process is independent of $\{U(t)\}; t \in (-\infty, \infty)\}$. The other subsystem is nonlinear, memoryless and has a characteristic $m(\bullet)$. Hence, $Y(t) = m(V(t))$. The characteristic is a Borel measurable function satisfying the following Lipschitz condition:

$$|m(x) - m(y)| \leq c_m |x - y| \tag{11.1}$$

for some c_m and all x, y in R, the same as (9.4) for discrete systems. Moreover, the characteristic $m(\bullet)$ is invertible and its inverse is denoted by $m^{-1}(\bullet)$.

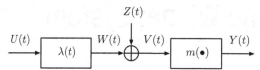

Figure 11.1 The continuous-time Wiener system.

We identify both subsystems, that is, estimate the characteristic $m(\bullet)$ of the nonlinear part, and the impulse response $\lambda(\bullet)$ of the linear subsystem from observations taken at input and output of the whole system, that is, from $\{U(t), Y(t); \; t \in (0, \infty)\}$.

11.2   Nonlinear subsystem

Obviously $V(t)$ has zero mean and variance $\sigma_V^2 = \sigma_U^2 \int_0^\infty \lambda^2(t)dt + \sigma_Z^2$. Observe now that the pair $(U(t - \tau), V(t))$ has a Gaussian distribution with marginal variances σ_U^2, σ_V^2, and the correlation coefficient $(\sigma_U/\sigma_V)\lambda(\tau)$. Therefore, $E\{U(t - \tau)|V(t)\} = (\sigma_U^2/\sigma_V^2)\lambda(\tau)V(t)$. In this way we have verified the following result:

LEMMA 11.1 *In the Wiener system,*

$$E\{U(t - \tau)|Y(t) = y\} = \alpha_\tau m^{-1}(y),$$

where $\alpha_\tau = (\sigma_U^2/\sigma_V^2)\lambda(\tau)$.

The discussion in Section 9.2.1 concerning the existence of a density $f(\bullet)$ of Y in a discrete-time system also applies in this situation. Since some results also require the density to exist, we assume sometimes that

$$m(\bullet) \text{ is strictly monotonous and has a bounded derivative.} \qquad (11.2)$$

Lemma 11.1 suggests the following estimate of $\alpha_\tau m^{-1}(y)$:

$$\hat{v}(y) = \frac{\int_0^t U(\xi - \tau)K\left(\dfrac{y - Y(\xi)}{h(t)}\right) d\xi}{\int_0^t K\left(\dfrac{y - Y(\xi)}{h(t)}\right) d\xi},$$

where $K(\bullet)$ is a suitably selected Borel measurable kernel function and $h(\bullet)$ is a positive bandwidth function, respectively. The choice of the kernel depends on the fact that $f(\bullet)$ exists or not.

Semirecursive forms of the estimate are as follows:

$$\tilde{v}(y) = \frac{\int_0^t U(\xi - \tau)\dfrac{1}{h(\xi)}K\left(\dfrac{y - Y(\xi)}{h(\xi)}\right) d\xi}{\int_0^t \dfrac{1}{h(\xi)}K\left(\dfrac{y - Y(\xi)}{h(\xi)}\right) d\xi}$$

$$\bar{v}(y) = \frac{\int_0^t U(\xi - \tau) K K \left(\frac{y - Y(\xi)}{h(\xi)} \right) d\xi}{\int_0^t K \left(\frac{y - Y(\xi)}{h(\xi)} \right) d\xi}$$

THEOREM 11.1 *Let $m(\bullet)$ satisfy (11.2). Let the kernel satisfy the following restrictions:*

$$\sup_y |K(y)| < \infty,$$

$$\int |K(y)| dy < \infty,$$

$$y K(y) \to 0 \ as \ |y| \to \infty,$$

and, moreover,

$$|K(y) - K(x)| \le c_K |y - x|$$

for some c_K, for all $x, y \in (-\infty, \infty)$. If

$$h(t) \to 0 \ as \ t \to \infty \tag{11.3}$$

and

$$t h^2(t) \to \infty \ as \ t \to \infty, \tag{11.4}$$

then

$$\hat{v}(y) \to v(y) \ as \ t \to \infty \ in \ probability \tag{11.5}$$

at every $y \in m(R)$ where $f(\bullet)$ is continuous and $f(y) > 0$.

THEOREM 11.2 *Let the kernel satisfy the appropriate restrictions of Theorem 11.1. If the bandwidth function $h(\bullet)$ satisfies (11.3) and*

$$t \int_0^t h(\tau) d\tau \to \infty \ as \ t \to \infty,$$

then

$$\bar{v}(y) \to v(y) \ as \ n \to \infty \ in \ probability$$

at every $y \in m(R)$ where $f(\bullet)$ is continuous and $f(y) > 0$.

THEOREM 11.3 *Let the kernel satisfy the appropriate restrictions of Theorem 11.1. If the positive bandwidth function $h(\bullet)$ satisfies (11.3) and*

$$\frac{1}{t^2} \int_0^t \frac{1}{h^2(\tau)} d\tau \to 0 \ as \ t \to \infty,$$

then

$$\tilde{v}(y) \to v(y) \ as \ n \to \infty \ in \ probability$$

at every $y \in m(R)$ where $f(\bullet)$ is continuous and $f(y) > 0$.

11.3   Dynamic subsystem

Since $E\{U(t-\tau)Y(t)\} = E\{Y(t)E\{U(t-\tau)|Y(t)\}\}$, applying Lemma 11.1, we find the quantity equal to $\alpha_\tau E\{Y(t)V(t)\}$. This verifies the following result.

LEMMA 11.2 *In the system,*

$$E\{U(t)Y(\tau+t)\} = \beta\lambda(\tau),$$

where $\beta = (\sigma_U/\sigma_V)E\{V(t)Y(t)\}.$

Having proved the lemma, we propose the following algorithm to estimate $\kappa(\tau) = \beta\lambda(\tau)$:

$$\hat\kappa(\tau) = \frac{1}{t}\int_0^t U(\xi)Y(\tau+\xi)d\xi.$$

THEOREM 11.4 *Let* $m(\bullet)$ *satisfy (11.1). For any* $\tau \in [0, \infty),$

$$E(\hat\kappa(\tau) - \kappa(\tau))^2 \to 0 \text{ as } t \to \infty.$$

Proof. We have, $E\{\hat\kappa(\tau)\} = E\{U(0)Y(t)\}$, which, by Lemma 11.2, is equal to $\kappa(\tau)$. Thus, $\hat\kappa(\tau)$ is an unbiased estimate of $\beta\lambda(t)$ and its variance equals

$$\frac{1}{t^2}\int_0^t\int_0^t \text{cov}[U(\xi)Y(\tau+\xi), U(\eta)Y(\tau+\eta)]d\xi d\eta$$
$$= \frac{1}{t^2}\int_0^t (t-\xi)\text{cov}[U(0)Y(\tau), U(\xi)Y(\tau+\xi)]d\xi,$$

which, by Lemma 11.3, is bounded from above by

$$\frac{1}{t^2}\gamma\int_0^t (t-\xi)\left\|e^{-A\xi}\right\|d\xi \le \frac{1}{t}\gamma\int_0^\infty \left\|e^{-A\xi}\right\|d\xi.$$

The proof is complete. ∎

11.4   Lemmas

LEMMA 11.3 *If (11.1) holds, then, for* $0 < \tau < t,$

$$|\text{cov}\left[U(0)m(V(\tau)), U(t)m(V(\tau+t))\right]| \le \gamma\left\|e^{-At}\right\|$$

with a finite $\gamma.$

Proof. Let $0 < \tau < t$. Clearly, $V(\tau + t) = c^T e^{-At} X(\tau) + \xi(\tau + t)$, where $\xi(\tau + t) = \int_\tau^{\tau+t} c^T e^{-At} b U(\tau) d\tau + Z(\tau + t)$. Since pairs $(U(0), V(\tau))$ and $(U(t), \xi(\tau + t))$ are independent, $\mathrm{cov}\,[U(0)m(V(\tau)), U(t)m(\xi(\tau + t))] = 0$. Hence, the covariance in the assertion is equal to

$$\mathrm{cov}\,[U(0)m(V(\tau))U(t)[m(V(\tau + t)) - m(\xi(\tau + t))]]$$
$$= E\,\{U(0)U(t)m(V(\tau))[m(V(\tau + t)) - m(\xi(\tau + t))]\}$$
$$- E\,\{U(0)m(V(\tau))\}\,E\,\{U(0)[m(V(\tau + t)) - m(\xi(\tau + t))]\}\,.$$

Applying (11.1), we get

$$|m(V(\tau + t)) - m(\xi(\tau + t))| \le c_m\,|V(\tau + t) - \xi(\tau + t)| \le c_m\,\left|c^T e^{-At} X(\tau)\right|$$
$$\le c_m\,\|c\|\,\left\|e^{-At}\right\|\,\|X(\tau)\|\,.$$

Thus, the absolute value of the variance is not greater than

$$c_m\,\|c\|\,\left\|e^{-At}\right\|\,(E\,|U(0)U(t)m(V(\tau))\,\|X(\tau)\|\|$$
$$+\,|E\,\{U(0)m(V(\tau))\}|\,E\,|U(0)\,\|X(\tau)\|\|)$$
$$=\,c_m\,\|c\|\,\left\|e^{-At}\right\|\,(E\,|U(0)m(V(\tau))\,\|X(\tau)\|\|\,E\,|U|$$
$$+\,|E\,\{U(0)m(V(\tau))\}|\,E\,|U(0)\,\|X(\tau)\|\|)$$
$$=\,\gamma\,\left\|e^{-At}\right\|$$

with a finite γ. ∎

Our next lemmas are given without proofs because they are very similar to those presented in Section 10.4.

LEMMA 11.4 *Let the density $f(\bullet)$ of Y exist. If the kernel satisfies the appropriate restrictions of Theorem 10.2, then,*

$$\left|\mathrm{cov}\left[U(\xi)K\left(\frac{y - Y(\tau + \xi)}{h(\xi)}\right), U(\eta)K\left(\frac{y - Y(\tau + \eta)}{h(\eta)}\right)\right]\right|$$
$$\le \begin{cases} \sqrt{h(\xi)h(\eta)}\omega(y), & \text{for } 0 \le |\xi - \eta| \le \tau \\ \dfrac{h(\eta)}{h(\xi)}\left\|e^{A(\xi - \eta)}\right\|\eta(y) & \text{for } \tau < \xi - \eta \\ \dfrac{h(\xi)}{h(\eta)}\left\|e^{A(\eta - \xi)}\right\|\eta(y) & \text{for } \tau < \eta - \xi \end{cases}$$

for some $\omega(y)$ finite at every $y \in m(R)$ where both $m^{-1}(\bullet)$ and $f(\bullet)$ are continuous.

COROLLARY 11.1 *Let the density* $f(\bullet)$ *of Y exist. If the kernel satisfies the appropriate restrictions of Theorem 10.1, then,*

$$\left| \mathrm{cov} \left[U(\xi)K \left(\frac{y - Y(\tau + \xi)}{h} \right), U(\eta)K \left(\frac{y - Y(\tau + \eta)}{h} \right) \right] \right|$$

$$\leq \begin{cases} h\eta(y), & \text{for } 0 \leq |\xi - \eta| \leq \tau \\ \left\| e^{-A|\xi - \eta|} \right\| \eta(y) & \text{for } \tau < |\xi - \eta| \end{cases}$$

for some $\eta(y)$ *finite at every* $y \in m(R)$ *where both* $m^{-1}(\bullet)$ *and* $f(\bullet)$ *are continuous.*

11.5   Bibliographic notes

Offline and semirecursive kernel estimates have been examined in Greblicki [112, 113].

12 Other block-oriented nonlinear systems

Thus far we have examined block-oriented systems of the cascade form, namely the Hammerstein and Wiener systems. The main tool that was used to recover the characteristics of the systems was based on the theory of nonparametric regression and correlation analysis. In this chapter, we show that this approach can be successfully extended to a class of block-oriented systems of the series-parallel form as well as systems with nonlinear dynamics. The latter case includes generalized Hammerstein and Wiener models as well as the sandwich system. We highlight some of these systems and present identification algorithms that can use various nonparametric regression estimates. In particular, Section 12.1 develops nonparametric algorithms for parallel, series-parallel, and generalized nonlinear block-oriented systems. Section 12.2 is devoted to a new class of nonlinear systems with nonlinear dynamics. This includes the important sandwich system as a special case.

12.1 Series-parallel, block-oriented systems

The cascade nonlinear systems presented in the previous chapters define the fundamental building blocks for defining general models of series-parallel forms. Together, all of these models may create a useful class of structures for modeling various physical processes. The choice of a particular model depends crucially on physical constraints and needs.

In this section, we present a number of nonlinear models of series-parallel forms for which we can relatively easily develop identification algorithms based on the regression approach used throughout the book.

12.1.1 Parallel nonlinear system

Our first example concerns a model in which the nonlinear memoryless element and the linear dynamic system are connected in parallel (see Figure 12.1). That is, we have the following input–output relationship:

$$Y_n = m(U_n) + V_n + Z_n, \tag{12.1}$$

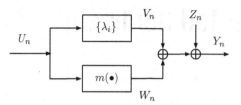

Figure 12.1 The parallel nonlinear system.

where V_n is the output of the linear subsystem. In general, we have

$$V_n = \sum_{i=0}^{\infty} \lambda_i U_{n-i}, \tag{12.2}$$

where the impulse-response function $\{\lambda_i\}$ satisfies $\sum_{i=0}^{\infty} \lambda_i^2 < \infty$.

More specifically, the state-space equation for the linear subsystem can be used, that is,

$$\begin{cases} X_{n+1} = AX_n + bU_n \\ \quad V_n = c^T X_n + dU_n \end{cases} \tag{12.3}$$

This corresponds to (12.2) with $\lambda_0 = d$ and $\lambda_j = c^T A^{j-1} b$, $j \geq 1$. The noise process $\{Z_i\}$ in (12.1) is white with zero mean and finite variance σ_Z^2. Let the input process $\{U_n\}$ be *iid* with the density $f(\bullet)$ and let, additionally, $EU_n^2 < \infty$ and $Em^2(U_n) < \infty$. These conditions can assure that the output signal $\{Y_n\}$ is a second-order stationary stochastic process. The parallel nonlinear system can be interpreted as a system where the linear part plays the role of the load or nuisance parameter that influences the signal-to-noise ratio of the output signal. Indeed, (12.1) can be written as follows:

$$Y_n = \mu(U_n) + \xi_n + Z_n, \tag{12.4}$$

where

$$\mu(U_n) = m(U_n) + \lambda_0 U_n + E\{U_0\} \sum_{i=1}^{\infty} \lambda_i \tag{12.5}$$

and

$$\xi_n = \sum_{i=1}^{\infty} \lambda_i [U_{n-i} - EU_{n-i}]. \tag{12.6}$$

Thus, the parallel system can be expressed in the signal-plus-noise form with the signal $\mu(u)$ and the noise $\xi_n + Z_n$, where ξ_n can be called the internal system noise. The total noise $\{\xi_n + Z_n\}$ has zero mean and finite variance

$$\text{var}(U_0) \sum_{i=1}^{\infty} \lambda_i^2 + \sigma_Z^2, \tag{12.7}$$

but it is not white anymore because ξ_i and ξ_j are correlated for $i \neq j$.

From (12.4) we can readily obtain that $E\{Y_n|U_n = u\} = \mu(u)$, and one can recover the nonlinearity $m(u)$ from $\mu(u)$ up to an unknown linear function. However, if $\lambda_0 = 1$ and $E\{U_0\} = 0$, then

$$m(u) = \mu(u) - u \tag{12.8}$$

and a full recovery of $m(u)$ from the regression function $\mu(u)$ is possible. Indeed, due to (12.8) an estimate of $m(u)$ can be defined as

$$\hat{m}(u) = \hat{\mu}(u) - u, \tag{12.9}$$

where as $\hat{\mu}(u)$ we can use any nonparametric regression estimate (see Chapters 3–7), derived from the input–output data $\{(U_1, Y_1), \ldots, (U_n, Y_n)\}$ generated from the model in (12.1).

REMARK 12.1 *In Chapter 2, we observed that the cascade Hammerstein system can also be written in the form of (12.4). In this case, we have*

$$\mu(U_n) = m(U_n) + E\{m(U_0)\} \sum_{i=1}^{\infty} \lambda_i,$$

and

$$\xi_n = \sum_{i=1}^{\infty} \lambda_i [m(U_{n-i}) - Em(U_{n-i})].$$

We can now compare the influence of noise on both systems. Since the measurement noise is the same for both systems it suffices to consider $\mathrm{var}(\xi_n)$. *In fact, we have*

$$\mathrm{var}(\xi_n) = \begin{cases} \sum_{i=1}^{\infty} \lambda_i^2 \, \mathrm{var}(U_0) & \text{for parallel system} \\ \sum_{i=1}^{\infty} \lambda_i^2 \, \mathrm{var}(m(U_0)) & \text{for cascade system} \end{cases}.$$

Thus, the influence of the noise is greater for the parallel system than for the cascade one if

$$\mathrm{var}(m(U_0)) \le \mathrm{var}(U_0). \tag{12.10}$$

Since it is known that $\mathrm{var}(m(U_0))$ *can be closely approximated by* $(m^{(1)}(EU_0))^2 \, \mathrm{var}(U_0)$ *then we can conclude that (12.10) holds for all the nonlinearities when* $|m^{(1)}(EU_0)| \le 1$. *Since we often have* $EU_0 = 0$ *then the latter criterion is equivalent to* $|m^{(1)}(0)| \le 1$. *A large class of practical nonlinearities meets the following constraint:*

$$c_1 u \le m(u) \le c_2 u.$$

If $c_2 \le 1$, *we can conclude that (12.10) holds. On the other hand, if* $c_1 > 1$, *then the opposite inequality to (12.10) is true and we may conclude that the cascade Hammerstein system is more influenced by the noise than the corresponding parallel connection.*

The generic formula in (12.9) allows us to form different nonparametric estimates of the nonlinearity $m(u)$. To focus our attention, let us consider the generalized kernel estimate

$$\hat{m}(u) = \frac{n^{-1} \sum_{i=1}^{n} Y_i \, k_b(u, U_i)}{n^{-1} \sum_{i=1}^{n} k_b(u, U_i)} - u, \tag{12.11}$$

where $k_b(u, v)$ is the kernel function indexed by the smoothing parameter $b = b(n)$.

REMARK 12.2 *It is worth noting that $k_b(u, v) = bk(b(u - v))$ gives the classical convolution kernel estimate with the kernel $k(u)$ and with the window width b^{-1}. On the other hand $k_b(u, v) = \sum_{\ell=0}^{b} \varphi_\ell(u)\varphi_\ell(v)$ defines the orthogonal series estimate with respect to the orthonormal basis $\{\varphi_\ell(u)\}$ and with the truncation parameter b. Next, $k_b(u, v) = bk(bu, bv)$ defines the multiresolution class of kernel functions. Here $k(u, v) = \sum_{j=-\infty}^{\infty} \phi(u - j)\phi(v - j)$ is the kernel corresponding to the scaling function $\phi(u)$.*

To establish the asymptotic properties of the estimate $\hat{m}(u)$ let us observe that under the condition $b(n) \to \infty$ as $n \to \infty$, and for a large class of kernel functions (see Chapters 3 and 6) we have

$$E\left\{ n^{-1} \sum_{i=1}^{n} Y_i \, k_b(u, U_i) \right\} = \int_{-\infty}^{\infty} \mu(x) f(x) \, k_b(u, x) dx \to \mu(u) f(u), \tag{12.12}$$

at every point u where both $\mu(u)$ and $f(u)$ are continuous. It is worth noting that the smoothness of $\mu(u)$ is determined by the smoothness of the nonlinearity $m(u)$. In fact, if, for example, $m(u)$ is Lipschitz continuous, that is,

$$|m(u) - m(v)| \leq L|u - v|,$$

then

$$|\mu(u) - \mu(v)| \leq (L + 1)|u - v|.$$

Regarding the variance of $n^{-1} \sum_{i=1}^{n} Y_i \, k_b(u, U_i)$, we have

$$\text{var}\left\{ n^{-1} \sum_{i=1}^{n} Y_i \, k_b(u, U_i) \right\} = n^{-1} \, \text{var}\{Y_n \, k_b(u, U_n)\}$$
$$+ 2n^{-1} \sum_{s=1}^{n-1} \left(1 - \frac{s}{n}\right) \text{cov}\{Y_s \, k_b(u, U_s), Y_0 \, k_b(u, U_0)\}. \tag{12.13}$$

The first term in (12.13) is bounded by

$$n^{-1} \int_{-\infty}^{\infty} \psi(x) f(x) \, k_b^2(u, x) dx,$$

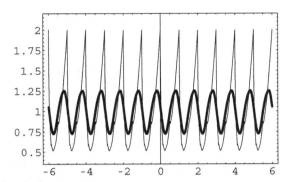

Figure 12.2 Function $c(u)$ for the Daubechies scaling functions of order $p = 2$ (thin line) and $p = 6$ (thick line).

where $\psi(u) = E\{Y_n^2 \mid U_n = u\}$ is a finite function if $E\{m^2(U_n)\} < \infty$ and $E\{U_n^2\} < \infty$. Also, for a large class of kernel functions, we can assume that

$$\int_{-\infty}^{\infty} g(x)k_b^2(u, x)dx \leq c(u, b)b, \qquad (12.14)$$

at every point u, where $g(u)$ is continuous.

REMARK 12.3 *The function $c(u, b)$ appearing in (12.14) is bounded and its particular form depends on the particular kernel function. For the convolution kernel, $c(u, b)$ is independent of b. On the other hand, for the multiresolution kernels (see Remark 12.2), $c(u, b) = c(ub)$, where $c(u)$ is given as follows:*

$$c(u) = \sum_{j=-\infty}^{\infty} \phi^2(u - j).$$

Thus, $c(u)$ is a periodic function with the period equal to one. Note that for the Haar scaling function, we have $c(u) = 1$. Figure 12.2 depicts $c(u)$ for the Daubechies scaling functions of order $p \geq 2$. Note that the amplitude of $c(u)$ decreases as p increases.

Regarding the second term in (12.13), we note that the covariance in this term can be written as follows:

$$\text{cov}\{Y_s k_b(u, U_s), Y_0 k_b(u, U_0)\}$$
$$= \lambda_s E\{k_b(u, U_0)\}E\{U_0\mu(U_0)k_b(u, U_0)\}$$
$$+ \sum_{i=-\infty}^{s-1}\sum_{j=-\infty}^{-1} \lambda_{s-i}\lambda_{-j} \text{cov}\{U_i k_b(u, U_s), U_j k_b(u, U_0)\}.$$

The first term in the previous formula tends to $\lambda_s f^2(u)\mu(u)u$ as $b \to \infty$. The second term can be decomposed as follows:

$$\sum_{i=-\infty}^{-1} \sum_{j=-\infty}^{-1} \lambda_{s-i}\lambda_{-j} \operatorname{cov}\{U_i k_b(u, U_s), U_j k_b(u, U_0)\}$$

$$+ \sum_{i=0}^{s-1} \sum_{j=-\infty}^{-1} \lambda_{s-i}\lambda_{-j} \operatorname{cov}\{U_i k_b(u, U_s), U_j k_b(u, U_0)\}. \qquad (12.15)$$

Note that the covariance in the second term in (12.15) is equal to zero. In turn, the first term in (12.15) converges to $E\{U_0^2\}f^2(u)\sum_{j=1}^{\infty}\lambda_{s+j}\lambda_j$. All of these considerations show that the second term in (12.13) is not greater than

$$\frac{a(u)}{n}\left\{\sum_{s=1}^{\infty}|\lambda_s| + \sum_{s=1}^{\infty}\sum_{j=1}^{\infty}|\lambda_{s+j}|\,|\lambda_j|\right\},$$

where $a(u)$ is a finite function independent of n. Note that the second term in the brackets is finite if $\sum_{s=1}^{\infty}|\lambda_s| < \infty$. The denominator in (12.11) represents a classical kernel density estimate that is known to converge to $f(u)$ as $n \to \infty$ in probability. The convergence in probability will be abbreviated as (P), whereas the rate in the probability sense (see Appendix C) as $O_P(\bullet)$. We can summarize the aforementioned discussion by stating the following theorem.

THEOREM 12.1 *Let $\hat{m}(u)$ be the generalized kernel estimator defined in (12.11) with the kernel function, which meets (12.14). Let $E\{m^2(U_0)\} < \infty$ and $E\{U_0^2\} < \infty$. Suppose that $\sum_{s=1}^{\infty}|\lambda_s| < \infty$. If*

$$b(n) \to \infty \quad and \quad \frac{b(n)}{n} \to 0 \quad as\ n \to \infty$$

then,

$$\hat{m}(u) \to m(u)(P) \quad as\ n \to \infty,$$

at every point u where $m(u)$ and $f(u)$ are continuous.

Arguing in the similar fashion as in Chapter 3, we can also show that for twice differentiable nonlinearities, we have the usual convergence rate

$$\hat{m}(u) = m(u) + O_P(n^{-2/5}).$$

Regarding the problem of identification of the linear subsystem (we assume that $\lambda_0 = 1$) of the parallel model, let us observe that

$$R_{YU}(s) = \tau^2 \lambda_s, \quad s = 1, 2, \ldots, \qquad (12.16)$$

where $R_{YU}(s) = \operatorname{cov}(Y_n, U_{n-s})$ and $\tau^2 = \operatorname{var}(U_n)$. Now λ_s can be estimated by

$$\hat{\lambda}_s = \frac{\hat{R}_{YU}(s)}{\hat{\tau}^2}, \qquad (12.17)$$

where $\hat{R}_{YU}(s)$ and $\hat{\tau}^2$ are the standard estimates of $R_{YU}(s)$ and τ^2, respectively; see Section 2.3.

The above results allow us to form a nonparametric estimate of the linear subsystem in the frequency domain. Indeed, formation of the Fourier transform of formula (12.16) yields

$$S_{YU}(\omega) = \tau^2 \Lambda(\omega), \quad |\omega| \leq \pi, \tag{12.18}$$

where $S_{YU}(\omega) = \sum_{s=-\infty}^{\infty} R_{YU}(s)e^{-is\omega}$ is the cross-spectral density function of the processes $\{Y_n\}$ and $\{U_n\}$. Moreover,

$$\Lambda(\omega) = \sum_{s=0}^{\infty} \lambda_s e^{-is\omega} \tag{12.19}$$

is the transfer function of the linear subsystem.

Because of (12.18) and the standard spectral estimation theory, we can propose the following generic nonparametric estimate of $\Lambda(\omega)$:

$$\hat{\Lambda}(\omega) = \hat{\tau}^{-2} \sum_{|s| \leq N} w(s/N)\hat{R}_{YU}(s)e^{-is\omega}, \tag{12.20}$$

where $w(t)$ is called a convergence factor (data window) that is a function defined on $[-1, 1]$ and usually satisfies $w(0) = 1$, and $|w(t)| \leq 1$. The popular choices are:

$$w(t) = \{0.54 + 0.46\cos(t)\}\mathbf{1}(|t| \leq 1) \quad \text{(Tukey–Hamming window)}$$

and

$$w(t) = \{1 - 6|t|^2 + 6|t|^3\}\mathbf{1}(|t| \leq 1/2) + 2(1 - |t|)^3\mathbf{1}(1/2 \leq |t| \leq 1)\,\text{(Parzen window)}.$$

In (12.20), N is the truncation parameter that can be chosen by the user or more objectively by the asymptotic analysis of the estimate $\hat{\Lambda}(\omega)$. In fact, such analysis shows that for a large class of data windows and if $N = N(n) \to \infty$ with $N(n)/n \to 0$ then the estimate $\hat{\Lambda}(\omega)$ converges to $\Lambda(\omega)$. The data-driven choice of N in (12.20), and generally in the context of nonlinear system identification, would be an interesting issue to pursue.

To gain insights on the above introduced estimation techniques, let us consider the following example.

Example 12.1 Consider the parallel model with the first-order stable autoregressive dynamic subsystem

$$X_n = aX_{n-1} + U_n,$$
$$Y_n = X_n + m(U_n) + Z_n, \quad n = 0, \pm 1, \pm 2, \ldots,$$

where $|a| < 1$. In our simulations, we assume that the noise Z_n is uniformly distributed over $[-0.1, 0.1]$ and the input U_n is uniform on $[0, 10]$. The nonlinearity

$m(u) = (5u^3 - 2u^2 + u)e^{-u}$. Simple algebra shows that the regression function $\mu(u) = E\{Y_n \mid U_n = u\}$ is equal to

$$\mu(u) = m(u) + u + E(U)\gamma,$$

where $\gamma = a(1-a)^{-1}$. Hence, $m(u)$ can be estimated from an estimate $\hat{\mu}(u)$ of the regression function $\mu(u)$ as follows:

$$\hat{m}(u) = \hat{\mu}(u) - u - \bar{U}_n\hat{\gamma}, \tag{12.21}$$

where $\bar{U}_n$ is the standard empirical mean estimate of $E(U)$, and $\hat{\gamma}$ is an estimate of the factor γ. There are two possible estimates of γ. In fact, note first that $\lambda_j = a^j$ and, therefore, $a = \lambda_1$. Using the estimate proposed in (12.17) we can estimate the parameter a by $\hat{a} = \frac{\hat{R}_{YU}(1)}{\hat{\tau}^2}$. Then $\hat{\gamma} = \hat{a}(1-\hat{a})^{-1}$ is the plug-in estimate of γ. For values of a close to ± 1 this can be an unstable estimate. Another estimate can be constructed first based on the fact that $\gamma = \sum_{i=1}^{\infty} \lambda_i$ and then on the observation that, see (12.19),

$$\gamma = \Lambda(0) - 1.$$

Hence, we have $\hat{\gamma} = \hat{\Lambda}(0) - 1$, where $\hat{\Lambda}(\omega)$ is defined in (12.20).

To assess the performance of $\hat{m}(u)$, we use the following criterion measure:

$$Error = E\left\{ n^{-1} \sum_{j=1}^{n} |\hat{m}(U_j) - m(U_j)|^2 \right\},$$

where thirty repetitions of the input–output data were used to evaluate the average operator. The generalized kernel estimate of $\mu(u)$ (see (12.11)) was used and applied in formula (12.21). The kernel $k_b(u, v)$ was generated by the Laguerre orthonormal polynomials, that is,

$$k_b(u, v) = \sum_{i=0}^{b} \ell_i(u)\,\ell_i(v),$$

where $\{\ell_i(u)\}$ is a Laguerre orthonormal polynomials basis. The property of such an estimate was extensively studied in Section 6.4, where it was shown that for s differentiable nonlinearities and input densities we can choose the truncation parameter b as $b = cn^{1/s}$ with the corresponding mean-squared error being of order $O(n^{-(2s-1)/2s})$. Figure 12.3 depicts the Error versus n for the parallel system with $a = 0.1$. The plug-in estimate of γ in (12.21) was used.

For each n the value of b minimizing the $Error$ was chosen. The selection of b is an important issue and Figure 12.4 shows the Error versus b for $n = 100$. The optimal $b^* = 19$ with the corresponding $Error = 0.16$ was observed. Figure 12.5 displays the plot of $m(u)$ and $\hat{m}(u)$ for $n = 100$, $b = 19$. The behavior of the estimate at the boundary $u = 0$ is clearly revealed. The relatively large bias of the estimate for $u \to 0$ is due to the fact that the Laguerre series does not generally converge at $u = 0$. To fix this problem, we can consider the Cesàro summation modification of our expansion which is known to converge at $u = 0$, see [288]. If the value of $m(0+)$ is known then we can use

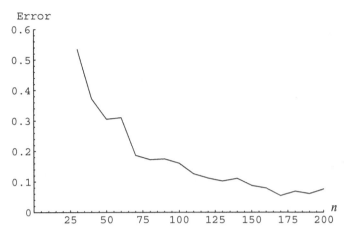

Figure 12.3 Error versus n for the parallel system.

the boundary corrected version of (12.21), that is, $\hat{m}(u) = m(0+) + \hat{\mu}(u) - u - \hat{\mu}(0)$, which converges to $m(u)$ for $u > 0$ and to $m(0+)$ for $u = 0$.

Analogous experiments were performed for the Hammerstein system with the parameters identical to those of the parallel model. The optimal $b^* = 23$ with the corresponding Error $= 0.083$ was observed. Hence, the Error for the parallel model is larger than that of the cascade connection. This phenomenon was discussed in Remark 12.1 and in this particular case, we can verify that the condition in (12.10) is satisfied.

12.1.2 Series-parallel models with nuisance characteristics

The Hammerstein cascade plays a fundamental role in numerous applications. Nevertheless, the departure from its basic structure can often appear in practice. In this section, we consider block-oriented systems that are combinations of series and parallel

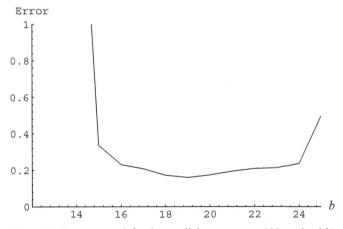

Figure 12.4 Error versus b for the parallel system, $n = 100$, optimal $b^* = 19$.

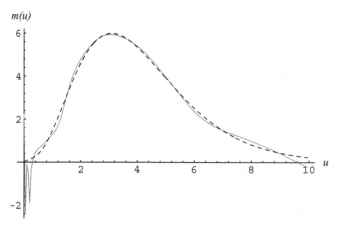

Figure 12.5 The characteristic $m(u) = (5u^3 - 2u^2 + u)e^{-u}$, $u \geq 0$ (dashed line) and its estimate $\hat{m}(u)$ (solid line) for the parallel system, $n = 100$, $b = 19$.

connections. We consider either a linear dynamical subsystem or a nonlinear element as nuisance parameters that disturb standard identification procedures of the Hammerstein system.

Let us first begin with the Hammerstein system with known nuisance dynamical subsystem $\{\xi_i\}$ connected in parallel with the nonlinear element $m(\bullet)$. This system is depicted in Figure 12.6. The input–output relationship of this system is given by

$$Y_n = \sum_{i=0}^{\infty} \lambda_i V_{n-i} + Z_n,$$ (12.22)

where

$$V_n = m(U_n) + \sum_{j=0}^{\infty} \xi_j U_{n-j}.$$ (12.23)

Under assumptions that are identical to those in the previous section, that is, that $\{U_n\}$ is an *iid* input process with $EU_n = 0$ and that $\lambda_0 = 1$, we can obtain the following formula for the regression function $\mu(u) = E\{Y_n \mid U_n = u\}$:

$$\mu(u) = m(u) + u + \gamma,$$ (12.24)

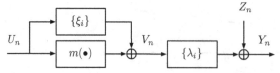

Figure 12.6 Parallel-series system with nuisance dynamics.

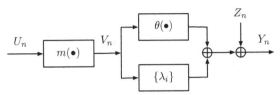

Figure 12.7 Parallel-series system with nuisance nonlinearity.

where $\gamma = E\{m(U_0)\} \sum_{j=1}^{\infty} \lambda_j$ is the constant, which has already appeared in the Hammerstein system. Formula (12.24) suggests that the nuisance dynamics $\{\xi_i\}$ has a limited influence on the problem of recovering the nonlinearity $m(u)$.

On the other hand, the dynamical system $\{\xi_i\}$ has a critical influence on recovery of the impulse response $\{\lambda_i\}$ of the second dynamical subsystem. Indeed, it can be easily shown that the correlation $R_{YU}(s) = \text{cov}(Y_n, U_{n-s})$ can be expressed in terms of $\{\xi_i\}$ and $\{\lambda_i\}$ by the following formula:

$$R_{YU}(s) = a\lambda_s + \tau^2 \sum_{j=0}^{s} \lambda_j \xi_{s-j}, \qquad (12.25)$$

where $\tau^2 = \text{var}(U_0)$ and $a = \text{cov}(m(U_0), U_0)$. Since $\lambda_0 = 1$ and τ^2 can be estimated from the input signal, the constant a can be determined as $a = R_{YU}(0) - \tau^2$.

Since $\{\xi_i\}$ is known, the formula in (12.25) gives a set of linear equations that can be solved with respect to $\{\lambda_i\}$. Plugging standard estimates for τ^2 and $R_{YU}(s)$ we can easily form the corresponding estimate of $\{\lambda_i\}$. The convolution structure of (12.25) makes the solution of the identification problem particularly simple in the frequency domain. Hence, formation of the Fourier transform of (12.25) yields

$$S_{YU}(\omega) = a\Lambda(\omega) + \tau^2 \Lambda(\omega) \, \Xi(\omega), \qquad (12.26)$$

where $\Xi(\omega) = \sum_{s=0}^{\infty} \xi_s e^{-is\omega}$ is the transfer function of the linear subsystem $\{\xi_i\}$. Solving (12.26) for $\Lambda(\omega)$ and then replacing $S_{YU}(\omega)$ by a lag window estimator (see (12.20)) we can form a nonparametric estimate of $\Lambda(\omega)$.

A system that is a complement to the one examined above is the parallel–series system with nuisance nonlinearity $\theta(\bullet)$ (see Figure 12.7). Here, we have

$$Y_n = \sum_{i=0}^{\infty} \lambda_i V_{n-i} + \theta(V_n) + Z_n, \qquad (12.27)$$

where $V_n = m(U_n)$. In this case, the regression function $\mu(u) = E\{Y_n \mid U_n = u\}$ is given by

$$\mu(u) = m(u) + \theta(m(u)) + \alpha, \qquad (12.28)$$

where $\alpha = E\{m(U_0)\} \sum_{i=1}^{\infty} \lambda_i$. Hence, the presence of $\theta(\bullet)$ critically influences the recovery of the system nonlinearity $m(\bullet)$. An approximate solution of (12.28) (with respect to $m(u)$) for given $\theta(u)$ and $\mu(u)$ is required in order to form an estimate of $m(u)$. For instance, the method of steepest descent can be used with the initial solution

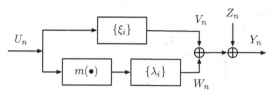

Figure 12.8 Parallel-series system with two linear subsystems.

$m_{(0)}(u) = \mu(u)$ and the consecutive iterations

$$m_{(\ell+1)}(u) = m_{(\ell)}(u) - \gamma_\ell \left\{ m_{(\ell)}(u) + \theta \left(m_{(\ell)}(u) \right) - \mu(u) \right\}$$

for $\ell = 0, 1, \ldots$. Here, $\{\gamma_\ell\}$ is the sequence controlling the step size in the iteration process. Concerning the linear subsystem, we can easily conclude from (12.27) that for $s \geq 1$, we have

$$\operatorname{cov}(Y_n, U_{n-s}) = \lambda_s \operatorname{cov}(m(U_0), U_0).$$

This with $\lambda_0 = 1$ allows us to recover the linear subsystem $\{\lambda_i\}$ in the presence of the nuisance nonlinearity in the identical way as for the standard Hammerstein system.

12.1.3   Parallel-series models

A system of that finds application in mechanical engineering, for example, engine transmission modeling, is shown in Figure 12.8. The system consists of the Hammerstein series model connected in parallel with another dynamical system having the characteristic $\{\xi_i\}$. Unlike in the previous section both linear subsystems are unknown.

The input-output description of the system is the following:

$$Y_n = W_n + V_n + Z_n, \tag{12.29}$$

where

$$W_n = \sum_{i=0}^{\infty} \lambda_i m(U_{n-i}) \quad \text{and} \quad V_n = \sum_{i=0}^{\infty} \xi_i U_{n-i}. \tag{12.30}$$

It is readily seen that the regression $\mu(u) = E\{Y_n \mid U_n = u\}$ is given in this case by

$$\mu(u) = m(u) + u + \alpha, \tag{12.31}$$

where $\alpha = E\{m(U_0)\} \sum_{j=1}^{\infty} \lambda_j + E\{U_0\} \sum_{j=1}^{\infty} \xi_j$. Hence the standard nonparametric regression approach is applicable in this case with an easily formulated estimate of the nonlinearity $m(\bullet)$.

Concerning the linear subsystems we must use the higher order covariance functions (cumulants) (see Section 12.2), in order to recover $\{\lambda_i\}$ and $\{\xi_i\}$. Let us assume, without loss of generality that the density function of the input signal is symmetric around zero

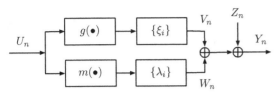

Figure 12.9 Uryson parallel-series system.

and that $m(\bullet)$ is an odd function. Then, we have

$$E\{Y_n U_{n-s}\} = \lambda_s E\{U_0\, m(U_0)\} + \xi_s E\left\{U_0^2\right\} \qquad (12.32)$$

and

$$E\left\{Y_n U_{n-s}^3\right\} = \lambda_s E\left\{U_0^3\, m(U_0)\right\} + \xi_s E\left\{U_0^4\right\} \qquad (12.33)$$

for $s = 1, 2, \ldots$. Note that under our assumptions $E\{U_0\, m(U_0)\} \neq 0$ and $E\{U_0^3\, m(U_0)\} \neq 0$. In addition, since $\lambda_0 = \xi_0 = 1$ then we have $E\{U_0\, m(U_0)\} = E\{Y_n U_n\} - E\{U_0^2\}$ and $E\{U_0^3 m(U_0)\} = E\{Y_n U_n^3\} - E\{U_0^4\}$. This combined with (12.32) and (12.33) allows us to find closed-form formulas for λ_s and ξ_s in terms of the cross covariance and the third-order cross cumulants of the stochastic processes $\{U_n\}$ and $\{Y_n\}$. See Section 12.2 for a discussion on properties of cumulants.

The extension of the system shown in Figure 12.8 to the case of a parallel model that consists of Hammerstein models in each path is named the Uryson model. This connection is shown in Figure 12.9 and is described by the following input–output relationship:

$$Y_n = W_n + V_n + Z_n, \qquad (12.34)$$

where

$$W_n = \sum_{i=0}^{\infty} \lambda_i\, m(U_{n-i}) \quad \text{and} \quad V_n = \sum_{i=0}^{\infty} \xi_i\, g(U_{n-i}). \qquad (12.35)$$

It is clear that this system is not identifiable and one must assume that one of the components of the system is known. For instance, if $g(\bullet)$ is given, then analogous to (12.32) and (12.33), we have

$$E\{Y_n U_{n-s}\} = \lambda_s E\{U_0 m(U_0)\} + \xi_s E\{U_0 g(U_0)\} \qquad (12.36)$$

and

$$E\left\{Y_n U_{n-s}^3\right\} = \lambda_s E\left\{U_0^3 m(U_0)\right\} + \xi_s E\left\{U_0^3 g(U_0)\right\}. \qquad (12.37)$$

The normalization $\lambda_0 = \xi_0 = 1$ and the knowledge of $g(\bullet)$ yield the closed-form solution for $\{\lambda_i\}$ and $\{\xi_i\}$. Furthermore, we can represent the nonlinearity $m(u)$ in terms of the regression function $\mu(u) = E\{Y_n \mid U_n = u\}$ as follows:

$$m(u) = \mu(u) - g(u) + \alpha, \qquad (12.38)$$

where $\alpha = -E\{m(U_0)\} \sum_{j=1}^{\infty} \lambda_j - E\{g(U_0)\} \sum_{j=1}^{\infty} \xi_i$.

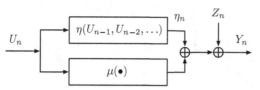

Figure 12.10 Generalized block-oriented system.

Hence, the formulas in (12.36), (12.37), and (12.38) provide building blocks in forming nonparametric estimates of the characteristics of the Uryson system.

12.1.4   Generalized nonlinear block-oriented models

It has already been observed that the majority of systems introduced in the previous sections (see, for example, (12.4)) can be written in the signal-plus-noise form. The signal part characterizes the nonlinearity we wish to estimate, whereas the noise part includes all dynamics present in the system plus other nuisance nonlinearities. This observation motivates the introduction of a more general class of nonlinear composite models that includes most of the previously defined connections. The system is characterized by a nonlinearity, which is embedded in a block-oriented structure containing dynamic linear subsystems and other "nuisance" nonlinearities.

The structure of the system is depicted in Figure 12.10 and is given by:

$$Y_n = \mu(U_n) + \eta(U_{n-1}, U_{n-2}, \ldots) + Z_n, \tag{12.39}$$

where $\mu(\bullet)$ is a nonlinearity to be identified. The process $\{\eta_n = \eta(U_{n-1}, U_{n-2}, \ldots)\}$ is the "system noise" that is induced by the system dynamics, that is, by its past input signals $(U_{n-1}, U_{n-2}, \ldots)$. For most practical systems, it is sufficient to assume that the system noise has the following additive form:

$$\eta_n = \eta_{1,n} + \eta_{2,n} + \cdots + \eta_{J,n}, \tag{12.40}$$

for some finite J. Furthermore, an individual component $\eta_{r,n}$ of η_n is of the form of a Hammerstein system, that is,

$$\eta_{r,n} = \sum_{t=1}^{\infty} \lambda_{r,t}\, s_r(U_{n-t}),$$

where $s_r(\bullet)$ are measurable functions such that $Es_r(U_0) = 0$, $\text{var}\{s_r(U_0)\} < \infty$, and $\sum_{t=1}^{\infty} \lambda_{r,t}^2 < \infty$, $r = 1, \ldots, J$. Figure 12.11 depicts the structure of the system noise. Let us note that under the above conditions on $s_r(\bullet)$ and $\{\lambda_{r,t}\}$, $r = 1, \ldots, J$, the system noise $\{\eta_n\}$ is a well-defined stationary stochastic process with $E\eta_n = 0$ and $\text{var}\{\eta_n\} < \infty$. The following remark illustrates some of the aforementioned concepts.

REMARK 12.4 *The linear form of the system noise allows us to put all the previously studied structures within the framework of the system in (12.39) and (12.40). In fact, for the Hammerstein system, we have $J = 1$ with $\{\lambda_{1,j}\} = \{\lambda_j\}$ and $s_1(u) = m(u) - E\{m(U_0)\}$, and for the parallel system, $s_1(u) = u - E\{U_0\}$. Also for these systems,*

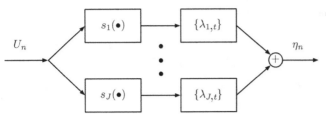

Figure 12.11 The structure of the system noise process $\{\eta_n\}$.

we have $\mu(u) = m(u) + E\{m(U_0)\} \sum_{i=1}^{\infty} \lambda_i$ and $\mu(u) = m(u) + u + E\{U_0\} \sum_{i=1}^{\infty} \lambda_i$, correspondingly.

For the parallel-series system depicted in Figure 12.8, we note that (see (12.29) and (12.30)) $\mu(u)$ is given by (12.31). The system noise is defined by two components, that is, $J = 2$ with $\{\lambda_{1j}\} = \{\lambda_j\}$, $\{\lambda_{2j}\} = \{\xi_j\}$, and $s_1(u) = m(u) - E\{m(U_0)\}$, $s_2(u) = u - E\{U_0\}$. Hence, the integer J reflects the number of dynamical subsystems which are present in the given composite nonlinear structure.

The structure of the system in (12.39) and (12.40) reveals that

$$\mu(u) = E\{Y_n \mid U_n = u\}. \tag{12.41}$$

Hence, one can recover the nonlinearity $\mu(u)$ from the regression function of Y_n on U_n. Let us apply the generalized kernel method (see (12.11)) for estimating $\mu(u)$, that is, we have

$$\hat{\mu}(u) = \frac{n^{-1} \sum_{i=1}^{n} Y_i k_b(u, U_i)}{n^{-1} \sum_{i=1}^{n} k_b(u, U_i)}. \tag{12.42}$$

We have already noted that in order to study the asymptotic behavior of $\hat{\mu}(u)$ it suffices to examine the expression $\hat{g}(u) = n^{-1} \sum_{i=1}^{n} Y_i k_b(u, U_i)$. We know that $E\hat{g}(u) \to \mu(u)f(u)$ as $b \to \infty$. The covariance structure of $\hat{g}(u)$ is more involved and the following lemma helps to tackle this problem:

LEMMA 12.1 *Let $\{Y_n\}$ be an output process of the system defined in (12.39) and (12.40). Then, for any measurable function $a(\bullet)$ and for $m < n$, we have*

$$\text{cov}(Y_n a(U_n), Y_m a(U_m)) = E\{a(U_0)\} \sum_{r=1}^{J} \lambda_{r,n-m} \, \text{cov}(s_r(U_0), \mu(U_0)a(U_0))$$

$$+ E^2\{a(U_0)\} \sum_{r=1}^{J} \sum_{\ell=1}^{\infty} \lambda_{r,\ell} \lambda_{r,\ell+n-m} E\left\{s_r^2(U_0)\right\}.$$

The proof of this lemma is straightforward and is left to the reader. It is worth noting that the covariance depends only on the difference $n - m$ and this plays an important role in the evaluation of the covariance structure of the estimate $\hat{\mu}(u)$.

Let us now evaluate $\text{var}\{\hat{g}(u)\}$. It is clear that

$$\text{var}\{\hat{g}(u)\} = n^{-1}\,\text{var}\{Y_n k_b(u, U_n)\} + 2n^{-2}\sum_{i=2}^{n}\sum_{\ell=1}^{i-1}\text{cov}(Y_i k_b(u, U_i), Y_\ell k_b(u, U_\ell))$$
$$= T_1(u) + T_2(u). \tag{12.43}$$

Using (12.39), we obtain

$$T_1(u) = n^{-1}\,\text{var}\{\mu(U_n)k_b(u, U_n)\} + n^{-1}\left(\text{var}\{\eta_n\} + \sigma_Z^2\right) E\left\{k_b^2(u, U_n)\right\}, \tag{12.44}$$

where $\sigma_Z^2 = \text{var}\{Z_0\}$. Thus by virtue of the assumption in (12.14) the term $T_1(u)$ is not greater than

$$c(u, b)n^{-1}b\left(\mu^2(u) + \text{var}\{\eta_n\} + \sigma_Z^2\right) f(u), \tag{12.45}$$

where the role of the factor $c(u, b)$ was discussed in Remark 12.3. Application of Lemma 12.1 to the covariance in $T_2(u)$ (see (12.43)) yields

$$\text{cov}(Y_i k_b(u, U_i), Y_\ell k_b(u, U_i))$$
$$= E\{k_b(u, U_0)\}\sum_{r=1}^{J}\lambda_{r,i-\ell}\,\text{cov}(s_r(U_0), \mu(U_0)k_b(u, U_0))$$
$$+ E^2\{k_b(u, U_0)\}\sum_{r=1}^{J}\sum_{v=1}^{\infty}\lambda_{r,v}\lambda_{r,v+i-\ell}E\left\{s_r^2(U_0)\right\}.$$

The right-hand side converges (as $b \to \infty$) to

$$f^2(u)\mu(u)\sum_{r=1}^{J}\lambda_{r,i-\ell}s_r(u) + f^2(u)\sum_{r=1}^{J}\sum_{v=1}^{\infty}\lambda_{r,v}\lambda_{r,v+i-\ell}E\left\{s_r^2(U_0)\right\} = \psi_{i-\ell}(u).$$

Then because

$$\sum_{i=2}^{n}\sum_{\ell=1}^{i-1}\psi_{i-\ell}(u) = n\sum_{i=1}^{n-1}\left(1 - \frac{i}{n}\right)\psi_i(u),$$

we find that the term $T_2(u)$ does not exceed

$$2n^{-1}\sum_{i=1}^{n-1}\left(1 - \frac{i}{n}\right)|\psi_i(u)|.$$

The Cesàro summation formula and the fact that $\sum_{i=1}^{\infty}|\psi_i(u)| < \infty$ imply that $T_2(u) = O(n^{-1})$.

All these considerations prove the following theorem.

THEOREM 12.2 *Let $\hat{\mu}(u)$ be the generalized kernel estimator defined in (12.42) with the kernel function, which satisfies (12.14). Let $E\{s_r^2(U_0)\} < \infty$ and $\sum_{s=1}^{\infty}|\lambda_{r,s}| < \infty$, $r = 1, \ldots, J$. If*

$$b(n) \to \infty \quad \text{and} \quad \frac{b(n)}{n} \to 0 \quad \text{as } n \to \infty$$

then

$$\hat{\mu}(u) \to \mu(u)(P) \quad as\ n \to \infty,$$

at every point u where $\mu(u)$ and $f(u)$ are continuous.

Writing the estimate $\hat{\mu}(u)$ in (12.42) in the ratio form $\hat{g}(u)/\hat{f}(u)$ and using the "ratio trick", that is, writing

$$\hat{\mu}(u) = \mu(u) + \frac{\hat{g}(u) - \mu(u)\hat{f}(u)}{f(u)} + (\hat{\mu}(u) - \mu(u))\frac{f(u) - \hat{f}(u)}{f(u)}, \tag{12.46}$$

we can see that the asymptotic behavior of $\hat{\mu}(u)$ is controlled by the second term in the above representation. Then the evaluation of the variance of this term in a similar fashion as in the proof of Theorem 12.2 shows that

$$\mathrm{var}\{\hat{\mu}(u)\} = c(u, b)n^{-1}b\frac{(\mathrm{var}\{\eta_n\} + \sigma_Z^2)}{f(u)} + O\left(n^{-1}\right). \tag{12.47}$$

The Taylor's expansion argument applied to the term $E\left\{\frac{\hat{g}(u) - \mu(u)\hat{f}(u)}{f(u)}\right\}$ (for twice differentiable nonlinearities and input densities) gives $E\hat{\mu}(u) = \mu(u) + O(b^{-2})$. This combined with (12.47) yields the usual optimal rate

$$\hat{\mu}(u) = \mu(u) + O_P(n^{-2/5}),$$

where the asymptotically optimal choice of the bandwidth $b(n)$ is $b^*(n) = an^{1/5}$.

It is worth noting that this asymptotic $b^*(n)$ does not depend on the correlation structure of the system noise $\{\eta_n\}$. In fact, owing to (12.47) it depends only on the overall noise variance $\mathrm{var}\{\eta_n\} + \sigma_Z^2$. In practice, however, one must specify $b(n)$ based only on the available training data. Such an automatic choice of the bandwidth can be conducted by two main strategies. The first one (often called the plug-in technique) involves utilizing an asymptotic formula for the optimal $b^*(n)$ and then estimating unknown quantities that appear in the formula by some pilot estimates. To be more specific, let us consider a version of (12.42) as a standard kernel estimate. Hence, let

$$\hat{\mu}(u) = \sum_{j=1}^{n} Y_j K_h(u - U_j) / \sum_{j=1}^{n} K_h(u - U_j), \tag{12.48}$$

where $K_h(u) = h^{-1}K(h^{-1}u)$. Here, h is the bandwidth and $K(\bullet)$ is an admissible kernel function (see Chapter 3). Proceeding as in the proof of Theorem 12.2 (see also Chapter 3 and Appendix A), we can show that

$$\mathrm{var}\{\hat{\mu}(u)\} = (nh)^{-1}\frac{(\mathrm{var}\{\eta_n\} + \sigma_Z^2)}{f(u)}\int_{-\infty}^{\infty} K^2(u)du(1 + o(1))$$

and

$$E\hat{\mu}(u) - \mu(u) = h^2\frac{\varphi(u)}{2}\int_{-\infty}^{\infty} u^2 K(u)du(1 + o(1)),$$

where $\varphi(u) = \mu^{(2)}(u) + 2\frac{\mu^{(1)}(u)f^{(1)}(u)}{f(u)}$. Hence, by forming the mean-squared error $\mathrm{var}\{\hat{\mu}(u)\} + (E\hat{\mu}(u) - \mu(u))^2$, we can readily obtain an expression for the asymptotically best local bandwidth that minimizes the error, that is,

$$h^*(u) = C(K) \left\{ \frac{\mathrm{var}\{\eta_n\} + \sigma_Z^2}{n\varphi^2(u)f(u)} \right\}^{1/5}, \tag{12.49}$$

where $C(K) = \{\int_{-\infty}^{\infty} K^2(u)du(\int_{-\infty}^{\infty} u^2 K(u)du)^{-2}\}^{1/5}$ is the factor that depends only on the kernel $K(u)$. For example, the standard Gaussian kernel gives $C(K) = 0.7763 \ldots$.

The prescription in (12.49) gives a local choice of the bandwidth which could be difficult to implement in practice as one must recalculate $h^*(u)$ at every point within the support of $\mu(u)$. Nevertheless, in a similar way (see Appendix A), one can derive a formula for the asymptotically best bandwidth that minimizes the global integrated mean-squared error, that is, we have

$$h^* = C(K) \left\{ \frac{\mathrm{var}\{\eta_n\} + \sigma_Z^2}{n\int_{-\infty}^{\infty} \varphi^2(u)f(u)du} \right\}^{1/5}. \tag{12.50}$$

In order to use (12.50), we must determine two quantities, $\sigma^2 = \mathrm{var}\{\eta_n\} + \sigma_Z^2$ and $\theta = \int_{-\infty}^{\infty} \varphi^2(u)f(u)du = E\{\varphi^2(U_0)\}$. To estimate θ we can first simplify the formula for $\varphi(u)$ by using a standard family of distributions of the input signal $\{U_n\}$ and a simple parametric model for $\mu(u)$. For example, let the density function of the input signal be Gaussian $N(0, \tau^2)$. Then since $f^{(1)}(u)/f(u) = -u/\tau^2$, we can replace $\varphi(u)$ by

$$\psi(u) = \mu^{(2)}(u) - 2\mu^{(1)}(u)u/\tau^2.$$

Next a quick parametric model (for instance the fourth-order polynomial) can be applied to obtain an estimate of $\mu(u)$ and its first two derivatives. A standard estimate of τ^2 along with the estimates of $\mu^{(1)}(u)$ and $\mu^{(2)}(u)$ give an estimate $\hat{\psi}(u)$ of $\psi(u)$. As a result,

$$\hat{\theta} = n^{-1} \sum_{i=1}^{n} \hat{\psi}^2(U_t)$$

can serve as an estimate of θ.

Regarding the variance σ^2 we can use the approximate residuals $\hat{e}_t = Y_t - \tilde{\mu}(U_t)$ to form a standard estimate of σ^2. Here $\tilde{\mu}(u)$ is a version of (12.48) using a preliminary bandwidth value, for example, $h_p = n^{-1/5}$. Yet another method is to estimate σ^2 directly from the output data using difference-type methods. This requires ordering the input data. Hence, let us use the rearranged training set

$$(U_{(1)}, Y_{[1]}), (U_{(2)}, Y_{[2]}), \ldots, (U_{(n)}, Y_{[n]}),$$

where $U_{(1)} < U_{(2)} < \cdots < U_{(n)}$, and $Y_{[i]}$s are paired with $U_{(i)}$s; see Chapter 7 for such concepts. The simplest difference-type estimates of the variance σ^2 are

$$\hat{\sigma}^2 = \frac{1}{2(n-1)} \sum_{t=1}^{n-1} \left(Y_{[t+1]} - Y_{[t]} \right)^2$$

and

$$\tilde{\sigma}^2 = \frac{1}{6(n-2)} \sum_{t=2}^{n-1} \left(Y_{[t+1]} - 2Y_{[t]} + Y_{[t-1]} \right)^2.$$

A detailed overview of difference-based methods for inference in nonparametric regression with correlated errors is given in [147, 214]. Note, however, that the difference techniques have been mostly studied in the fixed design estimation framework, that is, when the variable U_i is a fixed deterministic point.

All the aforementioned considerations and (12.50) lead to the following plug-in estimate of the bandwidth

$$\hat{h}_{PI} = C(K) \left\{ \frac{\hat{\sigma}^2}{n\hat{\theta}} \right\}^{1/5}. \tag{12.51}$$

The plug-in methods are tailored to specific estimation algorithms and they require a large amount of prior information about the smoothness of unknown nonlinearities. They, however, reveal a good performance in many studies concerning nonparametric regression with correlated errors.

Fully automatic methods for selecting smoothing parameters in nonparametric regression are based on various resampling strategies. A popular choice is a cross-validation principle that seeks the bandwidth $\hat{h}_{CV}$ for (12.48) that minimizes

$$CV(h) = n^{-1} \sum_{i=1}^{n} \{Y_i - \hat{\mu}_{-i}(U_i)\}^2 w(U_i), \tag{12.52}$$

where $\hat{\mu}_{-i}(u)$ denotes the version of (12.48) obtained by omitting the sample (U_i, Y_i). The weight function $w(u)$ is used to alleviate the behavior of the estimate at boundary points. It is known that the accuracy of $\hat{h}_{CV}$ is acceptable if data are independent. In the case of correlated errors the standard cross-validation rule breaks down as it tends to select a value of h which is much smaller than the optimal one. The criterion $CV(h)$ can be modified to adapt to correlated errors by dropping out blocks of samples instead of single samples. Yet another method is to construct a preliminary estimate $\tilde{\mu}(u)$ with a large value of h. Then, we form the residuals $\hat{e}_t = Y_t - \tilde{\mu}(U_t)$ in order to obtain a modified version of the cross-validation criterion. A minimizer of this criterion is, in turn, used as a next value of h. Further references and additional details on the problem of selecting smoothing parameters in the presence of correlated errors can be found in [224]. It is worth noting these contributions are dealing mostly with a specific parametric class of correlated errors (like ARMA-type errors) being additionally independent of the regression function. In the case of nonparametric inference for block-oriented systems, the errors are defined by a general linear process with the correlation structure depending on the unknown nonlinearity (see Remark 12.4).

Example 12.2 To evaluate the accuracy of our identification algorithms for small and moderate sample sizes we perform some simulation studies. In all our experiments the input signal $\{U_n\}$ is uniformly distributed over the interval $[-3, 3]$. The measurement noise $\{Z_n\}$ is also uniformly distributed in $[-0.1, 0.1]$. We apply the generalized kernel estimate as in Example 12.1.1 with the multiresolution kernel $k_b(u, v) = \sum_{j=-\infty}^{\infty} \phi(u - j)\phi(v - j)$, where the scaling function $\phi(u)$ is obtained from the Haar multiresolution analysis, that is, $\phi(u) = \mathbf{1}(0 \leq u < 1)$. The range of the input signal implies that we can specify the truncation parameter b as $b = 3 \times 2^J$, where J is the resolution level. The efficacy of the identification procedure $\hat{\mu}(u)$ is measured by the criterion *Error* introduced in Example 12.1.1.

In the first experiment a nonlinearity (see Figure 12.12 (a)),

$$m(u) = \begin{cases} 1 & \text{if } u \in [0.125, 0.25] \cup [0.375, 0.5] \cup [0.875, 1] \\ -1 & \text{if } u \in [-0.125, -0.25] \cup [-0.375, -0.5] \cup [-0.875, -1] \\ 0 & \text{otherwise} \end{cases} \quad (12.53)$$

is used in three different settings, that is, for the memoryless, cascade and parallel models. Since $m(u)$ in (12.53) is piecewise constant, this is an example of the nonlinearity well adapted to the Haar multiresolution basis.

The cascade and parallel models are described by the following state equations, respectively:

$$\begin{cases} X_n = -0.2X_{n-1} + m(U_n) \\ Y_n = X_n + Z_n \end{cases}, \quad (12.54)$$

$$\begin{cases} X_n = -0.2X_{n-1} + U_n \\ Y_n = X_n + m(U_n) + Z_n \end{cases}. \quad (12.55)$$

It is worth nothing that we have $Em(U_n) = EU_n = 0$ and therefore $\hat{\mu}(u)$ and $\hat{\mu}(u) - u$ are consistent estimates of the nonlinearity $m(u)$ in the cascade and parallel models, respectively. Figure 12.13 depicts the *Error* as a function of the sample size n. It is seen that the *Error* for the memoryless model is the smallest. Surprisingly the *Error* for the cascade model is about two to three times smaller than that of the parallel structure.

The value of the resolution level J has been set to 3 in all three cases. This is due to the fact that this value minimizes the *Error* for a small and moderate number of observations. In fact, Figure 12.14 displays the *Error* versus J for $n = 150$ observations. A clear global minimum at $J = 3$ is seen. Hence, the optimal partition of the u-axis is $1/2^J = 1/8$. Note that this agrees with the structure of the nonlinearity in (12.53), which is constant on the intervals of the size $1/8$.

In the second experiment a nonlinearity in (12.54), (12.55) that is not well adapted to the Haar basis has been selected, that is,

$$m(u) = \begin{cases} 10\exp(-10u) + 2 & \text{for } 0 \leq u \leq 0.5 \\ 3\cos(10\pi u) & \text{for } u > 0.5, \end{cases} \quad (12.56)$$

and $-\theta(-u)$ for $u < 0$. Figure 12.12 (b) displays this nonlinearity. Figure 12.15 depicts the *Error* versus J for the cascade and parallel structures based on $n = 150$ observations. Since the nonlinearity is not well suited for the Haar basis, larger values of J are required; the optimal J equals 5 for the cascade model and 6 for the parallel model.

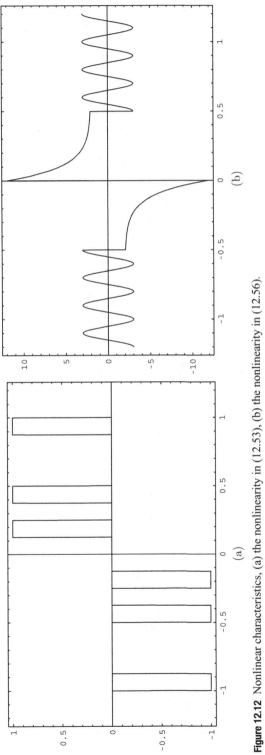

Figure 12.12 Nonlinear characteristics, (a) the nonlinearity in (12.53), (b) the nonlinearity in (12.56).

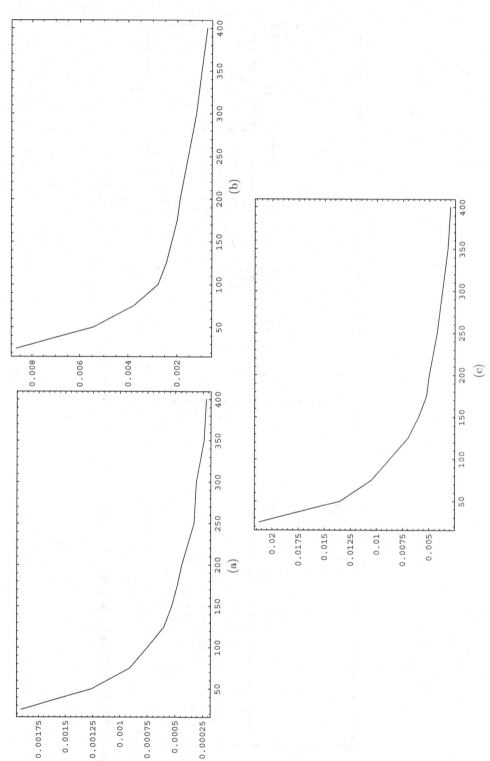

Figure 12.13 *Error* versus n for the nonlinearity in (12.53), (a) memoryless system, (b) cascade system, (c) parallel system.

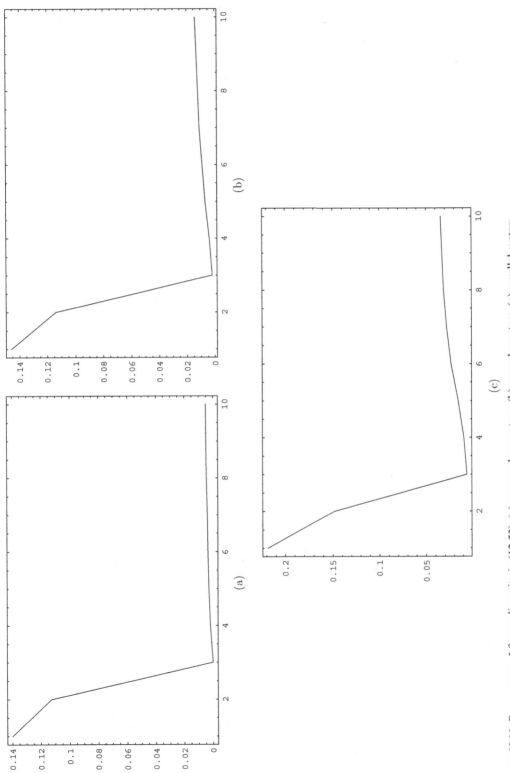

Figure 12.14 *Error* versus J for nonlinearity in (12.53), (a) memoryless system, (b) cascade system, (c) parallel system.

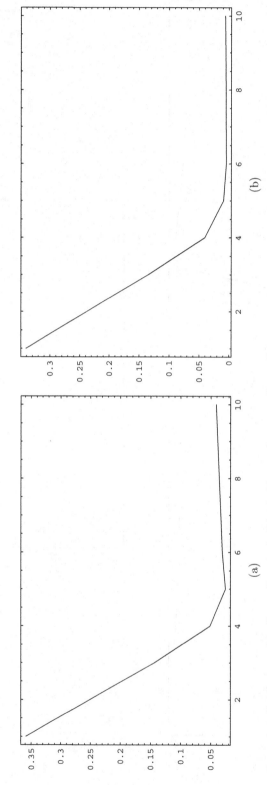

Figure 12.15 *Error* versus J for the nonlinearity in (12.56), (a) cascade system, (b) parallel system.

For the memoryless model the value of J is even larger and equals 10. The overall performance of $\hat{\mu}(u)$ is now considerably poorer.

Finally the model with the nuisance nonlinearity has been taken into account, that is, the model represented by the following equation:

$$\begin{cases} X_n = -0.2X_{n-1} + m(U_n) + m_0(U_{n-1}) \\ Y_n = X_n + Z_n \end{cases}, \tag{12.57}$$

where $m_0(u) = 0.1u^3$ is the nuisance nonlinearity and $m(u)$ is defined as in (12.56). Figure 12.16 displays the *Error* versus J. An optimal resolution level is equal to 5 for $n = 150$ observations. In the same figure (Figure 12.16 (b)), we show the *Error* of the version of (12.57) where the dynamical subsystem is set to zero (that is, the value -0.2 in (12.57) is replaced by 0). Let us observe that the optimal J is now considerably greater and equals 9. This reveals that the presence of dynamical subsystems in composite models to some extent helps in identification and it greatly influences the accuracy of identification algorithms for recovering nonlinear elements.

12.2   Block-oriented systems with nonlinear dynamics

Thus far, we have considered nonlinear systems where the dynamics and the nonlinearity were clearly separated. In particular, we have extensively studied the cascade models of the Hammerstein type. There are, however, certain qualitative limitations to the dynamics of cascade models with memoryless nonlinearities, just as with linear models. For example, the chemical reactor considered in [82] exhibits an underdamped oscillatory response to positive step perturbations. The static nonlinearity in the Hammerstein model can only influence the magnitude of the step input to the linear subsystem and not its dynamic character. Thus, positive-going and negative-going step responses for Hammerstein models may be different in magnitude, but they will be the same in character.

In this section, we propose a new model structure that generalizes the Hammerstein model by including a nonlinearity that has its own dynamics with a finite memory in front of a linear dynamical system. Hence, a nonlinear element is no longer memoryless and it possesses its own "local" memory typically of a short length; the linear dynamic system is in charge of long-range effects. The primary advantage of our model is that it is very flexible while still retaining considerable structural simplicity. This model structure is described in detail in the next section and important special cases are discussed to illustrate its relationship to the other structures mentioned and examined in the previous chapters. By relating the identification problem to a certain equation involving regression functions we present a class of nonparametric identification algorithms for determining the dynamic nonlinearity from input–output data. The issue of linear subsystem identification is also addressed. Here, the theory of cumulants is utilized. An important case with a Gaussian input process is also examined. Special attention is given to structures

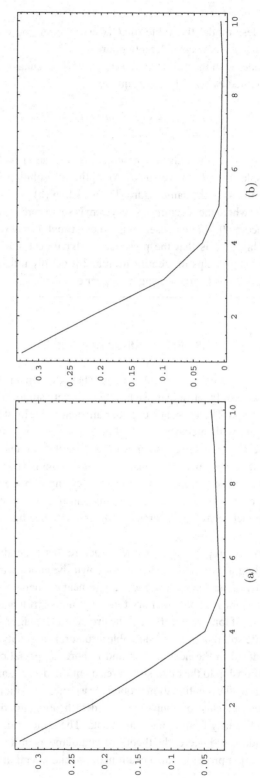

Figure 12.16 *Error* versus *J* for the system with two nonlinearities.

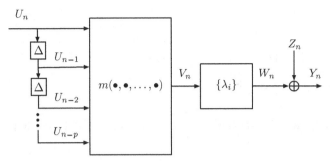

Figure 12.17 The Hammerstein system with the nonlinear dynamics of order p.

which play an important role in selected applications. These structures include the additive Hammerstein model with the nonlinear memory, and the sandwich system with the input linear subsystem possessing a finite memory. Consistent nonparametric estimates of these system nonlinearities and linear subsystems are developed.

12.2.1 Nonlinear models

Let us begin by recalling the Hammertein system with the nonlinearity $m(\bullet)$ and the linear subsystem $\{\lambda_i\}$, that is,

$$Y_n = \sum_{i=0}^{\infty} \lambda_i\, m(U_{n-i}) + Z_n. \tag{12.58}$$

As previously mentioned, the Hammerstein model has a number of shortcomings in capturing important dynamic features of the dynamical nonlinear system. In order to alleviate these problems, we propose the nonlinear dynamic model of the form shown in Figure 12.17, where Δ is the unit-time delay operator. Hence, the model has a cascade structure and it consists of a dynamic nonlinear element with a memory of length p followed by a linear dynamic system with the impulse response function $\{\lambda_i\}$. The characteristic of the first subsystem is denoted by $m(\bullet, \bullet, \ldots, \bullet)$ and it maps $p+1$ consecutive input variables $(U_n, U_{n-1}, \ldots, U_{n-p})$ into the scalar output V_n, that is,

$$V_n = m(U_n, U_{n-1}, \ldots, U_{n-p}), \tag{12.59}$$

where $m(\bullet, \bullet, \ldots, \bullet)$ is a measurable function of $p+1$ variables. The output signal of the linear subsystem $\{\lambda_i\}$ is disturbed by a stationary white random noise sequence Z_n yielding

$$Y_n = \sum_{i=0}^{\infty} \lambda_i V_{n-i} + Z_n. \tag{12.60}$$

Throughout this section, we shall assume that the linear subsystem is stable, that is, $\sum_{i=0}^{\infty} |\lambda_i| < \infty$. Moreover, we assume that $E m^2(U_n, U_{n-1}, \ldots, U_{n-p}) < \infty$, $\sigma_Z^2 = \operatorname{var} Z_n < \infty$, and the input signal $\{U_n\}$ is a sequence of *iid* random variables. By virtue

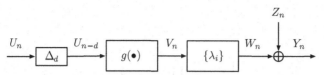

Figure 12.18 The Hammerstein system with the delayed input.

of Lemma 12.5 (see Section 12.2.6) the output $\{Y_n\}$ is a stationary stochastic process with

$$\text{var}\{Y_n\} \leq (p+1)\,\text{var}\{V_n\}\sum_{i=0}^{\infty}|\lambda_i|^2 + \sigma_Z^2. \tag{12.61}$$

The nonparametric identification problem for the system in (12.59) and (12.60) consists of estimating the nonlinearity $m(\bullet, \bullet, \ldots, \bullet)$ and the linear subsystem $\{\lambda_i\}$ without any further assumptions on the system characteristics. It is worth noting that the memory length of the nonlinear dynamics, that is, p, should also be estimated. A large value of p is, however, not advisable due to the curse of dimensionality, that is, the precision of estimating the $p+1$ function $m(\bullet, \bullet, \ldots, \bullet)$ is generally very poor (see Chapter 13 for a discussion of the problem of estimating multivariate functions).

There is a large class of practical block-oriented nonlinear models, which fall into the description given in (12.59) and (12.60), and we will illustrate this by a series of specific examples.

Example 12.3 (Hammerstein system with a delay). This is the most obvious special case of our model that has already been extensively discussed in previous chapters. However, using formula (12.59) and defining

$$m(U_n, U_{n-1}, \ldots, U_{n-p}) = g(U_{n-d}), \quad 0 \leq d \leq p, \tag{12.62}$$

we obtain the Hammerstein model with the static nonlinearity $g(\bullet)$ applied to the input signal delayed by d-time units. This is depicted in Figure 12.18, where the Δ_d is the d-time units delay operator, that is, $\Delta_d U_n = U_{n-d}$. The standard identification problem for this model is to recover $g(\bullet)$ and $\{\lambda_i\}$ from the input–output data $\{(U_1, Y_1), \ldots, (U_n, Y_n)\}$. For the *iid* input we can follow (recall that $\lambda_0 = 1$) the developments of Chapter 2 and obtain the following relationships

$$\lambda_j = \frac{\text{cov}(Y_n, U_{n-d-j})}{\text{cov}(Y_n, U_{n-d})}, \quad j = 1, 2, \ldots, \tag{12.63}$$

and

$$E\{Y_n | U_{n-d} = u\} = g(u) + c_0, \tag{12.64}$$

where $c_0 = E\{g(U_0)\}\sum_{j=1}^{\infty}\lambda_j$.

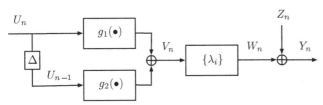

Figure 12.19 The Hammerstein system with the nonlinear dynamics of the additive form.

Formula (12.63) allows us to form a correlation-type estimate of the impulse response sequence of the dynamical subsystem and the corresponding transfer function. On the other hand (12.64) leads to the regression-type nonparametric estimate of $g(u)$.

It is worth noting that such obtained estimates of the system characteristics assume knowledge of the time delay d. If this is not the case, one should estimate d from the training data. For a system without the linear part, that is, when $Y_n = V_{n-d} + Z_n$, we can use the classical correlation theory and estimate d as

$$\hat{d} = \arg \max_{0 \leq s \leq p} \left\{ \left| n^{-1} \sum_{j=s+1}^{n} Y_j U_{j-s} \right| \right\}.$$

The expression under the absolute value is the consistent estimate of $\text{cov}(Y_n, U_{n-s})$, which is equal to zero if $s \neq d$. It is worth noting that the issue of estimating the time delay in discrete time stochastic nonlinear systems has rarely been addressed in the literature.

Example 12.4 (Additive model). Power amplifiers utilized in communication systems work at the saturation point with a nonflat gain over the frequency band and hence must be modeled as a nonlinear dynamical system with a short memory depth. A popular model [258] for this is described by (12.59) and (12.60) with $p = 1$ and

$$m(U_n, U_{n-1}) = g_1(U_n) + g_2(U_{n-1}), \tag{12.65}$$

where the static nonlinearities $g_1(\bullet)$ and $g_2(\bullet)$ are to be estimated from the training data. The system is depicted in Figure 12.19.

In [331] a special case of (12.65) of the following semiparametric form

$$m(U_n, U_{n-1}) = a U_n + g(U_{n-1}) \tag{12.66}$$

has been studied in the context of modeling chemical reactors. Here, one wishes to estimate the parameter a as well as the nonparametric nonlinearity $g(\bullet)$. Note that in [331] the nonlinearity $g(\bullet)$ has been specified to have the parametric polynomial form.

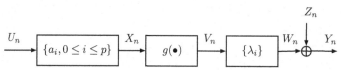

Figure 12.20 The sandwich system with the input linear subsystem of the memory length p.

Example 12.5 (Multiplicative model). An alternative to the additive structure introduced in the previous example is the multiplicative model

$$m(U_n, U_{n-1}) = g_1(U_n) g_2(U_{n-1}). \tag{12.67}$$

This model can be more suited in applications where the inputs are multiplied (like mixing and modulation processes) rather than added. From the approximation theory point of view, the additive function $g_1(u) + g_2(v)$ defines a linear approximation of the function $m(u, v)$, whereas $g_1(u)g_2(v)$ is a bilinear, hence, nonlinear approximation. The latter can often be more accurate.

Example 12.6 (Sandwich model). The model corresponding to (12.59) and (12.60) with

$$m(U_n, U_{n-1}, \ldots, U_{n-p}) = g\left(\sum_{i=0}^{p} a_i U_{n-i}\right) \tag{12.68}$$

is commonly referred to as a sandwich model. In fact, in this model the univariate nonlinearity $g(\bullet)$ is sandwiched between two linear systems. This system plays an important role in a number of applications ranging from communication technology to biological systems. In our case, the first linear subsystem is of a finite memory with the characteristic $\{a_i, 0 \le i \le p\}$, whereas the second one is of an infinite order with the characteristic $\{\lambda_i\}$. This system is depicted in Figure 12.20. The nonparametric identification problem here is to identify the linear subsystems and the nonlinearity $g(\bullet)$ without any parametric assumptions on the system characteristics. It is clear that if $g(\bullet)$ is a linear function then one cannot individually identify each linear subsystem. As we will see in this section the presence of the nonlinearity in the sandwich system helps us to identify all the components of the internal structure from input–output measurements alone.

In the next section, we describe a method for nonparametric identification of the system in (12.59) and (12.60). For simplicity of notation we will mostly focus on the case with $p = 1$. This case is advisable since systems with short input memory depth are probably the most useful due to the already mentioned curse of dimensionality.

As well, in the next two sections, we also describe the basic strategies for recovering the nonlinearity $m(\bullet, \bullet, \ldots, \bullet)$ and the impulse response function $\{\lambda_i\}$ of the Hammerstein system with nonlinear dynamics. We begin with the problem of recovering the nonlinearity followed by the linear subsystem identification.

12.2.2 Identification algorithms: Nonlinear system identification

Let us begin with the case $p = 1$, that is, we wish to recover the function $m(u, v)$ from the statistical characteristics of the input–output process $\{(U_n, Y_n)\}$. To do so let us define the following regression functions

$$r_1(u) = E\{Y_n \mid U_n = u\}, \tag{12.69}$$

and

$$r_2(u, v) = E\{Y_n \mid U_n = u, U_{n-1} = v\}. \tag{12.70}$$

The following theorem gives a fundamental relationship between the nonlinearity $m(u, v)$ appearing in the system defined in (12.59) and (12.60) and the regression functions $r_1(u)$ and $r_2(u, v)$.

THEOREM 12.3 *Let us consider the system in (12.59) and (12.60) with $p = 1$. Let the input signal $\{U_n\}$ be a sequence of iid random variables. Then we have*

$$m(u, v) = r_2(u, v) - \lambda_1 r_1(v) + c, \tag{12.71}$$

where

$$c = E\{m(U_1, U_0)\} \left\{ \lambda_1 + (\lambda_1 - 1) \sum_{i=1}^{\infty} \lambda_i \right\}. \tag{12.72}$$

Proof. By (12.60) and the fact that $\{U_n\}$ is a sequence of independent random variables, we have

$$E\{Y_n \mid U_n = u, U_{n-1} = v\} = m(u, v) + \lambda_1 E\{m(v, U_0)\} + E\{m(U_1, U_0)\} \sum_{i=2}^{\infty} \lambda_i$$

and

$$E\{Y_n \mid U_n = v\} = E\{m(v, U_0)\} + E\{m(U_1, U_0)\} \sum_{i=1}^{\infty} \lambda_i.$$

Eliminating the common term $E\{m(v, U_0)\}$ from the above equations yields the formula in (12.71). The proof of Theorem 12.3 is thus complete. ∎

REMARK 12.5 *Similar derivations as in the proof of Theorem 12.3 can be, in principle, carried out for the system with a longer input memory. Indeed, some algebra shows that for $p = 2$, we have*

$$m(u_0, u_1, u_2) = r_3(u_0, u_1, u_2) - \lambda_1 r_2(u_1, u_2) - \left(\lambda_2 - \lambda_1^2\right) r_1(u_2) + c_1,$$

where $r_3(u_0, u_1, u_2) = E\{Y_n \mid U_n = u_0, U_{n-1} = u_1, U_{n-2} = u_2\}$ and

$$c_1 = E\{m(U_2, U_1, U_0)\}\left\{1 + (\lambda_1(1 - \lambda_1) + \lambda_2 - 1)\sum_{i=0}^{\infty}\lambda_i\right\}.$$

The determination (up to a constant) of the system nonlinearity $m(u_0, u_1, u_2)$ requires the evaluation of three regression functions and the knowledge of λ_1 and λ_2. Generally, one can conjecture that for the system with the memory length p, we have

$$m(u_0, u_1, \ldots, u_p) = r_{p+1}(u_0, u_1, \ldots, u_p) + \sum_{j=1}^{p}\delta_j r_j(u_{p-j+1}, \ldots, u_{p-1}, u_p) + c_p,$$

where δ_j depends on $\lambda_1, \ldots, \lambda_j$ and $r_j(u_0, u_1, \ldots, u_j)$ is the j-dimensional regression function of Y_n on $(U_n, U_{n-1}, \ldots, U_{n-j+1})$.

The result of Theorem 12.3 explains that $m(u, v)$ can be determined from a linear combination of two regression functions. The knowledge of λ_1 and the constant c is also required. This result (which is similar to the requirements of the standard Hammerstein system) is a simple consequence of the cascade nature of the system and the fact that the signal $\{V_n\}$ interconnecting the subsystems is not accessible. The constant c is equal to zero if $E\{m(U_1, U_0)\} = 0$. This takes place if the density of U_n is even, whereas $m(u, v)$ is an odd function with respect to one of the variables, that is, either $m(-u, v) = -m(u, v)$ or $m(u, -v) = -m(u, v)$. Furthermore, often in practice we have $m(0, 0) = 0$ and then we can eliminate the constant c in (12.71), that is, we have

$$m(u, v) = \{r_2(u, v) - r_2(0, 0)\} - \lambda_1\{r_1(v) - r_1(0)\}.$$

Hence, we now need to calculate the regression functions in two points, that is, (u, v) and $(0, 0)$. It is also informative to evaluate the factor

$$\gamma = \lambda_1 + (\lambda_1 - 1)\sum_{i=1}^{\infty}\lambda_i, \tag{12.73}$$

that determines the constant c in (12.72), for some simple dynamical subsystems. Hence, for the AR(1) model

$$W_n = aW_{n-1} + V_n, \quad |a| < 1$$

we have $\lambda_i = a^i$ and $\sum_{i=1}^{\infty}\lambda_i = a(1 - a)^{-1}$ giving $\gamma = 0$. As a result, the constant c is zero. On the other hand, for the AR(2) model

$$W_n = aW_{n-1} + bW_{n-2} + V_n, \quad |a| + |b| < 1,$$

we have $\gamma = -b(1 - a - b)^{-1}$. Similar calculations for the ARMA(1, 1) process

$$W_n = aW_{n-1} + V_n + bV_{n-1}, \quad |a| < 1$$

yield $\gamma = (a + b)b(1 - a)^{-1}$.

Let us now reexamine the formula in (12.71) for the special cases discussed in Examples 12.3–12.6 in Section 12.2.1. In Example 12.3, we introduced the Hammerstein system with the delayed input. Formulas (12.63) and (12.64) provide an explicit prescription for recovering the system characteristics from the cross covariance and the regression function of the output and input signals. This readily yields consistent non-parametric estimates of the system characteristics. The details of these developments can easily be found by the reader. Let us now turn to Example 12.4, concerning an additive nonlinearity.

Example 12.7 (Additive model). Owing to Theorem 12.3 and (12.65), we can write

$$g_1(u) + g_2(v) = r_2(u, v) - \lambda_1 r_1(v) + c, \tag{12.74}$$

where $c = \{Eg_1(U_0) + Eg_2(U_0)\}\gamma$, γ being defined in (12.73). The problem which we face is how to extract the individual nonlinearities in (12.74) from the knowledge of the right-hand side of (12.74). In Chapter 13, we shall introduce the general concept of marginal integration for additive models. Nevertheless, the simplicity of the method allows us to apply it in the present situation. Hence, denoting by $f(\bullet)$ the density of the input signal $\{U_n\}$, the marginal integration strategy begins with integrating formula (12.74) with respect to $f(v)$, that is, we have

$$\int_{-\infty}^{\infty} \{g_1(u) + g_2(v)\} f(v) dv = \int_{-\infty}^{\infty} \{r_2(u, v) - \lambda_1 r_1(v) + c\} f(v) dv. \tag{12.75}$$

By this and noting that $\int_{-\infty}^{\infty} r_2(u, v) f(v) dv = E\{Y_n \mid U_n = u\}$, and $\int_{-\infty}^{\infty} r_1(v) f(v) dv = E\{Y_n\} = \{Eg_1(U_0) + Eg_2(U_0)\} \sum_{i=0}^{\infty} \lambda_i$, we obtain

$$g_1(u) = E\{Y_n \mid U_n = u\} + \alpha_1, \tag{12.76}$$

where

$$\alpha_1 = -\left\{ Eg_1(U_0) \sum_{i=1}^{\infty} \lambda_i + Eg_2(U_0) \left(\sum_{i=1}^{\infty} \lambda_i - 1 \right) \right\}.$$

Integration, as in (12.75), with respect to $f(u)$, and the observation that $\int_{-\infty}^{\infty} r_2(u, v) f(u) du = E\{Y_n \mid U_{n-1} = v\}$ yield

$$g_2(v) = E\{Y_n \mid U_{n-1} = v\} - \lambda_1 E\{Y_n \mid U_n = v\} + \alpha_2, \tag{12.77}$$

where

$$\alpha_2 = Eg_1(U_0)(\gamma - 1) + Eg_2(U_0)\gamma,$$

and γ is defined in (12.73).

Thus, (12.76) and (12.77) provide a simple prescription for recovering the system nonlinearities from the regression functions $E\{Y_n \mid U_n = u\}$ and $E\{Y_n \mid U_{n-1} = v\}$. The constants α_1 and α_2 can be set equal to zero if, for example, $Eg_1(U_0) = Eg_2(U_0) = 0$. The latter holds under some symmetry properties of $g_1(\bullet)$, $g_2(\bullet)$, and $f(\bullet)$. Also if $g_1(0) = g_2(0) = 0$ (the case often met in practice) then we can eliminate the constants in (12.76) and (12.77) and write

$$g_1(u) = E\{Y_n \mid U_n = u\} - E\{Y_n \mid U_n = 0\}, \tag{12.78}$$

and

$$g_2(v) = E\{Y_n \mid U_{n-1} = v\} - E\{Y_n \mid U_{n-1} = 0\}$$
$$-\lambda_1[E\{Y_n \mid U_n = u\} - E\{Y_n \mid U_n = 0\}]. \tag{12.79}$$

The estimates for $g_1(u)$ and $g_2(v)$ are now straightforward to obtain: estimate the regression functions $E\{Y_n \mid U_n = u\}$ and $E\{Y_n \mid U_{n-1} = v\}$ by using, for example, a kernel type estimator, and then plug them into (12.78) and (12.79).

An interesting semiparametric case of the additive system is defined in (12.66), that is, here $g_1(u) = au$ and $g_2(u) = g(v)$. Let $EU_0 = 0$ and $Eg(U_0) = 0$ without loss of generality. Then $g(v)$ can be recovered by (12.77), where $\alpha_2 = 0$. Also

$$au = r_1(u),$$

where we recall that $r_1(u) = E\{Y_n \mid U_n = u\}$. This readily implies that

$$a = \frac{E\{r_1(U_0)U_0\}}{E\{U_0^2\}}. \tag{12.80}$$

This gives the following generic estimate of a

$$\hat{a} = \frac{n^{-1}\sum_{i=1}^{n} U_i \hat{r}_1(U_i)}{n^{-1}\sum_{i=1}^{n} U_i^2}, \tag{12.81}$$

where $\hat{r}_1(u)$ is a nonparametric estimate (for example, the kernel estimate) of $r_1(u)$. Note that the averaging in (12.81) drastically reduces the variance of the nonparametric estimate and allows us to show that $\hat{a}$ converges to the true a with the optimal parametric $\sqrt{n}$ rate.

Example 12.8 (Multiplicative model). For the multiplicative model introduced in Example 12.5, formula (12.71) of Theorem 12.3 takes the following form:

$$g_1(u)g_2(v) = r_2(u, v) - \lambda_1 r_1(v) + c.$$

Reasoning as in Example 12.7, that is, by taking the marginal integration of the above equation with respect to $f(v)$ and $f(u)$, respectively, we can easily obtain

$$g_1(u) = a_1 E\{Y_n \mid U_n = u\} + \beta_1,$$
$$g_2(v) = a_2[E\{Y_n \mid U_{n-1} = v\} - \lambda_1 E\{Y_n \mid U_n = v\}] + \beta_2,$$

where $a_1 = 1/Eg_2(U_0)$, $a_2 = 1/Eg_1(U_0)$, $\beta_1 = -Eg_1(U_0)\sum_{i=1}^{\infty}\lambda_i$, $\beta_2 = Eg_2(U_0)\gamma$. Here, we must assume that $Eg_1(U_0) \neq 0$ and $Eg_2(U_0) \neq 0$.

Hence, we recover the nonlinearities up to multiplicative and additive constants. Some a priori knowledge about $g_1(u)$ and $g_2(v)$ allows us to eliminate some of these constants.

Example 12.9 (Sandwich model). For the sandwich model in (12.68) with $p = 1$ and $a_0 = 1$, we can write formula (12.71) of Theorem 12.3 in the following form

$$g(u + av) = r_2(u, v) - \lambda_1 r_1(v) + c$$

or equivalently

$$g(u) = r_2(u - av, v) - \lambda_1 r_1(v) + c, \tag{12.82}$$

where $c = \gamma E\{g(U_1 + aU_0)\}$, γ being defined in (12.73).

The formula in (12.82) holds for any v and one could choose $v = 0$, that is, we have

$$g(u) = r_2(u, 0) - \lambda_1 r_1(0) + c.$$

Under the commonly met condition

$$g(0) = 0, \tag{12.83}$$

this gives the following reconstruction formula

$$g(u) = r_2(u, 0) - r_2(0, 0). \tag{12.84}$$

This seems to be an attractive and simple way of estimating $g(u)$ via the two-dimensional regression function $r_2(u, v)$. Nevertheless, the statistical efficiency of such an approach is reduced because we can show that for twice differentiable nonlinearities we have the rate $O_P\left(n^{-1/3}\right)$, instead of the optimal univariate rate $O_P\left(n^{-2/5}\right)$ (see Chapter 13 for a discussion of the problem of estimating multivariate functions). In order to obtain an improved rate, we must apply the average operation in (12.82). In fact, integrating (12.82) with respect to $f(v)$, we obtain

$$g(u) = E\{r_2(u - aU_0, U_0)\} - \lambda_1 E\{r_1(U_0)\} + c$$
$$= E\{r_2(u - aU_0, U_0)\} + c_1, \tag{12.85}$$

where $c_1 = -Eg(U_1 + aU_0) \sum_{i=1}^{\infty} \lambda_i$.

Note also that under condition (12.83), we can rewrite the formula in (12.85) as follows:

$$g(u) = E\{r_2(u - aU_0, U_0)\} - E\{r_2(-aU_0, U_0)\}. \tag{12.86}$$

The theoretical average operation appearing in (12.85) (or (12.86)) can easily be replaced by an empirical average, that is, $E\{r_2(u - aU_0, U_0)\}$ is estimated by

$$n^{-1} \sum_{i=1}^{n} \hat{r}_2(u - aU_i, U_i),$$

where $\hat{r}_2(u, v)$ is a certain nonparametric estimate of the regression function $r_2(u, v)$.

In Chapter 13, we will demonstrate that the average operation is able to reduce the variance of a nonparametric estimate of a multivariate regression function and therefore improve the rate of convergence. Summing up we can fully recover the nonlinearity in the sandwich system (under condition (12.83)) by applying the following estimate:

$$\hat{g}(u) = n^{-1} \sum_{i=1}^{n} \{\hat{r}_2(u - \hat{a}U_i, U_i) - \hat{r}_2(-\hat{a}U_i, U_i)\}, \tag{12.87}$$

where $\hat{a}$ is a consistent estimate of the parameter a. We shall discuss in Section 12.2.3 how to estimate the linear subsystems of the generalized Hammerstein system with dynamic nonlinearity. It is worth noting that the formula for $g(u)$ in (12.84) does not depend on the characteristics of the linear subsystems. We can conjecture that the estimate in (12.87) can have the proper $O_P\left(n^{-2/5}\right)$ rate of convergence.

Thus far we have considered a simple case of the sandwich model when the input dynamical subsystem is of the FIR(1) type. Remark 12.5 provides the solution which covers the FIR(2) type subsystem. In fact, we have that

$$g(u_0 + a_1 u_1 + a_2 u_2) = r_3(u_0, u_1, u_2) - \lambda_1 r_2(u_1, u_2) - \left(\lambda_2 - \lambda_1^2\right) r_1(u_2) + c_1, \tag{12.88}$$

where the constant c_1 is defined in Remark 12.5. Similarly as in (12.85) we can apply the average operation (with respect to $f(u_1)f(u_2)$) to (12.88), that is, we can obtain

$$g(u) = E\{r_3(u - a_1 U_1 - a_2 U_2, U_1, U_2)\}$$
$$- \lambda_1 E\{r_2(U_1, U_2)\} - \left(\lambda_2 - \lambda_1^2\right) E\{r_1(U_2)\} + c_1.$$

Next, using $E\{r_2(U_1, U_2)\} = E\{r_1(U_2)\} = E\{Y_n\}$ and simple algebra yield

$$g(u) = E\{r_3(u - a_1 U_1 - a_2 U_2, U_1, U_2)\} + c_2, \tag{12.89}$$

where $c_2 = -E\{g(U_0 + a_1 U_1 + a_2 U_2)\} \sum_{i=1}^{\infty} \lambda_i$. Hence, the solution is a direct analog of the case in which $p = 1$ given in (12.85). This allows us to form the following fundamental result concerning the recovery of the nonlinearity in the sandwich system with the input linear subsystem of the memory length p.

THEOREM 12.4 *Let us consider the sandwich system*

$$\begin{cases} X_n = U_n + \sum_{t=1}^{p} a_t U_{n-t}, \\ V_n = g(X_n), \\ Y_n = \sum_{j=0}^{\infty} \lambda_j V_{n-j} + Z_n \end{cases}$$

with $1 \le p < \infty$. Let the input signal $\{U_n\}$ be a sequence of iid random variables. Then, we have

$$g(v) = Er_{p+1}\left(v - \sum_{t=1}^{p} a_t U_{n-t}, U_{n-1}, \ldots, U_{n-p}\right) + c_p, \tag{12.90}$$

where

$$c_p = -E g\left(X_n\right) \sum_{i=1}^{\infty} \lambda_i$$

and

$$r_{p+1}(u_0, u_1, u_2, \ldots, u_p) = E\{Y_n \mid U_n = u_0, U_{n-1} = u_1, \ldots, U_{n-p} = u_p\}$$

is the $(p + 1)$-dimensional regression function.

The procedure for estimating $g(u)$ is now clear. Using (12.90) in Theorem 12.4 and assuming that condition (12.83) is met, we have the following analog of (12.87)

$$\hat{g}(u) = n^{-1} \sum_{\ell=p+1}^{n} \left\{ \hat{r}_{p+1}\left(u - \sum_{i=1}^{p} \hat{a}_i U_{\ell-i}, U_{\ell-1}, \ldots, U_{\ell-p}\right) \right.$$
$$\left. - \hat{r}_{p+1}\left(-\sum_{i=1}^{p} \hat{a}_i U_{\ell-i}, U_{\ell-1}, \ldots, U_{\ell-p}\right) \right\}, \qquad (12.91)$$

where $\hat{r}_{p+1}(u_0, u_1, \ldots, u_p)$ is a nonparametric estimate of the regression function $r_{p+1}(u_0, u_1, u_2, \ldots, u_p)$, and $\{\hat{a}_i, 1 \leq i \leq p\}$ is an estimate of the input linear subsystem $\{a_i, 1 \leq i \leq p\}$. Again, the average operation can alleviate the variance of $\hat{r}_{p+1}(u_0, u_1, \ldots, u_p)$ such that $\hat{g}(u)$ can have the univariate optimal rate $O_P\left(n^{-2/5}\right)$ independent of the input subsystem memory length p.

REMARK 12.6 *The special case of the sandwich system is the Wiener model. In fact, we have $\lambda_j = \delta_{j0}$ and the corresponding input–output relationship is the following:*

$$Y_n = g\left(U_n + \sum_{i=1}^{p} a_i U_{n-i}\right) + Z_n. \qquad (12.92)$$

In this case, the constant c_p in (12.90) is equal to zero and, due to (12.90), we have

$$g(u) = E r_{p+1}\left(u - \sum_{i=1}^{p} a_i U_i, U_1, U_2, \ldots, U_p\right).$$

It should be noted that this formula can be directly derived from (12.92). Consequently, the estimate of $g(u)$ is given (from (12.91)) by

$$\hat{g}(u) = n^{-1} \sum_{\ell=p+1}^{n} \hat{r}_{p+1}\left(u - \sum_{i=1}^{p} \hat{a}_i U_{\ell-i}, U_{\ell-1}, \ldots, U_{\ell-p}\right). \qquad (12.93)$$

This gives the estimate of the system nonlinearity that is believed to have the optimal rate of convergence. Note that the estimate in (12.93) does not assume that $g(u)$ is invertible, that the input signal is Gaussian, and that the measurement noise Z_n is zero. These are important assumptions which were employed in the inverse regression *strategy examined thoroughly in Chapters 9–11. Nevertheless, the inverse regression method was applied for a general linear subsystem, whereas the estimate in (12.93) works only for the pth order moving average process. In Chapter 14, we will propose much more efficient methods for estimation of the Wiener system being parametrized by an univariate nonparametric function $g(\bullet)$ and a finite dimensional vector of parameters $\{a_i, 1 \leq i \leq p\}$. This semiparametric nature of the system will be greatly exploited and extended to a large class of semiparametric nonlinear systems.*

12.2.3   Identification algorithms: Linear system identification

In this section, we wish to address the issue of identification of the linear subsystem $\{\lambda_j\}$ of the generalized Hammerstein model defined in (12.59) and (12.60). This is the part of our system that gives infinite memory effects and its recovery is an important problem. We will make use of the theory of cumulants, which is thoroughly examined in [32].

Let $\mathbf{X} = (X_1, \ldots, X_p)^T$ be a p-dimensional random vector. The pth order cumulant, $\mathrm{cum}(X_1, \ldots, X_p)$, of $\mathbf{X}$ is the coefficient of the term $i^p t_1 t_2 \cdots t_p$ in the Taylor series expansion of the cumulant-generating function $\log E\{e^{i t^T \mathbf{X}}\}$, where $\mathbf{t} = (t_1, t_2, \ldots, t_p)^T$. The cumulants can be viewed as a generalization of classical moments with a number of useful invariant properties which are not shared by classical moments. Here we list the properties that are the most useful and directly applicable for the nonlinear system under consideration.

1. If $a_1, \ldots, a_p$ are constants, then we have

$$\mathrm{cum}(a_1 X_1, \ldots, a_p X_p) = a_1 \cdots a_p \, \mathrm{cum}(X_1, \ldots, X_p).$$

2. If any subset of the vector $(X_1, \ldots, X_p)^T$ is independent of those remaining, then

$$\mathrm{cum}(X_1, \ldots, X_p) = 0.$$

3. Cumulants are additive in their arguments, that is, for any random variables U and V, we have

$$\mathrm{cum}(U + V, X_1, \ldots, X_p) = \mathrm{cum}(U, X_1, \ldots, X_p) + \mathrm{cum}(V, X_1, \ldots, X_p).$$

In particular:

$$\mathrm{cum}(X_1 + U, \ldots, X_p) = \mathrm{cum}(X_1, \ldots, X_p) + \mathrm{cum}(U, X_2, \ldots, X_p)$$

and

$$\mathrm{cum}(X_1 + a, \ldots, X_p) = \mathrm{cum}(X_1, \ldots, X_p),$$

where a is a constant.

4. If $(X_1, \ldots, X_p)^T$ is independent of $(Z_1, \ldots, Z_p)^T$, then

$$\mathrm{cum}(X_1 + Z_1, \ldots, X_p + Z_p) = \mathrm{cum}(X_1, \ldots, X_p) + \mathrm{cum}(Z_1, \ldots, Z_p).$$

Note that $\text{cum}(X_1, \ldots, X_p)$ is well defined if $E \mid X_1 \cdots X_p \mid < \infty$, which is implied by $E \mid X_i \mid^p < \infty, i = 1, \ldots, p$. It is also worth noting that $\text{cum}(X_1, \ldots, X_p)$ is symmetric in its arguments. The second-, third-, and fourth-order cumulants are given by

$$\text{cum}(X_1, X_2) = E\{(X_1 - EX_1)(X_2 - EX_2)\},$$
$$\text{cum}(X_1, X_2, X_3) = E\{(X_1 - EX_1)(X_2 - EX_2)(X_3 - EX_3)\}, \quad (12.94)$$
$$\text{cum}(X_1, X_2, X_3, X_4) = E\{(X_1 - EX_1)(X_2 - EX_2)(X_3 - EX_3)(X_4 - EX_4)\}$$
$$- \text{cum}(X_1, X_2)\,\text{cum}(X_3, X_4)$$
$$- \text{cum}(X_1, X_3)\,\text{cum}(X_2, X_4)$$
$$- \text{cum}(X_1, X_4)\,\text{cum}(X_2, X_3).$$

Hence, the second-order cumulant is just the covariance between two random variables, whereas the third-order cumulant is the third-order centered moment of three random variables.

An important special case of the aforementioned discussion occurs when $X_j = X$ for all j. It is also well known that cumulants can measure the degree of nonnormality since the cumulants of degree greater than two for a Gaussian random process vanish.

For the stationary random process $\{X_n\}$ recorded at the time instants $t, t - \tau_1, \ldots, t - \tau_p$, we have

$$\text{cum}(X_t, X_{t-\tau_1}, \ldots, X_{t-\tau_p}) = R_{X\ldots X}(\tau_1, \tau_2, \ldots, \tau_p). \quad (12.95)$$

Hence, the pth order cumulant of the stationary process is a function of $p - 1$ variables. In particular, we obtain

$$\text{cum}\,(X_t, X_{t-\tau}) = R_{XX}(\tau), \quad \text{cum}\,(X_t, X_{t-\tau_1}, X_{t-\tau_2}) = R_{XXX}(\tau_1, \tau_2), \quad (12.96)$$

where $R_{XX}(\tau)$ and $R_{XXX}(\tau_1, \tau_2)$ are the covariance function and the third-order cumulant function of the stationary stochastic process $\{X_n\}$, respectively.

Let us now apply (without loss of generality) the cumulant theory to our system of order $p = 1$, see (12.59) and (12.60). Hence let us consider $\text{cum}(Y_{n+\ell}, U_n, U_{n-1})$ for $\ell \geq 1$. Since $Y_n = W_n + Z_n$ and due to Properties 2 and 3 of cumulants, we have

$$\text{cum}(Y_{n+\ell}, U_n, U_{n-1}) = \text{cum}(W_{n+\ell}, U_n, U_{n-1}).$$

By this formula, Properties 2 and 3, and the fact that $W_n = \sum_{j=0}^{\infty} \lambda_j m(U_{n-j}, U_{n-j-1})$, we obtain

$$\text{cum}(Y_{n+\ell}, U_n, U_{n-1}) = \text{cum}(\lambda_{\ell-1} m(U_{n+1}, U_n), U_n, U_{n-1})$$
$$+ \text{cum}(\lambda_\ell m(U_n, U_{n-1}), U_n, U_{n-1})$$
$$+ \text{cum}(\lambda_{\ell+1} m(U_{n-1}, U_{n-2}), U_n, U_{n-1}). \quad (12.97)$$

Since U_{n-1} is independent of (U_n, U_{n+1}) then due to Property 2, the first term in (12.97) is equal to zero. Similarly, the independence of U_n of (U_{n-1}, U_{n-2}) sets the last term in (12.97) to zero. The application of Property 1 to the second term in (12.97) yields

$$\text{cum}(Y_{n+\ell}, U_n, U_{n-1}) = \lambda_\ell\,\text{cum}(m(U_n, U_{n-1}), U_n, U_{n-1}). \quad (12.98)$$

Note that the above formula is well defined if $E \mid U_0 U_1 m(U_1, U_0) \mid < \infty$, or $E \mid U_0 \mid^3 < \infty$ and $E \mid m(U_1, U_0) \mid^3 < \infty$, which is a stronger requirement.

In order to eliminate the multiplicative constant in (12.98) let us recall that $\lambda_0 = 1$. Then, by analogous considerations to those presented above, we get

$$\text{cum}(Y_n, U_n, U_{n-1}) = \text{cum}(m(U_n, U_{n-1}), U_n, U_{n-1}). \tag{12.99}$$

It is, however, important to note that under typical symmetry assumptions on $m(\bullet, \bullet)$ (if it is an odd function in both arguments) and $f(\bullet)$ (if it is an even function) the expression on the right-hand side of (12.99) is equal to zero. To circumvent this difficulty, we observe that (12.98) holds also for nonlinear transformations of U_{n-1} and U_n. For the most practical situations the quadratic transformation is sufficient. Hence, assuming that $E U_n^2 < \infty$, we have

$$\text{cum}(Y_{n+\ell}, U_n^2, U_{n-1}^2) = \lambda_\ell \, \text{cum}(m(U_n, U_{n-1}), U_n^2, U_{n-1}^2), \quad \ell \geq 0.$$

This gives the following formula for λ_ℓ:

$$\lambda_\ell = \frac{\text{cum}(Y_{n+\ell}, U_n^2, U_{n-1}^2)}{\text{cum}(Y_n, U_n^2, U_{n-1}^2)}. \tag{12.100}$$

For a general nonlinear subsystem with a memory length of order p, we have, by analogous considerations, the following formula for λ_ℓ:

$$\lambda_\ell = \frac{\text{cum}(Y_{n+\ell}, U_n^2, \ldots, U_{n-p}^2)}{\text{cum}(Y_n, U_n^2, \ldots, U_{n-p}^2)}. \tag{12.101}$$

The aforementioned developments are summarized in the following theorem.

THEOREM 12.5 *Let us consider the nonlinear system in (12.59) and (12.60). Let the input signal $\{U_n\}$ be a sequence of iid random variables. If*

$$E \left\{ \mid m(U_n, \ldots, U_{n-p}) \mid U_n^2 \cdots U_{n-p}^2 \right\} < \infty, \tag{12.102}$$

then we have

$$\lambda_\ell = \frac{\text{cum}(Y_{n+\ell}, U_n^2, \ldots, U_{n-p}^2)}{\text{cum}(Y_n, U_n^2, \ldots, U_{n-p}^2)}, \quad \ell = 1, 2, \ldots.$$

Note that the condition in (12.102) is implied by

$$E \left\{ \mid m(U_n, \ldots, U_{n-p}) \mid^{p+2} \right\} < \infty \quad \text{and} \quad E \mid U_n \mid^{2p+4} < \infty. \tag{12.103}$$

Theorems 12.3 and 12.5 give a complete solution for the identification problem of the nonlinear system defined in (12.59) and (12.60) in the case of $p = 1$. The extension to the nonlinearity with larger memory length is possible by combining the result of Theorems 12.5 and the result established in Remark 12.5. Nevertheless, as we have already mentioned, it is not recommended to use the nonlinearity with large memory length due to the curse of dimensionality. In most practical cases the use of $p = 1$ or $p = 2$ is sufficient to grasp the essential nonlinearity effects of the underlying physical system. In the case $p = 1$, see (12.71), the nonlinear recovery needs the knowledge of λ_1. With the result of Theorem 12.5, we have

$$\lambda_1 = \frac{\text{cum}(Y_{n+1}, U_n^2, U_{n-1}^2)}{\text{cum}(Y_n, U_n^2, U_{n-1}^2)}. \tag{12.104}$$

Note that due to (12.94)

$$\text{cum}(Y_{n+1}, U_n^2, U_{n-1}^2) = E\left\{(U_{n-1}^2 - EU_{n-1}^2)(U_n^2 - EU_n^2)(Y_{n+1} - EY_{n+1})\right\}.$$

Hence, to form an estimate of $\{\lambda_\ell\}$ based on the result of Theorem 12.5 is straightforward. Indeed, we ought to replace the moments appearing in the formula for cumulants by their empirical counterparts. In fact, an estimate of λ_ℓ resulting from Theorem 12.5 takes the following form:

$$\hat\lambda_\ell = \frac{n^{-1}\sum_{t=2}^{n-\ell}(U_{t-1}^2 - \bar U_n^2)(U_t^2 - \bar U_n^2)(Y_{t+\ell} - \bar Y_n)}{n^{-1}\sum_{t=2}^{n}(U_{t-1}^2 - \bar U_n^2)(U_t^2 - \bar U_n^2)(Y_t - \bar Y_n)},$$

where $\bar U_n^2$ and $\bar Y_n$ are empirical counterparts of the mean values EU_n^2 and EY_n, respectively.

Hence, Theorems 12.3 and 12.5 lead to fully nonparametric methods for estimating the system characteristics. Nevertheless, in the examples introduced in Section 12.2.1 we have faced the semiparametric problem of estimating additional parameters which specify the low dimensional form of the system nonlinearity. This is a particularly important issue for the case of the sandwich system and its special case, the Wiener system. An important case study that follows concerns the sandwich system. Here the $(p + 1)$-dimensional system nonlinearity, see (12.68), is characterized by the one dimensional function $g(\bullet)$ and the linear projection of the last p observations of the input signal, that is, $U_n + \sum_{j=1}^p a_j U_{n-j}$. The latter defines the input linear subsystem of the sandwich model with the memory length p.

Example 12.10 (Sandwich system). In Theorem 12.4, we provided a full characterization of the sandwich system nonlinearity in terms of the p-dimensional projection (obtained via the process of marginal integration) of the $(p + 1)$-dimensional regression function, see formula (12.90). This formula, however, depends on the characteristic of the input linear system $\{a_i, 0 \le i \le p\}$. The semiparametric nature of the problem allows us to use the methodology, which will be thoroughly studied in Chapter 14. The main idea of this approach, however, can now be easily illustrated for the sandwich system.

For simplicity of presentation, let us assume that $p = 1$ and $Eg(X_n) = 0$, that is, the constant c_1 in (12.85) is equal to zero. Then, we have (see (12.85))

$$g(v) = Er_2(v - a^*U_{n-1}, U_{n-1}),$$

where a^* denotes the true value of the impulse response of the input linear system, that is, $X_n = U_n + a^*U_{n-1}$. Also let us recall that $r_2(u, v) = E\{Y_n \mid U_n = u, U_{n-1} = v\}$.

Denoting

$$g(v, a) = Er_2(v - aU_{n-1}, U_{n-1}), \tag{12.105}$$

for some a we observe that $g(v, a^*) = g(v)$. Let also $X_n(a) = U_n + aU_{n-1}$ such that $X_n(a^*) = X_n$. The semiparametric strategy begins with forming a pilot estimate of the function $g(v, a)$ for a given a, that is, we pretend that the parameter a is known. A generic estimate of $g(v, a)$ can easily be written as follows (see (12.87)):

$$\hat{g}(v, a) = n^{-1} \sum_{i=2}^{n} \hat{r}_2(v - aU_{i-1}, U_{i-1}), \tag{12.106}$$

where $\hat{r}_2(u, v)$ is any consistent nonparametric estimate of the regression function $r_2(u, v)$.

The estimate $\hat{g}(v, a)$ in (12.106) can be used to form the prediction of the output of the system nonlinearity $g(\bullet)$, that is, the process $\{\hat{V}_t(a) = \hat{g}(X_t(a), a)\}$ predicts (for a given a) the signal $\{V_t\}$. Unfortunately, the process $\{V_t\}$ is not observed and in order to form the predictive error criterion we must pass $\{\hat{V}_t(a)\}$ through the linear system $\{\lambda_j\}$. This leads to the following definition of the predictive error:

$$\hat{Q}_n(a) = n^{-1} \sum_{t=2}^{n} \{Y_t - \hat{Y}_t(a)\}^2, \tag{12.107}$$

where

$$\hat{Y}_t(a) = \sum_{j=0}^{\infty} \lambda_j \hat{V}_{t-j}(a). \tag{12.108}$$

It is natural now to estimate a^* as follows:

$$\hat{a} = \arg\min_a \hat{Q}_n(a).$$

Note that in the definition of the predictive error in (12.107) we used the true value $\{\lambda_j\}$ of the impulse response function of the output linear system. Theorem 12.5 describes the cumulant based method for estimating $\{\lambda_j\}$ that is entirely independent of the remaining characteristics of the system. This could lead to the modified version of $\hat{Y}_t(a)$ in (12.108), that is,

$$\hat{Y}_t(a) = \sum_{j=0}^{N} \hat{\lambda}_j \hat{V}_{t-j}(a), \tag{12.109}$$

where N is the truncation parameter defining how many λ_js should be taken into account.

It is worth noting that the aforementioned semiparametric approach can easily be extended to the case of the input linear system with the memory length p. In fact, we must use the formula in (12.90) in Theorem 12.4 instead of (12.105), that is, we can write (12.90) in the following vector form:

$$g(v, \mathbf{a}) = Er_{p+1}(v - \mathbf{a}^T\mathbf{U}_n, \mathbf{U}_n), \tag{12.110}$$

where $\mathbf{U}_n = (U_{n-1}, \ldots, U_{n-p})^T$. Here the parameter $\mathbf{a} = (a_1, \ldots, a_p)^T$ characterizes the class of input linear systems such that $\mathbf{a}^*$ is the true value of the impulse response of the linear system. Consequently, we can form the prediction error as in (12.107) and (12.109) with the estimate in (12.106) taking the following form:

$$\hat{g}(v, \mathbf{a}) = n^{-1} \sum_{i=p+1}^{n} \hat{r}_{p+1}(v - \mathbf{a}^T \mathbf{U}_i, \mathbf{U}_i).$$

Also note that in this case the process $\{\hat{V}_t(\mathbf{a}), p+1 \leq t \leq n\}$ is given by $\hat{V}_t(\mathbf{a}) = \hat{g}(X_t(\mathbf{a}), \mathbf{a})$ with $X_t(\mathbf{a}) = U_t + \mathbf{a}^T \mathbf{U}_t$. An estimate of $\mathbf{a}^*$ is then obtained as the minimizer of the predictive error $\hat{Q}_n(\mathbf{a})$ defined analogously to (12.107). This type of the semiparametric least squares error estimate is generally difficult to calculate for a large value of p since we need a numerical optimization procedure for minimizing the criterion $\hat{Q}_n(\mathbf{a})$, which is typically not a convex function of $\mathbf{a}$.

In Section 14.7.3, we propose a direct estimate of $\mathbf{a}^*$ using the theory of the average derivative estimate of the multiple-regression function.

12.2.4 Identification algorithms: the Gaussian input signal

In Chapters 9–11 we have developed efficient nonparametric estimates of the Wiener system based on the assumption that the input signal is Gaussian. This restriction can also be efficiently used in the context of the Hammerstein system with nonlinear dynamics. Indeed, thus far we have assumed the white noise input signal and we have proposed the cumulant based approach to identify the linear part of the system. On the other hand, the nonlinear part was recovered by the regression method and has mostly been confined to the system with short memory length.

In this section, we wish to take an advantage of a restricted class of input signals, that is, we assume that the input signal $\{U_n\}$ to the system pictured in Figure 12.17 is a stationary Gaussian process with the covariance function $R_{UU}(\tau)$. We will examine both linear and nonlinear parts of our system and we start with the former.

Linear system identification

The development of identification algorithms for the linear part of the system in (12.59) and (12.60) will make use of several identities which relate the cross-covariance of input and output signals with the covariance of the input signal alone. We begin with the following important result due, originally, to Brillinger [31]. This result concerns an identity for the cumulant function of a Gaussian random vector and its nonlinear transformations.

LEMMA 12.2 *(Brillinger) Let* $(X_1, \ldots, X_J, Y_1, \ldots, Y_K)^T$ *be a nonsingular Gaussian random vector with* $X_1, \ldots, X_J$ *being independent of each other. Let* $G : R^J \rightarrow R$ *be a measurable function of J variables such that*

$$E\{| G(X_1, \ldots, X_J)Y_1 \cdots Y_K |\} < \infty.$$

Then, we have

$$\text{cum}(G(X_1, \ldots, X_J), Y_1, \ldots, Y_K) = \sum_{s=1}^{J} \alpha_s \frac{\text{cov}(X_s, Y_1) \cdots \text{cov}(X_s, Y_K)}{(\text{var}(X_s))^K}, \quad (12.111)$$

where $\alpha_s = \text{cum}(G(X_1, \ldots, X_J), X_s, \ldots, X_s)$.

Proof. We will make use of Properties 1–4 of cumulants listed at the beginning of this section. Also let us recall the following property of the multivariate normal distribution. For $\mathbf{Z} \sim N_d(\mathbf{m}, \Sigma)$ there exists a nonsingular $d \times d$ matrix $\mathbf{C}$ such that $\Sigma = \mathbf{C}\mathbf{C}^T$ and $\mathbf{Z} = \mathbf{C}\mathbf{e} + \mathbf{m}$, where $\mathbf{e} \sim N_d(0, \mathbf{I})$, that is, the components of $\mathbf{e}$ are independent with the $N(0, 1)$ distribution. Moreover, $\mathbf{C}$ can be specified as a lower triangular matrix.

First let us note that since the cumulants are blind to an additive constant (see Property 3), we can assume, without loss of generality, that $(X_1, \ldots, X_J, Y_1, \ldots, Y_K)^T$ is of a zero mean random vector with the $J + K \times J + K$ covariance matrix Σ. Then we can represent the random variables $X_r (r = 1, \ldots, J)$ and $Y_s (s = 1, \ldots, K)$ in terms of the $J + K \times J + K$ matrix $\mathbf{C} = \{c_{i,j}\}$ and the vector $(e_1, e_2, \ldots, e_{J+K})^T$ with independent $N(0, 1)$ elements, as follows:

$$X_r = c_{r,r} e_r \quad (12.112)$$

and

$$Y_s = \sum_{\ell=1}^{J+s} c_{J+s,\ell} e_\ell, \quad (12.113)$$

where formula (12.112) reflects the fact that X_i are independent. Owing to Properties 2 and 3 of cumulants, independence of $\{e_j\}$, and (12.112) and (12.113), we can write

$$\text{cum}(G(X_1, \ldots, X_J), Y_1, \ldots, Y_K)$$

$$= \sum_{s=1}^{J} \text{cum}(G(X_1, \ldots, X_J), c_{J+1,s} e_s, \ldots, c_{J+K,s} e_s)$$

$$= \sum_{s=1}^{J} \text{cum}\left(G(X_1, \ldots, X_J), \frac{c_{J+1,s}}{c_{s,s}} X_s, \ldots, \frac{c_{J+K,s}}{c_{s,s}} X_s\right). \quad (12.114)$$

Let us write the matrix $\mathbf{C}$ in an explicit form

$$\mathbf{C} = \begin{bmatrix} \mathbf{C}_1 & 0 \\ \mathbf{D} & \mathbf{C}_2 \end{bmatrix},$$

where $\mathbf{C}_1$ and $\mathbf{C}_2$ are $J \times J$ and $K \times K$ lower triangular matrices, whereas $\mathbf{D}$ is the $K \times J$ matrix. Then, we note that $\mathbf{C}_1 = \text{diag}\{c_{s,s}, 1 \leq s \leq J\}$ and due to the representation $\Sigma = \mathbf{C}\mathbf{C}^T$, we obtain

$$\text{cov}\{(X_1, \ldots, X_J), (Y_1, \ldots, Y_K)\} = \mathbf{C}_1 \mathbf{D}^T.$$

Thus, we have that $c_{J+1,s}c_{s,s} = \text{cov}(X_s, Y_1), \ldots, c_{J+K,s}c_{s,s} = \text{cov}(X_s, Y_K)$, and, consequently, we can rewrite (12.114) as follows:

$$\sum_{s=1}^{J} \text{cum}(G(X_1, \ldots, X_J), X_s, \ldots, X_s) \frac{\text{cov}(X_s, Y_1) \cdots \text{cov}(X_s, Y_K)}{(c_{s,s}^2)^K}.$$

Noting that $c_{s,s}^2 = \text{var}(X_s)$ we can arrive to the formula in (12.111).

The proof of Lemma 12.2 is thus complete. ∎

It is worth noting that the $K + 1$ order cumulant on the left-hand side of (12.111) measures the multi-linear dependence between the Gaussian vector $(Y_1, \ldots, Y_K)^T$ and the nonlinearity $G(X_1, \ldots, X_J)$, which is a transformation of the another Gaussian vector $(X_1, \ldots, X_J)^T$. Hence, the formula in (12.111) says that the cumulant is proportional to the product of the covariance functions between the Gaussian vectors. The sequence $\{\alpha_s\}$ appearing in (12.111) is a combination of the moments of the form $E\{G(X_1, \ldots X_J)X_s^q\}$, $1 \leq q \leq K$ and can be explicitly evaluated in terms of the average derivatives of $G(X_1, \ldots, X_J)$. In the context of the nonlinear system analysis, the result of Lemma 12.2 is particularly useful for block-oriented nonlinear systems which have single-variable nonlinearities. This will be explained in the remarks given below.

REMARK 12.7 *The result of Lemma 12.2, established by Brillinger [31], generalizes a number of known identities concerning correlation properties of a stochastic process observed as an output of a nonlinear system fed by a stationary Gaussian process. In this remark we give the most classical version of (12.111) often referred to as Bussgang's theorem.*

Hence, let $\{U_n\}$ be an univariate stationary Gaussian process with the covariance function $R_{UU}(\tau)$ and $EU_n = 0$. Let also

$$V_n = g(U_n) + Z_n,$$

where $\{Z_n\}$ is independent, but otherwise arbitrary, of $\{U_n\}$ stochastic process, and $g(\bullet)$ is a measurable function. Consider the cross-covariance of $\{V_n\}$ and $\{U_n\}$

$$R_{VU}(\tau) = \text{cov}(V_n, U_{n-\tau}) = \text{cov}(g(U_\tau), U_{n-\tau}). \tag{12.115}$$

Recalling that the second order cumulant is the covariance function, we recognize that the right-hand side of (12.115) is the simplest version of (12.111) corresponding to $J = K = 1$, where Y_1 is identified as $U_{n-\tau}$, whereas X_1 as U_n. Thus, we can readily obtain

$$R_{VU}(\tau) = \text{cov}(g(U_0), U_0,)\frac{R_{UU}(\tau)}{\text{var}(U_0)}. \tag{12.116}$$

Let $\{U_n\}$ have the marginal density $f(u) \sim N(0, \sigma_U^2)$. The following identity can be easily verified

$$uf(u) = -\sigma_U^2 f^{(1)}(u). \tag{12.117}$$

This allows us to evaluate the coefficient $\text{cov}(g(U_0), U_0) = E\{g(U_0)U_0)\}$ in (12.116). Indeed, assuming that the nonlinearity $g(\bullet)$ is differentiable, and using (12.117) along *with integration by parts, we obtain*

$$E\{g(U_0)U_0\} = \sigma_U^2 E\{g^{(1)}(U_0)\}.$$

Consequently, we have the following version of (12.116)

$$R_{VU}(\tau) = \alpha R_{UU}(\tau), \tag{12.118}$$

where $\alpha = E\{g^{(1)}(U_0)\}$.

   Formula (12.118) is a classic result due to Bussgang; see Papoulis and Pillai [225] for a direct proof of this identity. This is also called the invariance property, meaning that regardless of the nonlinearity applied, the input and input–output covariance functions are identical up to a scale factor. In Barrett and Lampard [15] and Leipnik [194] various extensions of (12.118) to non-Gaussian processes were given. This theory is based on the concept of diagonal expansions of bivariate densities with respect to an orthonormal complete system of functions. In the case of a bivariate Gaussian distribution the Hermite polynomials appear in the orthogonal representation. In Nuttall [222] the most general class of stochastic processes (separable processes) for which the invariance property holds was established.

REMARK 12.8 *In considerations concerning the sandwich system and other nonlinear problems based on the third order cumulants we need the version of Lemma 12.2 corresponding to $J = 1$ and $K = 2$. Thus (12.111) takes the form:*

$$\mathrm{cum}(G(X_1), Y_1, Y_2) = \mathrm{cum}(G(X_1), X_1, X_1)\frac{\mathrm{cov}(X_1, Y_1)\,\mathrm{cov}(X_1, Y_2)}{(\mathrm{var}(X_1))^2}. \tag{12.119}$$

Application of this identity to the static system $V_n = g(U_n) + Z_n$, in Remark 12.7 yields

$$\begin{aligned}
R_{VUU}(\tau_1, \tau_2) &= \mathrm{cum}(V_n, U_{n-\tau_1}, U_{n-\tau_2}) \\
&= \mathrm{cum}(g(U_0), U_0, U_0)\frac{\mathrm{cov}(U_n, U_{n-\tau_1})\,\mathrm{cov}(U_n, U_{n-\tau_2})}{(\mathrm{var}(U_0))^2}.
\end{aligned}$$

For $\{U_n\}$ being the zero mean stationary Gaussian process with $\mathrm{var}(U_0) = \sigma^2$, we can find that

$$\mathrm{cum}(g(U_0), U_0, U_0) = \sigma^4 E\left\{g^{(2)}(U_0)\right\},$$

provided that $g(u)$ is twice differentiable. In fact, the twice application of (12.117) and integration by parts gives

$$E\left\{g(U_0)U_0^2\right\} = \sigma^4 E\left\{g^{(2)}(U_0)\right\} + \sigma^2 E\left\{g(U_0)\right\}.$$

This provides the following generalization of Bussgang's theorem, see (12.118),

$$R_{VUU}(\tau_1, \tau_2) = \beta R_{UU}(\tau_1)R_{UU}(\tau_2),$$

where $\beta = E\{g^{(2)}(U_0)\}$.

   Hence, the third order cumulant is invariant with respect to nonlinear transformations possessing two derivatives. It is natural to conjecture that the $p + 1$ order cumulant

$\text{cum}(V_n, U_{n-\tau_1}, U_{n-\tau_2}, \ldots, U_{n-\tau_p})$ *will be invariant for all nonlinearities* $g(\bullet)$ *with* p *derivatives.*

The main shortcoming of the identity in (12.111) is that the random variables $X_1, \ldots, X_J$ are assumed to be independent. This prevents us, from using (12.111) directly for nonlinear systems with multivariate nonlinearities.

Nevertheless, in the case when the vector $\mathbf{X} = (X_1, \ldots, X_J)^T$ is dependent, the matrix $\mathbf{C}_1$ appearing in the decomposition of the covariance matrix is fully lower triangular; see the proof of Lemma 12.2. Therefore, instead of (12.112) we have $\mathbf{X} = \mathbf{C}_1 \mathbf{e}_J$, where $\mathbf{e}_J = (e_1, \ldots, e_J)^T$. Then reasoning as in the proof of Lemma 12.2, we obtain

$$\text{cum}(G(X_1, \ldots, X_J), Y_1, \ldots, Y_K) = \sum_{s=1}^{J} \text{cum}(G(\mathbf{X}), c_{J+1,s}e_s, \ldots, c_{J+K,s}e_s).$$

$$(12.120)$$

We can now substitute each e_s by using the solution $\mathbf{e}_J = \mathbf{C}_1^{-1}\mathbf{X}$. Since $\mathbf{C}_1$ is lower triangular the solution is easily written as

$$e_s = \frac{X_s - \sum_{\ell=1}^{s-1} c_{s,\ell}e_\ell}{c_{s,s}}, \quad s = 1, \ldots, J.$$

Then, recalling that

$$\mathbf{C}_1\mathbf{C}_1^T = \text{cov}(\mathbf{X}) \quad \text{and} \quad \mathbf{C}_1\mathbf{D}^T = \text{cov}(\mathbf{X}, \mathbf{Y})$$

we can find $c_{J+j,s}e_s, 1 \le j \le K$ appearing in (12.120) in terms of $\text{cov}(\mathbf{X})$ and $\text{cov}(\mathbf{X}, \mathbf{Y})$. The solution of these matrix equations gives a formula for $\text{cum}(G(X_1, \ldots, X_J), Y_1, \ldots, Y_K)$ in terms of

$$\{\text{cum}(G(X_1, \ldots, X_J), X_s, \ldots, X_s), 1 \le s \le J\},$$
$$\{\text{cov}(X_s, Y_j), 1 \le j \le K, 1 \le s \le J)\},$$

as well as $\{\text{cov}(X_s, X_r), 1 \le s, r \le J\}$. The last term, that is, $\text{cov}(X_s, X_r)$ makes the resulting formula more complicated than the one established in Lemma 12.2.

As an example of the procedure outlined here, let us consider the case $J = 2$ and $K = 1$ with $(X_1, X_2, Y)^T$ being a Gaussian vector with zero mean and covariance matrix Σ such that $\text{var}(X_1) = \text{var}(X_2) = \sigma^2$, and correlation $\rho = \text{cov}(X_1, X_2)/\sigma^2$. Then some algebra (left to the reader) gives

$$\text{cov}(G(X_1, X_2), Y) = \text{cov}(G(X_1, X_2), X_1)a_1 + \text{cov}(G(X_1, X_2), X_2)a_2, \quad (12.121)$$

where

$$a_1 = \sigma^{-2}\frac{\text{cov}(X_1, Y) - \rho\,\text{cov}(X_2, Y)}{1 - \rho^2} \quad \text{and} \quad a_2 = \sigma^{-2}\frac{\text{cov}(X_2, Y) - \rho\,\text{cov}(X_1, Y)}{1 - \rho^2}.$$

Note that if $\rho = 0$ we can arrive to the identity in (12.111).

There is a direct way of dealing with the case of the dependent random vector $\mathbf{X} = (X_1, \ldots, X_J)^T$ based on the covariance matrix decomposition and a formula for differentiation of composite functions. Since the case of $K = 1$ is the most common in applications let us examine the version of Lemma 12.2 with $J > 1$ and $K = 1$, that is, let us consider $\mathrm{cov}(G(X_1, \ldots, X_J), Y)$ with $(X_1, \ldots, X_J, Y)^T$ being the $(J + 1)$-dimensional nonsingular Gaussian vector with zero mean and covariance matrix Σ. The following lemma gives an expression for $\mathrm{cov}(G(X_1, \ldots, X_J), Y)$ in terms of the average derivative of the nonlinearity $G(\bullet, \ldots, \bullet)$ and covariance between $\mathbf{X}$ and Y.

LEMMA 12.3 *Let* $(X_1, \ldots, X_J, Y)^T$ *be a nonsingular Gaussian random vector with zero mean and covariance matrix* Σ. *Let* $G : R^J \to R$ *be a measurable function of* J *variables, such that*

$$E\{|\, G(X_1, \ldots, X_J)Y\, |\} < \infty$$

and

$$E\left\{|\, \frac{\partial G(X_1, \ldots, X_J)}{\partial X_s}\, |\right\} < \infty, \quad s = 1, \ldots, J.$$

Then, we have

$$\mathrm{cov}(G(X_1, \ldots, X_J), Y) = \sum_{s=1}^{J} b_s \, \mathrm{cov}(X_s, Y), \qquad (12.122)$$

where $b_s = E\{\frac{\partial G(X_1, \ldots, X_J)}{\partial X_s}\}$.

Proof. Similarly as in the proof of Lemma 12.2 let us start with the decomposition of the matrix Σ as $\Sigma = \mathbf{C}\mathbf{C}^T$, such that

$$\mathbf{C} = \begin{bmatrix} \mathbf{C}_1 & 0 \\ \mathbf{d}^T & c \end{bmatrix},$$

and $\mathbf{C}_1 \mathbf{d} = \mathrm{cov}(\mathbf{X}, Y)$. Here $\mathbf{C}_1$ is the $(J \times J)$–lower triangular matrix, $\mathbf{d}$ is the $J \times 1$ vector, and c is a number. Then, we have

$$\mathbf{X} = \mathbf{C}_1 \mathbf{e} \quad \text{and} \quad Y = \mathbf{d}^T \mathbf{e} + c\varepsilon,$$

where $\binom{\mathbf{e}}{\varepsilon} \sim N_{J+1}(\mathbf{0}, \mathbf{I})$, and $\mathbf{I}$ is the unit diagonal matrix. Hence, we obtain

$$\mathrm{cov}(G(X_1, \ldots, X_J), Y) = \mathrm{cov}(G(\mathbf{C}_1\mathbf{e}), \mathbf{d}^T\mathbf{e}) = \mathbf{d}^T E\{G(\mathbf{C}_1\mathbf{e})\mathbf{e}\}, \qquad (12.123)$$

where the expectation is taken with respect to the random vector $\mathbf{e}$. Without loss of generality, let us consider the first component of the $J \times 1$ vector $E\{G(\mathbf{C}_1\mathbf{e})\mathbf{e}\}$. Hence, we wish to evaluate $E\{G(\mathbf{C}_1\mathbf{e})e_1\}$. Since $\mathbf{e} \sim N_J(\mathbf{0}, \mathbf{I})$, it suffices to examine the integral

$$\int_{-\infty}^{\infty} G(\mathbf{C}_1\mathbf{z})z_1 f(z_1)dz_1, \qquad (12.124)$$

where $\mathbf{z} = (z_1, \ldots, z_J)^T$ and $f(z_1)$ is the standard normal density function. The fact is that $z_1 f(z_1) = -f^{(1)}(z_1)$ and integration by parts give the following version of the integral in (12.124):

$$\int_{-\infty}^{\infty} \left\{ \frac{\partial}{\partial z_1} G(\mathbf{C}_1\mathbf{z}) \right\} f(z_1)dz_1.$$

To evaluate $\frac{\partial}{\partial z_1} G(\mathbf{C}_1, \mathbf{z})$, note that due to the formula for differentiation of composite functions we have for $\mathbf{x} = \mathbf{C}_1\mathbf{z}$

$$\frac{\partial}{\partial z_1} G(\mathbf{C}_1\mathbf{z}) = \frac{\partial G(x_1, \ldots, x_J)}{\partial x_1} \frac{\partial x_1}{\partial z_1} + \cdots + \frac{\partial G(x_1, \ldots, x_J)}{\partial x_J} \frac{\partial x_J}{\partial z_1}$$

$$= \frac{\partial G(x_1, \ldots, x_J)}{\partial x_1} c_{1,1} + \cdots + \frac{\partial G(x_1, \ldots, x_J)}{\partial x_J} c_{J,1},$$

where $\mathbf{v}_1 = (c_{1,1}, \ldots, c_{J,1})^T$ is the first column of the matrix $\mathbf{C}_1$.

Denoting by $\nabla G(\mathbf{x})$ the gradient vector of $G(\mathbf{x})$, we have shown that

$$E\{G(\mathbf{C}_1\mathbf{e})e_1\} = E\{\nabla^T G(\mathbf{X})\}\mathbf{v}_1.$$

Repeating the above calculations for each component of the vector $E\{G(\mathbf{C}_1\mathbf{e})\mathbf{e}\}$, we obtain the following representation for (12.123):

$$\mathbf{d}^T (E\{\nabla^T G(\mathbf{X})\}\mathbf{v}_1, \ldots, E\{\nabla^T G(\mathbf{X})\}\mathbf{v}_J)^T, \tag{12.125}$$

where $\mathbf{v}_1, \ldots, \mathbf{v}_J$ are the column vectors of the matrix $\mathbf{C}_1$. By changing the order of summation we see that (12.125) is equal to

$$E\{\nabla^T G(\mathbf{X})\}(\mathbf{d}^T \mathbf{c}_1, \ldots, \mathbf{d}^T \mathbf{c}_J)^T,$$

where $\mathbf{c}_1, \ldots, \mathbf{c}_J$ are the $J \times 1$ vectors representing the rows of matrix $\mathbf{C}_1$. Noting finally that $\mathbf{d}^T \mathbf{c}_i = \text{cov}(X_i, Y)$, $1 \leq i \leq J$ we can complete the proof of Lemma 12.3. ∎

REMARK 12.9 *The version of (12.111) which can be useful for the generalized Hammerstein system corresponds to the case when $K = 1$ and $J > 1$. Hence, (12.111) reads as*

$$\text{cov}(G(X_1, \ldots, X_J), Y_1) = \sum_{s=1}^{J} \text{cov}(G(X_1, \ldots, X_J), X_s) \frac{\text{cov}(X_s, Y_1)}{\text{var}(X_s)}, \tag{12.126}$$

provided that $X_1, \ldots, X_J$ are uncorrelated Gaussian random variables.

In the context of nonlinear systems this formula can be interpreted as follows. Let

$$V_n = g(U_n, U_{n-1}, \ldots, U_{n-p}) + Z_n \tag{12.127}$$

be a nonlinear finite memory system, where $\{Z_n\}$ is independent of $\{U_n\}$–a zero mean white Gaussian process with the variance σ^2. Note that for the independent input the covariance $R_{VU}(\tau) = \text{cov}(V_n, U_{n-\tau})$ is nonzero only for $0 \leq \tau \leq p$. Then, according to (12.126) and for $0 \leq \tau \leq p$, we have

$$R_{VU}(\tau) = \text{cov}(g(U_n, U_{n-1}, \ldots, U_{n-p}), U_{n-\tau})$$

$$= \sum_{s=0}^{p} \text{cov}(g(U_n, U_{n-1}, \ldots, U_{n-p}), U_{n-s}) \frac{\text{cov}(U_{n-s}, U_{n-\tau})}{\text{var}(U_{n-s})}$$

$$= \text{cov}(g(U_n, U_{n-1}, \ldots, U_{n-p}), U_{n-\tau})$$

for $n + \tau - p \leq n \leq n + \tau$.

For differentiable $g(\bullet, \bullet, \ldots, \bullet)$, we can apply (12.117) and integration by parts, that is, we obtain

$$E\{g(U_n, U_{n-1}, \ldots, U_{n-p})U_{n-\tau}\} = \sigma^2 E\left\{\frac{\partial g(U_n, U_{n-1}, \ldots, U_{n-p})}{\partial U_{n-\tau}}\right\}.$$

Hence, we have

$$R_{VU}(\tau) = \sigma^2 E\left\{\frac{\partial g(U_n, U_{n-1}, \ldots, U_{n-p})}{\partial U_{n-\tau}}\right\} \tag{12.128}$$

for $0 \leq \tau \leq p$.

It is worth noting that there is no further information contained in higher order moments with respect to the input process. In fact, we have $\operatorname{cum}(V_n, U_{n-\tau_1}, U_{n-\tau_2}) = 0$ for all $\tau_1 \neq \tau_2$.

In order to generalize (12.128) to the correlated input process $\{U_n\}$, we can apply Lemma 12.3. Hence, by virtue of (12.122), we readily obtain

$$R_{VU}(\tau) = \sum_{s=0}^{p} b_s R_{UU}(\tau - s), \tag{12.129}$$

where

$$b_s = E\left\{\frac{\partial g(U_n, U_{n-1}, \ldots, U_{n-p})}{\partial U_{n-s}}\right\}, \quad 0 \leq s \leq p. \tag{12.130}$$

It is seen that for $R_{UU}(s) = \sigma^2 \delta_{s0}$, that is, when $\{U_n\}$ is independent we can confirm formula (12.128).

The formula in (12.129) can be interpreted as a linear approximation of the nonlinear system in (12.127). In fact, it is straightforward to verify that a solution of the following minimization problem

$$E\left\{V_n - \sum_{s=0}^{p} \lambda_s U_{n-s}\right\}^2 \to \min_{\{\lambda_s\}} \tag{12.131}$$

must satisfy the equations in (12.129). That is, the optimal mean-square error linear time-invariant system which approximates the nonlinear system in (12.127) is given by the FIR(p) model with the characteristic $\{b_s, 0 \leq s \leq p\}$, where b_s is defined in (12.130).

The results of Lemmas 12.2 and 12.3 and the discussion in Remark 12.9 allow us to establish a simple relationship between the correlation functions $R_{YU}(\tau)$, $R_{UU}(\tau)$ and the impulse response function $\{\lambda_j\}$ of the linear subsystem of the Hammerstein system with nonlinear dynamics defined in (12.59) and (12.60). Hence, let the input signal $\{U_n\}$ be a zero-mean stationary Gaussian process with the covariance function $R_{UU}(\tau)$.

Let us first note for the linear part of the system

$$Y_n = \sum_{j=0}^{\infty} \lambda_j V_{n-j} + Z_n,$$

we have the following convolution formula:

$$R_{YU}(\tau) = \text{cov}(Y_n, U_{n-\tau}) = \sum_{j=0}^{\infty} \lambda_j \, \text{cov}(V_{n-j}, U_{n-\tau}) = \sum_{j=0}^{\infty} \lambda_j R_{VU}(\tau - j). \quad (12.132)$$

We can now directly apply the result of Remark 12.9, see (12.129), and evaluate the cross-covariance $R_{VU}(\tau)$ in (12.132). Indeed, by virtue of (12.129), $R_{VU}(\tau)$ is also given in the form of the convolution between the sequence $\{b_s\}$ and $\{R_{UU}(s)\}$, that is, we have

$$R_{VU}(\tau) = \sum_{s=0}^{p} b_s R_{UU}(\tau - s).$$

This and (12.132) yield the following theorem:

THEOREM 12.6 *Let (12.59) and (12.60) define the Hammerstein system with the nonlinear dynamics $m(U_n, U_{n-1}, \ldots, U_{n-p})$. Let the input signal $\{U_n\}$ be a zero-mean, stationary Gaussian process with the covariance function $R_{UU}(\tau)$. Suppose that*

$$E\{|\, m(U_n, U_{n-1}, \ldots, U_{n-p})U_{n-j}\,|\} < \infty, \quad j = 0, 1, \ldots, p,$$

and that $m(\bullet, \bullet, \ldots, \bullet)$ is a differentiable function. Then, we have the following convolution relationship

$$R_{YU}(\tau) = \{\lambda * b * R_{UU}\}(\tau), \quad (12.133)$$

where

$$b_j = E\left\{ \frac{\partial m(U_n, U_{n-1}, \ldots, U_{n-p})}{\partial U_{n-j}} \right\}, \quad j = 0, 1, \ldots, p.$$

Analogous to Remark 12.9 we can argue (see (12.131)) that $\{\lambda_s * b_s\}$ defines the impulse response function of the optimal mean-square error, linear time-invariant system, which approximates the Hammerstein system with nonlinear dynamics. The problem of best linear approximations of nonlinear systems and their application to system identification has been an active area in recent years; see [203], and [273] for recent results and further references on this important issue.

Formation of the Fourier transform of (12.133) gives

$$S_{YU}(\omega) = \Lambda(\omega)B(\omega)S_{UU}(\omega). \quad (12.134)$$

This formula reveals that one can recover (up to a multiplicative factor) the spectral characteristic of the linear subsystem from the spectral densities $S_{YU}(\omega)$ and $S_{UU}(\omega)$.

In an important case of $p = 1$, we can explicitly write

$$S_{YU}(\omega) = \Lambda(\omega)S_{UU}(\omega)\{b_0 + b_1 e^{-i\omega}\}, \quad (12.135)$$

where

$$b_0 = E\left\{ \frac{\partial m(U_n, U_{n-1})}{\partial U_n} \right\} \quad \text{and} \quad b_1 = E\left\{ \frac{\partial m(U_n, U_{n-1})}{\partial U_{n-1}} \right\}.$$

It is quite straightforward to evaluate the coefficients (b_0, b_1) for some specific nonlinearities. Hence, for the polynomial nonlinearity of order not greater than 3, that is, when

$$m(u, v) = \sum_{0 \le k+\ell \le 3} a_{k\ell} u^k v^\ell,$$

one can calculate that

$$b_0 = a_{10} + (a_{12} + 3a_{30})\sigma^2 + 2a_{21} R_{UU}(1)$$

and

$$b_1 = a_{01} + (a_{21} + 3a_{03})\sigma^2 + 2a_{12} R_{UU}(1),$$

where $\sigma^2 = \mathrm{var}(U_0)$.

For the cubic nonlinearity $\sum_{k+\ell=3} a_{k\ell} u^k v^\ell$ the terms a_{10}, a_{01} in the above formula should be set to zero. If there exists an identification procedure which can recover the nonlinearity $m(u, v)$ independently of the linear subsystem then one can estimate the numbers (b_0, b_1). This is, however, a rare case since the cascade nature of the system allows us to recover its characteristics only up to some unknown constants. Nevertheless, in Theorem 12.5 we have described the $(p + 2)$-order cumulant based method for recovering the linear subsystem characteristic $\{\lambda_j\}$. This method is very inefficient and complicated to implement. Indeed, the formula for λ_ℓ given in Theorem 12.5 requires the $(p + 2)$ – order cumulants and becomes highly inefficient for ℓ greater than p.

The main assumption employed in Theorem 12.5 is that the input signal is white but not necessarily Gaussian. In the following remark, we wish to find out what can be gained if the input is a white Gaussian process.

REMARK 12.10 *For the input signal being a white Gaussian process, we can write formula (12.133) as follows:*

$$R_{YU}(\tau) = \sigma^2 \sum_{\ell=0}^{\tau} \lambda_{\tau-\ell} b_\ell, \tag{12.136}$$

where $\sigma^2 = \mathrm{var}(U_0)$.

Without loss of generality let us consider the second order nonlinearity ($p = 1$). Then, due to (12.136), we readily have

$$b_0 = \sigma^{-2} R_{YU}(0) \quad and \quad b_1 = \sigma^{-2} R_{YU}(1) - \lambda_1 b_0.$$

These explicit formulas can lead to a direct estimation technique for the numbers (b_0, b_1), and consequently a nonparametric estimate of $\Lambda(\omega)$. In fact, by virtue of (12.135) and that $S_{UU}(\omega) = \sigma^2$, we have

$$\Lambda(\omega) = \frac{S_{YU}(\omega)}{a_0 + a_1 e^{-i\omega}}, \tag{12.137}$$

where $a_0 = R_{YU}(0)$ and $a_1 = R_{YU}(1) - \lambda_1 R_{YU}(0)$.

Note that in the formula for a_1, we still need to know λ_1. Here, we can recall expression (12.104), which determines λ_1 from the third-order cumulant of the input-output signals, that is, λ_1 can be obtained from the following formula:

$$\lambda_1 = \frac{\mathrm{cum}(Y_{n+1}, U_n^2, U_{n-1}^2)}{\mathrm{cum}(Y_n, U_n^2, U_{n-1}^2)}.$$

Hence, a nonparametric estimate of $\Lambda(\omega)$ that can be derived from the cross-spectra $S_{YU}(\omega)$ requires limited additional help from higher order statistics. Generally, for the p-order nonlinearity the nonparametric recovery of $\Lambda(\omega)$ in terms of $S_{YU}(\omega)$ needs the values of $\lambda_1, \ldots, \lambda_p$, that can be obtained from the $(p+2)$-order cumulants given in Theorem 12.5.

Thus, the computational saving in estimating the linear subsystem in the case of an uncorrelated Gaussian input is very essential for small and moderate values of p. In typical cases when $p = 1, 2$ the aforementioned considerations lead to much more efficient identification algorithms which essentially employ only the second order statistics of input–output data.

The aforementioned considerations point out that there can be some benefits in using higher order spectra even in the case of independent Gaussian input. If the input is correlated and there are several linear systems present in the overall nonlinear structure then it is necessary to go beyond the second order spectra. This will be transparent in the case of the sandwich system (see Example 12.6), where we have two linear subsystems separated by the nonlinearity. This structure plays an important role in many applications and it is treated separately in Section 12.2.5.

Nonlinear system identification

For Gaussian inputs, it is natural to employ Hermite polynomials to represent the system nonlinearities. In fact, as we mentioned in Section 6.5 (see also Appendix B.5) the Hermite polynomials constitute an orthonormal basis with respect to the weight function e^{-x^2}. In order to obtain Hermite polynomials orthogonal with respect to the standard normal density function, we need to slightly modify the definition of the Hermite polynomial. Hence, let

$$\bar{H}_k(x) = (-1)^k e^{-x^2/2} \left(e^{-x^2/2}\right)^{(k)}$$

be the kth modified Hermite polynomial. Then, it is easy to show that $\bar{H}(x)$ is related to the classical Hermite polynomial $H_k(x)$ (see Appendix B.5) by the following formula:

$$\bar{H}_k(x) = 2^{-k/2} H_k(x/\sqrt{2}).$$

Consequently, the system $\{\bar{h}_k(x) = (2^k k!)^{-1/2} H_k(x/\sqrt{2})\}$ defines the orthonormal basis with respect to the standard normal density. Here is a list of a few Hermite polynomials $\bar{h}_k(x)$:

$$\bar{h}_0(x) = 1,$$
$$\bar{h}_1(x) = x,$$
$$\bar{h}_2(x) = -\frac{1}{\sqrt{2}} + \frac{1}{\sqrt{2}}x^2,$$
$$\bar{h}_3(x) = -\sqrt{\frac{3}{2}}x + \frac{1}{\sqrt{6}}x^3,$$
$$\bar{h}_4(x) = \frac{3}{2\sqrt{6}} - \frac{3}{\sqrt{6}}x^2 + \frac{1}{2\sqrt{6}}x^4, \ldots.$$

It is also worth noting that the scaled version $\bar{h}_k(x/\sigma)$ of $\bar{h}_k(x)$ defines the orthonormal Hermite polynomials with respect to the $N(0, \sigma^2)$ density function. We wish to propose an estimate of the nonlinearity $m(u, v)$ that defines the system in (12.59) and (12.60) using the polynomials $\{\bar{h}_k(x)\}$.

Hence, without loss of generality, let us assume that the input signal $\{U_n\}$, exciting the system depicted in Figure 12.17, is a zero-mean, uncorrelated Gaussian process with unit variance. Let us also consider the system in (12.59) and (12.60) corresponding to $p = 1$. Then, by virtue of the result of Theorem 12.3, we have

$$m(u, v) = r_2(u, v) - \lambda_1 r_1(v), \tag{12.138}$$

where, additionally, we have assumed that $E\{m(U_1, U_0)\} = 0$. Since $EY_n^2 < \infty$ (see (12.61)), then we can expand the regression functions $r_2(u, v)$ and $r_1(v)$ in terms of the Hermite polynomials $\{\bar{h}_k(x)\}$. In the case of $r_1(v)$ this is standard orthogonal functions representation, whereas for $r_2(u, v)$ we need a two-dimensional extension. The latter can be easily done by using a product basis $\{\bar{h}_k(u)\bar{h}_\ell(v) : k, \ell = 0, 1, \ldots\}$; see also Chapter 13 for representations of multivariate functions. Hence, we have the following representations:

$$r_2(u, v) \approx \sum_{k=0}^{\infty} \sum_{\ell=0}^{k} a_{k-\ell,\ell} \bar{h}_{k-\ell}(u) \bar{h}_\ell(v) \tag{12.139}$$

and

$$r_1(v) \approx \sum_{k=0}^{\infty} b_k \bar{h}_k(v), \tag{12.140}$$

where

$$a_{k,\ell} = \int_{-\infty}^{\infty} \int_{-\infty}^{\infty} r_2(u, v) \bar{h}_k(u) \bar{h}_\ell(v) f(u) f(v) du\, dv$$

and

$$b_k = \int_{-\infty}^{\infty} r_1(v)\bar{h}_k(v)f(v)dv,$$

respectively. Noting that $f(u)$ is the density of U_n it is important to observe that

$$a_{k,\ell} = E\left\{Y_n\bar{h}_k(U_n)\bar{h}_\ell(U_{n-1})\right\} \quad \text{and} \quad b_k = E\left\{Y_n\bar{h}_k(U_n)\right\}. \tag{12.141}$$

This and (12.138)–(12.140) readily lead to the following orthogonal series estimate of $m(u, v)$:

$$\hat{m}(u, v) = \hat{r}_2(u, v) - \hat{\lambda}_1\hat{r}_1(v), \tag{12.142}$$

where $\hat{\lambda}_1$ is estimated via the cumulant formula in (12.104) and

$$\hat{r}_2(u, v) = \sum_{k=0}^{N}\sum_{\ell=0}^{k} \hat{a}_{k-\ell,\ell}\bar{h}_{k-\ell}(u)\bar{h}_\ell(v), \quad \hat{r}_1(v) = \sum_{k=0}^{N} \hat{b}_k\bar{h}_k(v) \tag{12.143}$$

with

$$\hat{a}_{k,\ell} = n^{-1}\sum_{j=2}^{n} Y_j\bar{h}_k(U_j)\bar{h}_\ell(U_{j-1}), \quad \hat{b}_k = n^{-1}\sum_{j=1}^{n} Y_j\bar{h}(U_j). \tag{12.144}$$

The formulas (12.143) and (12.144) define standard orthogonal series estimates with the truncation parameter N. Note that $\hat{a}_{k,\ell}$ and $\hat{b}_k$ are unbiased estimates of $a_{k,\ell}$ and b_k, respectively. We have already noted (see Chapter 6) that N controls the bias-variance tradeoff. In fact, let us consider the estimate $\hat{r}_2(u, v)$. By virtue of Parseval's formula the mean integrated squared error (MISE) of $\hat{r}_2(u, v)$ can be decomposed as follows:

$$\begin{aligned}
\text{MISE}(\hat{r}_2) &= E\int_{-\infty}^{\infty}\int_{-\infty}^{\infty} (\hat{r}_2(u, v) - r_2(u, v))^2 f(u)f(v)dudv \\
&= \sum_{k=0}^{N}\sum_{\ell=0}^{k} \text{var}(\hat{a}_{k-\ell,\ell}) + \sum_{k=N+1}^{\infty}\sum_{\ell=0}^{k} a_{k-\ell,\ell}^2. \tag{12.145}
\end{aligned}$$

The first term represents the integrated variance of $\hat{r}_2(u, v)$ and is an increasing function of N. The second term in (12.145) is an integrated bias of $\hat{r}_2(u, v)$ and it decreases with N. Arguing as in Section 12.2.4 (see also Section 6.5) we can show that the first term in (12.145) is of order $O\{\frac{N^{5/3}}{n}\}$. On the other hand, the second term in (12.145) tends to zero as $N \to \infty$ for all nonlinearities that satisfy

$$E\{m^2(U_1, U_0)\} < \infty. \tag{12.146}$$

An analogous analysis for the estimate $\hat{r}_1(v)$ shows that $\text{MISE}(\hat{r}_1) \to 0$ if $N \to \infty$ and $\frac{N^{5/6}}{n} \to 0$. Since $\hat{\lambda}_1 \to \lambda_1(P)$, then we have the following theorem:

THEOREM 12.7 *Let us consider the nonlinear system in (12.59) and (12.60). Let the input signal $\{U_n\}$ be a zero-mean, uncorrelated Gaussian process with unit variance. If (12.146) holds and if*

$$N(n) \to \infty \quad and \quad \frac{N^{5/3}(n)}{n} \to 0$$

then for the estimate $\hat{m}(u, v)$ in (12.142), we have

$$\mathrm{MISE}(\hat{m}) \to 0$$

as $n \to \infty$.

It should be also noted that if $m(u, v)$ is defined on a finite interval, then we can replace the condition $\frac{N^{5/3}(n)}{n} \to 0$ by $\frac{N(n)}{n} \to 0$.

Regarding the rate of convergence it suffices to examine the bias terms, that is, we wish to evaluate the rate at which $\sum_{k=N+1}^{\infty} \sum_{\ell=0}^{k} a_{k-\ell,\ell}^2$ and $\sum_{k=N+1}^{\infty} b_k^2$ tend to zero as $N \to \infty$. This analysis is independent of the dependence structure of data generated by our system and can be done along the lines of the analysis in Appendix B.5, see also [132].

Hence, let

$$m^{(i,j)}(u, v) = \frac{\partial^{i+j} m(u, v)}{\partial u^i v^j}$$

be the partial derivative of $m(u, v)$ of order $i + j$. We can show that if

$$E\left\{\left(m^{(i,j)}(U_1, U_0)\right)^2\right\} < \infty, \quad \text{for all } 0 \le i + j \le s \tag{12.147}$$

holds, then we have

$$\sum_{k=N+1}^{\infty} \sum_{\ell=0}^{k} a_{k-\ell,\ell}^2 = O(N^{-s}) \quad and \quad \sum_{k=N+1}^{\infty} b_k^2 = O(N^{-s}).$$

Combining this with the result of Theorem 12.7, we can conclude that under condition (12.147) and if

$$N(n) = n^{\frac{3}{3s+5}}, \tag{12.148}$$

then we have

$$\mathrm{MISE}(\hat{m}) = O(n^{-\frac{3s}{3s+5}}). \tag{12.149}$$

For the nonlinearity $m(u, v)$, which is defined on a finite interval and meets (12.147) we have a faster rate, that is,

$$\mathrm{MISE}(\hat{m}) = O(n^{-\frac{s}{s+1}}) \tag{12.150}$$

with the asymptotically optimal selection of the truncation parameter $N(n) = n^{\frac{1}{s+1}}$. For $s = 1$, the rate in (12.149) is $O(n^{-\frac{3}{8}})$, whereas the rate in (12.150) is $O(n^{-\frac{1}{2}})$.

It is worth noting that the rate in (12.150) is known to be the best possible rate of convergence for recovering a bivariate function possessing s derivatives.

12.2.5   Sandwich systems with a Gaussian input signal

In this section, we illustrate the theory developed in the previous section in the important case of the sandwich system. We shall see that the restriction put on the input signal to be a Gaussian process allows us to relax the assumption that the input linear system is of a finite memory.

Let us consider the sandwich system introduced in Example 12.6 (see also Figure 12.20). The system has the input linear subsystem with the impulse response function $\{a_j, 0 \le j \le p\}$, and is excited by a zero-mean, stationary Gaussian process. Recalling that

$$m(U_n, \ldots, U_{n-p}) = g\left(U_n + \sum_{j=1}^{p} a_j U_{n-j} \right),$$

we can easily find that the sequence $\{b_j\}$ in Theorem 12.6 is given by

$$b_j = a_j E\{g^{(1)}(X_n)\},$$

where $X_n = U_n + \sum_{j=1}^{p} a_j U_{n-j}$. This yields

$$R_{VU}(\tau) = \alpha \sum_{j=0}^{p} a_j R_{UU}(\tau - j), \tag{12.151}$$

where $\alpha = E\{g^{(1)}(X_0)\}$. This applied to formula (12.133) gives the following counterpart of (12.134):

$$S_{YU}(\omega) = \alpha A(\omega)\Lambda(\omega)S_{UU}(\omega). \tag{12.152}$$

Since we wish to identify both $A(\omega)$ and $\Lambda(\omega)$, then it is seen that this single equation is not sufficient to construct individual estimates of $A(\omega)$ and $\Lambda(\omega)$. Nevertheless, with the help of the third-order cumulants we can form another equation involving the unknown $A(\omega)$ and $\Lambda(\omega)$.

Hence, let us consider

$$R_{YUU}(\tau_1, \tau_2) = \text{cum}(Y_n, U_{n-\tau_1}, U_{n-\tau_2})$$

$$= \sum_{\ell=0}^{\infty} \lambda_\ell \, \text{cum}(g(X_{n-\ell}), U_{n-\tau_1}, U_{n-\tau_2}),$$

where Properties 1 and 3 of cumulants have been used. By virtue of Lemma 12.2 (applied with $J = 1$ and $K = 2$), we can obtain

$$\text{cum}(g(X_{n-\ell}), U_{n-\tau_1}, U_{n-\tau_2})$$
$$= \frac{\text{cum}(g(X_0), X_0, X_0)}{\{\text{var}(X_0)\}^2} \text{cov}(X_{n-\ell}, U_{n-\tau_1}) \, \text{cov}(X_{n-\ell}, U_{n-\tau_2}).$$

This combined with the fact that the cross-covariance function $R_{XU}(\tau)$ between $\{X_n\}$ and $\{U_n\}$ is given by

$$R_{XU}(\tau) = \sum_{\ell=0}^{p} a_\ell R_{UU}(\tau - \ell) \tag{12.153}$$

allows us to write the following relationship:

$$R_{YUU}(\tau_1, \tau_2) = \beta \sum_{\ell=0}^{\infty} \lambda_\ell R_{XU}(\tau_1 - \ell) R_{XU}(\tau_2 - \ell), \tag{12.154}$$

where $\beta = \mathrm{cum}(g(X_0), X_0, X_0)/\{\mathrm{var}(X_0)\}^2$.

Let us now define the cross-bispectrum of the processes $\{Y_n\}$ and $\{U_n\}$ as the two-dimensional Fourier transform of $R_{YUU}(\tau_1, \tau_2)$, that is,

$$S_{YUU}(\omega_1, \omega_2) = \sum_{\tau_1=-\infty}^{\infty} \sum_{\tau_2=-\infty}^{\infty} R_{YUU}(\tau_1, \tau_2) e^{-i\{\omega_1 \tau_1 + \omega_2 \tau_2\}}.$$

Hence, formation of the two-dimensional Fourier transform of the formulas (12.154) and (12.153) gives

$$S_{YUU}(\omega_1, \omega_2) = \beta A(\omega_1) A(\omega_2) \Lambda(\omega_1 + \omega_2) S_{UU}(\omega_1) S_{UU}(\omega_2). \tag{12.155}$$

This is the second expression besides (12.152), which involves the unknown characteristics $A(\omega)$ and $\Lambda(\omega)$ expressed in terms of the spectral characteristics of the input–output signals of the sandwich system.

It is worth noting that formula (12.152) was derived from the general result described in Theorem 12.6. This result restricts the input nonlinearity to be of finite memory. Direct calculations easily show that we can admit an infinite memory input linear subsystem, that is, the formula in (12.153) takes the following form:

$$R_{XU}(\tau) = \sum_{\ell=0}^{\infty} a_\ell R_{UU}(\tau - \ell).$$

Next, we have $R_{YU}(\tau) = \sum_{\ell=0}^{\infty} \lambda_\ell R_{VU}(\tau - \ell)$, where, due to the Bussgang identity (see Remark 12.7), we obtain $R_{VU}(\tau) = \alpha' R_{XU}(\tau)$, $\alpha' = \mathrm{cov}(g(X_0), X_0)/\mathrm{var}(X_0)$. Reasoning as in Remark 12.7 we can show $\alpha' = E\{g^{(1)}(X_0)\}$, provided that $g(\bullet)$ is a differentiable function, that is, $\alpha' = \alpha$, where α was introduced in (12.151).

All these considerations lead to the following theorem:

THEOREM 12.8 *Let us consider the sandwich system*

$$\begin{cases} X_n = U_n + \sum_{t=1}^{\infty} a_t U_{n-t}, \\ V_n = g(X_n), \\ Y_n = \sum_{j=0}^{\infty} \lambda_j V_{n-j} + Z_n, \end{cases}$$

where the input signal $\{U_n\}$ is a stationary Gaussian process with the covariance function $R_{UU}(\tau)$. Suppose that

$$E\{|X_0^2 g(X_0)|\} < \infty. \tag{12.156}$$

Then, we have the following relationships:

$$S_{YU}(\omega) = \alpha A(\omega)\Lambda(\omega)S_{UU}(\omega) \tag{12.157}$$

and

$$S_{YUU}(\omega_1, \omega_2) = \beta A(\omega_1)A(\omega_2)\Lambda(\omega_1 + \omega_2)S_{UU}(\omega_1)S_{UU}(\omega_2), \tag{12.158}$$

where $\alpha = \mathrm{cov}(g(X_0), X_0)/\mathrm{var}(X_0)$ *and* $\beta = \mathrm{cum}(g(X_0), X_0, X_0)/\{\mathrm{var}(X_0)\}^2$.

We should note that if $g(\bullet)$ is a twice-differentiable function then the condition in (12.156) is equivalent to $E\{| g^{(2)}(X_0) |\} < \infty$. In fact (see Remarks 12.7 and 12.8), we have

$$\alpha = E\{g^{(1)}(X_0)\} \quad \text{and} \quad \beta = E\{g^{(2)}(X_0)\}. \tag{12.159}$$

It is worth noting that the formulas in (12.159) hold for not necessarily zero-mean Gaussian input process. Indeed, this is an important issue since it is required that $\alpha, \beta \neq 0$ for the result in Theorem 12.8 to be meaningful. For the zero mean input process, the Gaussian density of X_0 is an even function and consequently we have $\alpha \neq 0$ and $\beta = 0$ for the nonlinearity $g(\bullet)$ that is an odd function, or vice versa $\alpha = 0$ and $\beta \neq 0$ for the nonlinearity $g(\bullet)$ that is an even function. Hence, a practical recommendation is to use the input Gaussian process with $EU_n = \mu \neq 0$. Then we have $EX_n = \mu \sum_{\ell=0}^{\infty} a_\ell \neq 0$, provided that $\sum_{\ell=0}^{\infty} a_\ell \neq 0$. The latter is equivalent to $A(0) \neq 0$, and this is met for a large number of linear systems. It should be noted, however, that for high-pass filters we have $A(0) = 0$; see [177] for a detailed discussion of this issue. Furthermore, as we have already noted if $g(\bullet)$ is a linear function, then there is no chance of separately identifying $A(\omega)$ and $\Lambda(\omega)$.

The formulas in (12.157) and (12.158) allow us to develop a fully nonparametric strategy for estimating the linear subsystems of the sandwich model. In fact, owing to (12.157) and (12.158), we have

$$\frac{S_{YUU}(\omega_1, \omega_2)S_{UU}(\omega_1 + \omega_2)}{S_{YU}(\omega_1 + \omega_2)S_{UU}(\omega_1)S_{UU}(\omega_2)} = \frac{\beta}{\alpha} \frac{A(\omega_1)A(\omega_2)}{A(\omega_1 + \omega_2)}. \tag{12.160}$$

By noting that $A(\omega)A(-\omega) =| A(\omega) |^2$ and then by setting $\omega_2 = -\omega_1 = -\omega$, we have

$$| A(\omega) |= L \frac{| S_{YUU}(\omega, -\omega) |^{1/2}}{S_{UU}(\omega)}, \tag{12.161}$$

where $L =| \frac{\alpha A(0)S_{UU}(0)}{\beta S_{YU}(0)} |^{1/2}$, and it is known that $S_{UU}(\omega) \geq 0$ for all ω. Next (12.157) yields

$$\Lambda(\omega) = \frac{1}{\alpha} \frac{S_{YU}(\omega)}{A(\omega)S_{UU}(\omega)}. \tag{12.162}$$

Formula (12.161) gives a prescription for the recovery of the amplitude of $A(\omega)$. Since $S_{UU}(\omega) \geq 0$, and (12.160) the phase of $A(\omega)$ is determined by the phase of $\frac{S_{YUU}(\omega_1, \omega_2)}{S_{YU}(\omega_1 + \omega_2)}$. Hence, let

$$\psi(\omega_1, \omega_2) = \arg\left\{ \frac{S_{YUU}(\omega_1, \omega_2)}{S_{YU}(\omega_1 + \omega_2)} \right\} \quad \text{and} \quad \theta(\omega) = \arg\{A(\omega)\}.$$

Then due to (12.160), we readily obtain

$$\theta(\omega_1) + \theta(\omega_2) - \theta(\omega_1 + \omega_2) = \psi(\omega_1, \omega_2). \tag{12.163}$$

This is a functional equation with respect to $\theta(\omega)$ for a given $\psi(\omega_1, \omega_2)$. Note that if $\psi(\omega_1, \omega_2) = 0$, then (12.163) is the classical Cauchy functional equation with a general unique solution $\theta(\omega) = c\omega$, where c is an arbitrary constant. Differentiating (12.163) with respect to ω_1 and ω_2, we get

$$\theta^{(2)}(\omega) = -\partial^2 \psi(\omega_1, \omega_2)/\partial\omega_1\partial\omega_2,$$

where $\omega = \omega_1 + \omega_2$. For many practical systems, $\theta(\omega)$ is linear for $\omega \to 0+$. Hence, we can assume that $\theta(0+) = 0$ and $\theta^{(1)}(0+) = -n_0$, which is given. These initial conditions allow us to express $\theta(\omega)$ in terms of $\psi(\omega_1, \omega_2)$. It is worth noting that for a class of systems commonly referred to as minimum-phase systems, the phase $\theta(\omega)$ can be determined from the magnitude $| A(\omega) |$.

In summary, we have obtained the following procedure for recovering the transfer functions $A(\omega)$ and $\Lambda(\omega)$ of the linear subsystems. First, we determine the amplitude of $A(\omega)$ from formula (12.161). This is followed by the recovery of the phase of $A(\omega)$ using (12.163) and the above outlined procedure for extracting $\theta(\omega) = \arg\{A(\omega)\}$. Having $A(\omega) = | A(\omega) | e^{i\theta(\omega)}$, we can finally get $\Lambda(\omega)$ from (12.162). This gives the procedure for reconstructing $A(\omega)$ and $\Lambda(\omega)$ being virtually independent on the form of the non-linearity $g(\bullet)$. The multiplicative constants appearing in (12.161) and (12.162) can be determined if some a priori knowledge about $A(\omega)$ and $\Lambda(\omega)$ is known, for example, that the gain factors $A(0)$, $\Lambda(0)$ are known.

Formulas (12.157) and (12.158) allow us to discriminate between two important special cases of the sandwich system. Indeed, if $\Lambda(\omega)$ is all-pass filter, that is, $\Lambda(\omega) = 1$, we obtain the Wiener system, and then (12.157) and (12.158) yield

$$\frac{S_{YUU}(\omega_1, \omega_2)}{S_{YU}(\omega_1)S_{YU}(\omega_2)} = \frac{\beta}{\alpha^2}. \tag{12.164}$$

On the other hand, if $A(\omega) = 1$, that is, we have the Hammerstein system, and from (12.157) and (12.158)

$$\frac{S_{YUU}(\omega_1, \omega_2)S_{UU}(\omega_1 + \omega_2)}{S_{YU}(\omega_1 + \omega_2)S_{UU}(\omega_1)S_{UU}(\omega_2)} = \frac{\beta}{\alpha}. \tag{12.165}$$

Hence, the constancy of the expressions on the left-hand side of (12.164) and (12.165) can be used (once the formulas are estimated from training data) to form test statistics to verify the plausibilities of the respective systems.

Regarding the nonlinear part of the sandwich system, we can combine the general solution given in Theorem 12.4 and the estimation technique from Section 12.2.4 that employs the Hermite orthogonal polynomials. Unfortunately, the result of Theorem 12.4 forces us to assume that the input signal is white Gaussian. Moreover, the input linear system must be of finite memory. Hence, without loss of generality, let $p = 1$, that is, we have $X_n = U_n + aU_{n-1}$. Then, an estimate of $g(\bullet)$ can be obtained by averaging an estimate of the regression function $r_2(u_0, u_1) = E\{Y_n \mid U_n = u_0, U_{n-1} = u_1\}$. Thus, we have (see Theorem 12.4)

$$\hat{g}(x) = n^{-1} \sum_{i=1}^{n} \{\hat{r}_2(x - \hat{a}U_i, U_i) - \hat{r}_2(-\hat{a}U_i, U_i)\}, \tag{12.166}$$

where $\hat{a}$ is a certain estimate of the parameter a of the input linear system. Note that in this case $A(\omega) = 1 + ae^{i\omega}$, that is, $a = A(0) - 1$ and a can easily be determined from the aforementioned algorithm for recovering $A(\omega)$; see the discussion following (12.161).

Furthermore, because the input is Gaussian, we can apply in (12.166) the estimate $\hat{r}_2(u_0, u_1)$ defined in (12.143), that is, we have

$$\hat{r}_2(u_0, u_1) = \sum_{k=0}^{N} \sum_{\ell=0}^{k} \hat{a}_{k-\ell,\ell} \bar{h}_{k-\ell}(u_0) \bar{h}_\ell(u_1) \tag{12.167}$$

and

$$\hat{a}_{k,\ell} = n^{-1} \sum_{j=2}^{n} Y_j \bar{h}_k(U_j) \bar{h}_\ell(U_{j-1}).$$

Because the averaging operation reduces the variance of a nonparametric estimate (see Chapter 13 for further discussion of this issue), we can conclude (see the discussion after Theorem 12.7) that estimate $\hat{g}(x)$ in (12.166) can reach the optimal rate of convergence $O_p(n^{-2/5})$ for twice differentiable nonlinearities defined on a finite interval. The extension of the estimate in (12.166) to the input linear system with the memory length $p > 1$ is straightforward. In fact, we can apply the result of Theorem 12.4 and then an orthogonal series estimate of the $p + 1$–dimensional regression function $r_{p+1}(u_0, \ldots, u_p) = E\{Y_n \mid U_n = u_0, \ldots, U_{n-p} = u_p\}$. The latter is a multivariate extension of (12.167), that is,

$$\hat{r}_{p+1}(u_0, \ldots, u_p) = \sum_{0 \leq \mathbf{k} \leq N} \hat{a}_{\mathbf{k}} \bar{h}_{k_0}(u_0) \cdots \bar{h}_{k_p}(u_p)$$

with

$$\hat{a}_{\mathbf{k}} = n^{-1} \sum_{j=p+1}^{n} Y_j \bar{h}_{k_0}(U_j) \cdots \bar{h}_{k_p}(U_{j-p}),$$

where we used the multiindex notation $\mathbf{k} = (k_0, \ldots, k_p)$.

12.2.6 Convergence of identification algorithms

In this section, we establish conditions for convergence of a nonparametric identification algorithm for recovering the system nonlinearity $m(\bullet, \bullet)$ of the system depicted in Figure 12.17. We use the result of Theorem 12.3, which allows us to propose the following plug-in estimate of $m(u, v)$:

$$\hat{m}(u, v) = \hat{r}_2(u, v) - \hat{\lambda}_1 \hat{r}_1(v) + c, \tag{12.168}$$

where the constant c is specified in Theorem 12.3. Moreover, $\hat{r}_2(u, v)$ and $\hat{r}_1(v)$ are certain nonparametric estimates of the regression functions $r_2(u, v)$ and $r_1(v)$, respectively. Next, $\hat{\lambda}_1$ is the cumulant based estimate of λ_1 discussed in Section 12.2.3, see formula (12.104). The constant c in (12.168) can be eliminated if either $E\{m(U_1, U_0)\} = 0$ or $m(0, 0) = 0$. In the latter case one must modify the estimate $\hat{m}(u, v)$ in (12.168) as follows:

$$\hat{m}(u, v) = \{\hat{r}_2(u, v) - \hat{r}_2(0, 0)\} - \hat{\lambda}_1 \{\hat{r}_1(v) - \hat{r}_1(0)\} . \qquad (12.169)$$

In this section, we make use of the standard kernel regression estimate (see Chapter 3) although other nonparametric techniques can be easily applied. Hence, we define

$$\hat{r}_2(u, v) = \frac{n^{-1}\sum_{j=2}^{n} Y_j K_h(u - U_j)K_h(v - U_{j-1})}{n^{-1}\sum_{j=2}^{n} K_h(u - U_j)K_h(v - U_{j-1})}, \qquad (12.170)$$

and

$$\hat{r}_1(v) = \frac{n^{-1}\sum_{j=1}^{n} Y_j K_h(v - U_j)}{n^{-1}\sum_{j=1}^{n} K_h(v - U_j)}, \qquad (12.171)$$

as the kernel estimates of the regression functions $r_2(u, v)$ and $r_1(v)$, respectively. Here $K_h(u) = h^{-1}K(h^{-1}u)$, where h is the bandwidth (smoothing parameter) and $K(u)$ is an admissible kernel function (see Chapter 3). Note that $\hat{r}_2(u, v)$ in (12.170) is the two-dimensional version of the kernel estimate, where the bivariate kernel $K_h(u, v)$ is of the product form, that is, $K_h(u, v) = K_h(u)K_h(v)$. See Chapter 13 for a further discussion on the issue of estimating multivariate functions.

The goal of this section is to provide a detailed proof of the convergence of the estimate $\hat{m}(u, v)$ to the true nonlinearity $m(u, v)$. We also examine the issue of the convergence rate and show that the kernel estimate $\hat{m}(u, v)$ tends to $m(u, v)$ with the optimal rate. Let $f(\bullet)$ denote the density function of the input signal $\{U_n\}$. For our future developments, we need the following technical lemma.

LEMMA 12.4 *Let $\{(V_i, K_i)\}$ be a sequence of independent and identically distributed random vectors in R^2 such that $E(V_0^2) + E(K_0^2) < \infty$. Let*

$$W_n = \sum_{i=-\infty}^{n} \chi_{n-i} V_i$$

be a linear process with $\sum_{i=0}^{\infty} |\chi_i| < \infty$. Then, for $t \geq 1$,

$$\mathrm{cov}(W_t K_t, W_0 K_0) = E^2(K_0)\,\mathrm{var}(V_0)\sum_{i=1}^{\infty} \chi_i \chi_{i+t}$$

$$+ E(K_0)E(V_0)\,\mathrm{cov}(K_0, V_0)\chi_t \sum_{i=1}^{\infty} \chi_i$$

$$+ E(K_0)\,\mathrm{cov}(V_0 K_0, V_0)\chi_0 \chi_t.$$

The proof of this result can be obtained by using straightforward algebra.

We will also employ the general fact that for a stationary stochastic process $\{X_i\}$ with a finite second moment, we have

$$\text{var}(X_1 + \cdots + X_n) = n\,\text{var}(X_1) + 2\sum_{j=1}^{n-1}(n-j)\,\text{cov}(X_0, X_j). \tag{12.172}$$

The next result provides the way of handling the variance of the sum of a p-dependent random sequence, that is, any two terms of the sequence are independent if the difference between their subscripts is larger than p.

LEMMA 12.5 *Let $\xi_1, \xi_2, \ldots$ be a sequence of random variables such that $E\xi_i = 0$, $E\xi_i^2 < \infty$, $i = 1, 2, \ldots$. Assume that for a nonnegative integer p the random variables ξ_i and ξ_j are independent whenever $\mid i - j \mid > p$. Then, for $n > p$*

$$E\left(\sum_{j=1}^{n} \xi_j\right)^2 \le (p+1)\sum_{j=1}^{n} E\xi_j^2.$$

Proof. Let us block the process $\{\xi_j\}$ into groups of independent random variables, that is,

$$\sum_{j=1}^{n} \xi_j = \{\xi_1 + \xi_{p+2} + \cdots\} + \{\xi_2 + \xi_{p+3} + \cdots\} + \cdots \{\xi_{p+1} + \xi_{2p+2} + \cdots\},$$

where each block has $\lfloor \frac{n}{p+1} \rfloor$ elements. Then,

$$E\left(\sum_{j=1}^{n} \xi_j\right)^2 \le (p+1)[E\left(\xi_1 + \xi_{p+2} + \cdots\right)^2$$

$$+ E\left(\xi_2 + \xi_{p+3} + \cdots\right)^2 + \cdots E\left(\xi_{p+1} + \xi_{2p+2} + \cdots\right)^2]$$

$$= (p+1)\sum_{j=1}^{n} E\xi_j^2.$$

This concludes the proof of Lemma 12.5. ∎

We need the following assumptions on the system nonlinearity $m(\bullet, \bullet)$ and the linear subsystem characteristic $\{\lambda_j\}$:

$$E\{m^2(U_1, U_0)\} < \infty, \tag{12.173}$$

$$\sum_{j=0}^{\infty} \lambda_j^2 < \infty. \tag{12.174}$$

Furthermore, we employ a class of bounded kernels that meets the following condition (see Chapter 3 for more facts about admissible kernels):

$$\int_{-\infty}^{\infty} \mid K(u) \mid du < \infty, \quad \mid u \mid\mid K(u) \mid \to 0 \quad \text{as } \mid u \mid \to \infty. \tag{12.175}$$

For such kernels we can establish, as shown in Appendix A (see also Chapter 13 for a further discussion on multivariate kernels), that if $\iint_{R^2} |\varphi(u, v)| \, du \, dv < \infty$ then

$$h^{-2} \iint_{R^2} \varphi(\xi_1, \xi_2) K \left(\frac{u - \xi_1}{h} \right) K \left(\frac{v - \xi_2}{h} \right) d\xi_1 \, d\xi_2 \to \varphi(u, v)$$

$$\times \left\{ \int_{-\infty}^{\infty} K(u) du \right\}^2 \quad \text{as } h \to 0, \tag{12.176}$$

at every point (u, v) of continuity of $\varphi(\bullet, \bullet)$.

Now we are in a position to form the theorem concerning the consistency of the kernel estimate $\hat{m}(u, v)$ of $m(u, v)$.

THEOREM 12.9 *Let the assumptions of Theorem 12.3 be satisfied. Suppose that conditions (12.173)–(12.175) hold. Let $m(\bullet, \bullet)$ and $f(\bullet)$ be bounded and continuous functions. If*

$$h \to 0 \quad \text{and} \quad nh^2 \to \infty, \tag{12.177}$$

then

$$\hat{m}(u, v) \to m(u, v), (P) \quad \text{as } n \to \infty,$$

for every point (u, v) at which $f(u) f(v) > 0$.

Proof. Considering Theorem 12.3, let us first note that

$$r_2(u, v) = m(u, v) + \lambda_1 \int_{-\infty}^{\infty} m(u, z) f(z) dz + c_1 \tag{12.178}$$

and

$$r_1(v) = \int_{-\infty}^{\infty} m(v, z) f(z) dz + c_2, \tag{12.179}$$

where the constants c_1 and c_2 can easily be derived from the proof of Theorem 12.3. Hence by (12.178) and (12.179) the smoothness of the regression functions $r_2(u, v)$ and $r_1(v)$ is determined by the smoothness of the nonlinearity $m(u, v)$ and the input density $f(u)$. In fact, if $m(u, v)$ is a continuous and bounded function on R^2 then by Lebesgue's dominated convergence theorem we can conclude that the integral

$$\int_{-\infty}^{\infty} m(v, z) f(z) dz$$

is a continuous function in v. Consequently, both regression functions $r_2(u, v)$ and $r_1(v)$ are continuous. By virtue of (12.168) it suffices to consider the convergence of the estimate $r_2(u, v)$ in (12.170). The estimate $\hat{r}_2(u, v)$ is of the ratio form, that is,

$$\hat{r}_2(u, v) = \frac{a_n(u, v)}{b_n(u, v)}, \tag{12.180}$$

where

$$a_n(u, v) = n^{-1} \sum_{j=2}^{n} Y_j K_h(u - U_j) K_h(v - U_{j-1}),$$

and $b_n(u, v)$ is defined by the denominator of (12.170). First, note that

$$E\{a_n(u, v)\}$$
$$= E\{Y_n K_h(u - U_n) K_h(v - U_{n-1})\}$$
$$= \iint_{R^2} r_2(u_1, u_0) K_h(u - u_1) K_h(v - u_0) f(u_0) f(u_1) du_0 du_1. \quad (12.181)$$

By (12.176), we can readily conclude that the integral in (12.181) tends to

$$r_2(u, v) f(u) f(v) \left\{ \int_{-\infty}^{\infty} K(u) du \right\}^2 \quad \text{as } h \to 0. \quad (12.182)$$

Let us express $a_n(u, v)$ in the following form:

$$a_n(u, v) = n^{-1} \sum_{j=2}^{n} Y_j K_j,$$

where $K_j = K_h(u - U_j) K_h(v - U_{j-1})$. We need to show that var $a_n(u, v) \to 0$. First, it follows from $Y_n = W_n + Z_n$ that

$$a_n(u, v) - E a_n(u, v) = n^{-1} \sum_{j=2}^{n} \{W_j K_j - E(W_j K_j)\} + n^{-1} \sum_{j=2}^{n} Z_j K_j. \quad (12.183)$$

The second term in (12.183) has a zero mean. Next, the second moment of this term is equal to

$$n^{-2}(n - 1) E(Z_1^2) E(K_1^2) \leq E(Z_1^2)(nh^2)^{-1} h^{-2} E\left\{ K^2\left(\frac{u - U_1}{h}\right) K^2\left(\frac{v - U_0}{h}\right) \right\}.$$

By virtue of (12.176), we have

$$h^{-2} E\left\{ K^2\left(\frac{u - U_1}{h}\right) K^2\left(\frac{v - U_0}{h}\right) \right\} \to f(u) f(v) \left\{ \int_{-\infty}^{\infty} K^2(u) du \right\}^2 \quad \text{as } h \to 0.$$

Thus, under the condition (12.177), we have shown that the second term in (12.183) tends (P) to zero as $n \to \infty$.

Let us now consider the first term in (12.183). We begin with the observation that $W_j = \sum_{t=0}^{\infty} \lambda_t V_{j-t}$ defines the linear stationary process with 1-dependent residuals $\{V_t = m(U_t, U_{t-1})\}$. Let us assume, without loss of generality, that $E V_t = 0$. Then, due to Lemma 12.5, we get

$$E W_j^2 \leq 2 E V_1^2 \sum_{t=0}^{\infty} \lambda_t^2.$$

By this, the stationarity of $\{W_j\}$, and (12.172), we can express the second moment of the first term in (12.183) in the following way:

$$n^{-2}(n-1)\,\mathrm{var}(W_n K_n) + n^{-2}\sum_{t=1}^{n-2}(n-1-t)\,\mathrm{cov}(W_t K_t, W_0 K_0) = P_n(u,v) + R_n(u,v).$$

(12.184)

To evaluate the term $P_n(u,v)$, let us denote $\eta_n = \sum_{j=2}^{\infty}\lambda_j V_{n-j}$. Then, we have

$$P_n(u,v) \leq 2n^{-1}\,\mathrm{var}\{(\lambda_0 V_n + \lambda_1 V_{n-2})K_n\} + 2n^{-1}\,\mathrm{var}\{\eta_n K_n\}$$
$$\leq 4\lambda_0^2 n^{-1}\,\mathrm{var}\{V_n K_n\} + 4\lambda_1^2 n^{-1}\,\mathrm{var}\{V_{n-1}K_n\} + 2n^{-1}\,\mathrm{var}\{\eta_n K_n\}$$
$$\leq 4\lambda_0^2 n^{-1}E\{V_n^2 K_n^2\} + 4\lambda_1^2 n^{-1}E\{V_{n-1}^2 K_n^2\} + 2n^{-1}\,\mathrm{var}\{\eta_n\}EK_n^2,$$

(12.185)

where we note that η_n is independent of K_n. Because of (12.176), we have

$$n^{-1}E\{V_n^2 K_n^2\} = (nh^2)^{-1}h^{-2}E\left\{m^2(U_n, U_{n-1})K^2\left(\frac{u-U_n}{h}\right)K^2\left(\frac{u-U_{n-1}}{h}\right)\right\}$$
$$\approx (nh^2)^{-1}m^2(u,v)f(u)f(v)\left\{\int_{-\infty}^{\infty}K^2(u)du\right\}^2 \quad \text{as } h \to 0.$$

By analogous derivations the second and third terms in (12.185) are of the orders

$$(nh^2)^{-1}f(u)f(v)E\{m^2(v, U_0)\}\left\{\int_{-\infty}^{\infty}K^2(u)du\right\}^2$$

and

$$(nh^2)^{-1}f(u)f(v)\left\{\int_{-\infty}^{\infty}K^2(u)du\right\}^2,$$

respectively. Hence, we have shown that

$$P_n(u,v) = O\left((nh^2)^{-1}\right).$$

(12.186)

In order to evaluate the term $R_n(u,v)$, in (12.184), let us write

$$W_t = W_t^e + W_t^o,$$

(12.187)

where $W_t^e = \sum_{j=-\infty}^{t}\lambda_{t-2j}V_{2j}$ and $W_t^o = \sum_{j=-\infty}^{t}\lambda_{t-(2j+1)}V_{2j+1}$. Analogously, let $W_0 = W_0^e + W_0^o$. It is important to note that W_t^e, W_t^o, W_0^e, and W_0^o are linear processes with independent residuals. Using the decomposition in (12.187), we can represent the covariance $\mathrm{cov}(W_t K_t, W_0 K_0)$ in (12.184) as follows:

$$\mathrm{cov}(W_t K_t, W_0 K_0) = \mathrm{cov}(W_t^e K_t, W_0^e K_0) + \mathrm{cov}(W_t^e K_t, W_0^o K_0)$$
$$+ \mathrm{cov}(W_t^o K_t, W_0^e K_0) + \mathrm{cov}(W_t^o K_t, W_0^e K_0).$$

(12.188)

The terms in the (12.188) formula can be directly evaluated by the application of Lemma 12.4. In fact, for $\text{cov}(W_t^e K_t, W_0^e K_0)$, we have

$$\text{cov}(W_t^e K_t, W_0^e K_0) = E(K_0)E(V_0^2 K_0)\xi_0\xi_t + E^2(K_0)E(V_0^2)\sum_{i=1}^{\infty}\xi_i\xi_{i+t},$$

where $\xi_i = \lambda_i$ for $i = 0, 2, 4, \ldots$, and $\xi_i = 0$ otherwise. Hence, the covariance $\text{cov}(W_t^e K_t, W_0^e K_0)$ is a function of the time lag t. Moreover, by virtue of (12.176)

$$E(V_0^2 K_0) \rightarrow m^2(u, v)f(u)f(v)\left\{\int_{-\infty}^{\infty} K^2(u)du\right\}^2 \quad \text{as } h \rightarrow 0$$

and

$$E(K_0) \rightarrow f(u)f(v)\left\{\int_{-\infty}^{\infty} K^2(u)du\right\}^2 \quad \text{as } h \rightarrow 0.$$

The remaining terms in (12.188) can be treated in the identical way. As a result,

$$R_n(u, v) = n^{-2}(n - 1)\sum_{t=1}^{n-2}\left(1 - \frac{t}{n - 1}\right)\gamma(t),$$

where the sequence $\gamma(t)$ can be written in an explicit way and due to (12.174) $\sum_{t=1}^{\infty} |\gamma(t)| < \infty$. Then, by Cesàro's summability theorem, we can conclude that $R_n(u, v) = O(n^{-1})$. Thus, we have shown that

$$a_n(u, v) \rightarrow r_2(u, v)f(u)f(v)\left\{\int_{-\infty}^{\infty} K(u)du\right\}^2 (P) \quad \text{as } n \rightarrow \infty.$$

Moreover, we have

$$\text{var}\{a_n(u, v)\} = O\left((nh^2)^{-1}\right). \tag{12.189}$$

By analogous considerations, we can prove that the term $b_n(u, v)$ in (12.180) tends (P) to

$$f(u)f(v)\left\{\int_{-\infty}^{\infty} K(u)du\right\}^2.$$

This proves that

$$\hat{r}_2(u, v) \rightarrow r_2(u, v)(P) \quad \text{as } n \rightarrow \infty.$$

In the identical way, we can obtain $\hat{r}_1(v) \rightarrow r_1(v)(P)$ as $n \rightarrow \infty$. Furthermore, we know that $\hat{\lambda}_1 \rightarrow \lambda_1(P)$ as $n \rightarrow \infty$. Then, recalling (12.168) and Theorem 12.3 we can conclude the proof of Theorem 12.9. ∎

The result of Theorem 12.9 can be used to evaluate the rate at which $\hat{m}(u, v)$ converges to $m(u, v)$. Indeed, by virtue of (12.168), we note that

$$\hat{m}(u, v) - m(u, v) = (\hat{r}_2(u, v) - r_2(u, v)) - \lambda_1(\hat{r}_1(v) - r_1(v)) - (\hat{\lambda}_1 - \lambda_1)\hat{r}_1(v). \tag{12.190}$$

Since $\hat{\lambda}_1 - \lambda$ tends to zero with the parametric rate $O_P(n^{-1/2})$, it is sufficient to examine the first two terms in (12.190). The most critical is the first term because it represents the error of estimating a function of two variables, whereas the second term concerns a function of a single variable. Hence, let us evaluate the rate at which $\hat{r}_2(u, v) - r_2(u, v) \to 0$. To do so, we use the representation of $\hat{r}_2(u, v)$ in (12.180) and the following identity (we called this the "ratio-trick" in (12.46)):

$$\hat{r}_2(u, v) - r_2(u, v) = \frac{a_n(u, v) - r_2(u, v)b_n(u, v)}{f(u)f(v)}$$

$$+ (\hat{r}_2(u, v) - r_2(u, v))\frac{f(u)f(v) - b_n(u, v)}{f(u)f(v)}, \quad (12.191)$$

where, without loss of generality, we assumed that $\int_{-\infty}^{\infty} K(u)du = 1$. Owing to Theorem 12.9, the rate at which $\hat{r}_2(u, v) - r_2(u, v) \to 0$ is controlled by the first term in (12.191). First, the proof of Theorem 12.9 reveals (see (12.189)) that the variance of the first term is of order $O((nh^2)^{-1})$. Regarding the bias we know that some smoothness conditions on $m(\bullet, \bullet)$ and $f(\bullet)$ must be assumed. Hence, let $m(\bullet, \bullet)$ and $f(\bullet)$ have two continuous and bounded derivatives on their support sets. Let for a general kernel function $K(\bullet, \bullet)$ on R^2 consider the following convolution integral:

$$h^{-2} \iint_{R^2} \varphi(\xi_1, \xi_2) K\left(\frac{u - \xi_1}{h}, \frac{v - \xi_2}{h}\right) d\xi_1\, d\xi_2, \quad (12.192)$$

where $\varphi(\bullet, \bullet)$ has two continuous and bounded derivatives. The kernel $K(\bullet, \bullet)$ is a symmetric function with respect to both arguments and $\iint_{R^2} K(u, v)du\, dv = 1$, $\iint_{R^2}(u^2 + v^2)K(u, v)du\, dv < \infty$. Then, by Taylor's formula the convolution in (12.192) is given by

$$\varphi(u, v) + \frac{h^2}{2}\left[\frac{\partial^2\varphi(u, v)}{\partial u^2}\iint_{R^2}\xi_1^2 K(\xi_1, \xi_2)d\xi_1\, d\xi_2\right.$$

$$\left. + \frac{\partial^2\varphi(u, v)}{\partial v^2}\iint_{R^2}\xi_2^2 K(\xi_1, \xi_2)d\xi_1\, d\xi_2\right] + o(h^2). \quad (12.193)$$

If $K(\bullet, \bullet)$ is a product kernel, then (12.193) takes the following form:

$$\varphi(u, v) + \frac{h^2}{2}\left[\frac{\partial^2\varphi(u, v)}{\partial u^2} + \frac{\partial^2\varphi(u, v)}{\partial v^2}\right]\int_{-\infty}^{\infty}\xi^2 K(\xi)d\xi + o(h^2). \quad (12.194)$$

Application of (12.194) to the first term in (12.191) with $\varphi(u, v) = r_2(u, v)f(u)f(v)$ (for the evaluation of the term $a_n(u, v)$) and $\varphi(u, v) = f(u)f(v)$ (for the evaluation of the term $b_n(u, v)$) yields

$$E\hat{r}_2(u, v) - r_2(u, v) = O(h^2). \quad (12.195)$$

Hence, by the aforementioned discussion, (12.190), and the fact that $\text{var}\,\hat{r}_2(u, v) = O((nh^2)^{-1})$, we obtain the next theorem on the convergence rate of the estimate $\hat{m}(u, v)$.

THEOREM 12.10 *Let all the assumptions of Theorem 12.9 hold. Let $m(\bullet, \bullet)$ and $f(\bullet)$ have two continuous and bounded derivatives on their support sets. Suppose that the* kernel function $K(\bullet)$ is symmetric and $\int_{-\infty}^{\infty} u^2 K(u)du < \infty$. If

$$h = cn^{-1/6}$$

then

$$\hat{m}(u, v) = m(u, v) + O_P(n^{-1/3}), \tag{12.196}$$

for every point (u, v) at which $f(u)f(v) > 0$.

It is worth noting that the rate in (12.196) is known to be optimal for estimating functions of two variables possessing two continuous derivatives. Furthermore, our method of proof is based on the "ratio-trick", which immediately gives the mean squared error rate, that is, we have

$$E\{\hat{m}(u, v) - m(u, v)\}^2 = O(n^{-2/3}).$$

The result in (12.196) can be extended to the case when $m(\bullet, \bullet)$ and $f(\bullet)$ have $s, s \geq 1$, continuous and bounded derivatives. To do so we need (in addition to the conditions of Theorem 12.9) the kernel function $K(\bullet)$ of the sth order, that is, $K(\bullet)$ meets the following restrictions:

$$\int_{-\infty}^{\infty} u^i K(u)du = 0, \quad i = 1, 2, \ldots, s - 1,$$

and

$$\int_{-\infty}^{\infty} |u^s K(u)| \, du < \infty.$$

In Chapter 3 (see also Appendix A), we provide a simple method for obtaining kernel functions of higher order. Under the above conditions we can show (using the technique identical to that used in the proof of Theorem 12.10) that if

$$h = cn^{-\frac{1}{2(s+1)}}$$

then

$$\hat{m}(u, v) = m(u, v) + O_P\left(n^{-\frac{s}{2(s+1)}}\right) \tag{12.197}$$

for every point (u, v) at which $f(u)f(v) > 0$. Clearly, the mean-squared error is $O(n^{-\frac{2s}{2(s+1)}})$. It is also worth noting that the rate $O_P(n^{-1/4})$ for $s = 1$ corresponds to the case when $m(\bullet, \bullet)$ and $f(\bullet)$ are Lipschitz continuous functions.

The technical arguments used in proving Theorems 12.9 and 12.10 can be carried over to the case of the system with nonlinear dynamics of order p. Then (see Remark 12.5) one must estimate the $(p + 1)$-dimensional regression function resulting in the following rate for recovering the system nonlinearity (for twice differentiable nonlinearities and input density functions):

$$O_P\left(n^{-\frac{2}{p+5}}\right). \tag{12.198}$$

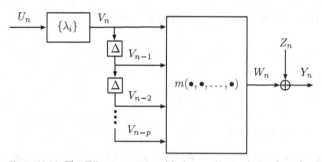

Figure 12.21 The Wiener system with the nonlinear dynamics of order p.

This rate reveals an apparent increase of the estimation error with p. This is the well-known curse of dimensionality that is present in any multivariate estimation problem. To appreciate this phenomenon, let us calculate the sample size $N \geq n$ required for the system with nonlinear dynamics of order $p > 1$ to have the identical accuracy as the system of order $p = 1$. Hence, because of (12.196) and (12.198), we have $N^{-\frac{2}{p+5}} = n^{-\frac{1}{3}}$, that is, $N = n^{\alpha p}$, where $\alpha = (p+5)/6p$. Since $1/6 < \alpha \leq 1$, we can conclude that to obtain a given degree of precision of a nonparametric estimate, the sample size must grow exponentially with the system order p. Hence, as we have already mentioned, the system with the input memory length $p = 1, 2$ is sufficient for most practical cases. The long memory effects are taken care of by the linear subsystem, which we allow to have an infinite memory.

12.3   Concluding remarks

In Section 12.2, we examined the generalization of the Hammerstein system that includes the nonlinearity possessing its own memory. This model has been shown to be very fruitful because it generalizes a number of important structures, for example, the Wiener and sandwich systems. One can also propose a generalized Wiener model being a counterpart of the system in Figure 12.17, that is, the model depicted in Figure 12.21. Here, the input–output description of the system is given by

$$Y_n = m(V_n, \ldots, V_{n-p}) + Z_n, \tag{12.199}$$

where $V_n = \sum_{j=0}^{\infty} \lambda_j U_{n-j}$.

This seems to be an attractive alternative to the system in Figure 12.17 with the input nonlinear dynamics. Unfortunately, the system in (12.199) is not identifiable for most common cases. Indeed, by taking

$$m(V_n, V_{n-1}, \ldots, V_{n-p}) = g\left(V_n + \sum_{j=1}^{p} a_j V_{n-j}\right),$$

we have

$$Y_n = g(a_n * \lambda_n * U_n) + Z_n,$$

that is, there are two linear systems in a series connection and this defines a nonidentifiable system.

There exists a broad class of nonlinear systems which can be tackled by the nonparametric regression approach. These systems do not have a block-oriented structure and their interpretability and flexibility to use in practice can be very limited. In particular, one can consider the following nonlinear counterpart of ARMA models:

$$Y_n = m(Y_{n-1}, Y_{n-2}, \dots, Y_{n-q}, U_n, U_{n-1}, \dots, U_{n-p}) + Z_n, \qquad (12.200)$$

where $m(\bullet, \dots, \bullet)$ is a $(p + q + 1)$ - dimensional function and $\{Z_n\}$ is a measurement noise. The nonlinearity $m(\bullet, \dots, \bullet)$ is a regression function of Y_n on the past outputs $Y_{n-1}, Y_{n-2}, \dots, Y_{n-q}$ and the current and past inputs $U_n, U_{n-1}, \dots, U_{n-p}$. Thus, it is a straightforward task to form a multivariate nonparametric regression estimate of $m(\bullet, \dots, \bullet)$, see (12.170) and Chapter 13. The convergence analysis of such an estimate will strongly depend on the stability conditions of the nonlinear recursive difference equation:

$$y_n = m(y_{n-1}, y_{n-2}, \dots, y_{n-q}, u_n, u_{n-1}, \dots, u_{n-p}).$$

With this respect, a fading-memory type assumption along with the Lipschitz continuity of $m(\bullet, \dots, \bullet)$ seem to be sufficient for the consistency of nonparametric regression estimates. Nevertheless, the accuracy of nonparametric regression estimates will be greatly limited by the apparent curse of dimensionality. In fact, for $m(\bullet, \dots, \bullet)$ being a Lipschitz continuous function the best possible rate can be

$$O_P\left(n^{-\frac{1}{3+p+q}}\right).$$

For the second order system $(q = p = 2)$ this gives a very slow rate of $O_P(n^{-1/7})$. To improve the accuracy we can apply the system introduced in Section 12.2, which can be viewed as a parsimonious approximation of the system (12.200). An alternative approach could be based on an additive approximation of (12.200), that is, one can consider the following additive system:

$$\begin{aligned} Y_n = m_1(Y_{n-1}) + m_2(Y_{n-2}) + \cdots + m_q(Y_{n-q}) \\ + g_0(U_n) + g_1(U_{n-1}) + \cdots + g_p(U_{n-p}) + Z_n, \qquad (12.201) \end{aligned}$$

where $m_i(\bullet)$'s and $g_j(\bullet)$'s are univariate functions. The estimation of such a system is much simpler than the model in (12.200) and consists of estimating the individual one-dimensional components $\{m_i(\bullet), g_j(\bullet)\}$. In fact, we can expect (assuming that all of the nonlinearities in (12.201) are Lipschitz) the optimal rate $O_P(n^{-1/3})$ (independent of p and q) for recovering the additive system. Note also that in (12.201) the nonlinear contribution of input signals $U_n, U_{n-1}, \dots, U_{n-p}$ and the past outputs $Y_{n-1}, Y_{n-2}, \dots, Y_{n-q}$ can easily be monitored and displayed. Nonlinear autoregressive time series models have been extensively studied in the statistical literature; see Fan and Yao [89], Tjøstheim [290], and Tong [291] for a review of this topic. In particular, nonparametric inference for these models has been initiated by Robinson [255]. Nonlinear additive time series structures have been considered in Chen and Tsay [46]. We examine additive systems of the block-oriented form in Chapter 13.

Thus far, it has been assumed that during the time of data measuring, the set of generating units and the structure of receivers do not change, hence nonlinearities and dynamic characteristics of the underlying system remain unchanged. An interesting extension of the aforementioned systems would relax this assumption and examine time-varying counterparts of block-oriented systems.

12.4   Bibliographical notes

Series-parallel connections introduced in Section 12.1 have been examined in the parametric framework in a number of contributions, see Bendat [16], Billings [19], Chen [45], Marmarelis and Marmarelis [207], and Westwick and Kearney [316], and the references cited therein. The nonparametric regression approach to identifying the parallel system was proposed in Greblicki and Pawlak [134] and next studied in Greblicki and Pawlak [135]. Nonparametric kernel and orthogonal series algorithms were examined, and their convergence properties were established. Nonparametric techniques for series-parallel models with nuisance characteristics were studied in Hasiewicz, Pawlak, and Śliwiński [158], see also Hasiewicz and Śliwiński [160]. In Ralston, Zoubir, and Boashash [249] a parallel-series model with polynomial nonlinearities was investigated, whereas the Uryson model was discussed in Gallman [91].

The generalized nonlinear block-oriented model of Section 12.1.4 was first introduced in Pawlak and Hasiewicz [234], see also Hasiewicz, Pawlak, and Śliwiński [158], Hasiewicz and Śliwiński [160], [159], and Pawlak and Hasiewicz [235] for further studies.

Block-oriented systems with nonlinear dynamics examined in Section 12.2 generalize a number of previously examined connections and have been mentioned and applied by Eskinat, Johnson, and Luyben [82], Greblicki and Pawlak [135], Pawlak, Pearson, Ogunnaike, and Doyle [237], Rozario and Papoulis [258], Zhu and Seborg [331], and recently re-examined in Enqvist and Ljung [81]. No consistent estimates of the system characteristics have been established. A fully nonparametric approach to the identification of these generalized models has been developed and thoroughly examined in Pawlak [232]. The theory of cumulants and its use in nonlinear system analysis and nonlinear signal processing is presented in Brillinger [32], Mathews and Sicuranza [208], and Nikias and Petropulu [219]. Bussgang's invariance theorem, originally proved for a single input nonlinear static element excited by a stationary Gaussian process, was extended to the multivariate case by Brillinger [31]. This seminal paper has been largely overlooked in works that followed, see Enqvist and Ljung [81], and Scarano, Caggiati, and Jacovitti [271]. The extension of Bussgang's theorem to non-Gaussian processes has been presented in Barrett and Lampard [15], Leipnik [194], and Nuttall [222].

The sandwich system plays an important role in numerous applications, see Quatieri, Reynold, and O'Leary [244], and Westwick and Kearney [316], and has been examined in Billings [19], Brillinger [31], Korenberg and Hunter [177], and Zhu [333]. A parametric class of nonlinearities has been assumed. The basic theorem assuring that from input-output measurements alone, we can extract information about the internal structure of the sandwich system, was proved by Boyd and Chua [28]. A fully nonparametric method for recovering the system nonlinearity is presented in Pawlak [231].

The problem of finding optimal linear approximations of nonlinear systems has been recently tackled in Enqvist and Ljung [81], Mäkilä [203], and Schoukens, Pintelon, Dobrowiecki, and Rolain [273].

The class of nonlinear systems defined in (12.200) has been examined in the control engineering literature under the name NARMAX models, see Chen and Billings [47] and Chen, Billings, Cowan, and Grant [48]. Mostly parametric estimation techniques have been investigated, see, however, Portier and Oulidi [239] for a nonparametric approach to such systems. The nonparametric inference of nonlinear autoregressive type time series models has been thoroughly studied in the statistical literature, see Chen and Tsay [46], Fan and Yao [89], Robinson [255], Tjøstheim [290], and Tong [291].

13 Multivariate nonlinear block-oriented systems

In all of the preceding chapters, we have examined the identification problem for block-oriented systems of various forms, that are characterized by a one-dimensional input process. In numerous applications, we confront the problem of identifying a system that has multiple inputs and multiple interconnecting signals. The theory and practical algorithms for identification of multivariate linear systems have been thoroughly examined in the literature [332]. On the other hand, the theory of identification of multivariate nonlinear systems has been far less explored. This is mainly due to the mathematical and computational difficulties appearing in multivariate problems. In this chapter, we examine some selected multivariate nonlinear models that are natural generalizations of the previously introduced block-oriented connections. An apparent curse of dimensionality that takes place in high-dimensional estimation problems forces us to focus on low-dimensional counterparts of the classical block-oriented structures. In particular, we examine a class of additive models, which provides a parsimonious representation for multivariate systems. Indeed, we show that the additive systems provide simple and interpretable structures, which also give a reasonable trade-off between the systematic modeling error and the estimation error of an identification algorithm. The theory of finding an optimal additive model is examined.

13.1   Multivariate nonparametric regression

As in all of the previous chapters, we will make use of the notion of a regression function. We need the extension of this concept to the multidimensional case. Hence, let $(\mathbf{U}, Y)$ be a random vector with values in $R^d \times R$. Then the conditional moment $m(\mathbf{u}) = E\{Y|\mathbf{U} = \mathbf{u}\}$ is referred to as the regression function of Y on the random vector $\mathbf{U}$. If $E|Y| < \infty$ then $m(\mathbf{u})$ is a well defined unique measurable function such that $E|m(\mathbf{U})| < \infty$. If, in addition, $EY^2 < \infty$ then $m(\mathbf{U})$ can be interpreted as the orthogonal projection of Y onto the closed subspace of all random variables $Z = \psi(\mathbf{U})$ such that $EZ^2 < \infty$. In particular, $m(\mathbf{u})$ solves the problem of the minimum mean-squared error prediction of Y given $\psi(\mathbf{U})$, that is,

$$m(\bullet) = \arg\min_{\psi} E(Y - \psi(\mathbf{U}))^2,$$

where the minimum is taken with respect to all measurable functions $\psi(\bullet)$ from R^d to R, such that $E\psi^2(\mathbf{U}) < \infty$. If there is not any a priori information about the form of $m(\bullet)$, then we are dealing with the case of nonparametric regression, which can be estimated by a number of local smoothing techniques. Hence, let $\{(\mathbf{U}_1, Y_1), \ldots, (\mathbf{U}_n, Y_n)\}$ be a training set drawn from the same distribution as $(\mathbf{U}, Y)$. Throughout this chapter, we assume that the process $\{\mathbf{U}_n, Y_n\}$ is stationary and therefore the regression function $E\{Y_n|\mathbf{U}_n = \mathbf{u}\}$ does not vary with time.

A generic form of a general class of kernel methods for estimating $m(\mathbf{u})$ can be written as follows:

$$\hat{m}(u) = \frac{n^{-1}\sum_{i=1}^{n} Y_i K_{\mathbf{B}}(\mathbf{u}, \mathbf{U}_i)}{n^{-1}\sum_{i=1}^{n} K_{\mathbf{B}}(\mathbf{u}, \mathbf{U}_i)}, \tag{13.1}$$

where $K_{\mathbf{B}}(\mathbf{u}, \mathbf{v})$ is a kernel function defined on $R^d \times R^d$ and $\mathbf{B}$ is a smoothing "parameter" that scales the kernel. The parameter $\mathbf{B}$ can have various multidimensional forms, that is, $\mathbf{B}$ can be a symmetric positive-definite matrix that allows us to adapt to the linear dependence structure between the coordinates of $\mathbf{u}$ and $\{\mathbf{U}_i\}$. A simpler form of the bandwidth assumes that $\mathbf{B}$ is a diagonal matrix with elements $(b_1, \ldots, b_d)$. Such a choice provides adjustment for different scales in the coordinates of $\mathbf{u}$ and $\{\mathbf{U}_i\}$. The simplest choice specifies a single global bandwidth b for all of the coordinates.

There are numerous forms of kernel functions in the d-dimensional space. A common choice is the convolution kernel

$$K_{\mathbf{B}}(\mathbf{u}, \mathbf{v}) = K_{\mathbf{B}}(\mathbf{u} - \mathbf{v}), \tag{13.2}$$

where the scaling provided by the bandwidth matrix $\mathbf{B}$, is such that $K_{\mathbf{B}}(\mathbf{u} - \mathbf{v}) = |\mathbf{B}|K(\mathbf{B}(\mathbf{u} - \mathbf{v}))$, where $K(\bullet)$ is the "scale-free" kernel function. Here, $|\mathbf{B}|$ denotes the determinant of $\mathbf{B}$. The convolution kernels are invariant to translations in an input space and they define the classical kernel-regression estimate.

Two common forms of $K(\bullet)$ in (13.2) are radial and product types of kernels. In the former type, a kernel depends on the distance between the arguments, that is, $K(\mathbf{u} - \mathbf{v}) = k(\|\mathbf{u} - \mathbf{v}\|)$, where $k(\bullet)$ is a single-variable kernel function. In the latter, a kernel is a product of univariate kernels applied to the individual coordinates of the vector $\mathbf{u} = (u_1, \ldots, u_d)^T$, that is,

$$K(\mathbf{u}) = K_1(u_1) \cdots K_d(u_d), \tag{13.3}$$

where $K_1(\bullet), \ldots, K_d(\bullet)$ are various admissible kernel functions (see Chapter 3 and Appendix A). It is worth mentioning that the kernel estimate $\hat{m}(\mathbf{u})$ in (13.1) equipped in the convolution kernel of the product form (13.3) is invariant under different scaling factors of the input random vector $\mathbf{U}$. This property is desirable from the practical point of view because the components of $\mathbf{U}$ may represent various incommensurable characteristics.

Figure 13.1 shows the bivariate kernel of the product form $K_{\mathbf{h}}(u_1, u_2) = h_1^{-1}K_1(h_1^{-1}u_1)h_2^{-1}K_2(h_2^{-1}u_2)$ scaled by the bandwidth vector $\mathbf{h} = (h_1, h_2)$. Here, $K_1(u) = (2\pi)^{-1/2}e^{-u^2/2}$ is the Gaussian kernel and $K_2(u) = \frac{3}{4}(1 - u^2)\mathbf{1}(|u| \leq 1)$ is the

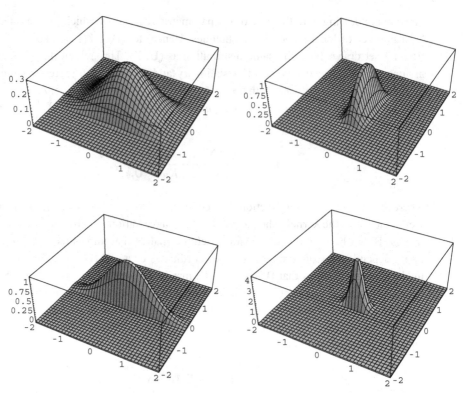

Figure 13.1 Two-dimensional product kernels for $(h_1, h_2) \in \{(1, 1), (1, 0.25); (0.25, 1), (0.25, 0.25)\}$.

optimal Epanechnikov kernel (see Chapter 3). The bandwidth vector is set to the following values:

$$(h_1, h_2) = \{(1, 1), (1, 0.25); (0.25, 1), (0.25, 0.25)\}.$$

A rich class of non-convolution kernels can be obtained from orthogonal representations of multivariate functions; see Appendix B for a related discussion on the one-dimensional case. The kernel function $K_{\mathbf{B}}(\mathbf{u}, \mathbf{v})$ generated by an orthonormal system $\{w_{\mathbf{k}}(\mathbf{u})\}$ is of the form

$$K_{\mathbf{B}}(\mathbf{u}, \mathbf{v}) = \sum_{|\mathbf{k} \leq \mathbf{B}} w_{\mathbf{k}}(\mathbf{u}) w_{\mathbf{k}}(\mathbf{v}), \qquad (13.4)$$

where $\mathbf{B}$ is the truncation parameter vector, and $\{w_{\mathbf{k}}(\mathbf{u}), \mathbf{k} \geq \mathbf{0}\}$ is an orthonormal basis with respect to some weight function $\lambda(\mathbf{u})$ defined on the set $\Omega \subseteq R^d$. Hence,

$$\int_{\Omega} w_{\mathbf{k}}(\mathbf{u}) w_{\mathbf{r}}(\mathbf{u}) \lambda(\mathbf{u}) d\mathbf{u} = \delta_{\mathbf{kr}}.$$

The notation $|\mathbf{k}|$ for the multiindex $\mathbf{k} = (k_1, \ldots, k_d)$ denotes the norm. The vector summation in (13.4) may have different meanings depending on the notion of the norm used for the multiindex $\mathbf{k}$. For example, the norm $|\mathbf{k}| = \max(k_1, \ldots, k_d)$ yields the

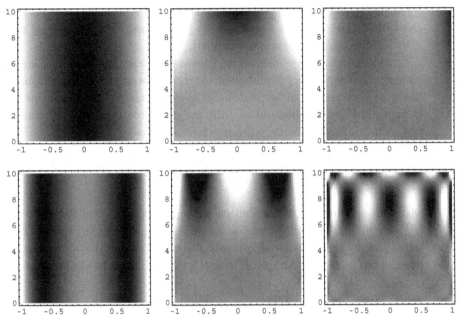

Figure 13.2 Density plots of the polynomials $p_{k_1}(u_1)l_{k_2}(u_2)$, $-1 \leq u_1 \leq 1$, $0 \leq u_2 \leq 10$.

cubic summation with a single truncation parameter B. Other summation types may need several truncation parameters. The choice of the weight function $\lambda(\mathbf{u})$ is essential for generating various forms of orthonormal systems in $\Omega \subseteq R^d$. The simplest choice is to use a product weight, that is,

$$\lambda(\mathbf{u}) = \lambda_1(u_1) \cdots \lambda_d(u_d), \tag{13.5}$$

where $\{\lambda_i(u), 1 \leq i \leq d\}$ are well-defined univariate weight functions. This yields the concept of tensor product orthonormal systems, that is, then we have

$$w_{\mathbf{k}}(\mathbf{u}) = w_{k_1}^1(u_1) \cdots w_{k_d}^d(u_d), \tag{13.6}$$

where $\{w_k^i(u), k \geq 0\}$ defines an orthonormal basis on $\Omega_i \subseteq R$, $i = 1, \ldots, d$ (with respect to the weight $\lambda_i(u)$), such that $\Omega = \Omega_1 \times \cdots \times \Omega_d$. This is a particularly simple way of generating multivariate orthonormal bases and corresponding kernel functions. This also allows us to design an appropriate support for the input variable $\mathbf{u}$. For example, in two dimensions, we can define the following orthonormal system:

$$w_{k_1 k_2}(u_1, u_2) = p_{k_1}(u_1)l_{k_2}(u_2), \quad k_1, k_2 \geq 0, \tag{13.7}$$

where $\{p_k(u)\}$ and $\{l_k(u)\}$ are Legendre and Laguerre orthonormal polynomials, respectively (see Appendix B). This defines the orthonormal basis on the set $\Omega = [-1, 1] \times [0, \infty)$. In Figure 13.2, we depict a collection of the polynomials in (13.7) for $(k_1, k_2) \in \{(2, 0), (2, 2), (3, 1); (4, 0), (4, 2), (8, 4)\}$.

It is also possible to construct nonproduct orthogonal systems by using a general weight function. For example, a radial weight function of the form $w_{\mathbf{k}}(\mathbf{u}) =$

$c(1 - \|\mathbf{u}\|^2)\mathbf{1}(\|\mathbf{u}\| \leq 1)$ yields a class of orthogonal Appell polynomials within the unit ball $\{\mathbf{u} : \|\mathbf{u}\| \leq 1\}$; see Dunkl and Xu [76] for an extensive study of the theory of multidimensional orthogonal polynomials.

Let us now turn to the problem of the statistical accuracy of nonparametric regression estimates. The consistency of nonparametric regression estimates can be obtained under very general conditions. In fact, there exists an elegant and powerful result due originally to Stone [284], which allows us to verify the global consistency of a general class of nonparametric local averaging regression estimates, that is,

$$\hat{m}(\mathbf{u}) = \sum_{i=1}^{n} v_{in}(\mathbf{u})Y_i, \tag{13.8}$$

where the weights $\{v_{in}(\mathbf{u})\}$ depend on the input data $\{\mathbf{U}_i\}$ and are often such that $\sum_{i=1}^{n} v_{in}(\mathbf{u}) = 1$. The weights provide local smoothing and can be generated by various techniques, for example, kernel methods, partition type estimates or nearest neighbor rules. In fact, the kernel estimate $\hat{m}(\mathbf{u})$ in (13.1) takes the form (13.8) with the weights

$$v_{in}(\mathbf{u}) = \frac{K_\mathbf{B}(\mathbf{u}, \mathbf{U}_i)}{\sum_{i=1}^{n} K_\mathbf{B}(\mathbf{u}, \mathbf{U}_i)}, \quad i = 1, \ldots, n.$$

Assuming that the training set $\{(\mathbf{U}_1, Y_1), \ldots, (\mathbf{U}_n, Y_n)\}$ is a sequence of independent random variables one can show that under some general conditions on $\{v_{in}(\mathbf{u})\}$ the estimate (13.8) can converge to $m(\mathbf{u})$ in the L_2 sense, that is,

$$\lim_{n \to \infty} E\{(\hat{m}(\mathbf{U}) - m(\mathbf{U}))^2\} = 0. \tag{13.9}$$

This is a deep result since the consistency holds for all possible distributions of $(\mathbf{U}, Y)$ and no smoothness of the regression function is required. Such distribution-free (universal) results have been verified for various nonparametric estimates. As well, the pointwise universal properties have been established; see [140] for extensive studies on the distribution-free theory of nonparametric regression. The situation becomes less ideal if one relaxes the assumption of data independence; this case is important for dynamical system identification. Here, it is known (see [140]) that if $\{(\mathbf{U}_n, Y_n)\}$ is a stationary and ergodic process then there is no universally consistent regression estimate (in the sense as in (13.9)).

Yet another issue in the multivariate-regression problem is the inverse relationship between the accuracy of nonparametric estimates and the dimensionality of the input variable $\mathbf{U}$. To illustrate this point, let us consider the problem of the convergence rate for a classical kernel estimate with product-type kernel functions. Let

$$\hat{m}(\mathbf{u}) = \frac{n^{-1} \sum_{i=1}^{n} Y_i K_\mathbf{h}(\mathbf{u} - \mathbf{U}_i)}{n^{-1} \sum_{i=1}^{n} K_\mathbf{h}(\mathbf{u} - \mathbf{U}_i)}, \tag{13.10}$$

where

$$K_\mathbf{h}(\mathbf{u}) = h_1^{-1} K(h_1^{-1} u_1) \cdots h_d^{-1} K(h_d^{-1} u_d). \tag{13.11}$$

Thus, we use a single kernel function $K(u)$ for all coordinates and the smoothing factor varies in each direction. We will find this kind of kernel function very versatile in our additive modeling of nonlinear systems (see Section 13.2).

For our subsequent developments concerning the convergence rate of $\hat{m}(\bullet)$ in (13.10), we need the following multidimensional counterpart of the result in Appendix A (see Section A.2.2).

LEMMA 13.1 *Let*

$$\varphi_{\mathbf{h}}(\mathbf{u}) = \int_{R^d} K_{\mathbf{h}}(\mathbf{u} - \mathbf{v})\varphi(\mathbf{v})d\mathbf{v}$$

be the convolution of $\varphi(\mathbf{u})$ with $K_{\mathbf{h}}(\mathbf{u}) = (\prod_{i=1}^d h_i)^{-1}K(h_1^{-1}u_1, \ldots, h_d^{-1}u_d)$, where $K(u_1, \ldots, u_d)$ is the d-dimensional kernel function, such that

$$\int_{R^d} u_i K(\mathbf{u})d\mathbf{u} = 0, \quad i = 1, \ldots, d$$

and

$$\int_{R^d} u_i u_j K(\mathbf{u})d\mathbf{u} = \begin{cases} 0 & \text{if } i \neq j \\ \mu_i(K) & \text{if } i = j \end{cases},$$

with $\mu_i(K) = \int_{R^d} u_i^2 K(\mathbf{u})d\mathbf{u} < \infty$.

Suppose that $\varphi(\mathbf{u})$ has two bounded derivatives. Then we have

$$\varphi_{\mathbf{h}}(\mathbf{u}) = \varphi(\mathbf{u}) + \frac{1}{2}\sum_{i=1}^d \frac{\partial^2 \varphi(\mathbf{u})}{\partial u_i^2} h_i^2 \mu_i(K) + o\left(\max_{1 \leq i \leq d}(h_i^2)\right). \quad (13.12)$$

The proof of this result is standard and is based on Taylor's expansion. Note that for the product kernel (13.11), the factor $\mu_i(K)$ is equal to $\int_R u^2 K(u)du < \infty$ for all $i = 1, \ldots, d$.

Let the kernel function $K(u)$ in (13.11) be a symmetric function which satisfies the conditions of Lemma A.1 in Appendix A. Let us assume that the data $\{(\mathbf{U}_1, Y_1), \ldots, (\mathbf{U}_n, Y_n)\}$ are *iid*. Arguing along the lines of the analysis given in Chapter 3 (see also the proof of Theorem 12.10) we can show that if the regression function $m(\mathbf{u})$ and the input density $f(\mathbf{u})$ are continuous then

$$\text{var}\{\hat{m}(\mathbf{u})\} = \frac{1}{nh_1 \cdots h_d} v(K)\frac{\sigma^2(\mathbf{u})}{f(\mathbf{u})}(1 + o(1)) \quad (13.13)$$

at every point $\mathbf{u}$ where $f(\mathbf{u}) > 0$. Here $\sigma^2(\mathbf{u}) = \text{var}\{Y|\mathbf{U} = \mathbf{u}\}$ and $v(K) = \left(\int_R K^2(z)dz\right)^d$.

To evaluate the bias of $\hat{m}(\mathbf{u})$ we can decompose $\hat{m}(\mathbf{u})$ as in (12.46) in Chapter 12 and then apply Lemma 13.1. Hence, assuming that $m(\mathbf{u})$ and $f(\mathbf{u})$ have two bounded derivatives, then at every point $\mathbf{u}$, where $f(\mathbf{u}) > 0$ we have

$$E\hat{m}(\mathbf{u}) = m(\mathbf{u}) + \frac{1}{2}\mu(K)\sum_{i=1}^d \psi_i(\mathbf{u})h_i^2 + o\left(\max_{1 \leq i \leq d}(h_i^2)\right), \quad (13.14)$$

where $\mu(K) = \int_R u^2 K(u)du$ and $\psi_i(\mathbf{u})$, $i = 1, \ldots, d$, are some bounded functions.

Combination of (13.13) and (13.14) can readily lead to a formula for the mean-squared error. This can be minimized with respect to the bandwidth vector $\mathbf{h} = (h_1, \ldots, h_d)$ yielding the optimal choice

$$h_i^* = a_i n^{-1/(4+d)}, \tag{13.15}$$

for some positive constants a_i, $i = 1, \ldots, d$.

Hence, the asymptotic choice of the different bandwidths is almost identical (up to the multiplicative constant). Plugging (13.15) into the mean-squared error gives the rate

$$E\{\hat{m}(\mathbf{u}) - m(\mathbf{u})\}^2 = O\left(n^{-\frac{4}{4+d}}\right). \tag{13.16}$$

This is a rather pessimistic result since the convergence rate is inversely proportional to the dimensionality of the input variable. In the next section, we examine low-dimensional approximations of fully nonparametric regression in order to mitigate the dimensionality problem.

13.2　Additive modeling and regression analysis

Fully nonparametric regression models in multivariate systems suffer from the curse of dimensionality. In order to ease this phenomenon, we consider a class of nonparametric additive models where a multivariate-regression function is approximated by the sum of univariate functions. We begin with a function approximation result that describes how to find an orthogonal projection of a function of several variables onto the subspace of additive functions.

13.2.1　Approximation by additive functions

Let us consider a class of measurable functions $m(\bullet)$ defined on R^d and such that

$$\int_{R^d} m^2(\mathbf{u})w(\mathbf{u})d\mathbf{u} < \infty, \tag{13.17}$$

where $w(\bullet)$ is a positive and integrable function on R^d. Let us denote a class of functions satisfying (13.17) as $L_2(w)$. We wish to characterize a low-complexity approximation of $m(\bullet) \in L_2(w)$ by means of functions of the additive form, that is, we seek the orthogonal projection of $m(\bullet)$ onto the following linear subspace:

$$\mathcal{A} = \{g \in L_2(w) : g(\mathbf{u}) = g_1(u_1) + \cdots + g_d(u_d)\}, \tag{13.18}$$

where the additive components $\{g_i(u_i), 1 \le i \le d\}$ are univariate functions. It is clear that the additive representation is not unique because the model $g(\mathbf{u}) = g_1(u_1) + g_2(u_2) + \cdots + g_d(u_d)$ is not distinguishable from $\tilde{g}(\mathbf{u}) = \{g_1(u_1) - c\} + g_2(u_2) + \cdots + \{g_d(u_d) + c\}$ for any constant c. Thus, we need some identifiability conditions on

$\{g_i(u_i), 1 \leq i \leq d\}$ in order to find a unique additive approximation of $m(\bullet) \in L_2(w)$. For identifiability, we may assume that

$$\int_R g_i(u_i)w_i(u_i)du_i = 0, \quad i = 1, \ldots, d. \tag{13.19}$$

Here, $w_i(\bullet)$ is the marginal function of $w(\bullet)$, that is,

$$w_i(u_i) = \int_{R^{d-1}} w(\mathbf{u})d\mathbf{u}_{-i}$$

where $d\mathbf{u}_{-i}$ denotes the integration with respect to all variables but the ith. Yet another way of making the additive approximation unique is to require that the weighted area under the additive function $g(\mathbf{u})$ is equal to the weighted area of $m(\mathbf{u})$, that is,

$$\int_{R^d} m(\mathbf{u})w(\mathbf{u})d\mathbf{u} = \int_{R^d} g(\mathbf{u})w(\mathbf{u})d\mathbf{u}. \tag{13.20}$$

It is worth noting that (13.20) makes the choice of $g(\mathbf{u})$ unique but not the choice of the individual components $\{g_i(u_i), 1 \leq i \leq d\}$. The constraints in (13.19) provide a unique choice of the individual components.

The error of the additive approximation is the squared distance in the $L_2(w)$ space. Hence, let

$$J(m; g) = \int_{R^d} [m(\mathbf{u}) - g(\mathbf{u})]^2 w(\mathbf{u})d\mathbf{u}, \tag{13.21}$$

for $g(\bullet) \in \mathcal{A}$, be the measure of the additive approximation of $m(\bullet)$. We wish to find $g(\bullet) \in \mathcal{A}$, which minimizes $J(m; g)$. Such a solution is the orthogonal projection of $m(\bullet)$ onto the subspace $\mathcal{A}$. The following theorem is fundamental for our subsequent considerations as it characterizes the optimal additive model.

THEOREM 13.1 *Let $m(\bullet) \in L_2(w)$. Suppose that $w(\bullet)$ is a positive and integrable function on R^d. The unique additive function $g^*(\mathbf{u}) = g_1^*(u_1) + \cdots + g_d^*(u_d)$ which minimizes $J(m; g)$ is characterized by the following system of equations:*

$$g_r^*(u_r) = \int_{R^{d-1}} m(\mathbf{u})\frac{w(\mathbf{u})}{w_r(u_r)}d\mathbf{u}_{-r} - \sum_{i \neq r}^{d} \int_{R^{d-1}} g_i^*(u_i)\frac{w(\mathbf{u})}{w_r(u_r)}d\mathbf{u}_{-r}, \tag{13.22}$$

subject to the constraints

$$\int_R g_r^*(u_r)w_r(u_r)du_r = 0, \tag{13.23}$$

for $r = 1, \ldots, d$.

It is worth noting that because $g^*(\bullet)$ is the orthogonal projection of $m(\bullet)$ onto the subspace $\mathcal{A}$, therefore

$$J(m; g^*) = \int_{R^d} m^2(\mathbf{u})w(\mathbf{u})d\mathbf{u} - \int_{R^d} g^{*2}(\mathbf{u})w(\mathbf{u})d\mathbf{u}. \tag{13.24}$$

The optimal additive approximation of $m(\bullet)$ is thus given by

$$m(\mathbf{u}) \approx g_1^*(u_1) + \cdots + g_d^*(u_d),$$

where $\{g_r^*(u_r)\}$ are characterized by the solution given in (13.22) and (13.23).

It is informative to write the system of equations in (13.22) for functions of two variables. Hence, we have

$$g_1^*(u_1) = \int_R m(u_1, u_2)\frac{w(u_1, u_2)}{w_1(u_1)}du_2 - \int_R g_2^*(u_2)\frac{w(u_1, u_2)}{w_1(u_1)}du_2,$$

$$g_2^*(u_2) = \int_R m(u_1, u_2)\frac{w(u_1, u_2)}{w_2(u_2)}du_1 - \int_R g_1^*(u_1)\frac{w(u_1, u_2)}{w_2(u_2)}du_1,$$

with the identifiability condition $\int_R g_r^*(u_r)w_r(u_r)du_r = 0, r = 1, 2.$

Proof of Theorem 13.1: In the proof, we make use of the calculus of variations. Hence, let us consider variations in the components $\{g_r(u_r)\}$ of the additive function $g(\mathbf{u})$:

$$g_r(u_r) = g_r^*(u_r) + \varepsilon_r \delta_r(u_r), \quad r = 1, \ldots, d, \tag{13.25}$$

where $\{\varepsilon_r\}$ are small positive numbers and $\{\delta_r(u_r)\}$ are arbitrary functions. By inserting (13.25) in $J(m; g)$ we get a function $J(\varepsilon_1, \ldots, \varepsilon_d)$ dependent on $(\varepsilon_1, \ldots, \varepsilon_d)$. The functional $J(m; g)$ has a minimum value for $g^*(\bullet)$ if

$$\frac{\partial J(\varepsilon_1, \ldots, \varepsilon_d)}{\partial \varepsilon_r} = 0 \quad \text{for } (\varepsilon_1, \ldots, \varepsilon_d) = (0, \ldots, 0), r = 1, \ldots, d.$$

This is equivalent to

$$\int_{R^d} \left(m(\mathbf{u}) - \sum_{i=1}^d g_i^*(u_i) \right) \delta_r(u_r)w(\mathbf{u})d\mathbf{u} = 0, \quad r = 1, \ldots, d.$$

Because this must hold for an arbitrary choice of functions $\{\delta_r(u_r), r = 1, \ldots, d\}$, it follows that

$$\int_{R^{d-1}} m(\mathbf{u})w(\mathbf{u})d\mathbf{u}_{-r} = \sum_{i=1}^d \int_{R^{d-1}} g_i^*(u_i)w(\mathbf{u})d\mathbf{u}_{-r}, \quad r = 1, \ldots, d. \tag{13.26}$$

Noting that the right-hand side of (13.26) is equal to

$$g_r^*(u_r) \int_{R^{d-1}} w(\mathbf{u})d\mathbf{u}_{-r} + \sum_{i \neq r}^d \int_{R^{d-1}} g_i^*(u_i)w(\mathbf{u})d\mathbf{u}_{-r}$$

we can conclude the proof of Theorem 13.1. ∎

A particularly simple solution of (13.22) and (13.23) takes place for the weight function $w(\mathbf{u})$ of the product form. Hence, if $w(\mathbf{u}) = \prod_{i=1}^d w_i(u_i)$, then we can observe that due to the identifiability condition (13.23) the integral

$$\int_{R^{d-1}} g_i^*(u_i)\frac{w(\mathbf{u})}{w_r(u_r)}d\mathbf{u}_{-r}$$

appearing in (13.22) is equal to zero. This readily yields the following important result.

THEOREM 13.2 *Let all the conditions of Theorem 13.1 hold. Moreover, let the weight function $w(\mathbf{u})$ be of the product form $w(\mathbf{u}) = \prod_{i=1}^{d} w_i(u_i)$. Then, a unique additive function $g^*(\mathbf{u}) = g_1^*(u_1) + \cdots + g_d^*(u_d)$, which minimizes $J(m; g)$ is given by the following equations:*

$$g_r^*(u_r) = \int_{R^{d-1}} m(\mathbf{u}) w_{-r}(\mathbf{u}) d\mathbf{u}_{-r}, \quad r = 1, \ldots, d, \qquad (13.27)$$

where $w_{-r}(\mathbf{u}) = \prod_{i \neq r}^{d} w_i(u_i)$.

The solution in (13.27) can be interpreted as the weighted marginal integration of the function $m(\mathbf{u})$ with respect to $d - 1$ variables. The identifiability condition (13.23) required for the solution (13.27) is satisfied if the weight function $w(\mathbf{u})$ is even and all of the additive components are odd. This is often unrealistic for functions $m(\mathbf{u})$, which are positive on the support of $w(\mathbf{u})$. Then, we can modify the solution in (13.27) by subtracting a constant such that the global constraint in (13.20) is met. Hence, we can use

$$g_r^*(u_r) = \int_{R^{d-1}} m(\mathbf{u}) w_{-r}(\mathbf{u}) d\mathbf{u}_{-r} - c, \qquad (13.28)$$

where $c = (1 - d^{-1}) \int_{R^d} m(\mathbf{u}) w(\mathbf{u}) d\mathbf{u}$. It is easy to verify that for such a choice of c, the solution $g^*(\bullet)$ satisfies the condition (13.20). Note that we assumed, without loss of generality, that $\int_R w_r(v) dv = 1, r = 1, \ldots, d$. The following examples shed some light on further aspects of additive modeling.

Example 13.1 Let us consider the function $m(\mathbf{u}) = u_1 \cdots u_d$. Let $w(\mathbf{u}) = 1_{[0,1]^d}(\mathbf{u})$, that is, the weight function is supported on the hypercube $[0, 1]^d$. Simple algebra resulting from (13.27) shows that the following functions:

$$g_r^*(u_r) = 2^{-d}\{2u_r - 1\}, \quad r = 1, \ldots, d$$

give the unique solution of the equations (13.22) and (13.23). Note that the solution consists of linear functions symmetrical with respect to the point $u = 1/2$.

Due to (13.24), the corresponding error is $J(m; g^*) = 3^{-d} - \frac{d}{3} 2^{-2d}$. On the other hand, we can apply the strategy in (13.28), which leads to a nonsymmetrical solution

$$\bar{g}_r^*(u_r) = 2^{-d}\{2u_r - (1 - 1/d)\}.$$

For this solution, the corresponding error is $3^{-d} - 2^{-2d}\left(\frac{d}{3} + 1\right)$, which is smaller than that for the symmetrical additive approximation. Note that $m(\mathbf{u})$ is positive on the support of $w(\mathbf{u})$ and the solution $\bar{g}_r^*(u_r)$ is less negative than $g_r^*(u_r)$. Figure 13.3 plots both solutions for $d = 3$. Hence, we have the best additive approximation of $u_1 u_2 u_3$ given by $(u_1 + u_2 + u_3 - 1)/4$. In Figure 13.4, we depict the approximation error as a continuous curve that is a function of d. Surprisingly, the additive approximations improve exponentially fast with the dimensionality of the input vector $\mathbf{u}$.

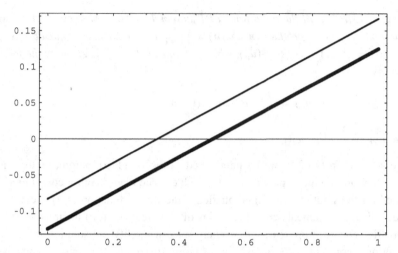

Figure 13.3 The components of the additive approximation of $m(\mathbf{u}) = u_1 u_2 u_3 : g_r^*(\bullet)$ (thick line) and $\bar{g}_r^*(\bullet)$.

Example 13.2 In this example we consider a concrete two-dimensional function of the form $m(u, v) = v^3 \arctan(u + v^3)$. The shape of $m(u, v)$ is shown in Figure 13.5. The weight function $w(u, v)$ is of the product form and is assumed to be a uniform density on $[-1, 1]^2$. A numerical integration algorithm has been applied to obtain the additive components corresponding to the identifiability condition (13.23), that is,

$$g_1^*(u) = \frac{1}{2} \int_{-1}^{1} m(u, v) dv \quad \text{and} \quad g_2^*(v) = \frac{1}{2} \int_{-1}^{1} m(u, v) du. \tag{13.29}$$

These functions are displayed in Figure 13.6. Interestingly, the functions are concave and convex, whereas the original nonlinearity $m(u, v)$ is not. The error $4^{-1} \int_{-1}^{1} (m(u, v) - g_1^*(u) - g_2^*(v))^2 du\, dv$ of the additive approximation was evaluated

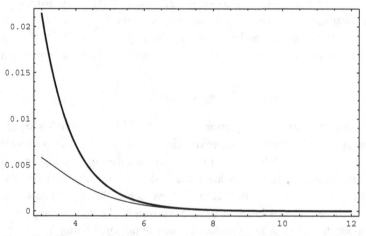

Figure 13.4 The error of additive approximations $g^*(\bullet)$ (thick line) and $\bar{g}^*(\bullet)$ versus d for $m(\mathbf{u}) = u_1 \cdots u_d$.

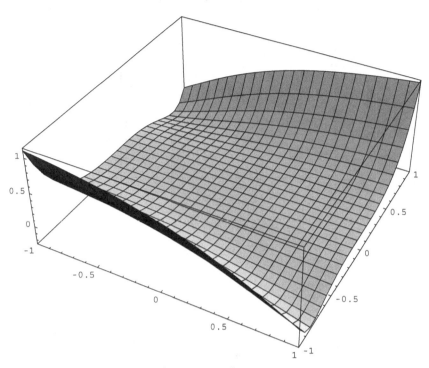

Figure 13.5 Function $m(u, v) = v^3 \arctan(u + v^3)$.

giving the value 0.0332693. The modified version of (13.28) (satisfying the global constraint (13.20)) was also evaluated. Hence, the functions $g_1^*(u) - \alpha$, $g_2^*(u) - \alpha$, where $\alpha = 8^{-1} \int_{-1}^{1} \int_{-1}^{1} m(u, v) du\, dv$, were generated and are plotted in Figure 13.7. The error of this approximation is 0.0220564 and is lower than that for the solution in (13.29).

13.2.2 Additive regression

The analysis in the previous section can be carried over to the problem of additive modeling of a multivariate regression function $m(\mathbf{u}) = E\{Y|\mathbf{U} = \mathbf{u}\}$. In this case, it seems to be natural to choose the weight function $w(\bullet)$ in (13.22) as the probability distribution of the random vector $\mathbf{U} = (U_1, \ldots, U_d)^T$.

Since our additive approximation has been established with respect to the squared error, we must assume that $EY^2 < \infty$. Suppose also that $\mathbf{U}$ has a density $f(\bullet)$. The approximation error (13.21) is now equal to

$$J(m; g) = E\{(m(\mathbf{U}) - g(\mathbf{U}))^2\}.$$

Let us examine the solution in (13.22) when $w(\mathbf{u}) = f(\mathbf{u})$. Note that the first term on the right-hand side of (13.22) has a simple statistical interpretation, that is, we have

$$\int_{R^{d-1}} m(\mathbf{u}) \frac{f(\mathbf{u})}{f_r(u_r)} d\mathbf{u}_{-r} = E\{m(\mathbf{U})|U_r = u_r\} = E\{Y|U_r = u_r\}, \quad (13.30)$$

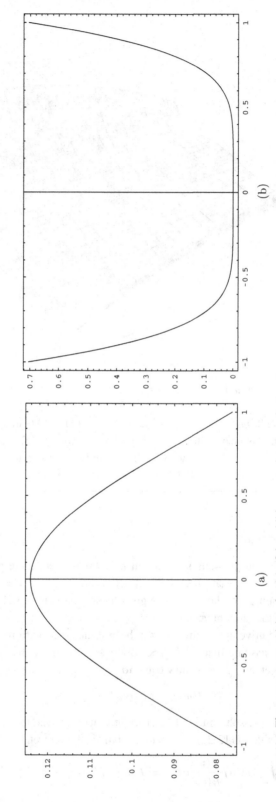

Figure 13.6 The components $g_1^*(u)$ and $g_2^*(v)$ of the additive approximation of $m(u, v)$.

(a)

(b)

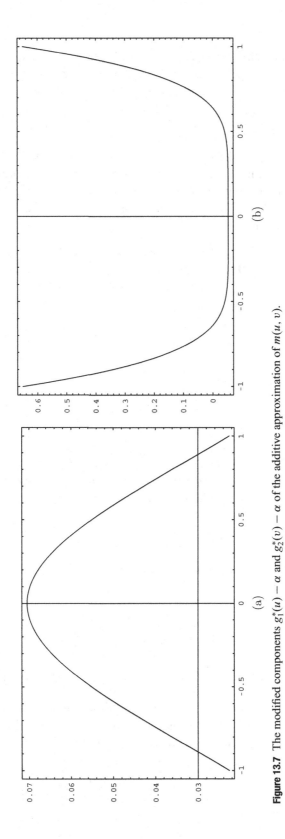

Figure 13.7 The modified components $g_1^*(u) - \alpha$ and $g_2^*(v) - \alpha$ of the additive approximation of $m(u, v)$.

where $f_r(u_r)$ is the marginal density for U_r. Hence, we have obtained a proper one-dimensional regression function of Y on the rth coordinate of the vector $\mathbf{U}$. Regarding the second term on the right-hand side of (13.22), we observe that

$$\int_{R^{d-1}} g_i^*(u_i)\frac{f(\mathbf{u})}{f_r(u_r)}d\mathbf{u}_{-r} = \int_R g_i^*(u_i)\frac{f_{ri}(u_r, u_i)}{f_r(u_r)}du_i = E\{g_i^*(U_i)|U_r = u_r\}, \quad (13.31)$$

where $f_{ri}(u_r, u_i)$ is the joint density function of (U_r, U_i). This equation shows that the second term in (13.22) is controlled entirely by two-dimensional densities of the marginals of the input vector $\mathbf{U}$. On the other hand, owing to (13.30) the first term in (13.22) needs a full d-dimensional density of $\mathbf{U}$. All these considerations yield the following theorem.

THEOREM 13.3 *Let* $(\mathbf{U}, Y)$ *be a random vector in* $R^d \times R$ *with the regression function* $m(\mathbf{u}) = E\{Y|\mathbf{U} = \mathbf{u}\}$. *Let* $E\{Y^2\} < \infty$. *The unique additive function* $g^*(\mathbf{u}) = g_1^*(u_1) + \cdots + g_d^*(u_d)$, *which minimizes* $E\{(m(\mathbf{U}) - g(\mathbf{U}))^2\}$, *is characterized by the following system of equations:*

$$g_r^*(u_r) = E\{Y|U_r = u_r\} - \sum_{i \neq r}^{d} E\{g_i^*(U_i)|U_r = u_r\}, \quad (13.32)$$

subject to the constraints

$$E\{g_r^*(U_r)\} = 0, \quad (13.33)$$

for $r = 1, \ldots, d$.

REMARK 13.1 *The solution given in Theorem 13.3 becomes particularly simple if the vector* $\mathbf{U}$ *has statistically independent components. Then, due to the identifiability condition (13.33), the second term on the right-hand side of (13.32) is equal to zero, and we obtain*

$$g_r^*(u_r) = E\{Y|U_r = u_r\}.$$

Hence in this case the orthogonal projection of $m(\mathbf{u})$ *onto the subspace of additive functions* $g_1(u_1) + \cdots + g_d(u_d)$ *such that* $E\{g_r^2(U_r)\} < \infty$, $r = 1, \ldots, d$, *is given by the sum of univariate regression functions*

$$\sum_{r=1}^{d} E\{Y|U_r = u_r\}.$$

This fact is known in the statistical literature as the Hájek projection [295].

The case of independent components described in the above remark is very restrictive in practice and one would like to find a complete solution of the equations (13.32), (13.33). Observing that (13.32) can be written as

$$g_r^*(u_r) = E\left\{Y - \sum_{i \neq r}^{d} g_i^*(U_i)|U_r = u_r\right\},$$

we can propose an iterative procedure for finding $g_r^*(u_r)$. We start with initial guesses for $g_i(u_i)$, $i = 1, \ldots, d$ and then improve our choices by computing the regression function of $Y - \sum_{i \neq r}^d g_i(U_i)$ on U_r, $r = 1, \ldots, d$. This process is repeated until some convergence is reached. This is a basic idea for a simple, yet powerful algorithm for finding components of additive models. It is often referred to as the backfitting algorithm due originally to Breiman and Friedman [30] (see also [161]). Formally, we have the following iterative procedure:

- Set initial guesses $g_r^{(0)}(\bullet)$, $r = 1, \ldots, d$.
- Obtain new evaluations $g_r^{(t)}(\bullet)$ by computing the regression function of $Y - \sum_{i \neq r}^d g_i^{(t-1)}(U_i)$ on U_r, $r = 1, \ldots, d$.
- Iterate the above with respect to t until convergence.

The stopping rule for the backfitting algorithm can be defined by some objective fitting measure or just by monitoring changes in the individual components. In practical implementations of this algorithm, we use a certain nonparametric estimate of the regression function. In fact, we can use different estimates for different components of the additive model. The backfitting procedure is very general and can be applied to a large class of additively separable models. The main shortcoming of the backfitting method is that it has unknown basic statistical properties like consistency, convergence rates, and limit theorems.

An alternative approach for finding the individual components of the additive model can be based on Theorem 13.2. This result allows us to find an optimal solution with respect to the product weight function $w(\mathbf{u}) = \prod_{i=1}^d w_i(u_i)$. Hence, we can obtain the components of the optimal additive approximation of the regression function $m(\mathbf{u})$ via the marginal integration

$$g_r^*(u_r) = \int_{R^{d-1}} m(\mathbf{u})w_{-r}(\mathbf{u})d\mathbf{u}_{-r}, \quad r = 1, \ldots, d, \tag{13.34}$$

subject to the conditions

$$\int_R g_r^*(u_r)w_r(u_r)du_r = 0, \quad r = 1, \ldots, d. \tag{13.35}$$

Alternatively, we can ignore (13.35) (see (13.28)) and determine $g_r^*(u_r)$ by

$$g_r^*(u_r) = \int_{R^{d-1}} m(\mathbf{u})w_{-r}(\mathbf{u})d\mathbf{u}_{-r} - c, \tag{13.36}$$

with $c = (1 - d^{-1}) \int_{R^d} m(\mathbf{u})w(\mathbf{u})d\mathbf{u}$.

It is important to note that in order for the solutions (13.34) or (13.36) to be valid we need to assume that $m(\mathbf{u})$ satisfies the following condition:

$$\int_{R^d} m^2(\mathbf{u})w(\mathbf{u})d\mathbf{u} < \infty. \tag{13.37}$$

This is not a very restrictive requirement because the weight function $w(\bullet)$ can be specified by the user. Nevertheless, we should emphasize that natural weight for $m(\mathbf{u})$ is the density $f(\bullet)$ of $\mathbf{U}$ since the regression function "lives" in the space $L_1(f)$ or $L_2(f)$ if the mean-squared error theory is employed.

All of the aforementioned considerations have been based on the knowledge of true regression functions and input densities. This is an impractical approach because the only information we can use is the training set. Nevertheless, to construct estimates of the individual components of an additive model, we can use the fundamental result of Theorems 13.3. First of all, we could apply the empirical counterpart of the iterative backfitting algorithm with all conditional means replaced by their empirical counterparts. The convergence of such an estimate is rather difficult to establish. Yet another strategy can directly employ the formula (13.32) by writing

$$g_r^*(u_r) = m_r(u_r) - \sum_{i \neq r}^{d} \int_R g_i^*(v) f_{i|r}(v|u_r) dv, \qquad (13.38)$$

where $m_r(u_r)$ is the regression function on Y on U_r, and $f_{i|r}(v|u_r)$ is the conditional density of U_i on U_r. All of these functions can easily be estimated from the data set $\{(\mathbf{U}, Y_1), \ldots, (\mathbf{U}_n, Y_n)\}$. For example, the kernel estimates of these functions are:

$$\hat{m}_r(u_r) = \frac{n^{-1} \sum_{t=1}^{n} Y_t K_{h_r}(u_r - U_{r,t})}{n^{-1} \sum_{t=1}^{n} K_{h_r}(u_r - U_{r,t})},$$

$$\hat{f}_{i|r}(v|u_r) = \frac{n^{-1} \sum_{t=1}^{n} K_{h_i}(v - U_{i,t}) K_{h_r}(u_r - U_{r,t})}{n^{-1} \sum_{t=1}^{n} K_{h_r}(u_r - U_{r,t})}.$$

Here $U_{r,t}$ and $U_{i,t}$ denote the rth and ith components of the random vector $\mathbf{U}_t$, respectively. Note that we use a separate bandwidth for each coordinate, that is, we employ the vector bandwidth $\mathbf{h} = (h_1, \ldots, h_d)$. Plugging these estimates into (13.38) gives the following implicit estimates for $g_r^*(\bullet)$:

$$\hat{g}_r(u_r) = \hat{m}_r(u_r) - \sum_{i \neq r}^{d} \int_R \hat{g}_i(v) \hat{f}_{i|r}(v|u_r) dv, \qquad (13.39)$$

with the constraints $\int_R \hat{g}_r(v) \hat{f}_r(v) dv = 0$, $r = 1, \ldots, d$, where $\hat{f}_r(v)$ is the kernel estimate of the marginal density function of U_r. A practical solution of (13.39) can be obtained through some iteration process. Note also that the solution in (13.39) needs one-dimensional integration, and this can be done accurately by some numerical formulas.

Statistical properties of the estimate in (13.39) (often referred to as the smooth backfitting algorithm) have been established in the case of *iid* data by Mammen, Linton and Nielsen [205]. The existing theory, however, assumes that the regression function $m(\mathbf{u})$ can be exactly expressed as the additive function, that is,

$$m(\mathbf{u}) = \sum_{r=1}^{d} g_r(u_r). \qquad (13.40)$$

Under this critical assumption, it can be shown that

$$\text{var}\{\hat{g}_r(u_r)\} = \frac{v_r(u_r)}{f_r(u_r)} \frac{1}{nh_r}(1 + o(1)),$$

where $v_r(u_r) = \text{var}(Y|U_r = u_r) \int_R K^2(z)dz$. Furthermore, for the twice differentiable additive components $\{g_r(\bullet)\}$ in (13.40) and the input density, we have

$$E\{\hat{g}_r(u_r)\} = g_r^*(u_r) + \beta(u_r)h_r^2 + o(h_r^2),$$

where the function $\beta(u_r)$ depends on the derivatives $g_r^{(1)}(u_r)$, $g_r^{(2)}(u_r)$, and $\partial f(\mathbf{u})/\partial u_r$, $r = 1, \ldots, d$. This result readily leads to the fact that the kernel estimate of the additive model, that is,

$$\hat{g}(\mathbf{u}) = \sum_{r=1}^{d} \hat{g}_r(u_r), \tag{13.41}$$

converges to $m(\mathbf{u})$ in (13.40) with the one-dimensional optimal rate $O_P\left(n^{-2/5}\right)$. Although this is a desired property, the estimate in (13.39) must be obtained by some numerical algorithm.

A competitive approach can use the result of Theorem 13.2. This is based on the subjective choice of the weight function of the product form $w(\mathbf{u}) = \prod_{i=1}^{d} w_i(u_i)$. This yields the marginal integration. strategy for obtaining the individual components of the optimal additive approximation. Hence, taking the modification (13.36) into account we can estimate $g_r^*(\bullet)$ via

$$\hat{g}_r(u_r) = \int_{R^{d-1}} \hat{m}(\mathbf{u})w_{-r}(\mathbf{u})d\mathbf{u}_{-r} - \hat{c}, \tag{13.42}$$

where $\hat{c} = (1 - d^{-1}) \int_{R^d} \hat{m}(\mathbf{u})w(\mathbf{u})d\mathbf{u}$, and we assume that $w(\mathbf{u})$ is a density function on R^d. In (13.42) $\hat{m}(\mathbf{u})$ is a pilot nonparametric estimate of the d-dimensional regression function. Although any such estimate has a common d-dimensional behavior, that is, its accuracy deteriorates with d, the marginal integration process reduces the estimate variance and "maps" its accuracy to the one-dimensional situation. In fact, the integration process projects the d-dimensional estimate onto the one-dimensional subspace.

The implementation of (13.42) requires multidimensional integration and this can be done by various methods. The simplest method is Monte Carlo integration, that is, determine $\hat{g}_r(u_r)$ via

$$\tilde{g}_r(u_r) = \frac{1}{L}\sum_{j=1}^{L} \hat{m}(u_r, \mathbf{V}_j^{-r}) - (1 - d^{-1})\frac{1}{L}\sum_{j=1}^{L} \hat{m}(\mathbf{V}_j), \tag{13.43}$$

where $\{\mathbf{V}_j\}$ are random vectors generated from the density $w(\mathbf{u})$, whereas $\{\mathbf{V}_j^{-r}\}$ are simulated from the density $w_{-r}(\mathbf{u})$. Since $w(\mathbf{u})$ is of the product form, simulation of these vectors can be performed very efficiently. It is known that this simple Monte Carlo method gives an unbiased estimate of $\hat{g}_r(u_r)$ with the variance of order $O(L^{-1})$. There are improved sampling methods that can achieve a variance of order $O(L^{-4})$ [327]. The complexity of a direct evaluation of the integral in (13.42) originates from the fact that common nonparametric regression estimates are in the ratio form. In fact, we can write the kernel estimate in (13.10) as follows:

$$\hat{m}(\mathbf{u}) = \frac{n^{-1}\sum_{i=1}^{n} Y_i K_{\mathbf{h}}(\mathbf{u} - \mathbf{U}_i)}{\hat{f}(\mathbf{u})}, \tag{13.44}$$

where $\hat{f}(\mathbf{u}) = n^{-1}\sum_{i=1}^{n} K_{\mathbf{h}}(\mathbf{u} - \mathbf{U}_i)$ is the kernel density estimate. In order to eliminate the dependence of $\hat{m}(\mathbf{u})$ on the argument in the denominator, we can move $\hat{f}(\mathbf{u})$ inside the summation sign by replacing $\mathbf{u}$ by $\mathbf{U}_i$. Thus, we can define

$$\tilde{m}(\mathbf{u}) = n^{-1}\sum_{i=1}^{n} \frac{Y_i}{\hat{f}(\mathbf{U}_i)} K_{\mathbf{h}}(\mathbf{u} - \mathbf{U}_i). \tag{13.45}$$

This estimate (often called the internally corrected kernel estimate) was originally proposed by Mack and Müller [202] as a flexible method for estimating derivatives of $m(\mathbf{u})$. In fact, the derivative of $\tilde{m}(\mathbf{u})$ has a much simpler form than the derivative of $\hat{m}(\mathbf{u})$. See also [172, 201, 212] for a further discussion on the internally corrected regression estimates. In the context of estimating the components of the additive model via the marginal integration in (13.42), the estimate (13.45) is especially useful.

Let us try to express the formula in (13.42) in some more explicit form for both types of kernel estimates, that is, the classical kernel method $\hat{m}(\mathbf{u})$ in (13.44) and the modified one in (13.45). Note that throughout this chapter we use the product kernel, that is, $K_{\mathbf{h}}(\mathbf{u}) = K_{h_1}(u_1)\cdots K_{h_d}(u_d)$. Hence, without loss of generality, let us focus on the first component of the additive function. The first step is to partition each d-dimensional vector $\mathbf{u}$ as $\mathbf{u} = (x, \mathbf{v})$, $\mathbf{v} \in R^{d-1}$. Then, we can express (13.42) in the following way:

$$\hat{g}_1(x) = \int_{R^{d-1}} \hat{m}(x, \mathbf{v})\bar{w}(\mathbf{v})d\mathbf{v} - \hat{c}, \tag{13.46}$$

where $\bar{w}(\mathbf{v}) = \prod_{i=2}^{d} w_i(u_i)$. Also, the kernel function can be written as:

$$K_{\mathbf{h}}(\mathbf{u}) = K_{h_1}(x)K_{\mathbf{b}}(\mathbf{v}), \tag{13.47}$$

where $\mathbf{b} = (h_2, \ldots, h_d)$. Plugging formula (13.44) into (13.46) and using (13.47), we obtain

$$\hat{g}_1(x) = \frac{n^{-1}\sum_{i=1}^{n} O_i(x)K_{h_1}(x - U_{1,i})}{n^{-1}\sum_{i=1}^{n} K_{h_1}(x - U_{1,i})} - \hat{c}. \tag{13.48}$$

where

$$O_i(x) = Y_i \int_{R^{d-1}} \frac{K_{\mathbf{b}}(\mathbf{v} - \mathbf{V}_i)}{\hat{f}(\mathbf{v}|x)} \bar{w}(\mathbf{v})d\mathbf{v}. \tag{13.49}$$

Here

$$\hat{f}(\mathbf{v}|x) = \frac{n^{-1}\sum_{i=1}^{n} K_{h_1}(x - U_{1,i})K_{\mathbf{b}}(\mathbf{v} - \mathbf{V}_i)}{n^{-1}\sum_{i=1}^{n} K_{h_1}(x - U_{1,i})}$$

is the kernel estimate of the conditional density of $\mathbf{V}$ on U_1.

Thus, we can interpret (13.48) as the standard univariate kernel estimate with the output variable transformed from Y_i to $O_i(x)$. The fact that we have a $(d-1)$-dimensional function $\hat{f}(\mathbf{v}|x)$ appearing in the integral formula in (13.49) makes this method difficult to use. This is not the case when the internally modified estimate in (13.45) is used. Indeed, plugging (13.45) into (13.46), we have

$$\tilde{g}_1(x) = n^{-1} \sum_{i=1}^{n} Q_i K_{h_1}(x - U_{1,i}) - \tilde{c}, \qquad (13.50)$$

where

$$Q_i = \frac{Y_i}{\hat{f}(\mathbf{U}_i)} \int_{R^{d-1}} K_{\mathbf{b}}(\mathbf{v} - \mathbf{V}_i) \bar{w}(\mathbf{v}) d\mathbf{v} \qquad (13.51)$$

is independent of x. Here $\tilde{c}$ is an estimate of the constant c in (13.36) using the estimate $\tilde{m}(\mathbf{u})$.

The integral in (13.51) is completely defined in terms of two functions that we can control, that is, the kernel and weight functions. This integral is easy to evaluate in many cases. For a small value of the bandwidth $\mathbf{b}$, we can approximate the integral in (13.51) as $\bar{w}(\mathbf{V}_i)$. The estimate may then be computed as, simply,

$$\tilde{g}_1(x) = n^{-1} \sum_{i=1}^{n} \frac{Y_i}{\hat{f}(\mathbf{U}_i)} \bar{w}(\mathbf{V}_i) K_{h_1}(x - U_{1,i}) - \tilde{c}. \qquad (13.52)$$

The constant $\tilde{c}$ in (13.52) can be determined by

$$\tilde{c} = (1 - d^{-1})n^{-1} \sum_{i=1}^{n} \frac{Y_i}{\hat{f}(\mathbf{U}_i)} w(\mathbf{U}_i).$$

An exact formula for the integral in (13.51) can also be derived. For example, let us assume that both $K(u)$ and $w_i(u_i)$ are Gaussian. Hence, let $K(u)$ be the $N(0, 1)$ density function, and $w_i(u)$ be $N(0, \tau_i^2)$. Then, because (13.51) is the convolution integral between the kernel and weight functions, we readily obtain

$$\int_{R^{d-1}} K_{\mathbf{b}}(\mathbf{v} - \mathbf{V}_i) \bar{w}(\mathbf{v}) d\mathbf{v} = \prod_{r=2}^{d} \frac{1}{a_r} \phi \left(\frac{V_{r,i}}{a_r} \right),$$

where $\phi(\bullet)$ is the $N(0, 1)$ density function and $a_r = \sqrt{h_r^2 + \tau_r^2}$, $r = 2, \ldots, d$. In the special case when all $a_r = a$ for $r = 2, \ldots, d$, we can write (13.50) in a very simple way:

$$\tilde{g}_1(x) = n^{-1} \sum_{i=1}^{n} Y_i \frac{K_{h_1}(x - U_{1,i}) \xi_i}{\hat{f}(\mathbf{U}_i)} - \tilde{c},$$

where $\xi_i = (\sqrt{2\pi}a)^{-1} \exp\left(-\sum_{r=2}^{d} V_{r,i}^2/2a^2\right)$.

In order to assess the statistical accuracy of the estimate $\tilde{g}_1(x)$ in (13.50) we need to establish some basic statistical properties of the internally normalized kernel estimate.

To do so, let us consider the idealized case when the input density $f(\bullet)$ is completely known. Then, (13.45) takes the form

$$\tilde{m}(\mathbf{u}) = n^{-1} \sum_{i=1}^{n} \frac{Y_i}{f(\mathbf{U}_i)} K_\mathbf{h}(\mathbf{u} - \mathbf{U}_i).$$

For *iid* samples, the variance of $\tilde{m}(u)$ is equal to

$$\text{var}\{\tilde{m}(\mathbf{u})\} = n^{-1} \text{var}\left\{\frac{Y_n}{f(\mathbf{U}_n)} K_\mathbf{h}(\mathbf{u} - \mathbf{U}_n)\right\}. \tag{13.53}$$

The convolution property of the kernel function and the standard analysis (see Chapters 3 and 12) reveal that the variance of $\tilde{m}(\mathbf{u})$ is of order

$$(nh^d)\frac{\sigma^2(\mathbf{u}) + m^2(\mathbf{u})}{f(\mathbf{u})} \int_{R^d} K^2(\mathbf{u}) d\mathbf{u},$$

where $\sigma^2(\mathbf{u}) = \text{var}\{Y|\mathbf{U} = \mathbf{u}\}$. This formula includes the term $m^2(\mathbf{u})$ which is absent in the expression for the variance of the standard kernel-regression estimate. Since an identical result can be obtained for the estimate $\tilde{m}(\mathbf{u})$ in (13.45), we can conclude that

$$\text{var}\{\tilde{m}(\mathbf{u})\} \geq \text{var}\{\hat{m}(\mathbf{u})\}. \tag{13.54}$$

Hence, the internally modified estimate has an increased variability.

Concerning the structure of the bias of $\tilde{m}(\mathbf{u})$ we can show, by some more involved analysis (use Lemma 13.1 and (13.14)), that

$$E\tilde{m}(\mathbf{u}) = m(\mathbf{u}) + \frac{1}{2}\mu(K) \sum_{i=1}^{d} \frac{\partial^2 m(\mathbf{u})}{\partial u_i^2} h_i^2 + o\left(\max_{1 \leq i \leq d}(h_i^2)\right). \tag{13.55}$$

This result is only possible if we use a different smoothing strategy for the term $\hat{f}(\bullet)$ appearing in (13.45). That is, one should use different bandwidths and kernel functions to form an estimate of the input density. The expression in (13.55) should be compared to the formula for the bias (see (13.14)) of the classical kernel estimate. It should be noted that the bias of $\tilde{m}(\mathbf{u})$ has a very simple structure as it merely depends on the assumed smoothness of the regression function.

An interesting issue that emerges is how the aforementioned properties of the classical and internally modified kernel estimates are preserved in the problem of estimating the components of additive models. We shed some light on this problem in Section 13.3 that deals with the question of recovering nonlinear characteristics of multichannel, block-oriented systems.

13.3   Multivariate systems

In all of the previous chapters, the main focus has been on the identification of single-input–single-output nonlinear systems. In Chapter 12, however, we applied some

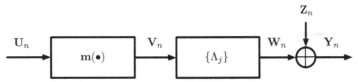

Figure 13.8 Multiple-input–multiple-output Hammerstein system.

multivariate nonparametric estimation concepts in order to capture the memory effect embedded in a nonlinear function. In this section we give a brief introduction to the problem of nonparametric identification of multivariate nonlinear systems. Such systems are characterized by the existence of several nonparametric characteristics. The theory of additive approximations examined in Section 13.2 will be employed. We should, however, emphasize that the problem of identification of multivariate nonlinear systems has received a little attention in the literature and many fundamental issues remain open for future research.

We begin with the multivariate counterpart of the Hammerstein system which is depicted in Figure 13.8. The input $\{\mathbf{U}_n\}$ is a d-dimensional random vector that is mapped to a d-dimensional interconnected signal $\mathbf{V}_n$ via the nonlinear vector-valued function $\mathbf{m}(\bullet)$, that is, $\mathbf{m}(\bullet) = (m_1(\bullet), \ldots, m_d(\bullet))^T$, where $m_i(\bullet)$, $i = 1, \ldots, d$ are measurable functions from R^d to R. The dynamical subsystem is characterized by the impulse response $(q \times d)$-matrix sequence $\{\Lambda_j\}$. Hence, the output process $\mathbf{Y}_n$ is the q-dimensional random vector disturbed by the noise process $\mathbf{Z}_n$. Thus, we have:

$$\begin{cases} \mathbf{Y}_n = \mathbf{W}_n + \mathbf{Z}_n \\ \mathbf{W}_n = \sum_{i=0}^{\infty} \Lambda_i \mathbf{V}_{n-i} \\ \mathbf{V}_n = \mathbf{m}(\mathbf{U}_n) \end{cases} . \tag{13.56}$$

Let us assume, similarly as in Chapter 3, that the input process is *iid*. Let the linear subsystem be stable, that is, $\sum_{j=0}^{\infty} \|\Lambda_j\| < \infty$, where $\|\Lambda_j\|$ denotes a norm of the matrix Λ_j, and let $\Lambda_0 = \Gamma$, be the $(q \times d)$-matrix having all elements equal to one. Then, arguing as in Chapter 3, we obtain

$$E\{\mathbf{Y}_n|\mathbf{U}_n = \mathbf{u}\} = \Gamma\mathbf{m}(\mathbf{u}) + \mathbf{C}, \tag{13.57}$$

where $\mathbf{C} = \sum_{i=1}^{\infty} \Lambda_i E\{\mathbf{m}(\mathbf{U}_0)\}$ is the q-dimensional vector.

It is clear from (13.57) that we cannot recover $\mathbf{m}(\mathbf{u})$ from the q regression functions defined on the left-hand side of (13.57). Therefore, we have to put further restrictions on the structure of the system in order to be able to recover its characteristics.

The simplest case is the multiple-input Hammerstein system (depicted in Figure 13.9) with the one dimensional interconnecting signal V_n. For this system, we can readily obtain

$$E\{Y_n|\mathbf{U}_n = \mathbf{u}\} = m(\mathbf{u}) + c, \tag{13.58}$$

where $c = E\{m(\mathbf{U}_0)\} \sum_{i=1}^{\infty} \lambda_i$. Hence, the multivariate nonlinearity $m(\mathbf{u})$ can be recovered from the d-dimensional regression function $E\{Y_n|\mathbf{U}_n = \mathbf{u}\}$. A kernel regression

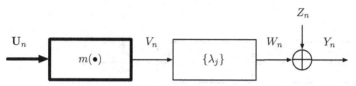

Figure 13.9 Multiple-input–single-output Hammerstein system.

estimate can be applied and then, by the analysis given in Chapter 3, combined with Lemma 13.1 (see (13.16)), we can demonstrate that the accuracy of such a method is $O_P(n^{-\frac{2}{4+d}})$ for a twice differentiable nonlinearity $m(\mathbf{u})$. For a large value of d, this is an unacceptably slow rate. By virtue of the theory discussed in Section 13.2 we may alleviate this problem by projecting $m(\bullet)$ onto the class of additive functions. This leads to an additive Hammerstein system depicted in Figure 13.10. The components of this additive structure can be selected in such a way that the system in Figure 13.10 can be viewed as the best additive approximation of the system in Figure 13.9. In fact, the formula in (13.58) proves that $m(\bullet)$ is the regression function of Y_n on $\mathbf{U}_n$. Then, a direct application of Theorem 13.3 shows that there is a unique solution minimizing

$$E\{(m(\mathbf{U}_n) - g(\mathbf{U}_n))^2\} \tag{13.59}$$

with respect to all functions $g(\mathbf{u}) = g_1(u_1) + \cdots + g_d(u_d)$ such that $E\{g_i(U_i)\} = 0$, $i = 1, \ldots, d$.

It is worth noting that the criterion in (13.59) can be equivalently expressed as $E\{(Y_n - Y_n^A)^2\}$, where Y_n is the output of the system in Figure 13.9, whereas Y_n^A is the output of the system in Figure 13.10. The solution minimizing (13.59) is characterized in (13.32) and can be efficiently obtained via the marginal integration strategy discussed thoroughly in Section 13.2.2.

For example, one can use the simplified solution described in (13.52) and estimate the nonlinearity $g_1(\bullet)$ in Figure 13.10. Hence, let us write $\mathbf{u}_{-1}$ for the version of the vector $\mathbf{u}$ with its first element dropped. Then, for a given product-type weight function $w(\mathbf{u})$ let $\bar{w}(\mathbf{u}_{-1}) = \prod_{i=2}^d w_i(u_i)$. Consequently, owing to (13.52), we have

$$\tilde{g}_1(x) = n^{-1} \sum_{i=1}^n \frac{Y_i}{\hat{f}(\mathbf{U}_i)} \bar{w}(\mathbf{U}_{-1,i}) K_{h_1}(x - U_{1,i}) - \tilde{c}, \tag{13.60}$$

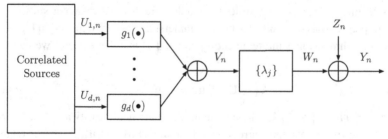

Figure 13.10 Additive Hammerstein system.

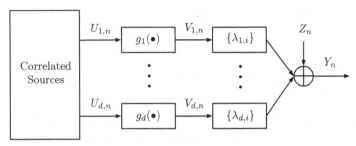

Figure 13.11 Multichannel Hammerstein system.

where the normalizing constant $\tilde{c}$ is defined in (13.52). In this algorithm the user must specify the bandwidth defining the estimate $\hat{f}(\bullet)$ of the input density $f(\bullet)$. Note that this factor has to be determined only once for all additive components of the system. On the other hand, the bandwidth h_1 must be tailored to the given additive element. In Chapter 12, we have discussed the problem of choosing the bandwidth for nonparametric regression estimation. Nevertheless, this issue requires further investigation in the present situation.

Further extension of the additive methodology can be carried over to the version of the multivariate Hammerstein system in Figure 13.8, which takes the additive form with respect to both nonlinear and linear subsystems. This structure is depicted in Figure 13.11 and is often referred to as the multichannel Hammerstein system. On one hand, the multichannel Hammerstein system can be viewed as a more powerful additive approximation of the system (13.56). On the other hand, this model has appeared as a natural structure for a number of applications like multi sensor systems, power systems, multiuser detection, and physiological and neural systems [35, 40, 149, 287, 293, 300, 304, 323]. Surprisingly there has been a very little research done for the problem of identification of multichannel nonlinear models; see [56] for an early contribution on a parametric identification of a two-channel Hammerstein model and its application to chemical engineering.

The nonparametric identification of the multichannel Hammerstein system was carried out in [235]. The kernel estimate of the internally modified form (see (13.45)) has been applied. Then the marginal integration approach was utilized to obtain nonparametric estimates of the individual components of the system in Figure 13.11. Thus, the estimate $\tilde{g}_1(x)$ in (13.50) (see also (13.60)) was obtained based on the training data $\{(\mathbf{U}_1, Y_1), \ldots, (\mathbf{U}_n, Y_n)\}$ generated by the truly additive model in Figure 13.11.

Under usual conditions on the kernel function and the proper choice of the bandwidth vector $\mathbf{h} = (h_1, \ldots, h_d)$, it can be shown [162, 235] that the asymptotic variance of the estimate $\tilde{g}_1(x)$ of the first component $g_1(x)$ is of the order

$$\frac{\varphi(x)}{f_1(x)} \frac{1}{nh_1} \int_R K^2(u)du, \tag{13.61}$$

where $\varphi(x)$ is the bounded function depending on the input density of the vector $\mathbf{U}$, the nonlinear and linear characteristics of all channels, and the weight function $w(\bullet)$. Here $f_1(\bullet)$ is the marginal density of the first component of $\mathbf{U}$. Regarding the bias term of the estimate $\tilde{g}_1(x)$ it can be demonstrated [162, 235] that

$$E\tilde{g}_1(x) = g_1(x) + c \left\{ \frac{g_1^{(2)}(x)}{f_1(x)} - \int_R g_1(t) w_1^{(2)}(t) dt \right\} h_1^2 + o(h_1^2),
\qquad (13.62)$$

where c is some positive constant.

This is a rather surprising result because the bias depends neither on the smoothness of the nonlinearities in other channels, nor on the input density of **U**. It merely depends on the smoothness (here expressed by the second derivative) of the nonlinearity being estimated, the marginal density of the input signal of the given channel, and the smoothness of the first component of the weight function. The latter is not a very restrictive requirement since the user can select a smooth weight function. The above useful property of the independence on the smoothness of the nonlinearities in other channels is not shared by standard kernel estimates when they are applied in the process of the marginal integration. The bias of such obtained estimates depends on the smoothness of the nonlinearities in other channels as well as on the input density of **U**. Hence the correlation structure of the input signal influences not only the statistical variability of the estimate but also its systematic error. This is clearly an undesirable property.

The results in (13.61) and (13.62) yield the optimal one-dimensional rate $O_P(n^{-2/5})$ for recovering the nonlinearity $g_1(x)$. It is important to note that the bandwidth h_1 used for estimating $g_1(x)$ must be specified as $h_1 = n^{-1/5}$, whereas the remaining bandwidths $h_2, \ldots, h_d$ must be selected larger than it is usually recommended; see (13.15). In fact, for the consistency of the kernel estimates the following condition (see (13.13)):

$$nh_1 \cdots h_d \to \infty \qquad (13.63)$$

is necessary. Therefore, if one decides to use a single bandwidth h for all d channels, then the desirable rate $O_P(n^{-2/5})$ holds if $d \leq 4$. Indeed, for $h_i = n^{-1/5}$ for all i, the condition (13.63) holds if $d \leq 4$. Such an improper choice of the bandwidth can drastically reduce the quality of the marginal integration method. This is illustrated in the following simulation example.

Example 13.3 Let us consider the two-channel Hammerstein system excited by the Gaussian input signal (U_n, V_n) with zero mean and the covariance matrix of the following form:

$$\begin{pmatrix} 1/4 & \rho/4 \\ \rho/4 & 1/4 \end{pmatrix},$$

where ρ is the correlation coefficient. The measurement noise Z_n is white Gaussian with zero mean and variance 0.01. The nonlinearity in the first channel is of the polynomial form, that is, $g_1(u) = \frac{4}{3}u^3 - \frac{1}{3}u$, whereas the linear subsystem is the FIR(4) model with the transfer function $L_1(z) = 1 + 0.2z^{-1} + 0.1z^{-2} + 0.05z^{-3} + 0.05z^{-4}$. The nonlinearity $g_1(u)$ is estimated by the marginal integration method with a pilot two-dimensional kernel-regression estimate using the Gaussian kernel and a single

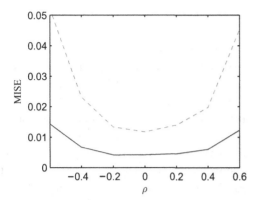

Figure 13.12 MISE versus ρ for recovering the polynomial characteristic in the two-channel Hammerstein system. The internal estimate (dotted line) and classical kernel estimate (solid line).

bandwidth for both variables. The product weight function of the Gaussian $N(0, 1/9)$ form is applied. In Figure 13.12, we depict the mean integrated squared error (MISE) versus ρ with the sample size $n = 320$. The second channel is assumed to have the same characteristics as the first one. The dotted line corresponds to the internally modified kernel estimate, whereas the solid line is the error of the classical kernel estimate. A loss in efficiency, due to increasing correlation between inputs, is clearly seen. Moreover, the accuracy of the internal estimate is drastically reduced compared to the classical kernel estimate. This is due to the use of a single bandwidth for both variables. In this case, the independence (see (13.62)) of the internal estimate on the correlation of the input signal is not valid anymore. In Figure 13.13, we show the same phenomenon, which is additionally amplified by the fact that the nonlinearity in the second channel is a quantizer with six levels. The lack of smoothness of this characteristic increases the error for both estimates.

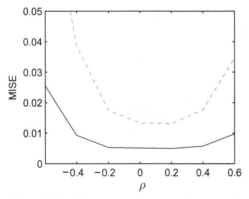

Figure 13.13 MISE versus ρ for recovering the polynomial characteristic in the two-channel Hammerstein system. The internal estimate (dotted line) and classical kernel estimate (solid line). The second channel nonlinearity is a quantizer.

13.4   Concluding remarks

In the considerations examined in this chapter we have touched on the difficult issue of low-dimensional representations of fully nonparametric nonlinear systems. The main tool we have introduced and examined was based on the theory of additive approximations of multivariate functions. This allows us to develop a powerful and general additive modeling strategy for nonparametric identification of multivariate nonlinear block-oriented systems. Hence, one should replace all multivariate functions present in the system by their additive approximations. An important remaining issue, however, is whether one can recover efficiently the individual components of these additive structures. We have demonstrated that this is the case for the multiple-input and multichannel Hammerstein systems. Furthermore, to achieve a complete low-dimensional representation of a nonlinear multivariate dynamical system, we must also incorporate similar representations for linear dynamical parts of the system.

There are further ways to improve the flexibility of the additive approximation by allowing terms with interactions, that is,

$$g(u_1, \dots, u_d) = g_{12}(u_1, u_2) + g_{23}(u_2, u_3) + \cdots + g_{d-1d}(u_{d-1}, u_d) + g_{d1}(u_d, u_1).$$
(13.64)

The marginal integration method in Section 13.2 can be extended to this case as well. Here, we must estimate bivariate functions and this can be done with reasonable accuracy. An interesting issue is to find optimal interaction pairs (u_{i_1}, u_{i_2}), $1 \leq i_1, i_2 \leq d$, in order to reduce the approximation error.

In this chapter, we have mostly focused on the Hammerstein system. It should be clear, however, that similar considerations can be carried over to series-parallel structures examined in Chapter 12. For example, one could consider the parallel nonlinear system (Section 12.1.1) with a multidimensional nonlinearity and its corresponding additive approximation.

Yet another class of low complexity representations can be based on the combination of parametric and nonparametric inference. For example, the following

$$g(\mathbf{u}) = \sum_{i=1}^{N} g_i(\mathbf{a}_i^T \mathbf{u})$$

can be used to approximate a multivariate function $m(\mathbf{u})$ in terms of univariate nonlinearities $g_i(\bullet)$ and vectors $\mathbf{a}_i$, $i = 1, \dots, N$. Such semiparametric structures will be examined in Chapter 14. This includes the multivariate counterparts of Wiener and sandwich systems. A finite-dimensional parametrization of dynamical subsystems will define semiparametric models for which we propose some efficient identification algorithms.

13.5   Bibliographic notes

A few studies exist concerning the identification problem of multivariate nonlinear systems. Early methods rely on Wiener and Volterra expansions and lead to complex algorithms. Moreover, smooth (parametric) nonlinearities have been assumed and a Gaussian input process was applied in most studies, see Westwick and Kearney [316, 317], Westwick and Verhaegen [315], and Zhu [332]. More recent contributions use parametric models for nonlinearities and apply the subspace methods for recovering linear subsystems, see Chan, Bao, and Whiten [43], and Gomez and Baeyens [104]. The nonparametric approach for estimating the multiple-input Hammerstein system was proposed by Greblicki and Pawlak [127]. An additive class of block-oriented nonlinear systems was studied by Pawlak and Hasiewicz [235], which includes a number of connections introduced in Section 12.1.

The use of the additive models in various problems of statistical inference was summarized in Hastie and Tibshirani [161], and Härdle, Müller, Sperlich, and Werwatz [152]. The backfitting algorithm was proposed by Breiman and Friedman [30] and further examined in Hastie and Tibshirani [161]. The marginal integration strategy for recovering individual components in the additive model was first proposed by Linton and Nielsen [197]; see also Fan and Yao [89] for further discussion of this method. Mammen, Linton, and Nielsen [205] introduced a smooth version of the backfitting algorithm. A fundamental issue of finding the best additive model has been addressed. The proposed theory uses a weight function that is a kernel estimate of the input variable. In Hengartner and Sperlich [162] a detailed analysis of the marginal integration method is given. The internally normalized kernel estimate is employed. All of the aforementioned studies deal with independent data. The use of additive models in time series analysis is discussed by Tjøstheim [290], and Fan and Yao [89]. The extension of additive modeling to nonlinear system identification has been examined by Pawlak and Hasiewicz [235].

The internally normalized kernel estimate was proposed by Mack and Müller [202] and further examined by Mack and Müller [201], Müller and Song [212], Jones, Davies, and Park [172], and Müller [211].

14 Semiparametric identification

In this chapter, we discuss the problem of identification of a class of semiparametric block-oriented systems. This class of block-oriented systems can be restricted to a parameterization that includes a finite-dimensional parameter and nonlinear characteristics that run through a nonparametric class of mostly univariate functions. The parametric part of a semiparametric model defines characteristics of linear dynamical subsystems and low-dimensional projections of multivariate nonlinearities. The nonparametric part of the model comprises all static nonlinearities defined by functions of a single variable. A general methodology for identifying semiparametric block-oriented systems is developed. This includes a semiparametric version of least squares and a direct method using the concept of the average derivative of a regression function. These general approaches are applied in cases of semiparametric versions of Wiener, Hammerstein, and parallel systems. Section 14.2 gives examples of semiparametric block-oriented systems. This includes the multivariate version of Hammerstein and Wiener systems. In Section 14.3, we give a general approach to semiparametric inference. Section 14.4 is devoted to an important case study concerning the semiparametric Wiener system. Sections 14.5 and 14.6 provide similar considerations for semiparametric Hammerstein and parallel systems. In Section 14.7, we derive direct estimation methods for semiparametric nonlinear systems.

14.1 Introduction

In all of the preceding chapters, we have examined various fully nonparametric block-oriented systems. Indeed, we have not imposed any a priori knowledge about characteristics of the nonlinear systems under study. Hence, the system was represented by the pair $(\lambda, \mathbf{m}(\bullet))$, where $\lambda \in R^{\infty}$ is an infinite-dimensional parameter representing impulse response sequences of linear dynamical subsystems, whereas $\mathbf{m}(\bullet)$ is a vector of nonparametric multidimensional functions describing nonlinear elements. Such a fully nonparametric model does not suffer from risk of misspecification. However, corresponding nonparametric estimators exhibit low convergence rates due to the complexity of an assumed block-oriented structure and the multidimensional nature of interconnecting signals. In contrast, a parametric model carries a risk of incorrect model specification, but if it is correct it will typically enjoy a fast $O(n^{-1})$ parametric rate of convergence.

In practice, we can accept an intermediate model which lies between parametric and fully nonparametric cases when a linear subsystem can be parameterized by a finite-dimensional parameter, whereas nonlinear characteristics run through a nonparametric class of low-dimensional functions. This semiparametric model allows one to design practical identification algorithms that share the efficiency of parametric modeling while preserving the high flexibility of the nonparametric case. In fact, in many cases, as it will be shown, we are able to identify linear and nonlinear parts of a block-oriented system under much weaker conditions on the system characteristics and on the probability distribution of the input signal.

An estimation strategy used for semiparametric systems is based on the concept of interchanging parametric and nonparametric estimation methods. The basic idea is to first analyze the parametric part of the block-oriented structure as if all static nonlinearities were known. To eliminate the dependence of a parametric fitting criterion on the nonlinearities, we form pilot nonparametric estimates of the nonlinearities being indexed by a finite-dimensional vector of the admissible value of the parameter. Then this is used to establish a parametric fitting criterion (such as least squares) with random functions representing all estimated nonparametric nonlinearities. On the other hand, nonparametric regression estimates of the nonlinearities use estimates of the parametric part. As a result of this interchange, we need some data resampling schemes in order to achieve some statistical independence between the estimators of parametric and nonparametric parts of the system. This improves the efficiency of the estimates and facilitates the mathematical analysis immensely.

In our studies on the identification of semiparametric systems, we use the least squares method with a data splitting strategy where the training set is divided into two nonoverlapping parts. The first part plays the role of a testing sequence that defines the least squares fitting function, whereas the other part is used as a training sequence to form preliminary nonparametric regression estimates.

We establish sufficient conditions for the convergence of our identification algorithms for a general class of semiparametric block-oriented systems. Then, this general theory is illustrated in the case of semiparametric Wiener, Hammerstein, and parallel systems of various forms. In the Wiener system case, the semiparametric approach leads to an identification algorithm that is consistent (with the optimal rate) for noninvertible nonlinearities. Furthermore the method can be applied in the presence of an output measurement noise and when an input signal is not necessarily Gaussian. These practical conditions were not applicable for the nonparametric algorithms presented in Chapters 9, 10, and 11. This two-stage semiparametric least squares method also has the advantage of a unified approach for a large class of input distributions and complex block-oriented systems.

In addition, we present an alternative approach for estimating the parametric part of specific semiparametric block-oriented systems which does not need any optimization procedures. This direct strategy relies on the concept of the average derivative estimation of a regression function. This may be an appealing method in many applications because it is simple and noniterative. However, it is important to note that the method requires smooth density functions of input signals and can be applicable to only a limited class of block-oriented nonlinear systems.

14.2 Semiparametric models

In all of the previous chapters, we have examined various block-oriented models with general linear subsystems and nonparametric nonlinearities. In such a setting, a number of nonparametric estimates of system nonlinearities have been proposed and their convergence rates have been evaluated. These estimates are very robust to any model misspecification, but unfortunately suffer low convergence rates due to the model complexity and the dimensionality of interconnecting signals. For example, in the problem of Wiener system identification, discussed in Chapter 9, our inverse regression-estimation strategy yielded the rate $O_P(n^{-1/3})$ instead of the optimal rate $O_P(n^{-2/5})$. As shown in Chapter 13, for the Hammerstein system with a d-dimensional input signal, a nonparametric estimate of the system nonlinearity has the optimal rate $O_P(n^{-2/(d+4)})$, that is, as d increases the convergence rate deteriorates. In this section, we list a number of semiparametric systems which provide a high degree of modeling flexibility while improving on the convergence rate. A semiparametric model has a natural parameterization $(\theta, \mathbf{g}(\bullet))$, where θ is a finite-dimensional parameter and $\mathbf{g}(\bullet)$ is a vector of univariate nonparametric functions. This defines a semiparametric model in which we aim at estimating θ and $\mathbf{g}(\bullet)$. In our future considerations, we will often denote by θ^* and $\mathbf{g}^*(\bullet)$ the true characteristics of the underlying semiparametric model.

14.2.1 Semiparametric Hammerstein models

In Chapter 13 (see Figure 13.9), we examined the Hammerstein model with a d-dimensional input, that is, we have the following input–output description:

$$Y_n = \sum_{l=0}^{\infty} \lambda_l m(\mathbf{U}_{n-l}) + Z_n, \tag{14.1}$$

where $m(\mathbf{u})$, $\mathbf{u} = (u_1, \ldots, u_d)^T \in R^d$, is the d-dimensional nonlinearity, whereas $\lambda = \{\lambda_l, 0 \leq l \leq \infty\}$ defines the impulse response function of the linear subsystem.

It has been demonstrated (under common smoothing conditions) (see Chapter 13) that the kernel regression estimate $\hat{m}(\mathbf{u})$ of $m(\mathbf{u})$ has the following rate:

$$\hat{m}(\mathbf{u}) = m(\mathbf{u}) + O_P(n^{-2/(d+4)}). \tag{14.2}$$

It is apparent that the rate $O_P(n^{-2/(d+4)})$ gets slower as d increases. In order to alleviate this curse of dimensionality, we have proposed to approximate the nonlinearity $m(\mathbf{u})$ by an additive function, that is,

$$m(\mathbf{u}) \simeq \sum_{i=1}^{d} m_i(u_i),$$

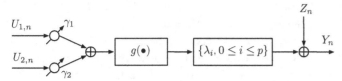

Figure 14.1 The semiparametric Hammerstein model with $d = 2$.

where $m_i(\bullet), i = 1, \ldots, d$ are univariate functions. This representation allows us to improve the rate in (14.2) to $O_P(n^{-2/5})$ independent of d, provided that sufficient over-smoothing of the additive components is made (see Chapter 13 for details). Although this approach partially eliminates the curse of dimensionality, it may introduce the involved problem of estimating all additive components.

To recover the linear part, λ, of the Hammerstein system we can use the correlation method (see Chapter 13). Hence, we can estimate λ by

$$\hat{\lambda}_i = \frac{n^{-1} \sum_{j=1}^{n-i} Y_{i+j} \eta(\mathbf{U}_j)}{n^{-1} \sum_{j=1}^{n} Y_j \eta(\mathbf{U}_j)}, \tag{14.3}$$

for $i = 1, \ldots$, where it is assumed that $\lambda_0 = 1$. In the above formula, $\eta : R^d \to R$ is a known function chosen by the user such that $E\eta(\mathbf{U}_n) = 0$ and $E\{m(\mathbf{U}_n)\eta(\mathbf{U}_n)\} \neq 0$.

It should be noted that the estimate in (14.3) is very inefficient for large values of i. In order to overcome the high dimensionality nature of the nonparametric Hammerstein system (14.1), we propose the following semiparametric low complexity approximation of the Hammerstein system

$$Y_n = \sum_{l=0}^{p} \lambda_l\, g(\gamma^T \mathbf{U}_{n-l}) + Z_n, \tag{14.4}$$

where $g(\bullet)$ is a univariate function; the parameter $\gamma = (\gamma_1, \ldots, \gamma_d) \in R^d$ defines a projection of the input signal $\mathbf{U}_n$ onto an one-dimensional subspace; and p defines the memory length of the model, which is assumed to be finite and often known. Hence, in this case the model is defined by the pair $(\theta, g(\bullet))$, where $\theta = (\lambda, \gamma) \in R^{p+d+1}$ is the parametric part of the semiparametric system. The objective is to estimate the parameters $\lambda = \{\lambda_l, 0 \leq l \leq p\}$, $\gamma = \{\gamma_i, 1 \leq i \leq d\}$ and the single, univariate function $g(\bullet)$. Figure 14.1 depicts this semiparametric version of the Hammerstein system for $d = 2$.

It is worth noting that some normalization of the model in (14.4) is necessary in order to uniquely identify $(\theta, g(\bullet))$. First of all, θ cannot be identified if $g(\bullet)$ is constant. Next, the scaling effect of the cascade Hammerstein model forces us to normalize γ in order to be able to recover $g(\bullet)$. Hence, we should assume, for example, that $\gamma_1 = 1$. We also require that the input signal $\{\mathbf{U}_n\}$ has a density function. This assures that the probability of the event $\{\delta^T \mathbf{U}_n = c\}$ is not equal to one for some constant c and $\delta \in R^d$. This fact is necessary for the identifiability of γ. By virtue of the discussions in Chapters 2 and 12 we also know that in order to uniquely recover $g(\bullet)$ we need $\lambda_0 = 1$ and $E\{g(\gamma^T \mathbf{U}_n)\} = 0$. The latter assumption can be replaced by the requirement that $g(0) = 0$, see Chapter 2.

The aforementioned conditions are sufficient for identification of (λ, γ) and $g(\bullet)$ and are likely to be satisfied in most practical applications.

The problem of estimating (λ, γ) and $g(\bullet)$ will be examined in the remaining part of this chapter. We should note, however, that one can recover the impulse response function $\{\lambda_l, 0 \leq l \leq p\}$ independent of γ and $g(\bullet)$. In fact, under the aforementioned normalization conditions and the assumption that the input signal $\{U_n\}$ is white, we have

$$\hat{\lambda}_i = \frac{n^{-1} \sum_{j=1}^{n-i} Y_{i+j}\, \eta(U_j)}{n^{-1} \sum_{j=1}^{n} Y_j\, \eta(U_j)}, \tag{14.5}$$

for $i = 1, \ldots, p$.

The consistency of $\hat{\lambda}_i$ can be easily established, see Chapter 13. Indeed, the variance of the estimate in (14.5) is of order $O((n-p)/n^2)$, which makes this estimate more efficient than (14.3) by the fact that p is finite. The problem of estimating γ and $g(\bullet)$ is nonstandard and this issue will be examined in Section 14.5.

14.2.2   Semiparametric Wiener models

In Chapter 9, we introduced and examined the identification problem of the Wiener system with an infinite memory, that is, the system described by the following input–output equation:

$$Y_n = m \left(\sum_{l=0}^{\infty} \lambda_l\, U_{n-l} \right). \tag{14.6}$$

Assuming that the input process is Gaussian (white or color), we have shown that the impulse response function $\{\lambda_l, 0 \leq l \leq \infty\}$ can be consistently estimated by the correlation method, resulting in the following estimate:

$$\hat{\lambda}_i = \frac{n^{-1} \sum_{j=1}^{n-i} Y_{i+j}\, U_j}{n^{-1} \sum_{j=1}^{n} Y_j\, U_j}, \tag{14.7}$$

for $i = 1, \ldots,$ where it is assumed that $\lambda_0 = 1$.

As noted for the estimate in (14.3), this estimate is very inefficient for large values of i. Furthermore, for invertible nonlinearities we have proved that several nonparametric regression estimates can consistently recover the inverse $m^{-1}(y)$. For twice differentiable nonlinearities the rate of convergence of the estimates was shown to be,

$$\widehat{m^{-1}}(y) = m^{-1}(y) + O_P(n^{-1/3}). \tag{14.8}$$

Since the optimal rate of convergence for twice differentiable nonlinearities is $O_P(n^{-2/5})$, the rate given in (14.8) is suboptimal. In Section 14.4, we will examine the following counterpart of the model given in (14.6),

$$Y_n = m \left(\sum_{l=0}^{p} \lambda_l\, U_{n-l} \right) + Z_n, \tag{14.9}$$

where the model order p is finite and known.

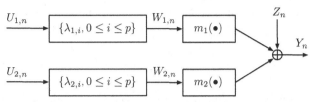

Figure 14.2 The additive Wiener model.

Note that in the above model we allow measurement noise which was absent in (14.6). The model in (14.9) is of the semiparametric form with the parameterization $(\lambda, m(\bullet))$. Using the semiparametric estimation methodology developed in this chapter we propose estimates of $\lambda = \{\lambda_l, 0 \le l \le p\}$ and $m(\bullet)$ which are consistent for a large class of not-necessarily Gaussian input signals and noninvertible nonlinearities. Also, estimates of $m(\bullet)$ exhibit the optimal rate $O_P(n^{-2/5})$.

The semiparametric restriction of the classical Wiener system allows us to examine the following two-channel version of this model:

$$Y_n = m_1 \left(\sum_{l=0}^{p_1} \lambda_{1,l} \, U_{1,n-l} \right) + m_2 \left(\sum_{l=0}^{p_2} \lambda_{2,l} \, U_{2,n-l} \right) + Z_n. \qquad (14.10)$$

Here $\{\lambda_{1,l}, 0 \le l \le p_1\}$, $\{\lambda_{2,l}, 0 \le l \le p_2\}$ are the impulse response functions of the linear subsystems, and $m_1(\bullet)$, $m_2(\bullet)$ are the characteristics of the nonlinear subsystems. The multichannel extension of the above system can also be easily derived. Figure 14.2 shows the structure of the two-channel Wiener system with the input signal $\{(U_{1,n}, U_{2,n})\}$ and $p_1 = p_2 = p$.

It is worth mentioning that the multichannel Wiener system has a number of important applications in biomedical systems and communication engineering. It is also known that this system can approximate any nonlinear system with fading memory. Other semiparametric models can be defined in a similar fashion, that is, models which are characterized by a finite dimensional parameter and nonparametric functions of a single variable.

14.3 Statistical inference for semiparametric models

In this section, we give a general approach to semiparametric statistical inference, which later we will apply to concrete nonlinear block-oriented systems. In particular, we will examine the Wiener system identification problem within the introduced framework.

Let (U, Y) be a random vector distributed according to the law $\mathbf{P}(\bullet, \bullet)$. In system identification U is usually identified with the input signal, and Y with the output signal. In parametric modeling the distribution $\mathbf{P}(\bullet, \bullet)$ is known up to a finite dimensional parameter θ^*, that is, $\mathbf{P}(\bullet, \bullet) = \mathbf{P}_{\theta^*}(\bullet, \bullet)$. Given a training set $T = \{(U_1, Y_1), \ldots, (U_n, Y_n)\}$ of the input–output signals, an estimator $\hat{\theta}$ of θ^* can be found by minimizing a criterion function

$$Q_n(\theta) = \frac{1}{n} \sum_{i=1}^{n} \Psi_\theta(U_i, Y_i), \tag{14.11}$$

where $\Psi_\theta(\bullet, \bullet)$ is a known function for every $\theta \in \Theta$, and Θ is a admissible set of parameters, such that $\theta^* \in \Theta$.

An example of $\Psi_\theta(\bullet, \bullet)$ is a maximum likelihood estimate for which

$$\Psi_\theta(U, Y) = -\log \frac{\mathbf{P}_\theta(U, Y)}{\mathbf{P}_{\theta^*}(U, Y)}. \tag{14.12}$$

In system identification the maximum likelihood estimate is usually difficult to implement. In fact, we usually deal with dependent data and, moreover, a distribution of the noise process is unknown making it difficult to find an analytical formula for $\mathbf{P}_\theta(\bullet, \bullet)$. Indeed, an estimation strategy depends critically on the selected model relating the output signal with past values of the input signals. Thus, for a given model the least-squares type estimators are more adequate. To illustrate this point let the input–output relationship of a nonlinear dynamical system be given by

$$Y_n = g(U_n, U_{n-1}, \ldots, U_{n-p}; \theta) + Z_n, \tag{14.13}$$

where Z_n is the measurement noise and p is the system memory. Then, one can choose

$$\Psi_\theta(\mathbf{U}_i, Y_i) = M(Y_i - g(U_i, U_{i-1}, \ldots, U_{i-p}; \theta)), \tag{14.14}$$

where $\mathbf{U}_i = (U_i, U_{i-1}, \ldots, U_{i-p})^T$ and $M(\bullet)$ is a criterion function.

The special cases $M(t) = t^2$ and $M(t) = |t|$ correspond to least squares and least absolute deviation estimators, respectively. Other choices are: $M(t; \alpha) = t^2 \mathbf{1}(|t| \leq \alpha) + (2\alpha|t| - \alpha^2)\mathbf{1}(|t| > \alpha)$, and $M(t; \beta) = -(2\pi\beta^2)e^{-t^2/2\beta^2}$. The criterion $M(t; \alpha)$ yields a class of robust estimation methods due to Huber, whereas $M(t; \beta)$ gives an alternative scheme to deal with outliers. Parametric estimates which are obtained with respect to the criterion function $M(\bullet)$ are often called M-estimators. These estimators play an important role in the parametric statistical inference as they define robust statistical methods.

In the nonparametric setting, the distribution $\mathbf{P}(\bullet, \bullet)$ is completely unknown, and therefore an input–output mapping such as (14.13) cannot be parameterized. Consequently, the minimization of the criterion in (14.11) is not well defined and the classical parametric M-estimators cannot be directly applied. In order to circumvent this difficulty one must constrain a class of admissible functions to a certain functional space, which size must be carefully controlled. This leads to the concept of the penalized M-estimators. However, these nonparametric estimators are rarely given in the explicit form and this issue will not be covered here. In all of the previous chapters, we have studied the nonparametric setup in the context of particular block-oriented systems with infinite memory. All of our nonparametric estimates have been given in the explicit form.

In many block-oriented models we have an intermediate situation between the parametric and fully nonparametric cases. As mentioned above, this takes place when linear subsystems can be parameterized by a finite dimensional parameter, whereas nonlinear characteristics run through a nonparametric class of functions. Furthermore, in the multivariate setting we would like to eliminate the curse of dimensionality and this can be achieved by projecting multivariate nonparametric functions onto a low dimensional nonparametric subspace. This introduces additional "projection" parameters, see Section 14.2 for examples of such a situation.

In this so-called semiparametric (intermediate) case, we have a natural parameterization

$$(\theta, m(\bullet)) \mapsto \mathbf{P}_{\theta, m(\bullet)}(\bullet, \bullet),$$ (14.15)

where $\theta \in \Theta \subset R^s$ is a finite-dimensional parameter representing impulse response sequences of linear dynamical subsystems and projection parameters. Here, Θ is a subset of R^s that defines a class of admissible parameters, and $m(\bullet)$ is a nonparametric function describing the characteristic of a nonlinear subsystem. Generally $m(\bullet) = (m_1(\bullet), \ldots, m_q(\bullet))$, that is, $m(\bullet)$ is a vector of functions representing all static nonlinearities of the underlying model. Typically, $m_i(\bullet)$ are required to be functions of a single variable. For our further considerations we denote by $(\theta^*, m^*(\bullet))$ the true characteristics of the underlying block-oriented system, such that $\theta^* \in \Theta$.

In semiparametric modeling, we first aim at estimating the parameter θ in the presence of an infinite dimensional "nuisance parameter" $m(\bullet)$ followed by a recovery of $m(\bullet)$. This two-step procedure can be efficiently implemented and yields high-accuracy estimation algorithms. In fact, the criterion in (14.11), due to the parameterization in (14.15), now takes the following form:

$$Q_n(\theta, m(\bullet)) = \frac{1}{n} \sum_{i=1}^{n} \Psi_{\theta, m(\bullet)}(U_i, Y_i).$$ (14.16)

The basic idea in finding an estimate $\hat{\theta}$ of the true value θ^* is to eliminate the dependence of $Q_n(\theta, m(\bullet))$ on the nonparametric functions $m(\bullet)$. Hence, we wish to analyze the parametric parts of the semiparametric system as if the nonparametric parts were given.

Let us assume that one can estimate $m(\bullet)$ for a given θ. This leads to a nonparametric estimate $\hat{m}(\bullet; \theta)$, which itself depends on θ. In the limit $\hat{m}(\bullet; \theta)$ is expected to tend to a function $m(\bullet; \theta)$ which is a regression function of the corresponding output and input signals, that is, it is an orthogonal projection of all system nonlinearities onto the space spanned by $\theta \in \Theta$. Clearly if $\theta = \theta^*$ then $m(\bullet; \theta^*) = m^*(\bullet)$. This important concept is illustrated in Figure 14.3 where $\varepsilon_n = O_n - E\{O_n | W_n(\theta)\}$ is the residual error. We can substitute $\hat{m}(\bullet; \theta)$ in $Q_n(\theta, m(\bullet))$ for $m(\bullet)$ to obtain the following criterion depending solely on θ:

$$\hat{Q}_n(\theta) = \frac{1}{n} \sum_{i=1}^{n} \Psi_{\theta, \hat{m}(\bullet; \theta)}(U_i, Y_i).$$ (14.17)

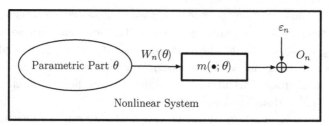

Figure 14.3 The regression function $m(\bullet; \theta) = E\{O_n | W_n(\theta)\}$ such that $m(\bullet; \theta^*) = m^*(\bullet)$.

It is worth noting that in the case of nonlinear dynamical systems with the memory size p, we ought modify our criterion $Q_n(\theta, m(\bullet))$ to the following form:

$$Q_n(\theta, m(\bullet)) = \frac{1}{n} \sum_{i=p+1}^{n} \Psi_{\theta, m(\bullet)}(\mathbf{U}_i, Y_i), \qquad (14.18)$$

where $\mathbf{U}_i = (U_i, U_{i-1}, \ldots, U_{i-p})^T$. The criterion $\hat{Q}_n(\theta)$ is then modified accordingly. It is now natural to define an estimate $\hat{\theta}$ of θ^* as the minimizer of $\hat{Q}_n(\theta)$, that is,

$$\hat{\theta} = \arg \min_{\theta \in \Theta} \hat{Q}_n(\theta). \qquad (14.19)$$

This approach may lead to an effective estimator of θ^* subject to two conditions. First, as we have already noted, we should be able to estimate the nonlinearities $m(\bullet; \theta)$ for a given $\theta \in \Theta$. The difficulty of this step depends on the complexity of the studied nonlinear system, that is, whether nonlinear components can be easily estimated as if the parametric part of the system were known. We will demonstrate that this is the case for a large class of block-oriented nonlinear systems. Second, we must minimize the criterion $\hat{Q}_n(\theta)$, which may be an expensive task mostly if θ is highly dimensional and if the gradient vector of $\hat{Q}_n(\theta)$ is difficult to evaluate. To partially overcome these computational difficulties we can use the following generic iterative method:

Step 1. Select an initial $\hat{\theta}^{(0)}$ and set $\hat{m}(\bullet; \hat{\theta}^{(0)})$.
Step 2. Minimize

$$\tilde{Q}_n(\theta) = \frac{1}{n} \sum_{i=1}^{n} \Psi_{\theta, \hat{m}(\bullet; \hat{\theta}^{(0)})}(U_i, Y_i)$$

with respect to θ and use the obtained value $\hat{\theta}^{(1)}$ to update $\hat{m}(\bullet; \theta)$, that is, go to Step 1 to get $\hat{m}(\bullet; \hat{\theta}^{(1)})$.
Step 3. Iterate between the above two steps until a certain stopping rule is satisfied.

Note that in the above algorithm the criterion $\tilde{Q}_n(\theta)$ has a weaker dependence on θ than the original criterion $\hat{Q}_n(\theta)$. In fact, in $\tilde{Q}_n(\theta)$ the nonlinearities $\hat{m}(\bullet; \theta)$ are already specified. Nevertheless, minimization of $\tilde{Q}_n(\theta)$ still requires an optimization algorithm and this can be obtained via one of many possible versions of the steepest descent method. We will illustrate the efficiency of this algorithm in Section 14.4.

Once the estimate $\hat{\theta}$ is specified one can plug it back into our pilot estimate $\hat{m}(\bullet; \theta)$ to obtain an estimate of $m^*(\bullet)$. Hence, let

$$\hat{m}(\bullet) = \hat{m}(\bullet; \hat{\theta}) \qquad (14.20)$$

be a nonparametric estimate of the system nonlinearities $m^*(\bullet)$.

It is worth noting that in the above two-step scheme the estimate $\hat{m}(\bullet; \theta)$ and the criterion function $\hat{Q}_n(\theta)$ which is used to obtain $\hat{\theta}$ share the same training data. This is usually not the recommended strategy since it may lead to estimates with unacceptably large variance. Indeed, some resampling schemes would be useful here which would partition the training data into the testing and training sequences. The former should be used to form the criterion in (14.17), whereas the latter to obtain the nonparametric estimate $\hat{m}(\bullet; \theta)$. This will facilitate the mathematical analysis of the estimation algorithms. This issue will be revisited in Section 14.4.

14.3.1   Consistency of semiparametric estimates

In the previous section, we described a general scheme of deriving estimates of characteristics of the linear dynamical parts and nonlinear elements of a class of semiparametric, block-oriented systems. In this section, we examine the issue of consistency of these estimates.

Let us recall that $(\theta^*, m^*(\bullet))$ stands for the true characteristics of the nonlinear system. Throughout this whole chapter we assume that θ^* is a finite dimensional vector, that is, that $\theta^* \in R^s$. On the other hand, we do not make any parametric assumptions about $m^*(\bullet)$.

Prior to outlining the problem of consistency of the estimates defined in (14.19) and (14.20), a fundamental issue that warrants further considerations is whether the underlying nonlinear system is identifiable. Because we have not assumed any specific structure of the system, this question is impossible to answer. In fact, the problem of identifiability can only be verified for specific classes of systems and this issue will addressed in Section 14.4.

The consistency of the estimates defined in (14.19) and (14.20) can be examined separately, that is, we can first determine the consistency of $\hat{\theta}$ in (14.19) and then of the resulting estimate $\hat{m}(\bullet)$ in (14.20). However, the dependence of $\hat{m}(\bullet; \theta)$ on θ plays a critical role for the consistency of $\hat{\theta}$. In turn, the accuracy of $\hat{m}(\bullet)$ also depends on the quality of the parametric estimate $\hat{\theta}$.

Ideally, we would like to estimate θ^* as efficiently as if the nonlinear nonparametric part $m^*(\bullet)$ were known and vice versa to estimate $m^*(\bullet)$ as accurately as if the parametric component θ^* were known. Estimates with such ideal efficiency are labeled as having an *oracle* property.

In the following two subsections, we shall establish sufficient conditions for the consistency of the parametric $\hat{\theta}$ and nonparametric $\hat{m}(\bullet)$ estimates. We recall the notation that $\eta_n \overset{n}{\to} \eta$, (P) stands for the convergence in probability of the random sequence $\{\eta_n\}$ to the limit random variable η as $n \to \infty$. Also $\eta_n = \eta + O_P(a_n)$, some $a_n \overset{n}{\to} 0$, stands for the fact that $\delta_n a_n^{-1}(\eta_n - \eta) \overset{n}{\to} 0$ (P) for $\delta_n \overset{n}{\to} 0$ arbitrarily slow. It is then said that $\eta_n - \eta$ tends to 0 (P) at the rate a_n.

A: Parametric estimation

The minimization of the random criterion $\hat{Q}_n(\theta)$ in (14.17) can be viewed as the problem of finding parametric M-estimators. This is, however, a nonstandard case because $\hat{Q}_n(\theta)$ depends on a random function $\hat{m}(\bullet; \theta)$. Hence, the consistency of $\hat{\theta}$ depends critically on the smoothness of the mapping:

$$\theta \mapsto \hat{m}(\bullet; \theta). \tag{14.21}$$

In fact, the consistency of the estimate $\hat{\theta}$ requires the continuity of the mapping in (14.21). This requirement can usually be easily verified if a specific nonparametric estimate $\hat{m}(\bullet; \theta)$ is used.

The first step required to establish the consistency of $\hat{\theta}$ is to determine the limit-deterministic criterion function $Q(\theta)$ such that,

$$\hat{Q}_n(\theta) \xrightarrow{n} Q(\theta), \quad (P), \quad \text{for every } \theta \in \Theta. \tag{14.22}$$

The above limit exists if the mapping $\hat{m}(\bullet; \theta) \mapsto \Psi_{\theta, \hat{m}(\bullet; \theta)}(\bullet, \bullet)$ is continuous, which is usually the case, and if,

$$\hat{m}(\bullet; \theta) \xrightarrow{n} m(\bullet; \theta), \quad (P), \quad \text{for every } \theta \in \Theta. \tag{14.23}$$

The limit function $m(\bullet; \theta)$ is a regression function of corresponding output and input signals of the underlying nonlinear system in which the parametric part has the value $\theta \in \Theta$, see Figure 14.3. It is clear that $m(\bullet; \theta^*) = m^*(\bullet)$, the true characteristics of the nonlinear elements.

By the above discussion, we may conclude that,

$$Q(\theta) = E\{\Psi_{\theta, m(\bullet; \theta)}(U, Y)\}. \tag{14.24}$$

It is also expected that the minimizer $\bar{\theta}$ of $Q(\theta)$ should well characterize the true value θ^*, and indeed in many cases we have

$$\theta^* = \arg \min_{\theta \in \Theta} Q(\theta). \tag{14.25}$$

Nevertheless, θ^* need not be a unique minimum of $Q(\theta)$.

The minimum in (14.25) exists if the set Θ is compact and if the mapping

$$\theta \mapsto \Psi_{\theta, m(\bullet; \theta)}(\bullet, \bullet)$$

is continuous. This, due to (14.24), is implied by the continuity of $\theta \mapsto \Psi_{\theta, \bullet}(\bullet, \bullet)$ and $\theta \mapsto m(\bullet; \theta)$. The latter properties can only be verified in concrete cases, and we will illustrate this in Section 14.4.

The above discussion reveals that it is reasonable to expect that the minimizer $\hat{\theta}$ of $\hat{Q}_n(\theta)$ converges to the minimizer $\bar{\theta}$ of $Q(\theta)$. As we have already noted that the value $\bar{\theta}$ can often be identified with the true value θ^*, and this will be assumed throughout this chapter.

Note that the criterion $\hat{Q}_n(\theta)$ is not a convex function of θ and therefore need not achieve a unique minimum. This, however, is of no serious importance for the consistency since we may weaken our requirement on the minimizer $\hat{\theta}$ and define $\hat{\theta}$ as any estimator that nearly minimizes $\hat{Q}_n(\theta)$, that is,

$$\hat{Q}_n(\hat{\theta}) \le \inf_{\theta \in \Theta} \hat{Q}_n(\theta) + \varepsilon_n, \tag{14.26}$$

for any random sequence $\{\varepsilon_n\}$, such that $\varepsilon_n \xrightarrow{n} 0$, (P). It is clear that (14.26) implies that $\hat{Q}_n(\hat{\theta}) \le \hat{Q}_n(\theta^*) + \varepsilon_n$ and this is sufficient for the convergence of $\hat{\theta}$ defined in (14.26) to θ^*.

Since $\hat{\theta}$ depends on the whole mapping $\theta \mapsto \hat{Q}_n(\theta)$, the convergence of $\hat{\theta}$ to θ^* requires uniform consistency of the corresponding criterion function, that is, we need

$$\sup_{\theta \in \Theta} |\hat{Q}_n(\theta) - Q(\theta)| \xrightarrow{n} 0, (P). \tag{14.27}$$

This uniform convergence is the most difficult step in proving the consistency result. First, we observe that (14.27) is implied by

$$\sup_{\theta \in \Theta} |\hat{Q}_n(\theta) - E\hat{Q}_n(\theta)| \xrightarrow{n} 0, (P), \tag{14.28}$$

and

$$\sup_{\theta \in \Theta} |E\hat{Q}_n(\theta) - Q(\theta)| \xrightarrow{n} 0. \tag{14.29}$$

The uniform convergence of $\hat{Q}_n(\theta)$ to its average required in (14.28) can be analyzed by the well established theory of empirical processes. The main requirement of this theory is that the following set of functions,

$$\{\Psi_{\theta, \hat{m}(\bullet; \theta)}(\bullet, \bullet); \theta \in \Theta\}, \tag{14.30}$$

defines the so-called Glivienko–Cantelli class. When Θ is compact, which is assumeed throughout this chapter, simple sufficient conditions for the class in (14.30) to be the Glivenko–Cantelli set of functions are:

(a) The mapping $\theta \mapsto \Psi_{\theta, \hat{m}(\bullet; \theta)}(U, Y)$ is continuous for every (U, Y).
(b) There exists a function $\alpha(U, Y)$, such that

$$\Psi_{\theta, \hat{m}(\bullet; \theta)}(U, Y) \le \alpha(U, Y) \text{ for every } \theta \in \Theta \tag{14.31}$$

and $E\{\alpha(U, Y)\} < \infty$.

Conditions (14.31) assure that the class of functions in (14.30) satisfy a uniform law of large numbers in (14.28). On the other hand, the convergence in (14.29) is equivalent to,

$$\sup_{\theta \in \Theta} E|\Psi_{\theta, \hat{m}(\bullet; \theta)}(U, Y) - \Psi_{\theta, m(\bullet; \theta)}(U, Y)| \xrightarrow{n} 0. \tag{14.32}$$

This property can be deduced from the continuity of the mappings $\theta \mapsto \Psi_{\theta, \hat{m}(\bullet, \theta)}(\bullet, \bullet)$, $\theta \mapsto \Psi_{\theta, m(\bullet; \theta)}(\bullet, \bullet)$, condition (14.31), and the compactness of Θ.

In summary, the uniform convergence in (14.27) is essential for establishing the convergence of $\hat{\theta}$, defined in (14.26), to θ^*. It should be noted, however, that this requirement is not necessary and weaker conditions can be formulated. The following theorem summarizes our discussion on the consistency of $\hat\theta$.

THEOREM 14.1 *Let $\hat{Q}_n(\theta)$ be a random criterion function given in (14.17) and defined for $\theta \in \Theta$, where Θ is a compact subset of R^s. Let the limit criterion $Q(\theta)$ given in (14.22) be a continuous function on Θ. Suppose that condition (14.31) is met. Then, for any sequence of estimators $\hat\theta$ that satisfy (14.26), we have*

$$\hat\theta \overset{n}{\to} \theta^*, (P).$$

Proof. By the definition of $\hat\theta$, we have $\hat{Q}_n(\hat\theta) \le \hat{Q}_n(\theta^*) + \varepsilon_n$. By virtue of (14.31), the compactness of Θ, and the continuity of $Q(\theta)$, we have (14.27), and consequently $\hat{Q}_n(\theta^*) \overset{n}{\to} Q(\theta^*)(P)$. As a result we get $\hat{Q}_n(\hat\theta) \le Q(\theta^*) + \tilde\varepsilon_n$ where $\tilde\varepsilon_n \overset{n}{\to} 0(P)$. This yields

$$Q(\hat\theta) - Q(\theta^*) \le Q(\hat\theta) - \hat{Q}_n(\hat\theta) + \tilde\varepsilon_n$$
$$\le \sup_{\theta\in\Theta} |\hat{Q}_n(\theta) - Q(\theta)| + \tilde\varepsilon_n \overset{n}{\to} 0, (P).$$

This result and the facts that Θ is compact and $Q(\theta)$ is a continuous function imply that there exists, for every $\delta > 0$, a number $\eta > 0$, such that if $\|\hat\theta - \theta^*\| \ge \delta$ then $Q(\hat\theta) - Q(\theta^*) > \eta$. Hence,

$$P(\|\hat\theta - \theta^*\| \ge \delta) \le P(Q(\hat\theta) - Q(\theta^*) > \eta) \overset{n}{\to} 0.$$

This completes the proof of Theorem 14.1. ■

Often the mapping $a : \theta \to \Psi_{\theta,m(\bullet)}$, which defines the criterion in (14.16), and the mapping $\theta \mapsto \hat{m}(\bullet; \theta)$ are differentiable. Then, it is more convenient to examine zeros of the criterion,

$$\hat{q}_n(\theta) = \frac{\partial}{\partial\theta} \hat{Q}_n(\theta) = \frac{1}{n} \sum_{i=1}^{n} \psi_{\theta,\hat{m}(\bullet;\theta)}(U_i, Y_i), \tag{14.33}$$

where $\psi_{\theta,\hat{m}(\bullet;\theta)}(\bullet, \bullet) = \frac{\partial}{\partial\theta} \Psi_{\theta,\hat{m}(\bullet;\theta)}(\bullet, \bullet)$. Note that we can efficiently evaluate this derivative in the following way:

$$\frac{\partial}{\partial\theta} \Psi_{\theta,\hat{m}(\bullet;\theta)}(\bullet, \bullet) = \frac{\partial\Psi_{\theta,\hat{m}(\bullet;\theta)}(\bullet, \bullet)}{\partial a} \mathbf{1} + \frac{\partial\Psi_{\theta,\hat{m}(\bullet;\theta)}(\bullet, \bullet)}{\partial\hat{m}(\bullet;\theta)} \frac{\partial\hat{m}(\bullet;\theta)}{\partial\theta}, \tag{14.34}$$

where $\mathbf{1} = (1, 1, \dots, 1)$ is the s-dimensional vector.

Similarly, as in Theorem 14.1 we can use the zeros of $\hat{q}_n(\theta)$ as a definition of our estimate $\hat\theta$. Then, the zeros of the limit criterion $q(\theta) = E\{\psi_{\theta,m(\bullet;\theta)}(U, Y)\}$ can be identified with the true value θ^* of the parameter.

The consistency $\hat\theta \overset{n}{\to} \theta^*$ (P) and differentiability of the mapping $\theta \mapsto \Psi_{\theta,\hat{m}(\bullet;\theta)}(\bullet, \bullet)$ allow us to consider the problem of the convergence rate. In fact, one can determine a formal Taylor series expansion of $\hat{Q}_n(\theta)$ around θ^*,

$$\hat{Q}_n(\hat\theta) = \hat{Q}_n(\theta^*) + \hat{q}_n(\theta^*)^T(\hat\theta - \theta^*) + \frac{1}{2}(\hat\theta - \theta^*)^T V_{\hat\theta}(\hat\theta - \theta^*), \tag{14.35}$$

where V_θ is the $s \times s$ matrix of second derivatives of $\hat{Q}_n(\theta)$ and $\bar{\theta}$ is a point between $\hat{\theta}$ and θ^*. By taking the derivative of the expansion in (14.35) and noting that $\hat{q}_n(\hat{\theta}) = 0$, we can obtain that

$$\sqrt{n}(\hat{\theta} - \theta^*) = -D_n(\theta^*)\sqrt{n}\hat{q}_n(\theta^*), \qquad (14.36)$$

for some matrix $D_n(\theta)$.

With regard to (14.36), note first that $\sqrt{n}\hat{q}_n(\theta^*) = n^{-1/2} \sum_{i=1}^n \psi_{\theta^*,\hat{m}(\bullet;\theta^*)}(U_i, Y_i)$. Next, we have $\psi_{\theta^*,\hat{m}(\bullet;\theta^*)}(\bullet, \bullet) \xrightarrow{n} \psi_{\theta^*,m^*(\bullet)}(\bullet, \bullet)$ (P) and then $E\psi_{\theta^*,m^*(\bullet)}(U, Y) = q(\theta^*)$. Since $q(\theta^*) = 0$ and $\mathrm{var}(\sqrt{n}\hat{q}_n(\theta^*))$ tends to a constant, we can conclude that $\sqrt{n}\hat{q}_n(\theta^*)$ is bounded (P). The examination of the asymptotic behavior of the matrix $D_n(\theta)$ is more delicate. It can be shown, however, that $D_n(\theta)$ tends (P) to some nonsingular matrix.

The following theorem gives some sufficient conditions for the $\sqrt{n}$ convergence rate of the estimate $\hat{\theta}$. We omit, however, technical details yielding the proof of this theorem.

THEOREM 14.2 *Let $\hat{Q}_n(\theta)$ be a random criterion function given in (14.17) defined for $\theta \in \Theta$, where Θ is a compact subset of R^s. Let the limit criterion $Q(\theta)$ given in (14.22) admit the second-order Taylor expansion at a point of minimum θ^* with a nonsingular symmetric second derivative matrix. Furthermore, assume that the mapping $\theta \mapsto \Psi_{\theta,\hat{m}(\bullet;\theta)}(\bullet, \bullet)$ is differentiable at θ^* with the derivative $\dot{\psi}_{\theta,\hat{m}(\bullet;\theta)}(\bullet, \bullet)$ satisfying,*

$$\| \dot{\psi}_{\theta,\hat{m}(\bullet;\theta)}(U, Y) \| \le \beta(U, Y),$$

for all $\theta \in \Theta$, with $E\beta^2(U, Y) < \infty$. Then, for any sequence of estimators $\hat{\theta}$ that satisfy (14.26) with $n\varepsilon_n \xrightarrow{n} 0$ (P) and such that $\hat{\theta} \xrightarrow{n} \theta^*$ (P), we have

$$\hat{\theta} = \theta^* + O_P(n^{-1/2}).$$

In Sections 14.4, 14.5, and 14.6, we willl illustrate the universality of the aforementioned approach to the problem of identifying semiparametric versions of the Wiener, Hammerstein, and parallel systems.

B: Nonparametric estimation

The result of Theorem 14.1 assures that $\hat{\theta} \xrightarrow{n} \theta^*$, (P). Therefore, it is reasonable to expect that the estimate $\hat{m}(\bullet)$ in (14.20) converges to $m(\bullet;\theta^*) = m^*(\bullet)$. The following decomposition will facilitate this claim

$$\hat{m}(\bullet) - m^*(\bullet) = \{\hat{m}(\bullet) - \hat{m}(\bullet;\theta^*)\} + \{\hat{m}(\bullet;\theta^*) - m^*(\bullet)\}. \qquad (14.37)$$

The convergence (P) of the second term to zero in the above decomposition represents a classical problem in nonparametric estimation. This convergence depends on the nature of the dependence between random signals within the underlying system. If the system has a finite memory, the random processes are outputs of finite impulse response filters, and the convergence property can easily be established. For general dependence structures, that is, for systems with infinite memory, the convergence can be a difficult issue to verify.

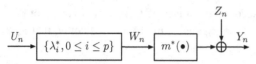

Figure 14.4 The identified semiparametric Wiener system.

Concerning the first term in (14.37), note that $\hat{m}(\bullet) = \hat{m}(\bullet; \hat{\theta})$. Then, we can apply the linearization technique, that is, we have

$$\hat{m}(\bullet; \hat{\theta}) - \hat{m}(\bullet; \theta^*) = \left\{ \frac{\partial}{\partial \theta} \hat{m}(\bullet; \theta)_{|\theta=\theta^*} \right\}^T \left(\hat{\theta} - \theta^* \right) + o \left(\|\hat{\theta} - \theta^*\|^2 \right). \quad (14.38)$$

To show the convergence (P) of the first term to zero it suffices to prove that $\frac{\partial}{\partial \theta} \hat{m}(\bullet; \theta)_{|\theta=\theta^*}$ has a finite limit (P) as $n \to \infty$. This fact can be directly verified for a specific estimate $\hat{m}(\bullet; \theta)$ of $m(\bullet; \theta)$. There exists a general result that allows one to determine the convergence of the derivatives of an estimate from the convergence of the estimate itself, see the Appendix.

The decomposition in (14.37) combined with the result of Theorem 14.2 allows us also to examine the problem of the convergence rate of the estimate $\hat{m}(\bullet)$. First of all, because the derivative term in (14.38) tends (P) to a finite value, then, due to Theorem 14.2, the first term on the right-hand side of (14.37) is of order $O_P(n^{-1/2})$. Next, for most nonparametric regression estimates $\hat{m}(\bullet; \theta^*)$, we usually have $\hat{m}(\bullet; \theta^*) = m^*(\bullet) + O_P(n^{-\alpha})$, where typically $1/3 \leq \alpha < 1$. Consequently, we obtain,

$$\hat{m}(\bullet) = m^*(\bullet) + O_P(n^{-\alpha}). \quad (14.39)$$

With respect to α, if $\hat{m}(\bullet)$ is a kernel regression estimate and $m^*(\bullet)$ is Lipschitz continuous then (14.39) holds with $\alpha = 1/3$. For twice differentiable nonlinearities, this takes place with $\alpha = 2/5$.

14.4   Statistical inference for semiparametric Wiener models

In this section, we will illustrate the semiparametric methodology developed in the previous sections by the examination of the constrained Wiener system introduced in Section 14.2. The system is shown in Figure 14.4 and is characterized by the pair $(\lambda^*, m^*(\bullet))$, where $\lambda^* \in R^{p+1}$ is the vector representing the impulse response function of the linear subsystem. The identification problem of the Wiener system examined in Chapters 9, 10, and 11 is based on the concept of the inverse regression and has certain apparent limitations. Indeed, it has been assumed that:

- The input signal $\{U_n\}$ is either a white or color Gaussian process.
- The measurement noise Z_n is absent, that is, $Z_n = 0$.
- The nonlinear characteristic $m^*(\bullet)$ is invertible, that is, $m^{*-1}(\bullet)$ exists.

On the other hand, in the previous algorithms it was possible to define the memory of the linear subsystem as infinite. On the contrary, in the semiparametric approach introduced in Section 14.3, we need a finite dimensional parameterization of the dynamical subsystem. Hence, let us consider the semiparametric case of the Wiener system which is parameterized by a finite memory linear subsystem. Thus, we have the following input–output relationship:

$$
\begin{cases}
W_n = \displaystyle\sum_{l=0}^{p} \lambda_l^* \, U_{n-l} \\[2mm]
Y_n = m^*(W_n) + Z_n
\end{cases}
\tag{14.40}
$$

where the order p of the dynamical subsystem is assumed to be known.

As we have already noted, the tandem nature of the Wiener system yields the scaling effect, that is, one can only estimate λ^* up to a multiplicative constant. Let us consider a Wiener system with the characteristics $\bar{m}^*(w) = m^*(w/c)$ and $\bar{\lambda}^* = c\lambda^*$, c being an arbitrary nonzero constant. Then it is easy to see that the new system is indistinguishable from the original one. Thus, in order to get around this identifiability problem we need some normalization of the sequence $\lambda^* = \{\lambda_l^*, 0 \leq l \leq p\}$. A simple normalization is to assume that $\lambda_0 = 1$.

To proceed further, it is necessary to introduce the space Λ of all admissible impulse response functions $\lambda = \{\lambda_l, 0 \leq l \leq p\}$ of order p which satisfy the normalization constraint $\lambda_0 = 1$. Hence, let $\Lambda = \{\lambda \in R^{p+1} : \lambda_0 = 1\}$ such that $\lambda^* \in \Lambda$. By virtue of the semiparametric methodology of Section 14.3, we first wish to characterize the system nonlinearity for a given $\lambda \in \Lambda$. Hence, let

$$
W_n(\lambda) = \sum_{l=0}^{p} \lambda_l \, U_{n-l},
\tag{14.41}
$$

be the interconnecting signal of the Wiener system corresponding to $\lambda \in \Lambda$. Consequently, it is natural to use the following regression function,

$$
m(w; \lambda) = E\{Y_n | W_n(\lambda) = w\},
\tag{14.42}
$$

as the best approximate of $m^*(w)$ for a given $\lambda \in \Lambda$. Note that $m(w; \lambda)$ is the best predictor of the output signal for a given $\lambda \in \Lambda$. It is clear that $W_n(\lambda^*) = W_n$ and $m(w; \lambda^*) = m^*(w)$. The smoothness of $m(w; \lambda)$ plays an important role in the statistical analysis of our identification algorithms. Since $m(w; \lambda) = E\{m^*(W_n) | W_n(\lambda) = w\}$, the smoothness of $m(w; \lambda)$ is controlled by the smoothness of $m^*(w)$ and the conditional distribution of W_n on $W_n(\lambda)$.

Example 14.1 To illustrate the dependence of $m(w; \lambda)$ on $m^*(w)$ in terms of smoothness, let $\{U_n\}$ be an *iid* sequence with a normal distribution $N(0, \sigma^2)$. Then, denoting by $\phi(\bullet)$ the $N(0, 1)$ density and after some algebra we have,

$$
m(w; \lambda) = \int_{-\infty}^{\infty} m^*(\mu(\lambda)w + v\sigma(\lambda))\phi(v)dv,
\tag{14.43}
$$

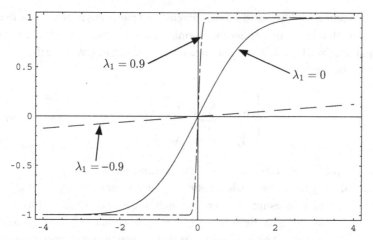

Figure 14.5 The regression function $m(w; \lambda)$ in (14.43) versus w, with $\lambda = (1, \lambda_1)^T$, $\lambda^* = (1, 1)^T$, $m^*(w) = \text{sgn}(w)$. Values $\lambda_1 = -0.9, 0, 0.9$.

where

$$\mu(\lambda) = \frac{\lambda^T \lambda^*}{\|\lambda\|^2}, \sigma^2(\lambda) = \sigma^2 \left[\|\lambda^*\|^2 - \frac{(\lambda^T \lambda^*)^2}{\|\lambda\|^2} \right].$$

In Figure 14.5 we plot $m(w; \lambda)$ in (14.43) as a function of w with $\lambda = (1, \lambda_1)^T$, $\lambda^* = (1, 1)^T, \sigma^2 = 1$, and the discontinuous nonlinearity $m^*(w) = \text{sgn}(w)$. Values $\lambda_1 = -0.9, 0, 0.9$ are used. Note that the value $\lambda_1 = 0$ indicates that there is no dynamical subsystem in the Wiener model. The continuity of $m(w; \lambda)$ is apparent.

In Figure 14.6, we plot $m(w; \lambda)$ versus λ for a few selected values of w, that is, $w = -1, -0.1, 0.1, 1$. The sensitivity of $m(w; \lambda)$ with respect to λ is small for points which lie far from the point of discontinuity $w = 0$. On the other hand, we observe a great influence of λ at the points which are close to the discontinuity.

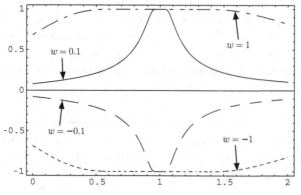

Figure 14.6 The regression function $m(w; \lambda)$ in (14.43) versus λ_1, with $\lambda = (1, \lambda_1)^T$, $\lambda^* = (1, 1)^T$, $m^*(w) = \text{sgn}(w)$. Values $w = -1, -0.1, 0.1, 1$.

In general, it can be observed that for a very large class of nonlinearities and distributions of $(W_n, W_n(\lambda))$ which are not necessarily continuous, $m(w; \lambda)$ is a continuous function provided that $\|\lambda - \lambda^*\| > 0$.

Our principle issue is to recover the pair $(\lambda^*, m^*(\bullet))$ from the training set

$$T = \{(U_1, Y_1), \ldots, (U_n, Y_n)\}, \tag{14.44}$$

where $\{U_i\}$ is an iid sequence of random variables with the density function $f_U(\bullet)$. Owing to the semiparametric methodology discussed in Section 14.3, we first must estimate (for a given λ) the regression function $m(w; \lambda)$. This can easily be done using any previously studied regression estimates applied to synthetic data parametrized by λ :

$$\{(W_{p+1}(\lambda), Y_{p+1}), \ldots, (W_n(\lambda), Y_n)\}. \tag{14.45}$$

This yields an estimate $\hat{m}(w; \lambda)$, which allows one to define a predictive error as a function of only the linear subsystem characteristic λ.

Let us choose the criterion function defined in (14.14). Then, we can write the prediction error as follows:

$$\hat{Q}_n(\lambda) = \frac{1}{n} \sum_{j=p+1}^{n} M(Y_j - \hat{m}(W_j(\lambda); \lambda)). \tag{14.46}$$

The strategy of estimating λ^* is now based on the minimization of $\hat{Q}_n(\lambda)$. In this respect, the choice $M(t) = t^2$ is the most popular, and this error function will be used throughout this section. It is worth noting that the formula in (14.46) is the counterpart of (14.17), which is now specialized to the Wiener system and the M-type criterion function.

14.4.1 Identification algorithms

The criterion $\hat{Q}_n(\lambda)$ in (14.46) uses the same data to form the pilot estimate $\hat{m}(w; \lambda)$ and to define $\hat{Q}_n(\lambda)$. This is not generally a good strategy and some form of resampling scheme should be applied in order to separate the data into the testing and training sequence. The former should be used to form the criterion in (14.17), whereas the latter to obtain the estimate $\hat{m}(w; \lambda)$. This will facilitate not only the mathematical analysis of the estimation algorithms but also gives a desirable separation of parametric and nonparametric estimation problems, which allows one to evaluate parametric and nonparametric estimates more efficiently.

Consider the partition strategy that reorganizes a set of training data T into two nonoverlapping subsets that are statistically independent. Owing to the fact that the observations Y_n and Y_{n+p+1} are statistically independent, we define T_1 as the subset of training set T consisting of n_1 observations after deleting the first p data points due to the memory effect. Similarly, let T_2 be the remaining part of T separated from T_1 by the distance of length p. By construction we note that T_1 and T_2 are independent random subsets of T. This is the key property which allows us to design efficient estimates of λ^*, $m(w; \lambda)$, and consequently $m^*(\bullet)$. We use the subset T_1 to estimate the regression

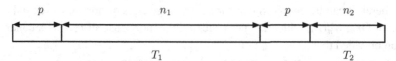

Figure 14.7 An example of the partition of the training set T into independent subsets T_1 and T_2.

function $m(w; \lambda)$ whereas T_2 is used as a testing sequence to form the least-squares criterion to recover the impulse response sequence λ^*. Also, let I_1 and I_2 denote the indices of data points $\{(U_i, Y_i), 1 \le i \le n\}$ which belong to T_1 and T_2, respectively. Figure 14.7 shows an example of the partition of T into T_1 and T_2. Here we have $n_2 = n - 2p - n_1$.

It is clear that there are other possible partitions of a training data set. In fact, the machine learning theory principle says the testing sequence T_2 should consists of independent observations, whereas the training sequence T_1 can be arbitrary. In our situation this strategy can be easily realized by choosing the testing observations that are $p + 1$ positions apart from each other. Figure 14.8 shows such a partition where only T_2 is indicated, with the remaining part of the data set (except for the first p observations) defining T_1. In this case the testing set T_2 is a sequence of independent random variables and $n_2 = n/p$. Note, however, that for a linear subsystem with a long memory size p, the testing sequence can be unacceptably small.

In the analysis that follows, we employ the partition which divides the training set into two statistically independent subsets, as shown in Figure 14.7.

A number of nonparametric estimates of the regression function $m(w; \lambda)$ can be proposed. In Chapter 3, we studied the classical Nadaraya–Watson estimate which, when applied to the subset T_1 of the data set in (14.45), takes the following form:

$$\hat{m}(w; \lambda) = \frac{\sum_{j \in I_1} Y_j K\left(\frac{w - W_j(\lambda)}{h}\right)}{\sum_{j \in I_1} K\left(\frac{w - W_j(\lambda)}{h}\right)}, \tag{14.47}$$

for a given $\lambda \in \Lambda$.

The recursive and semirecursive kernel estimates that were introduced in Chapters 4 and 5 can also be used to estimate $m(w; \lambda)$. The order statistics kernel estimate of Chapter 7 is particularly attractive under the present circumstances since it is not of the

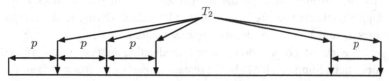

Figure 14.8 An example of the partition of the training set T into subsets T_1 and T_2, which provides the testing set T_2 as a sequence of *iid* random variables. Only the subset T_2 is shown; the remaining part of the data set (except for the first p observations) defines T_1.

ratio form. The order statistics kernel estimate is defined as follows:

$$\tilde{m}(w; \lambda) = \sum_{j \in I_1} Y_{[j]} h^{-1} \int_{W_{(j-1)}(\lambda)}^{W_{(j)}(\lambda)} K\left(\frac{w - v}{h}\right) dv, \tag{14.48}$$

where, for a given $\lambda \in \Lambda$, $\{W_{(j)}(\lambda), j \in I_1\}$ is the order statistic of the data set in (14.45) confined to observations from T_1. In addition, $\{Y_{[j]}, j \in I_1\}$ is a set of the corresponding output observations paired with $\{W_{(j)}(\lambda), j \in I_1\}$.

A less accurate, but computationally more attractive, version of $\tilde{m}(w; \lambda)$ is

$$\tilde{m}(w; \lambda) = \sum_{j \in I_1} Y_{[j]}(W_{(j)}(\lambda) - W_{(j-1)}(\lambda)) h^{-1} K\left(\frac{w - W_{(j)}(\lambda)}{h}\right). \tag{14.49}$$

The aforementioned pilot estimates of $m(w; \lambda)$ can now be used to form the least-squares approach to recover the impulse response λ^* of the Wiener system. Thus, the least-squares version of the criterion function in (14.17) or (14.46) confined to the data set T_2 takes the following form:

$$\hat{Q}_n(\lambda) = \frac{1}{n_2} \sum_{i \in I_2} \{Y_i - \hat{m}(W_i(\lambda); \lambda)\}^2, \tag{14.50}$$

where the classical kernel estimate (14.47) can be replaced by virtually any other non-parametric estimate, for example, the estimates in (14.48) and (14.49).

A natural estimate of λ^* is the minimizer of $\hat{Q}_n(\lambda)$, that is, let

$$\hat{\lambda} = \arg\min_{\lambda \in \Lambda} \hat{Q}_n(\lambda). \tag{14.51}$$

Once the estimate $\hat{\lambda}$ is obtained, one can define the following estimate of the nonlinear characteristic $m^*(\bullet)$ of the Wiener system

$$\hat{m}(w) = \hat{m}(w; \hat{\lambda}), \tag{14.52}$$

where $\hat{m}(w; \lambda)$ is virtually any nonparametric estimate of $m(w; \lambda)$. Let us note that we can use one nonparametric estimate in the definition of $\hat{Q}_n(\lambda)$ and another one (according to the prescription in (14.52)) to recover the nonlinearity.

It is clear that the criterion $\hat{Q}_n(\lambda)$ need not possess a unique minimum and, moreover, an efficient procedure to find the minimum of $\hat{Q}_n(\lambda)$ is required. One numerical method to determine $\hat{\lambda}$ is to evaluate the gradient vector of $\hat{Q}_n(\lambda)$, which is necessary for the application of steepest descent algorithms. Hence, let us define the n_2-dimensional vector $\varphi(\lambda) = (\varphi_i(\lambda), i \in I_2)^T$, where,

$$\varphi_i(\lambda) = Y_i - \hat{m}(W_i(\lambda); \lambda). \tag{14.53}$$

Because of our normalization imposed on $\lambda \in \Lambda$, we may identify λ as a vector in R^p. Then, direct differentiation of $\hat{Q}_n(\lambda)$ yields,

$$\frac{\partial \hat{Q}_n(\lambda)}{\partial \lambda} = 2n_2^{-1} G^T(\lambda)\varphi(\lambda), \tag{14.54}$$

where $G(\lambda)$ is the $n_2 \times p$ matrix with the (i, j) entry given by $\partial\varphi_i(\lambda)/\partial\lambda_j$, $i \in I_2$, $j = 1, \ldots, p$. The evaluation of the function $\varphi_i(\lambda)$ in (14.53) and its derivative,

$$\frac{\partial\varphi_i(\lambda)}{\partial\lambda_j} = -\frac{\partial\hat{m}(W_i(\lambda);\lambda)}{\partial\lambda_j},$$

can be conducted for specific nonparametric estimates $\hat{m}(w;\lambda)$. It is a relatively easy task to evaluate the derivative for the order statistics kernel estimate in (14.49) and it is more complicated for the classical kernel estimator in (14.47).

To ease the computational burden we can apply the iterative algorithm introduced Section 14.3. This requires determining for a given $\hat{\lambda}^{(old)}$ the minimum of the partially specified criterion,

$$\tilde{Q}_n(\lambda) = \frac{1}{n_2} \sum_{i \in I_2} \left\{ Y_i - \hat{m}\left(W_i(\lambda); \hat{\lambda}^{(old)}\right) \right\}^2. \tag{14.55}$$

The gradient of $\tilde{Q}_n(\lambda)$ is given by formula (14.54), where

$$\varphi_i(\lambda) = Y_i - \hat{m}\left(W_i(\lambda); \hat{\lambda}^{(old)}\right), \tag{14.56}$$

and

$$\frac{\partial\varphi_i(\lambda)}{\partial\lambda} = -\frac{\partial\hat{m}\left(W_i(\lambda); \hat{\lambda}^{(old)}\right)}{\partial\lambda}. \tag{14.57}$$

Calculation of the derivative of $\varphi_i(\lambda)$ is now straightforward. Indeed, for the estimate $\hat{m}(w;\lambda)$ in (14.47), we have

$$\frac{\partial\varphi_i(\lambda)}{\partial\lambda} = -\hat{m}^{(1)}\left(W_i(\lambda); \hat{\lambda}^{(old)}\right) \tilde{\mathbf{U}}_i, \tag{14.58}$$

where $\hat{m}^{(1)}(w;\lambda) = \partial\hat{m}(w;\lambda)/\partial w$ and $\tilde{\mathbf{U}}_i = (U_{i-1}, \ldots, U_{i-p})^T$.

Concerning the order statistics estimate $\tilde{m}(w;\lambda)$ in (14.49), we have,

$$\frac{\partial\varphi_i(\lambda)}{\partial\lambda} = -\tilde{m}^{(1)}\left(W_i(\lambda); \hat{\lambda}^{(old)}\right) \tilde{\mathbf{U}}_i,$$

where

$$\hat{m}^{(1)}(w;\lambda) = \frac{\partial\tilde{m}(w;\lambda)}{\partial w} = \sum_{j \in I_1} Y_{[j]}(W_{(j)}(\lambda) - W_{(j-1)}(\lambda))h^{-2}K^{(1)}\left(\frac{w - W_{(j)}(\lambda)}{h}\right). \tag{14.59}$$

In summary, we can propose the following efficient algorithm to minimize $\tilde{Q}_n(\lambda)$ that uses the simplified criterion $\tilde{Q}_n(\lambda)$:

Step 1: For a selected estimate $\hat{m}(w;\lambda)$ of the regression function $m(w;\lambda)$ specify an initial value $\hat{\lambda}^{(old)}$ and set $\hat{m}(w; \hat{\lambda}^{(old)})$.

Step 2: Choose $\lambda^{(0)}$ and iterate for $t = 0, 1, 2, \ldots$

$$\lambda^{(t+1)} = \lambda^{(t)} - \gamma_t G^T(\lambda^{(t)})\varphi(\lambda^{(t)}),$$

where $\varphi(\lambda)$ is the n_2-dimensional vector such that

$$\varphi_i(\lambda) = Y_i - \hat{m}(W_i(\lambda); \hat{\lambda}^{(\text{old})}), \quad i \in I_2$$

and $G(\lambda)$ is the $n_2 \times p$ matrix with the (i, j) entry given by

$$\frac{\partial \varphi_i(\lambda)}{\partial \lambda_j} = -\hat{m}^{(1)}(W_i(\lambda); \hat{\lambda}^{(\text{old})})U_{i-j}, \quad i \in I_2, j = 1, \ldots, p.$$

Stop iterations once two successive values of $\tilde{Q}_n(\lambda)$ in (14.55) differ insignificantly (quantified by a given small number $\epsilon = 10^{-4}$). This produces a new $\hat{\lambda}^{(\text{new})}$.

Step 3: Use the obtained $\hat{\lambda}^{(\text{new})}$ from Step 2 to update $\hat{m}(w; \hat{\lambda}^{(\text{old})})$, that is, we get $\hat{m}(w; \hat{\lambda}^{(\text{new})})$.

Step 4: Iterate between the above steps until the criterion function $\hat{Q}_n(\lambda)$ in (14.50) does not change significantly.

There are various ways to refine the above procedure, and some additional comments on the implementation are provided in Remarks 14.1, 14.2, 14.3.

REMARK 14.1 *We can speed up the algorithm by employing some preliminary estimate of λ^*, rather than selecting an arbitrary $\hat{\lambda}^{(\text{old})}$. For instance, we can use the following correlation estimate of λ^*, as shown in Chapter 9:*

$$\hat{\lambda}_s^{(\text{old})} = \frac{\sum_{j=s+1}^{n} U_{j-s} Y_j}{\sum_{j=1}^{n} U_j Y_j}, \quad s = 1, \ldots, p.$$

It is worth noting that $\hat{\lambda}_s^{(\text{old})}$ is a consistent estimate of λ_s^* provided that the input signal is a white Gaussian process (see Chapter 9). If the input process is at least close to being Gaussian, this choice for $\hat{\lambda}^{(\text{old})}$ may drastically reduce the number of iterations required in the algorithm.

REMARK 14.2 *The algorithm uses a kernel estimate which, in turn, needs the selection of the bandwidth parameter h. Due to the splitting strategy, our criterion $\hat{Q}_n(\lambda)$ or its simplified form $\tilde{Q}_n(\lambda)$ are already in the form of a predictive error, and we can incorporate the bandwidth into the definition of our criterion. Hence, we can use*

$$\tilde{Q}_n(\lambda; h) = \frac{1}{n_2} \sum_{i \in I_2} \left\{ Y_i - \hat{m}\left(W_i(\lambda); \hat{\lambda}^{(\text{old})} \right) \right\}^2$$

as the criterion for selecting both $\hat{\lambda}$ and h.

Then, in Step 1 of the algorithm, we should select an initial value of h, and in Step 2, we should apply a simple search algorithm (not necessarily based on the derivative $\partial \tilde{Q}_n(\lambda; h)/\partial h$) to find an improved value of h. A value of h obtained in this manner can serve as the bandwidth for the final estimate in (14.52) of the system nonlinearity.

REMARK 14.3 *The aforementioned procedure needs the derivative $\partial \hat{m}(w; \lambda)/\partial w$ of the kernel estimates. This can easily be obtained as, for example, in (14.59). Nevertheless,*

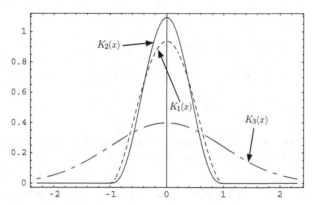

Figure 14.9 Smooth kernel functions.

smooth kernel functions are required here. The following are the second-order kernels (see Chapter 3), which have two, three, and infinity number of derivatives, respectively:

$$K_1(x) = \frac{15}{16}(1 - x^2)^2 \, \mathbf{1}(|x| \le 1),$$

$$K_2(x) = \frac{35}{32}(1 - 3x^2 + 3x^4 - x^6) \, \mathbf{1}(|x| \le 1),$$

$$K_3(x) = (2\pi)^{-1/2} \, e^{-x^2/2}.$$

Figure 14.9 depicts these kernels.

14.4.2   Convergence analysis

This section is concerned with the convergence analysis of the identification algorithms $\hat{\lambda}$ and $\hat{m}(\bullet)$ proposed in (14.51) and (14.52), respectively. We will employ the basic methodology established in Section 14.3.1.

Let $f_W(\bullet)$ and $f_W(\bullet; \lambda)$ be density functions of random processes $\{W_n\}$ and $\{W_n(\lambda)\}$, respectively. Note that $f_W(\bullet)$ and $f_W(\bullet; \lambda)$ always exist since are they are obtained by the $(p + 1)$-fold convolution of the scaled version of $f_U(\bullet)$ – the probability density function of the input process. In the subsequent sections of this chapter we give sufficient conditions for the convergence of $\hat{\lambda}$ and $\hat{m}(\bullet)$.

A: Parametric estimation

Owing to the results in Section 14.3.1, we can extend a definition of the least-squares estimate to a class of minimizers that nearly minimize $\hat{Q}_n(\lambda)$, that is, (see (14.26)),

$$\hat{Q}_n(\hat{\lambda}) \le \inf_{\lambda \in \Lambda} \hat{Q}_n(\lambda) + \varepsilon_n, \tag{14.60}$$

for any random sequence $\{\varepsilon_n\}$ such that $\varepsilon_n \overset{n}{\to} 0$, (P). As we have already noted, (14.60) implies that $\hat{Q}_n(\hat{\lambda}) \leq \hat{Q}_n(\lambda^*) + \varepsilon_n$ and this is sufficient for the convergence of $\hat{\lambda}$ in (14.60) to λ^*.

The decomposition of $\hat{Q}_n(\lambda)$ into the terms defined in (14.28) and (14.29) needs an average of $\hat{Q}_n(\lambda)$. Due to the independence of the sample sets T_1 and T_2 we have,

$$\bar{Q}(\lambda) = E\{\hat{Q}_n(\lambda)|T_1\} = E\{(Y_t - \hat{m}(W_t(\lambda); \lambda))^2|T_1\}, \tag{14.61}$$

where $(W_t(\lambda), Y_t)$ is a random vector, which is independent of T_1. The definition of $m(w; \lambda)$ in (14.42) and the fact that the noise is independent of $\{Y_n\}$ yield:

$$\bar{Q}(\lambda) = E Z_t^2 + E\{(m(W_t(\lambda); \lambda) - m^*(W_t))^2\}$$
$$+ E\{(\hat{m}(W_t(\lambda); \lambda) - m(W_t(\lambda); \lambda))^2|T_1\}. \tag{14.62}$$

The last term in the above decomposition represents the integrated squared error between the kernel estimate $\hat{m}(w; \lambda)$ and the regression function $m(w; \lambda)$. Using Lemma 14.5 of Section 14.9 and the techniques developed in Chapter 3 we can easily show that under the standard assumptions on the kernel function and the bandwidth sequence $\{h_n\}$ (see Assumptions **A4** and **A5** listed below), the last term in (14.62) tends (P) to zero. Since, moreover, $\hat{Q}_n(\lambda)$ converges (P) to its average $\bar{Q}(\lambda)$ for every $\lambda \in \Lambda$, then we may conclude that:

$$\hat{Q}_n(\lambda) \overset{n}{\to} Q(\lambda), (P) \quad \text{for every } \lambda \in \Lambda, \tag{14.63}$$

where,

$$Q(\lambda) = E Z_t^2 + E\{(m(W_t(\lambda); \lambda) - m^*(W_t))^2\}. \tag{14.64}$$

This asymptotic criterion can be now used to characterize the true impulse response function λ^*. In fact, since $Q(\lambda^*) = E Z_t^2$, we have $Q(\lambda^*) \leq Q(\lambda)$, $\lambda \in \Lambda$. Nevertheless, λ^* need not be a unique minimum of $Q(\lambda)$. Indeed, the second term in (14.64) is equal to zero for such λ values which belong to the following set:

$$S = \{\lambda \in \Lambda : P\{m^*(W_t) = E(m^*(W_t)|W_t(\lambda))\} = 1\}. \tag{14.65}$$

This set defines all possible values minimizing $Q(\lambda)$ and it is clear that $\lambda^* \in S$.

The property $P\{m^*(W_t) = E(m^*(W_t)|W_t(\lambda))\} = 1$ may hold for other λ values, but this happens in very rare cases. Note, however, that $S = R^{p+1}$ if $m^*(\bullet)$ is a constant function. Excluding this singular situation we may certainly assume that $Q(\lambda)$ has the unique global minimum at λ^*. This assumption will be applied throughout our convergence theory.

The following formal assumptions are required for consistency:

A1 Let the density $f_U(\bullet)$ of the input process be a continuous function bounded away from zero in some small neighborhood of the point $u = 0$.

A2 Let $m^*(\bullet)$ be a nonconstant continuous function defined on the support of the random process $\{W_n\}$ such that $E|m^*(W_n)|^2 < \infty$.

A3 Let the space $\Lambda = \{\lambda \in R^{p+1} : \lambda_0 = 1\}$ of all admissible impulse response functions be a compact subset of R^{p+1}.

A4 Let the kernel function $K(\bullet)$ be continuous and satisfy the following restriction:

$$k_1 \mathbf{1}_{[-r,r]}(w) \leq K(w) \leq k_2 \mathbf{1}_{[-R,R]}(w),$$

for some positive constants $r \leq R$, $k_1 \leq k_2$.

A5 Let the smoothing sequence $\{h_n\}$ be such that $h_n \to 0$ and $nh_n \to 0$ as $n \to \infty$.

The kernel function satisfying Assumption **A4** is called a boxed kernel and there is a large class of kernels that may be categorized as such.

The following theorem gives conditions for the convergence of the identification algorithm defined in (14.60) to the true impulse response function λ^*. The formal proof of this result relies on the verification of conditions (14.31) and can be found in Section 14.9.

THEOREM 14.3 *Let $\hat{\lambda}$ be any estimate defined in (14.60) and let λ^* be a unique minimizer of the limit criterion $Q(\lambda)$. Suppose that Assumptions A1–A5 hold. Then we have,*

$$\hat{\lambda} \overset{n}{\to} \lambda^*, (P).$$

The critical part in proving this theorem is to show the uniform convergence of $\hat{Q}_n(\lambda)$ to its average $\bar{Q}(\lambda)$, that is, that,

$$\sup_{\lambda \in \Lambda} |\hat{Q}_n(\lambda) - \bar{Q}(\lambda)| \to 0, (P) \quad \text{as } n \to \infty.$$

Such a property is often called a Glivienko–Cantelli property. This is the property of a set of functions,

$$\{(Y - \hat{m}(W(\lambda); \lambda))^2 : \lambda \in \Lambda\}, \tag{14.66}$$

which defines the criterion $\hat{Q}_n(\lambda)$. In (14.31), simple sufficient conditions for the class to be defined as Glivienko–Cantelli are given.

If stronger requirements are imposed on (14.66), for example, that the nonlinearity $m^*(\bullet)$ and the noise process $\{Z_n\}$ are bounded, then the set in (14.66) is of the Vapnik–Chervonenkis class. This allows one to show the following exponential inequality:

$$P\left\{\sup_{\lambda \in \Lambda} |\hat{Q}_n(\lambda) - \bar{Q}(\lambda)| \geq \delta | T_1\right\} \leq c(n)e^{-\alpha n_2 \delta^2}, \tag{14.67}$$

for every $\delta > 0$ and some $\alpha > 0$. The sequence $c(n)$ is known to not grow faster than a polynomial in n. It is worth noting that bound (14.67) holds uniformly over all training sequences T_1 of size n_1. The important consequence of this is that the accuracy of the estimate $\hat{\lambda}$ does not depend critically on the training sequence T_1. Hence, the training sequence can be quite arbitrary, whereas the testing part T_2 of the training set T should be as independent as possible. This observation would favour the resampling scheme shown in Figure 14.8.

Theorem 14.2 gives the general result concerning the rate of convergence of generalized least-squared type estimates of unknown parameters of a semiparametric system.

This, combined with the consistency $\hat{\lambda} \overset{n}{\to} \lambda^*(P)$ obtained in Theorem 14.3, allows us to evaluate the rate of convergence of the estimate $\hat{\lambda}$.

In fact, first we must verify that the asymptotic criterion $Q(\lambda)$ in (14.64) admits the second-order Taylor expansion at $\lambda = \lambda^*$. It is clear that by the optimality of λ^* we have $\partial Q(\lambda^*)/\partial\lambda = \mathbf{0}$. Then, the second derivative of $Q(\lambda)$ at $\lambda = \lambda^*$ exists if additional smoothness conditions are placed on the input density $f_U(\bullet)$ and nonlinearity $m^*(\bullet)$. This can be formally verified using Lemma 14.3 from Section 14.9. As a result, we must strengthen Conditions **A1** and **A2** and assume that $f_U(\bullet)$ and $m^*(\bullet)$ have two continuous, bounded derivatives.

Yet another requirement in Theorem 14.2 is that the mapping $\lambda \mapsto (Y_i - \hat{m}(W_i(\lambda); \lambda))^2, i \in I_2$ is differentiable at $\lambda = \lambda^*$. Note that the derivative of the mapping is given by

$$-2(Y_i - \hat{m}(W_i(\lambda); \lambda))\frac{\partial \hat{m}(W_i(\lambda); \lambda)}{\partial \lambda}. \tag{14.68}$$

A bound for the term $|Y_i - \hat{m}(W_i(\lambda); \lambda)|$ in (14.68) that is uniform with respect to λ can be found using the technique which was used in the proof of Theorem 14.3 (see Section 14.9.3).

Furthermore, to bound the term $\partial \hat{m}(W_i(\lambda); \lambda)/\partial\lambda$ in (14.68), we note that

$$\frac{\partial \hat{m}(W_i(\lambda); \lambda)}{\partial \lambda} = \hat{m}^{(1)}(W_i(\lambda); \lambda)\mathbf{U}_i + \sum_{j \in I_1} D_j(W_i(\lambda); \lambda)\mathbf{U}_j,$$

where $\hat{m}^{(1)}(w; \lambda) = \partial\hat{m}(w; \lambda)/\partial w$ and $\mathbf{U}_i = (U_i, U_{i-1}, \ldots, U_{i-p})^T$. The term $D_j(w; \lambda)$ is the derivative of $\hat{m}(W_j(\lambda); \lambda)$ with respect to $W_j(\lambda)$ and is given in (14.71) with λ^* and W_j replaced by λ and $W_j(\lambda)$, respectively. Then, we can proceed as in the proof of Theorem 14.3 and conclude that the conditions of Theorem 14.2 are met. All these informal considerations lead to the following theorem:

THEOREM 14.4 *Let all the assumptions of Theorem 14.3 be satisfied. Let the derivative $K^{(1)}(\bullet)$ of the kernel function exist and be bounded. Suppose that $f_U(\bullet)$ and $m^*(\bullet)$ have two continuous, bounded derivatives.*

Then for any sequence of estimators $\hat{\lambda}$ that satisfy (14.60) with $n\varepsilon_n \overset{n}{\to} 0(P)$ and such that $\hat{\lambda} \overset{n}{\to} \lambda^*(P)$ we have

$$\hat{\lambda} = \lambda^* + O_P(n^{-1/2}).$$

This result shows that the semiparametric least-squares estimation method can reach the usual $\sqrt{n}$ parametric rate of convergence. Nevertheless, additional smoothness conditions on the input density and system nonlinearity are required. On the contrary, the correlation type estimators of λ^* can reach the $\sqrt{n}$ rate without virtually any assumptions on the nonlinearity and the system memory. The critical assumption, however, was that the input signal is Gaussian. In Remark 14.1, we discuss how to combine these two estimation methods in order to reduce the computational cost of the least-squares estimate. It is also worth noting that the correlation estimate is given by an explicit formula whereas the least-squares method is not. In Section 14.7, we give another class of parameter estimators which can be given by direct formulas. They can be applied to only a limited class of semiparametric models but their ease of use make them an attractive alternative.

B: Nonparametric estimation

The estimate $\hat{\lambda}$ of the linear subsystem found in the preceding section allows one to define an estimate of $\hat{m}^*(\bullet)$ as in (14.52), that is, $\hat{m}(\bullet) = \hat{m}(\bullet; \hat{\lambda})$, where $\hat{m}(\bullet; \lambda)$ is the kernel estimate defined in (14.47). The first step in proving the consistency result for $\hat{m}(\bullet)$ is to apply the decomposition in (14.37). The convergence of the second term in this decomposition, that is,

$$\hat{m}(\bullet; \lambda^*) - m^*(\bullet) \xrightarrow{n} 0, (P),$$ (14.69)

represents the classical problem in nonparametric estimation. In our case the output process is p-dependent, that is, the random variables Y_i and Y_j are independent as long as $|i - j| > p$. This observation and the result of Lemma 14.6 in Section 14.9 yield (14.69); see the proof of Theorem 14.5 in Section 14.8.

Concerning the first term in (14.37), note that we wish to apply the linearization technique (see (14.38)) with respect to $\hat{\lambda} - \lambda^*$. To do so, let us write the kernel estimate $\hat{m}(w; \lambda)$ in (14.47) as follows:

$$\hat{m}(w; \lambda) = \frac{\hat{r}(w; \lambda)}{\hat{f}(w; \lambda)},$$ (14.70)

where $\hat{r}(w; \lambda) = n_1^{-1} h^{-1} \sum_{j \in I_1} Y_j K(\frac{w - W_j(\lambda)}{h})$ and $\hat{f}(w; \lambda) = n_1^{-1} h^{-1} \sum_{j \in I_1} K(\frac{w - W_j(\lambda)}{h})$. Note that $\hat{f}(w; \lambda)$ is the kernel estimate of the density function $f(w; \lambda)$, whereas $\hat{m}(w; \lambda)$ is the kernel estimate of $m(w; \lambda) f(w; \lambda)$.

Now using (14.70) and recalling that $W_j(\lambda^*) = W_j$, we can express the derivative of $\hat{m}(w; \lambda^*)$ with respect to W_j, $j \in I_1$ as follows:

$$D_j(w) = n_1^{-1} h^{-2} K^{(1)} \left(\frac{w - W_j}{h} \right) \cdot \frac{\hat{r}(w; \lambda^*) - Y_j \hat{f}(w; \lambda^*)}{\hat{f}^2(w; \lambda^*)},$$ (14.71)

where $\hat{r}(w; \lambda^*)$, $\hat{f}(w; \lambda^*)$ are defined as in (14.70) with $\lambda = \lambda^*$. Here, $K^{(1)}(w)$ denotes the derivative of the kernel function.

Next, let us note that

$$W_j(\hat{\lambda}) - W_j(\lambda^*) = \sum_{t=1}^{p} (\hat{\lambda}_t - \lambda_t^*) U_{j-t}, \quad j \in I_1.$$

Then, we can approximate $\hat{m}(w) - \hat{m}(w; \lambda^*)$ by the first term of Taylor's formula,

$$\sum_{j \in I_1} D_j(w)(W_j(\hat{\lambda}) - W_j(\lambda^*)) = \sum_{t=1}^{p} (\hat{\lambda}_t - \lambda_t^*) A_{t,n}(w),$$

where

$$A_{t,n}(w) = \sum_{j \in I_1} D_j(w) U_{j-t},$$

for $1 \le t \le p$.

Since, by Theorem 14.3, we have that $\hat{\lambda}_t - \lambda_t^* \overset{n}{\to} 0(P)$, it is sufficient to show that the stochastic term $A_{t,n}(w)$ tends (P) to a finite function as $n \to \infty$. Let us note that by the technical considerations (see Section 14.9) that lead to the consistency result in (14.69), we know that $\hat{f}(w; \lambda^*)$ and $\hat{r}(w; \lambda^*)$ converge (P) to $f_W(w)$ and $m^*(w) f_W(w)$, respectively. By these convergences and (14.71), we see that the term $A_{t,n}(w)$ is determined by the following two expressions:

$$J_1(w) = n_1^{-1} h^{-2} \sum_{j \in I_1} K^{(1)} \left(\frac{w - W_j}{h} \right) U_{j-t},$$

$$J_2(w) = n_1^{-1} h^{-2} \sum_{j \in I_1} K^{(1)} \left(\frac{w - W_j}{h} \right) Y_j U_{j-t}.$$

Since the term $J_2(w)$ is more general (setting $Y_j \equiv 1$ gives $J_1(w)$), it suffices to examine that term. Let us start by noting that

$$J_2(w) = \frac{\partial}{\partial w} \bar{J}_2(w), \tag{14.72}$$

where

$$\bar{J}_2(w) = n_1^{-1} h^{-1} \sum_{j \in I_1} K \left(\frac{w - W_j}{h} \right) Y_j X_{j-t}.$$

In Section 14.9, we show that

$$\bar{J}_2(w) \overset{n}{\to} m^*(w) a(w), (P), \tag{14.73}$$

where $a(w)$ is some finite function. The convergence (P) of $\bar{J}_2(w)$ implies the convergence (P) of the derivative due to the general result presented in Lemma 14.4, that is, (14.72) and (14.73) yield

$$J_2(w) \overset{n}{\to} \frac{\partial}{\partial w} \{ m^*(w) a(w) \}, (P).$$

The aforementioned discussion explains the main steps used to prove the convergence of the estimate $\hat{m}(w)$ defined in (14.52) to the true nonlinearity $m^*(w)$. More technical considerations necessary for establishing the results in (14.69) and (14.73) can be found in Section 14.9.

Note that the linearization technique requires some differentiability conditions both on the system characteristics and the kernel function. Hence, we need the following additional assumptions:

A6 Let $f_U(\bullet)$ have a bounded and continuous derivative.
A7 Let $m^*(\bullet)$ have a bounded and continuous derivative.
A8 Let the derivative $K^{(1)}(\bullet)$ of the kernel function exist and be bounded.

All these considerations lead to the following convergence result for the nonlinear subsystem identification algorithm.

THEOREM 14.5 *Let $\hat{m}(\bullet) = \hat{m}(\bullet; \hat{\lambda})$, where $\hat{m}(\bullet; \lambda)$ is the kernel regression estimate defined in (14.47). Let all of the assumptions of Theorem 14.3 hold. If, further, Assumptions A6, A7, and A8 are satisfied, then we have*

$$\hat{m}(w) \to m^*(w), (P), \quad as \ n \to \infty$$

at every point $w \in R$ where $f_W(w) > 0$.

The conditions imposed in Theorem 14.5 are by no means the weakest possible and it may be conjectured that the convergence holds at a point where $f_W(w)$ and $m^*(w)$ are continuous.

In the proof of Theorem 14.5, we have already shown that $\hat{m}(w) - \hat{m}(w, \lambda^*)$ is of order,

$$\sum_{t=1}^{p} (\hat{\lambda}_t - \lambda_t) A_{t,n}(w),$$

where $A_{t,n}(w) \overset{n}{\to} A_t(w)(P)$, some finite function $A_t(w)$. Then, due to Theorem 14.4, we have that

$$\hat{m}(w) - m^*(w) = \{\hat{m}(w; \lambda^*) - m^*(w)\} + O_P(n^{-1/2}). \tag{14.74}$$

Hence, the rate of convergence of $\hat{m}(w)$ to $m^*(w)$ depends merely on the speed at which the first term on the right-hand side of (14.74) tends to zero. This is, however, an usual problem in nonparametric estimation. Indeed, the rate is controlled by the smoothness of the nonlinearity $m^*(w)$ and density $f_W(\bullet)$. Note that the smoothness of $f_W(\bullet)$ can be inferred by the smoothness of $f_U(\bullet)$. The smoothness required to get the second term in (14.74) is described in Theorem 14.4. Since we have assumed that $f_U(\bullet)$ and $m^*(\bullet)$ have two continuous bounded derivatives, then by standard analysis described in Chapter 2 we may readily obtain that $\hat{m}(w; \lambda^*) - m^*(w) = O_P(n^{-2/5})$, provided that the kernel function is even. Consequently, we come to the following theorem.

THEOREM 14.6 *Let all the assumptions of Theorem 14.4 and Theorem 14.5 be satisfied. Suppose that the kernel function is even. Then we have*

$$\hat{m}(w) = m^*(w) + O_P(n^{-2/5}).$$

14.4.3   Simulation examples

In order to illustrate the practical usefulness of the theoretical results obtained in the preceding section, this section gives some simulation examples. We use the Wiener model in which the linear dynamic subsystem is given by

$$W_n = U_n + 0.8U_{n-1} - 0.6U_{n-2} + 0.4U_{n-3}. \tag{14.75}$$

The nonlinear part of the model is represented by the characteristics $m_1^*(w) = \arctan(2w)$ and $m_2^*(w) = \lfloor w \rfloor$, where $\lfloor \cdot \rfloor$ denotes the rounding function. Hence, two different Wiener models are taken into consideration, one with a smooth nonlinearity $(m_1^*(w))$ and the other with a nonlinearity having discontinuities $(m_2^*(w))$. The input excitation $\{U_n\}$ is a uniformly distributed random sequence on the interval $[-3, 3]$. The output is corrupted by a noise signal which is uniformly distributed on $[-c, c]$, where c is selected such that $c = 0.1 \cdot g_{\max}$, in which $g_{\max}$ is the maximum of the nonlinear characteristic. Both the pilot kernel estimate (14.47) and the final estimate in (14.52) use the Gaussian kernel $K(w) = (2\pi)^{-1/2} \exp(-w^2/2)$ and the bandwidth sequence $h = n_1^{-1/5}$. In order to assess the accuracy of our estimation algorithms, we employ two kinds of estimation errors:

• The error assessing the accuracy of the identification algorithm of the linear subsystem,

$$\text{Err}(\hat{\lambda}) = \frac{1}{L} \sum_{i=1}^{L} \frac{\left\| \hat{\lambda}^{[i]} - \lambda^* \right\|}{\|\lambda^*\|}, \qquad (14.76)$$

where L is the number of experimental runs and $\hat{\lambda}^{[i]}$ is the value of the estimate $\hat{\lambda}$ obtained in the ith experiment.

• The global estimation error of the nonlinear part

$$\text{MISE}(\hat{m}) = \frac{1}{LT} \sum_{i=1}^{L} \sum_{j=1}^{T} (m^*(w_j) - \hat{m}^{[i]}(w_j; \hat{\lambda}))^2, \qquad (14.77)$$

where $\{w_j : j = 1, \ldots, T\}$ are points equally spaced in the interval $[-8.4; 8.4]$. The number 8.4 is selected because $|W_n| \leq 8.4$. In all experiments we use $L = 10$ and $T = 1,000$. In all our numerical results, the error $\text{MISE}(\hat{m})$ is given using a logarithmic scale.

In the first simulation example we study the rate of convergence of both of the errors, that is, the speed at which $\text{Err}(\hat{\lambda})$ and $\text{MISE}(\hat{m})$ tend to zero as the sample size n increases. The results are shown in Figure 14.10 where n ranges from 10 to 200. The subsets T_1, T_2 of the training set T are of equal size, that is, they each have $0.5n - 3$ data points. It is apparent that the smoothness of the nonlinearity has a significant influence on the accuracy of the proposed identification algorithms. Indeed, the smoother the nonlinearity is the faster is the speed at which $\text{Err}(\hat{\lambda})$ and $\text{MISE}(\hat{m})$ tend to zero. It is worth noting that in the case of characteristic $m_2^*(w)$ the estimation error $\text{Err}(\hat{\lambda})$ for the linear part tends to zero faster than the error $\text{MISE}(\hat{m})$ assessing the nonlinearity estimate. This is consistent with our convergence results. In fact, $m_2^*(w)$ meets the conditions of Theorem 14.3 (see comments that were presented below Theorem 14.3), whereas it does not satisfy Assumption **A7** .

In the next experiment, we investigate the issue of data splitting, that is, how to choose n_1 and n_2 for a fixed value of n. The subsets T_1 and T_2 are arranged as in Figure 14.7. Actually, this is not an optimal partition of the training set and this interesting issue will be examined elsewhere. Figure 14.11 shows the errors $\text{Err}(\hat{\lambda})$ and $\text{MISE}(\hat{m})$ versus n_1,

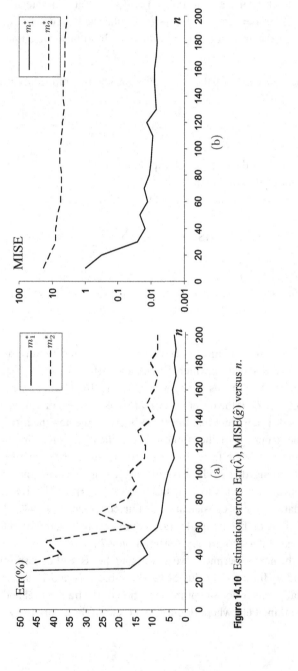

Figure 14.10 Estimation errors Err($\hat{\lambda}$), MISE($\hat{g}$) versus n.

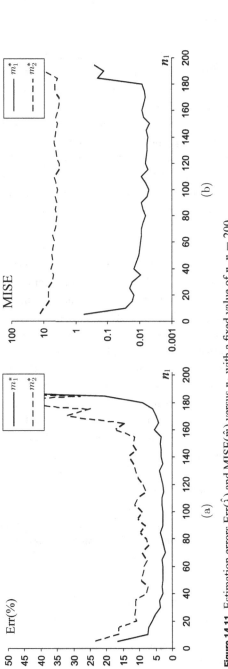

Figure 14.11 Estimation errors $\mathrm{Err}(\hat{\lambda})$ and $\mathrm{MISE}(\hat{m})$ versus n_1 with a fixed value of n, $n = 200$.

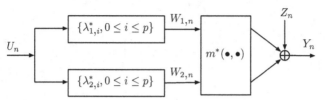

Figure 14.12 The generalized two-channel Wiener model.

the size of the training subset T_1, where n_1 ranges from 5 to 195. The size of the whole training set is $n = 200$. Note that $n_2 = n - 2p - n_1$, where $p = 3$ in our case. It is clear from our results that there is a wide range of n_1 for which the quality of the estimate $\hat{\lambda}$ does not change. On the other hand, a large value of n_1 can reduce the accuracy of $\hat{\lambda}$; a large n_1 improves the precision of the pilot kernel estimate in (14.47) but it results in the small value of n_2 yielding a poor least-squares estimate as defined in (14.51). A small n_1 also has a negative influence on the pilot kernel estimate used in (14.50) and as a result gives a large value of the criterion $\hat{Q}_n(\lambda)$. An analogous trade-off can be observed for the nonlinear characteristic estimation. As a practical recommendation, we can use $n_1 = n_2 = 0.5n - p$, which is equal to 97 in our experimental setup.

14.4.4   Extensions

Thus far, we have examined the one channel Wiener system with a finite memory and the univariate nonlinearity. We have employed the semiparametric approach to identify the parametric and nonparametric parts of the system. This strategy can be further extended to other types of Wiener systems. Among many possible alternatives we single out a multichannel model with separate dynamical parts and a common nonlinearity. A two-channel version of this particular class of Wiener systems is shown in Figure 14.12. This model is important since it can be shown that any general class of nonlinear systems, which satisfy the so-called fading memory property, can be approximated by the aforementioned multichannel Wiener model. A system has fading memory if two input signals which are close in the recent past, but not necessarily close in the remote past, yield present outputs that are close. This property aims at strengthening the concept of system continuity and assures us that the system "forgets" initial conditions. Clearly any system with a finite memory meets this property.

It is worth mentioning that the inverse regression approach examined in Chapters 9–11 would be difficult to apply to the multichannel Wiener model. On the other hand, this model can be easily identified within the semiparametric framework examined in Section 14.3. Hence, without loss of generality, let us consider the two-channel system depicted in Figure 14.12, that is, we have,

$$Y_n = m^* \left(\sum_{i=0}^{p} \lambda_{1,i}^* U_{n-i}, \sum_{j=0}^{p} \lambda_{2,j}^* U_{n-j} \right) + Z_n, \qquad (14.78)$$

where $\lambda_1^* = \{\lambda_{1,i}^*, 0 \le i \le p\}$ and $\lambda_2^* = \{\lambda_{2,i}^*, 0 \le i \le p\}$ are unknown parameters and $m^*(\bullet, \bullet)$ is an unknown nonlinearity.

The first important issue, similar to that studied for the single input Wiener model, is whether the parameter $\lambda^* = (\lambda_1^*, \lambda_2^*) \in R^s$, $s = 2p + 2$, is identifiable. The previous normalization $\lambda_{1,0} = \lambda_{2,0} = 1$ is not sufficient in this case; we must further restrict a class of admissible impulse response sequences and nonlinearities. Concerning the parameter space of all admissible impulse response functions we assume that $\lambda \in \Lambda \subset R^s$ for Λ being a compact subset of R^s, where $\lambda = (\lambda_1, \lambda_2)$.

In general, we can only identify a linear subspace spanned by $(\lambda_1^*, \lambda_2^*)$. To be able to identify the individual parameters we can assume that λ_1^* and λ_2^* are not collinear. Furthermore, assume that $m^*(\bullet, \bullet)$ is not a constant function and that the derivatives of $m^*(w_1, w_2)$ with respect to each of the variables are not linearly dependent. This assures us that the nonlinearity is sufficiently far from being constant and linear.

The solution of the identification problem for the model (14.78) is now straightforward. Indeed, we can follow the ideas developed in Section 14.3 starting with an important concept of the optimal predictor of the output signal for a given $\lambda \in \Lambda$,

$$m(w_1, w_2; \lambda) = E\{Y_n | W_{1,n}(\lambda_1) = w_1, W_{2,n}(\lambda_2) = w_2\}, \qquad (14.79)$$

where $W_{1,n}(\lambda_1) = \sum_{i=0}^{p} \lambda_{1,i} U_{n-i}$ and $W_{2,n}(\lambda_2) = \sum_{j=0}^{p} \lambda_{2,j} U_{n-j}$. We have the obvious constraints $W_{1,n}(\lambda_1^*) = W_{1,n}$, $W_{2,n}(\lambda_2^*) = W_{2,n}$ and $m(w_1, w_2; \lambda^*) = m^*(w_1, w_2)$. Next, using the partition strategy of the training set shown in Figure 14.7, the regression function $m(w_1, w_2; \lambda)$ can be estimated by the two-dimensional version of the kernel estimate,

$$\hat{m}(w_1, w_2; \lambda) = \frac{\sum_{j \in I_1} Y_j K\left(\frac{w_1 - W_{1,j}(\lambda_1)}{h}\right) K\left(\frac{w_2 - W_{2,j}(\lambda_2)}{h}\right)}{\sum_{j \in I_1} K\left(\frac{w_1 - W_{1,j}(\lambda_1)}{h}\right) K\left(\frac{w_2 - W_{2,j}(\lambda_2)}{h}\right)}, \qquad (14.80)$$

for a given $\lambda = (\lambda_1, \lambda_2) \in \Lambda$. See (14.47) for the one-dimensional version of this estimate. This allows us to form the least-squares criterion for the estimation of λ; hence, we have

$$\hat{\lambda} = \arg\min_{\lambda \in \Lambda} \frac{1}{n_2} \sum_{i \in I_2} \{Y_i - \hat{m}(W_{1,i}(\lambda_1), W_{2,i}(\lambda_2); \lambda)\}^2. \qquad (14.81)$$

The corresponding estimate of $m^*(w_1, w_2)$ is defined as follows:

$$\hat{m}(w_1, w_2) = \hat{m}(w_1, w_2; \hat{\lambda}). \qquad (14.82)$$

An efficient computational algorithm, identical to that in Section 14.4.1, for finding $\hat{\lambda}$ can be easily worked out.

The convergence analysis analogous to that given in the proof of Theorem 14.3 and Theorem 14.4 leads to the result

$$\hat{\lambda}_1 = \lambda_1^* + O_P(n^{-1/2}), \quad \hat{\lambda}_2 = \lambda_2^* + O_P(n^{-1/2}),$$

where we need to assume that $f_U(\bullet)$ and $m^*(\bullet, \bullet)$ are twice continuously differentiable. Then, the reasoning leading to the results of Theorem 14.5 and Theorem 14.6 readily yields,

$$\hat{m}(w_1, w_2) = m^*(w_1, w_2) + O_P(n^{-1/3}), \tag{14.83}$$

where $f_U(\bullet)$ and $m^*(\bullet, \bullet)$ are twice continuously differentiable. Note that the rate in (14.83) is slower than that for the one channel Wiener system, see Theorem 14.6. This is due to the fact that we are estimating a bivariate function for which the rate is slower than for an univariate one, see Chapter 13.

The following general d-channel Wiener system,

$$Y_n = m^* \left(\sum_{i=0}^{p} \lambda_{1,i}^* U_{n-i}, \ldots, \sum_{j=0}^{p} \lambda_{d,j}^* U_{n-j} \right) + Z_n, \tag{14.84}$$

can be tackled in the identical way. It is worth noting that the system in (14.84) corresponds to the network, which, as we have already mentioned, can approximate any nonlinear system with the fading memory property. In this theory, the choice of the order d is controlled by the fading memory assumption and generally, the larger value of d, the smaller the approximation error. However, this is true only in the deterministic theory, whereas in the case of random data we must estimate the parameters $(\lambda_1^*, \ldots, \lambda_d^*)$ and the d-dimensional function $m^*(\bullet, \ldots, \bullet)$. The model order d should also be a part of the estimation problem since an optimal order is expected to exist for training data coming from an arbitrary nonlinear system that meets the fading memory assumption. This is an example of the modeling-estimation tradeoff quantified by the balance between the approximation error and the variance of the estimates.

In fact, due to the curse of dimensionality, the general model (14.84) is unlikely to be useful when d is large. As a matter of fact, the function $m^*(\bullet, \ldots, \bullet)$ can be estimated at the rate $O_P\left(n^{-2/(4+d)}\right)$ (see Chapter 13), and this is a very slow rate for large d. A possible remedy to overcome this dimensionality problem is to examine an additive approximation of (14.84), that is,

$$Y_n = \sum_{r=1}^{d} m_r^* \left(\sum_{i=0}^{p} \lambda_{r,i}^* U_{n-i} \right) + Z_n, \tag{14.85}$$

where $\{m_r^*(\bullet), 1 \le r \le d\}$ are univariate functions.

The model in (14.85) is a dynamical counterpart of projection pursuit models extensively studied in the statistical literature for low dimensional representations of multivariate regression. In this static set up, it has been demonstrated that the projection pursuit regression can approximate any d-dimensional function [75], which is sufficiently smooth. To the best of the authors' knowledge, there is not such a result in the context of dynamical systems. The system in (14.85), however, is difficult to identify, that is, to estimate all $\{\lambda_r^*, 1 \le r \le d\}$ and $\{m_r^*(\bullet), 1 \le r \le d\}$. We explain this issue in the next generalization of the semiparametric Wiener model.

As we have already mentioned in Chapter 13, it is an interesting issue to examine various multivariate extensions of block-oriented models. Here, we consider the following counterpart of the Wiener model (14.78) with two-dimensional input $\{(U_{1,n}, U_{2,n})\}$,

$$Y_n = m^*(\sum_{i=0}^{p} \lambda_{1,i}^* U_{1,n-i}, \sum_{j=0}^{p} \lambda_{2,j}^* U_{2,n-j}) + Z_n. \tag{14.86}$$

An estimation method for recovering $(\lambda_1^*, \lambda_2^*)$ and $m^*(\bullet, \bullet)$ is identical to the algorithm outlined above.

The d-dimensional generalization of (14.86) can be easily written down. A practical strategy is to consider a one-dimensional additive approximation of (14.86) (or its d-dimensional version) depicted in Figure 14.2. Using our modified notation that emphasizes the true value of the model characteristics, we write the input–output formula for the additive system as follows:

$$Y_n = m_1^* \left(\sum_{i=0}^{p} \lambda_{1,i}^* U_{1,n-i} \right) + m_2^* \left(\sum_{j=0}^{p} \lambda_{2,j}^* U_{2,n-j} \right) + Z_n. \tag{14.87}$$

Now, we wish to estimate $(\lambda_1^*, \lambda_2^*)$ and $m_1^*(\bullet), m_2^*(\bullet)$.

This identification problem can be solved by mixing the semiparametric methodology and the marginal integration method used for additive models in Chapter 13. In fact, we begin with the projection function,

$$m(w_1, w_2; \lambda) = E\{Y_n | W_{1,n}(\lambda_1) = w_1, W_{2,n}(\lambda_2) = w_2\}, \tag{14.88}$$

where we used the standard notation,

$$W_{1,n}(\lambda_1) = \sum_{i=0}^{p} \lambda_{1,i} U_{1,n-i}, \quad W_{2,n}(\lambda_2) = \sum_{j=0}^{p} \lambda_{2,j} U_{2,n-j}. \tag{14.89}$$

It is worth noting that, due to the dependence between $\{U_{1,n}\}$ and $\{U_{2,n}\}$, we cannot write $m(w_1, w_2; \lambda)$ as an additive function. This, however, does not bother us because $W_{1,n}(\lambda_1^*) = W_{1,n}, W_{2,n}(\lambda_2^*) = W_{2,n}$, and therefore,

$$m(w_1, w_2; \lambda^*) = m_1^*(w_1) + m_2^*(w_2). \tag{14.90}$$

This fundamental identity allows us to recover all the components of the model in (14.87).

The parametric part of (14.87), that is, estimating $\lambda^* = (\lambda_1^*, \lambda_2^*)$, is done in exactly the same way as in (14.81), where the kernel estimate $\hat{m}(w_1, w_2; \lambda)$ of $m(w_1, w_2; \lambda)$ is defined in (14.80) with $W_{1,j}(\lambda_1)$ and $W_{2,j}(\lambda_2)$ being defined in (14.89). This yields the least squares estimate $\hat{\lambda}$ of λ^*.

The consistency $\hat{\lambda} \xrightarrow{n} \lambda^*$ (P) can be established along the lines of the proof of Theorem 14.3. If this is the case, we can anticipate (see the proof of Theorem 14.5) that,

$$\hat{m}(w_1, w_2; \hat{\lambda}) \xrightarrow{n} m(w_1, w_2; \lambda^*)(P). \tag{14.91}$$

In (14.90), the limit function is equal to $m_1^*(w_1) + m_2^*(w_2)$. Hence, it remains to recover the individual components $m_1^*(\bullet)$ and $m_2^*(\bullet)$ of the additive form. This can be achieved by the marginal integration method discussed in Chapter 13. We can define the following marginal integration estimates of $m_1^*(w_1)$ and $m_2^*(w_2)$,

$$\hat{m}_1(w_1) = \int_{-\infty}^{\infty} \hat{m}(w_1, w_2; \hat{\lambda}) \psi(w_2) dw_2,$$

$$\hat{m}_2(w_2) = \int_{-\infty}^{\infty} \hat{m}(w_1, w_2; \hat{\lambda}) \psi(w_1) dw_1, \qquad (14.92)$$

for a given weight function $\psi(w)$.

Using the results of Chapter 13 along with (14.90) and (14.91), we can prove that the estimates in (14.92) converge to $m_1^*(\bullet)$ and $m_2^*(\bullet)$, respectively. This leads to a complete solution of the identification problem of the additive Wiener model using the semiparametric least-squares method combined with the marginal integration technique for resolving additive models. Nevertheless, in Section 14.7.3 we present another identification method that gives direct estimates of λ^*.

14.5   Statistical inference for semiparametric Hammerstein models

In Section 14.2.1 we introduced a semiparametric low-dimensional version of the multiple-input Hammerstein system (see Figure 14.1). The input–output relationship for this d-dimensional system, see (14.4), is given by,

$$Y_n = \sum_{\ell=0}^{p} \lambda_\ell^* g^*(\gamma^{*T} \mathbf{U}_{n-\ell}) + Z_n, \qquad (14.93)$$

where $\lambda^* = \{\lambda_i^*, 0 \le i \le p\}$ and $\gamma^* = \{\gamma_j^*, 1 \le j \le d\}$ are unknown parameters, and $g^*(\bullet)$ is a single variable function representing the nonlinearity of the system. The identifiability of the system forces us to normalize the parameters (see Section 14.2.1) such that $\lambda_0^* = 1$ and $\gamma_1^* = 1$. Furthermore, $g^*(\bullet)$ need not be a constant function and $E\{g^*(\gamma^{*T} \mathbf{U}_n)\} = 0$. We also assume that the input signal $\{\mathbf{U}_n\}$ constitutes a sequence of iid random vectors with the density function $f_\mathbf{U}(\bullet)$ defined on R^d.

As we have already noted in Section 14.2.1 the impulse response function λ^* of the linear subsystem can be estimated via the correlation method (see (14.5)) independent of γ^* and $g^*(\bullet)$. On the other hand, the problem of recovering γ^* and $g^*(\bullet)$ can be treated by the semiparametric method.

Let $\Theta = \{(\lambda, \gamma) : \lambda_0 = 1, \gamma_1 = 1\}$ be a parameter space which is assumed to be a compact subset of R^{p+d+1}. For a given $(\lambda, \gamma) \in \Theta$, let us define the regression function,

$$g(w; \gamma) = E\{Y_n | \gamma^T \mathbf{U}_n = w\}. \qquad (14.94)$$

This, due our normalization, is equal to,

$$E\{g^*(\gamma^{*T} \mathbf{U}_n) | \gamma^T \mathbf{U}_n = w\}.$$

Note that $g(w; \gamma)$ is independent of λ and also that $g(w; \gamma^*) = g^*(w)$. Hence, for a given $(\lambda, \gamma) \in \Theta$, we can eliminate the dependence of the estimation problem on $g^*(\bullet)$

by estimating $g(w; \gamma)$ using the kernel regression method,

$$\hat{g}(w; \gamma) = \frac{\sum_{j \in I_1} Y_j K\left(\frac{w - \gamma^T U_j}{h}\right)}{\sum_{j \in I_1} K\left(\frac{w - \gamma^T U_j}{h}\right)}, \tag{14.95}$$

where we again used the partition scheme to divide the training set into two independent subsets (see Figure 14.7).

Employing the kernel-regression estimate allows us to form the least squares criterion for selecting the parameter $(\lambda, \gamma) \in \Theta$. Hence, we have,

$$\hat{Q}_n(\lambda, \gamma) = \frac{1}{n_2} \sum_{i \in I_2} \left(Y_i - \sum_{\ell=0}^{p} \lambda_\ell \hat{g}(\gamma^T U_i; \gamma) \right)^2. \tag{14.96}$$

Thus, we define the estimate of the system parameters as follows:

$$(\hat{\lambda}, \hat{\gamma}) = \arg \min_{(\lambda, \gamma) \in \Theta} \hat{Q}_n(\lambda, \gamma). \tag{14.97}$$

Once the estimates $(\hat{\lambda}, \hat{\gamma})$ are obtained we can define the following nonparametric estimate of the system nonlinearity:

$$\hat{g}(w) = \hat{g}(w; \hat{\gamma}).$$

Since we can show that $(\hat{\lambda}, \hat{\gamma})$ tends to $(\lambda^*, \gamma^*)(P)$ (see Section 14.3.1) it is expected that $\hat{g}(w)$ tends to $g^*(w)$. We leave it to the reader to fill in the missing points in the consistency proofs.

It is worth noting that if we use the correlation-based type of estimate for λ^*, then we can simplify $\hat{Q}_n(\lambda, \gamma)$ in (14.96), as follows:

$$\hat{Q}_n(\gamma) = \frac{1}{n_2} \sum_{i \in I_2} \left(Y_i - \sum_{\ell=0}^{p} \hat{\lambda}_\ell \, \hat{g}(\gamma^T U_i; \gamma) \right)^2,$$

where $\hat{\lambda}$ is the correlation estimate of λ^*. The estimate of γ^* is now defined as the minimizer of $\hat{Q}_n(\gamma)$.

14.6   Statistical inference for semiparametric parallel models

In our final example concerning the use of the semiparametric least-squares method, let us consider the parallel system introduced in Section 12.1.1. In the current framework, this system (shown in Figure 14.13) has the following input–output description:

$$Y_n = m^*(U_n) + \sum_{j=0}^{p} \lambda_j^* U_{n-j} + Z_n. \tag{14.98}$$

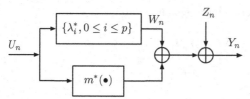

Figure 14.13 Semiparametric nonlinear parallel model.

The identifiability condition for this system is that $\lambda_0^* = 1$ (see Chapter 12). Hence, let $\Lambda = \{\lambda \in R^{p+1} : \lambda_0 = 1\}$ be a set of all admissible parameters that is assumed to be the compact subset of R^{p+1}.

The semiparametric least-squares strategy begins with the elimination of the nonlinear characteristic from the optimization process. To this end, let

$$W_n(\lambda) = \sum_{j=0}^{p} \lambda_j U_{n-j}, \qquad (14.99)$$

be the output of the linear subsystem for a given $\lambda \in \Lambda$. Clearly $W_n(\lambda^*) = W_n$. Next, let

$$m(u; \lambda) = E\{Y_n - W_n(\lambda)|U_n = u\} \qquad (14.100)$$

be the best model (regression function) of $m^*(u)$ for a given $\lambda \in \Lambda$. Indeed, the signal $Y_n - W_n(\lambda) - Z_n$ is the output of the nonlinear subsystem for $\lambda \in \Lambda$. Noting that,

$$m(u; \lambda) = m^*(u) + \sum_{j=0}^{p} (\lambda_j^* - \lambda_j) \, E\{U_{n-j}|U_n = u\},$$

we can conclude that $m(u; \lambda^*) = m^*(u)$. For a given training set $T = \{(U_1, Y_1), \ldots, (U_n, Y_n)\}$, we can easily form a nonparametric estimate of the regression function $m(u; \lambda)$. Hence, let

$$\hat{m}(u; \lambda) = \frac{(nh)^{-1} \sum_{t=p+1}^{n} (Y_t - W_t(\lambda)) K \left(\frac{u-U_t}{h} \right)}{(nh)^{-1} \sum_{t=1}^{n} K \left(\frac{u-U_t}{h} \right)}, \qquad (14.101)$$

be the kernel-regression estimate of $m(u; \lambda)$.

The mean-squared criterion for estimating λ^* can now be defined as follows:

$$\hat{Q}_n(\lambda) = n^{-1} \sum_{t=p+1}^{p} (Y_t - \hat{m}(U_t; \lambda) - W_t(\lambda))^2 . \qquad (14.102)$$

The minimizer of the prediction error $\hat{Q}_n(\lambda)$ defines an estimate $\hat{\lambda}$ of λ^*. As soon as $\hat{\lambda}$ is determined, we can estimate $m^*(u)$ by the two-stage process, that is, we have,

$$\hat{m}(u) = \hat{m}(u; \hat{\lambda}). \qquad (14.103)$$

Thus far we have used the same data for estimating the pilot regression estimate $\hat{m}(u; \lambda)$ and the criterion function $\hat{Q}_n(\lambda)$. This may lead to consistent estimates but the mathematical analysis of such algorithms is lengthy. In Section 14.4, we employed the partition resampling scheme, which gives a desirable separation of the training and testing data sets and reduces the mathematical complications. This strategy can be easily applied here, that is, we can use the subset T_1 of T to derive the kernel estimate in (14.101) and then use the remaining part of T for computing the criterion function $\hat{Q}_n(\lambda)$.

For estimates of $\hat{\lambda}$ and $\hat{m}(u)$ obtained as outlined above, we can follow the technical arguments given in Section 14.4.2 and show that $\hat{\lambda} \xrightarrow{n} \lambda^*(P)$ and consequently $\hat{m}(u; \hat{\lambda}) \xrightarrow{n} m(u; \lambda^*) = m^*(u)(P)$. We omit details of this convergence analysis but this should now be easily accomplished by the reader.

The minimization procedure required to obtain $\hat{\lambda}$ can be involved due to the highly nonlinear nature of $\hat{Q}_n(\lambda)$. A reduced complexity algorithm can be developed based on the general iterative scheme described in Section 14.3. Hence, for a given $\hat{\lambda}^{(\text{old})}$, set $\hat{m}(u; \hat{\lambda}^{(\text{old})})$. Then, we form the modified criterion,

$$\tilde{Q}_n(\lambda) = n^{-1} \sum_{t=p+1}^{p} \left(Y_t - \hat{m}(U_t; \hat{\lambda}^{(\text{old})}) - W_t(\lambda) \right)^2, \qquad (14.104)$$

and we find

$$\hat{\lambda}^{(\text{new})} = \arg \min_{\lambda \in \Lambda} \tilde{Q}_n(\lambda).$$

Next, we use $\hat{\lambda}^{(\text{new})}$ to get $\hat{m}(u; \hat{\lambda}^{(\text{new})})$ and iterate the above process until the original criterion $\hat{Q}_n(\lambda)$ does not change significantly.

It is worth noting that $W_t(\lambda)$ in (14.104) is a linear function of λ and therefore we can explicitly find $\hat{\lambda}^{(\text{new})}$ that minimizes $\tilde{Q}_n(\lambda)$. Indeed, this is the classical linear least-squares problem with the following solution:

$$\hat{\lambda}^{(\text{new})} = (\mathbf{U}^T \mathbf{U})^{-1} \mathbf{U}^T \mathbf{O}, \qquad (14.105)$$

where $\mathbf{O}$ is the $(n-p) \times 1$ vector with the tth coordinate being equal to $Y_t - \hat{m}(U_t; \hat{\lambda}^{(\text{old})})$, $t = p+1, \ldots, n$. $\mathbf{U}$ is a $(n-p) \times (p+1)$ matrix, $\mathbf{U} = (\mathbf{U}_{p+1}^T, \ldots, \mathbf{U}_n^T)^T$, where $\mathbf{U}_t = (U_t, \ldots, U_{t-p})^T$.

We should note that the above algorithm can work with the dependent input process $\{U_n\}$. However, if $\{U_n\}$ is a sequence of iid random variables, then the result of Chapter 12 gives a simple way of recovering λ^* and $m^*(u)$. Hence, we observed that,

$$\lambda_j^* = \frac{\text{cov}(Y_n, U_{n-j})}{\text{var}(U_0)}, \qquad j = 1, \ldots, p,$$

and

$$m^*(u) = E\{Y_n | U_n = u\} - u.$$

Empirical counterparts of $\text{cov}(Y_n, U_{n-j})$, $\text{var}(U_0)$, and the regression function $E\{Y_n | U_n = u\}$ define the estimates of the system characteristics (see Chapter 12). Although these are explicit estimates, they are often difficult to generalize in more complex cases. On the other hand, the semiparametric approach can easily be extended to a large class of interconnected complex systems.

14.7   Direct estimators for semiparametric systems

The least-squares estimates of the parametric part of a semiparametric model require solving highly nonlinear and multidimensional optimization problems. In addition, the estimates are not given in an explicit form and therefore are difficult to use. Nevertheless, the least squares semiparametric approach is very general and can be applied to a broad class of nonlinear systems. Furthermore, input signals with general distributions are admitted.

There is, however, a need for finding direct estimates of parameters of nonlinear systems, even at the expense of putting stronger assumptions on the underlying characteristics. In this section, we derive such explicit estimates based on the theory of *average derivative estimation*. We begin with, the fundamental concept of average derivative estimation, and then we will illustrate this approach in the case of some nonlinear semiparametric models.

14.7.1   Average derivative estimation

Given random variables $\mathbf{U} \in R^d$ and $Y \in R$ with the regression function $M(\mathbf{u}) = E\{Y|\mathbf{U} = \mathbf{u}\}$, the *average derivative* of $M(\mathbf{u})$ is the average slope of $M(\mathbf{u})$, that is,

$$\delta = E\{\mathcal{D}M(\mathbf{U})\}, \tag{14.106}$$

where $\mathcal{D}M(\mathbf{u}) = \frac{\partial M(\mathbf{u})}{\partial \mathbf{u}}$ is the gradient of $M(\mathbf{u})$.

The following elementary, yet very useful, identity relates the average derivative δ with the average of corresponding input–output signals and the input-density function. This fact will be very helpful in constructing direct estimates of the parametric part of semiparametric nonlinear models.

THEOREM 14.7 *Let $(\mathbf{U}, Y) \in R^d \times R$ be a pair of random vectors such that $\mathbf{U}$ has a density $f(\bullet)$ defined on the set $S \subseteq R^d$. Suppose that $f(\bullet)$ has a continuous derivative and $f(\bullet)$ is zero on the boundary ∂S of S. Assume that the regression function $M(\mathbf{u}) = E\{Y|\mathbf{U} = \mathbf{u}\}$ has a derivative $\mathcal{D}M(\mathbf{u})$. Then, we have,*

$$\delta = E\{\mathcal{D}M(\mathbf{U})\} = -E\left\{Y\frac{\mathcal{D}f(\mathbf{U})}{f(\mathbf{U})}\right\}. \tag{14.107}$$

Proof. Using integration by parts we have,

$$E\{\mathcal{D}M(\mathbf{U})\} = \int_S (\mathcal{D}M(\mathbf{u}))f(\mathbf{u})d\mathbf{u} = M(\mathbf{u})f(\mathbf{u})_{|\partial S} - \int_S M(\mathbf{u})\mathcal{D}f(\mathbf{u})d\mathbf{u}$$

$$= -\int_S M(\mathbf{u})\mathcal{D}f(\mathbf{u})d\mathbf{u}.$$

Then, we can note that,

$$\int_S M(\mathbf{u})\mathcal{D}f(\mathbf{u})d\mathbf{u} = E\left\{M(\mathbf{U})\frac{\mathcal{D}f(\mathbf{U})}{f(\mathbf{U})}\right\}$$

$$= E\left\{E\left\{Y\frac{\mathcal{D}f(\mathbf{U})}{f(\mathbf{U})}|U\right\}\right\} = E\left\{Y\frac{\mathcal{D}f(\mathbf{U})}{f(\mathbf{U})}\right\}.$$

This proves the claim of Theorem 14.7. ∎

One can easily generalize the result of Theorem 14.7 to the case of the weighted average derivative of $\delta_w = E\{w(\mathbf{U})\mathcal{D}M(\mathbf{U})\}$ of $M(\mathbf{u})$. Indeed, for a smooth weight function $w(\bullet)$ defined on S, we can show that

$$\delta_w = E\{w(\mathbf{U})\mathcal{D}M(\mathbf{U})\} = -E\{Y\mathcal{D}w(\mathbf{U})\} - E\left\{Yw(\mathbf{U})\frac{\mathcal{D}f(\mathbf{U})}{f(\mathbf{U})}\right\}. \qquad (14.108)$$

The choice $w(\mathbf{u}) = f(\mathbf{u})$ is particularly interesting because we obtain a formula for the weighted average derivative of $M(\mathbf{u})$ that is not in the ratio form. In fact, for $w(\mathbf{u}) = f(\mathbf{u})$ we have the following version of (14.108):

$$\delta_w = E\{f(\mathbf{U})\mathcal{D}M(\mathbf{U})\} = -2E\{Y\mathcal{D}f(\mathbf{U})\}. \qquad (14.109)$$

Nonparametric average derivative estimation tries to estimate δ (or δ_w) from the training data $T = \{(\mathbf{U}_1, Y_1), \dots, (\mathbf{U}_n, Y_n)\}$ without any knowledge of the regression function $M(\mathbf{u})$ and the input density $f(\mathbf{u})$. Owing to (14.107) it is natural to estimate the vector δ by,

$$\hat{\delta} = -\frac{1}{n}\sum_{i=1}^{n}\hat{Y}_i\hat{\ell}(\mathbf{U}_i), \qquad (14.110)$$

where $\hat{\ell}(\mathbf{u})$ is a certain nonparametric estimate of the following function:

$$\ell(\mathbf{u}) = \frac{\mathcal{D}f(\mathbf{u})}{f(\mathbf{u})}. \qquad (14.111)$$

A plug-in strategy for $\ell(\mathbf{u})$ would give,

$$\hat{\ell}(\mathbf{u}) = \frac{\mathcal{D}\hat{f}(\mathbf{u})}{\hat{f}(\mathbf{u})}, \qquad (14.112)$$

where $\hat{f}(\mathbf{u})$ and $\mathcal{D}\hat{f}(\mathbf{u})$ are nonparametric estimates of $f(\mathbf{u})$ and $\mathcal{D}f(\mathbf{u})$, respectively. For instance, one can use the kernel density estimate

$$\hat{f}(\mathbf{u}) = \frac{1}{nh^d}\sum_{j=1}^{n}K\left(\frac{\mathbf{u} - \mathbf{U}_j}{h}\right) \qquad (14.113)$$

with a d-dimensional kernel function $K(\mathbf{u})$. Then, the derivative of the kernel estimate $\hat{f}(\mathbf{u})$ is given by the following formula

$$\mathcal{D}\hat{f}(\mathbf{u}) = \frac{1}{nh^{d+1}}\sum_{j=1}^{n}\mathcal{D}K\left(\frac{\mathbf{u} - \mathbf{U}_j}{h}\right), \qquad (14.114)$$

where $\mathcal{D}K(\mathbf{u})$ stands for the derivative of $K(\mathbf{u})$. This yields the following kernel estimate of $\ell(\mathbf{u})$:

$$\hat{\ell}(\mathbf{u}) = \frac{\left(nh^{d+1}\right)^{-1} \sum_{j=1}^{n} \mathcal{D}K\left(\frac{\mathbf{u}-\mathbf{U}_j}{h}\right)}{\left(nh^{d}\right)^{-1} \sum_{j=1}^{n} K\left(\frac{\mathbf{u}-\mathbf{U}_j}{h}\right)}. \tag{14.115}$$

It should be emphasized that in (14.115), we use the plug-in estimate of $\mathcal{D}f(\mathbf{u})$, that is, the derivative $\mathcal{D}f(\mathbf{u})$ is estimated by $\mathcal{D}\hat{f}(\mathbf{u})$.

There are two fundamental problems with the estimate $\hat{\delta}$ defined in (14.110). The first problem is that, we use the same data both to compute the average and to evaluate the nonparametric estimate $\hat{\ell}(\mathbf{u})$. This issue can be overcome by using some resampling schemes. For instance, one can apply the partition strategy to the training data T (see Section 14.4.1), that is, we can write,

$$\hat{\delta} = -\frac{1}{n_2} \sum_{i\in I_2} Y_i \frac{\mathcal{D}\hat{f}(\mathbf{U}_i)}{\hat{f}(\mathbf{U}_i)}, \tag{14.116}$$

with $\hat{f}(\mathbf{u}) = \frac{1}{n_1 h^d} \sum_{j\in I_1} K\left(\frac{\mathbf{u}-\mathbf{U}_j}{h}\right)$. Hence, we calculate the average using the testing set T_2, whereas the kernel estimate is derived from the training set T_1. Since we can often choose T_2 to be independent of T_1, the estimate (14.116) has a simple bias, that is,

$$E\hat{\delta} = -E\left\{Y\frac{\mathcal{D}\hat{f}(\mathbf{U})}{\hat{f}(\mathbf{U})}\right\},$$

where $(\mathbf{U}, Y)$ is independent of the training set T_1.

The second problem concerning the estimate $\hat{\delta}$ in (14.110) as well as the one in (14.116) is that both involve dividing by $\hat{f}(\bullet)$ and a refined estimator employing the truncation term is advisable in practice. Thus, we may define the following modified version of $\hat{\delta}$:

$$\tilde{\delta} = -\frac{1}{n} \sum_{i=1}^{n} Y_i \hat{\ell}(\mathbf{U}_i) \mathbf{1}(\hat{f}(\mathbf{U}_i) > \vartheta), \tag{14.117}$$

where ϑ is a sequence of positive numbers (the cutoff parameter). In the asymptotic analysis of the estimate $\tilde{\delta}$ we should require that $\vartheta = \vartheta_n \to 0$ as $n \to \infty$.

Another option to eliminate the ratio problem is to use the weighted average derivative defined in (14.108). In particular, for $\delta_w = E\{w(\mathbf{U})\mathcal{D}M(\mathbf{U})\}$ with $w(\mathbf{u}) = f(\mathbf{u})$, we have (see (14.109)),

$$\delta_w = -2E\{Y\mathcal{D}f(\mathbf{U})\}.$$

This suggests the following estimate of δ_w:

$$\hat{\delta}_w = -\frac{2}{n} \sum_{i=1}^{n} Y_i \mathcal{D}\hat{f}(\mathbf{U}_i). \tag{14.118}$$

We should note again that some resampling scheme would be required here to improve the statistical accuracy of the estimate $\hat{\delta}_w$. An advantage of (14.118) over (14.110) is that the estimate $\hat{\delta}_w$ is not in the ratio form, yielding more stability and an easier to implement estimation algorithm. Nevertheless, when the average derivative estimates are applied to the block-oriented semiparametric systems, we find the estimate $\hat{\delta}$ in (14.110) to be more flexible. This is illustrated in the next three subsections.

Concerning the asymptotic behavior of the estimates $\hat{\delta}$ and $\hat{\delta}_w$, we first note that the expected limit values, that is, δ and δ_w represent linear functionals of the nonparametric function $M(\mathbf{u})$. Therefore, we can expect that the rate of convergence for $\hat{\delta}$ and $\hat{\delta}_w$ is faster than the pointwise rate of nonparametric estimates of $f(\mathbf{u})$ and $\mathcal{D}M(\mathbf{u})$. In fact, the parametric $O(n^{-1})$ rate can be anticipated. This is true depending on the smoothness of $f(\mathbf{u})$ and $M(\mathbf{u})$, and on the dimensionality of the input random process $\{\mathbf{U}_n\}$.

An involved analysis shows that the mean-squared error for the kernel-type estimates $\hat{\delta}$ and $\hat{\delta}_w$ of δ and δ_w, respectively, is of the following order:

$$E\|\hat{\delta} - \delta\|^2 = c_1 n^{-1} + c_2 n^{-2} h^{-d-2} + c_3 h^{2s} + o_2, \tag{14.119}$$

for some constants c_1, c_2, and c_3, with o_2 representing the smaller order terms. It is assumed that the training set $\{(\mathbf{U}, Y_1), \ldots, (\mathbf{U}, Y_n)\}$ comprises a sequence of iid random variables. Furthermore, the functions $\mathcal{D}f(\mathbf{u})$, $\mathcal{D}M(\mathbf{u})$ are s-times differentiable, $s > 1$, and the kernel $K(\mathbf{u})$ is of the sth order. See Chapter 13 for the definition of the order of multivariate kernels.

In (14.119), the first two terms represent an asymptotic variance. The first term describes the standard parametric $O(n^{-1})$ decay of the variance, whereas the second term depends on the dimensionality and reflects the semiparametric nature of the multi-dimensional estimation problem, that is, the fact that we estimate the finite dimensional parameter δ that is a linear functional of the nonparametric function $M(\mathbf{u})$. The last term in (14.119) is the standard evaluation of the bias for s differentiable functions.

Formula (14.119) allows us to specify the asymptotical optimal h and the corresponding rate. Indeed, balancing the second and third terms in (14.119), we obtain,

$$h^* = an^{-\frac{2}{2s+d+2}}, \tag{14.120}$$

and consequently,

$$E\|\hat{\delta} - \delta\|^2 = c_1 n^{-1} + c_4 n^{-\frac{4s}{2s+d+2}}. \tag{14.121}$$

Hence, for

$$d \leq 2(s - 1), \tag{14.122}$$

we obtain the parametric rate $E\|\hat{\delta} - \delta\|^2 = O(n^{-1})$; otherwise, the rate is slower and is equal to $E\|\hat{\delta} - \delta\|^2 = O\left(n^{-\frac{4s}{2s+d+2}}\right)$. Alternatively, the choice of the smoothing parameter as $h = an^{-\alpha}$ for $1/2s \leq \alpha \leq 1/(d+2)$ gives the parametric rate $O(n^{-1})$.

Thus, for all symmetric kernels (such as Gaussian) and for $f(\mathbf{u})$ and $M(\mathbf{u})$ being three-times differentiable we have $s = 2$ and consequently the $O(n^{-1})$ rate is only possible for a one- or two-dimensional regression function $M(\mathbf{u})$.

14.7.2   The average derivative estimate for the semiparametric Wiener model

In this section, we illustrate the theory developed in Section 14.7.1 to derive a direct estimate of the impulse response function of the single input semiparametric Wiener system. Recalling the notation of Section 14.4, we can write the input–output relationship of the Wiener system as follows:

$$Y_n = M(\mathbf{U}_n) + Z_n, \tag{14.123}$$

where $\mathbf{U}_n = (U_n, U_{n-1}, \ldots, U_{n-p})^T$, and

$$M(\mathbf{U}_n) = m^* \left(\sum_{j=0}^{p} \lambda_j^* U_{n-j} \right). \tag{14.124}$$

We can easily note that $M(\mathbf{u}) = E\{Y_n | \mathbf{U}_n = \mathbf{u}\}$ is the regression function of Y_n on $\mathbf{U}_n$. Let us now consider the average derivative $E\{\mathcal{D}M(\mathbf{U}_n)\}$ of $M(\mathbf{u})$. We begin with the following fundamental identity:

$$E\{\mathcal{D}M(\mathbf{U}_n)\} = \lambda^* E \left\{ m^{*(1)}(W_n) \right\}, \tag{14.125}$$

where $m^{*(1)}(\bullet)$ is the derivative of $m^*(\bullet)$ and $W_n = \sum_{j=0}^{p} \lambda_j^* U_{n-j}$. Owing to the result of Theorem 14.7, the left-hand side of (14.125) is equal to

$$-E \left\{ Y_n \frac{\mathcal{D}f(\mathbf{U}_n)}{f(\mathbf{U}_n)} \right\}, \tag{14.126}$$

where $f(\bullet)$ is the density function of the vector $\mathbf{U}_n$.

Consequently, by (14.125), (14.126), and the fact that $\lambda_0^* = 1$, we arrive at the following formula for λ_j^*:

$$\lambda_j^* = \frac{E \left\{ Y_n \frac{\frac{\partial}{\partial U_{n-j}} f(\mathbf{U}_n)}{f(\mathbf{U}_n)} \right\}}{E \left\{ Y_n \frac{\frac{\partial}{\partial U_n} f(\mathbf{U}_n)}{f(\mathbf{U}_n)} \right\}}, \qquad j = 1, 2, \ldots, p \tag{14.127}$$

provided that,

$$E\{m^{*(1)}(W_n)\} \neq 0. \tag{14.128}$$

The formula in (14.127) can be greatly simplified for the input process $\{U_n\}$, which is an *iid* random sequence. In fact, we have

$$f(\mathbf{U}_n) = \prod_{i=0}^{p} f_U(U_{n-i}),$$

and

$$\frac{\partial}{\partial U_{n-j}} f(\mathbf{U}_n) = f_U^{(1)}(U_{n-j}) \prod_{i=0, i \neq j}^{p} f_U(U_{n-i}). \tag{14.129}$$

Consequently, under condition (14.128), we can readily obtain the following basic formula for λ_j^*:

$$\lambda_j^* = \frac{E\left\{Y_n \ell(U_{n-j})\right\}}{E\left\{Y_n \ell(U_n)\right\}}, \qquad j = 1, 2, \ldots, p, \qquad (14.130)$$

where

$$\ell(u) = \frac{f_U^{(1)}(u)}{f_U(u)}. \qquad (14.131)$$

All of these considerations are summarized in the following theorem.

THEOREM 14.8 *Let the Wiener system be given by the input–output mapping in (14.123) and (14.124). Let the input signal $\{U_n\}$ be a sequence of iid random variables with the density $f_U(u)$ that is a differentiable function on its support. Assume that $E\{m^{*(1)}(W_n)\} \neq 0$. Then we have,*

$$\lambda_j^* = \frac{E\left\{Y_n \ell(U_{n-j})\right\}}{E\left\{Y_n \ell(U_n)\right\}}, \qquad j = 1, 2, \ldots, p,$$

where $\ell(u) = \frac{f_U^{(1)}(u)}{f_U(u)}$.

We can draw a number of interesting conclusions from the above theorem. First of all, note that if the input $\{U_n\}$ is *iid* Gaussian $N(\mu, \sigma^2)$, then we have

$$\ell(u) = -\frac{u - \mu}{\sigma^2}. \qquad (14.132)$$

This (for $\mu = 0$) readily yields the following formula for λ_j^*:

$$\lambda_j^* = \frac{E\left\{Y_n U_{n-j}\right\}}{E\left\{Y_n U_n\right\}}, \qquad j = 1, 2, \ldots, p. \qquad (14.133)$$

We easily recognize that this is the correlation method introduced in Chapter 9 for recovering the impulse response sequence of the fully nonparametric Wiener system excited by the *iid* Gaussian input signal. Hence, the formula presented in Theorem 14.8 can be viewed as the generalized correlation method defined by the correlation between the output process $\{Y_n\}$ and the process $\{\ell(U_n)\}$. The latter is a nonlinear function of the input signal $\{U_n\}$, where the nonlinearity $\ell(\bullet)$ is completely determined by the input density function $f_U(\bullet)$. It is worth noting that if $f_U(\bullet)$ is symmetric then $f_U^{(1)}(\bullet)$ is an odd function and consequently, we have

$$\ell(-u) = -\ell(u). \qquad (14.134)$$

Thus, the nonlinearity $\ell(\bullet)$ is an odd function.

The function $\ell(\bullet)$ plays an important role in our future developments and it is worth finding other general properties of $\ell(\bullet)$. This can be achieved by restricting a class of input densities, and we discuss this in the following remark:

REMARK 14.4 *A rich family of density functions is the class of log-concave densities, that is, densities for which* $\log f_U(u)$ *is a concave function on R. Noting that,*

$$\ell(u) = (\log f_U(u))^{(1)},$$

we can conclude that for log-concave densities $\ell(u)$ *is a nonincreasing function.*

Log-concave densities have all moments and are known to be unimodal. Furthermore, the convolution of two log-concave densities is again log-concave; see [9] for properties and applications of log-concave densities.

These facts have interesting consequences in the context of Wiener system identification. In fact, the density function $f_W(w)$ *of the interconnecting signal* $\{W_n\}$ *is the convolution of scaled versions of the input density* $f_U(u)$. *Thus, if* $f_U(u)$ *is log-concave then* $f_W(w)$ *is also log-concave and is therefore unimodal.*

The class of log-concave densities comprises many well-known parametric distributions such as Gaussian, exponential, Gamma, Gumbel, logistic, Laplace, and many others. The generalized Gaussian density,

$$f_U(u; \alpha) = a(\alpha) \exp\left(-|b(\alpha)u|^\alpha\right),$$

is log-concave if $\alpha \geq 1$. *Here,* $a(\alpha)$ *and* $b(\alpha)$ *are normalized constants, such that* $E\{U\} = 0$ *and* $\mathrm{var}\{U\} = 1$. *It can be demonstrated that for the density* $f_U(u; \alpha)$, *we have,*

$$\ell(u; \alpha) = -\alpha b^\alpha(\alpha)|u|^{p-1} \operatorname{sgn}(u),$$

where $b(\alpha) = (\Gamma(3/\alpha)/\Gamma(1/\alpha))^{1/2}$.

In Figure 14.14 we plot $\ell(u; \alpha)$ *for* $\alpha = 1$ *(Laplace distribution),* $\alpha = 1.2$, *and* $\alpha = 2$ *(Gaussian distribution). Also, we consider the normalized* $(\mathrm{var}\{U\} = 1)$ *logistic distribution, that is, when,*

$$f_U(u; \beta) = \frac{\beta e^{-\beta u}}{(1 + e^{-\beta u})^2},$$

with $\beta = \pi/\sqrt{3}$. *Here it can be shown that* $\ell(u; \beta) = \beta(e^{-\beta u} - 1)(1 + e^{-\beta u})^{-1}$. *We observe the monotonicity and continuity of* $\ell(u; \alpha)$ *for* $\alpha > 1$. *For* $\alpha = 1$ *we have the discontinues function* $\ell(u; 1) = -\operatorname{sgn}(u)$.

It is worth noting that the monotonicity property is lost for non-log-concave densities. For instance, the unimodal Gaussian mixture $0.8N(0, 1) + 0.2N(0, 0.04)$ *has the function* $\ell(u)$ *shown in Figure 14.15. The monotonicity property of* $\ell(u)$ *for log-concave densities suggests that it would be worthwhile to construct an estimate of* $\ell(u)$ *which is also monotonic. This seems to be workable by employing a penalized (with respect to the log-convexity property) maximum likelihood estimate of* $f_U(u)$ *[206].*

We should also note that the formula in (14.130) fails for input densities that are nonsmooth. In particular, if $f_U(u)$ *is uniform then (14.130) cannot be used. In this case, the least-squares method presented in Section 14.4 can still be applied.*

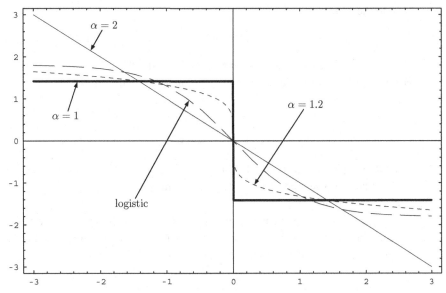

Figure 14.14 Function $\ell(u)$ for the generalized Gaussian density with $\alpha = 1, 1.2, 2$ and logistic distribution.

We can also obtain alternative formulas to that given in (14.127) using the concept of the weighted average derivative. Indeed, using (14.109), we can readily obtain that

$$\lambda_j^* = \frac{E\left\{Y_n \frac{\partial}{\partial U_{n-j}} f(\mathbf{U}_n)\right\}}{E\left\{Y_n \frac{\partial}{\partial U_n} f(\mathbf{U}_n)\right\}}, \qquad j = 1, 2, \ldots, p \qquad (14.135)$$

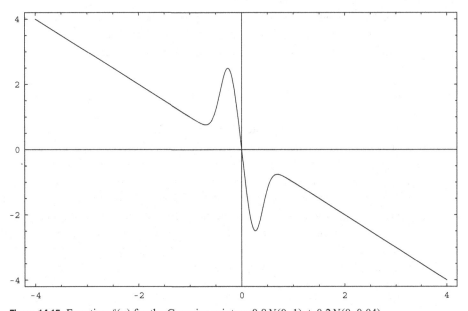

Figure 14.15 Function $\ell(u)$ for the Gaussian mixture $0.8N(0, 1) + 0.2N(0, 0.04)$.

subject to the condition,

$$E\{f(\mathbf{U}_n)m^{*(1)}(W_n)\} \neq 0. \tag{14.136}$$

Note that this is a weaker requirement than that in (14.128).

For the independent input signal, (14.135) can be written as follows:

$$\lambda_j^* = \frac{E\left\{Y_n f_U^{(1)}(U_{n-j}) \prod_{i=0, i\neq j}^p f_U(U_{n-i})\right\}}{E\left\{Y_n f_U^{(1)}(U_n) \prod_{i=1}^p f_U(U_{n-i})\right\}}. \tag{14.137}$$

No further simplification of this formula seems to be possible and therefore we can conclude that the weighted average derivative strategy is not very useful for Wiener-system identification.

The aforementioned considerations allow us to estimate λ_j^* based on a sample analogue of the equation in Theorem 14.8. This direct estimate does not require any computational techniques for minimization. A sample analog of (14.130) is

$$\hat{\lambda}_j = \frac{n^{-1}\sum_{t=j+1}^n Y_t \hat{\ell}(U_{t-j})}{n^{-1}\sum_{t=1}^n Y_t \hat{\ell}(U_t)}, \tag{14.138}$$

where $\hat{\ell}(u)$ is a nonparametric estimate of $\ell(u)$. Specifically, one can apply the kernel estimate, see (14.130), and obtain,

$$\hat{\ell}(u) = \frac{(nh^2)^{-1}\sum_{t=1}^n K^{(1)}\left(\frac{u-U_t}{h}\right)}{(nh)^{-1}\sum_{t=1}^n K\left(\frac{u-U_t}{h}\right)}. \tag{14.139}$$

Thus, (14.138) with (14.139) defines the kernel type density estimate of the impulse response sequence of the Wiener system.

As we have already mentioned, the ratio form of the estimate $\hat{\ell}(u)$ may lead to some instabilities. A simple correction would use the cutoff parameter $\vartheta > 0$; see (14.117). This would give the following numerically stable version of $\hat{\lambda}_j$:

$$\hat{\lambda}_j = \frac{n^{-1}\sum_{t=j+1}^n Y_t \hat{\ell}(U_{t-j}) \mathbf{1}(\hat{f}_U(U_{t-j}) > \vartheta)}{n^{-1}\sum_{t=1}^n Y_t \hat{\ell}(U_t) \mathbf{1}(\hat{f}_U(U_t) > \vartheta)}, \tag{14.140}$$

where $\hat{f}_U(u)$ is the kernel estimate of the input density $f_U(u)$.

To improve further small sample properties of the estimate $\hat{\lambda}_j$, we can use some resampling schemes that aim at separating data used to estimate $\hat{\ell}(u)$ and to determine an empirical mean for $E\{Y_n\ell(U_{n-j})\}$. We have already examined the partition strategy that, due to its simplicity, can be universally used in many nonparametric/semiparametric circumstances, and allows us to reduce mathematical analysis immensely. The data splitting, however, does not utilize data in an optimal way and can result in reduced accuracy for small sample sizes. A more efficient strategy is the leave-one-out resampling scheme that results in the following estimate:

$$\tilde{\lambda}_j = \frac{n^{-1}\sum_{t=j+1}^n Y_t \hat{\ell}_{-(t-j)}(U_{t-j})}{n^{-1}\sum_{t=1}^n Y_t \hat{\ell}_{-t}(U_t)},$$

where,

$$\hat{\ell}_{-t}(u) = \frac{((n-1)h^2)^{-1} \sum_{j=1,j\neq t}^{n} K^{(1)}\left(\frac{u-U_j}{h}\right)}{((n-1)h)^{-1} \sum_{j=1,j\neq t}^{n} K\left(\frac{u-U_j}{h}\right)},$$

is the version of $\hat{\ell}(u)$ with the tth observation deleted. This method may have better small sample properties than the partition scheme but at the expense of increased computational cost and more involved mathematical analysis.

Asymptotically, all of the aforementioned methods are equivalent, and we can use the result presented in (14.119). The main difference in proving this result is that we now have to cope with dependent data. However, the final asymptotic formula for the mean-squared error remains the same. Hence, noting that we now have a single input, that is, $d = 1$, we obtain from (14.119),

$$E(\hat{\lambda}_j - \lambda_j^*)^2 = c_1 n^{-1} + c_2 n^{-2} h^{-3} + c_3 h^{2s} + o_2, \quad j = 1, \dots, p.$$

This readily leads to the following theorem:

THEOREM 14.9 *Let all of the assumptions of Theorem 14.8 be met. Let $f_U^{(1)}(u)$ and $m^{*(1)}(w)$ be s-times differentiable, $s > 1$, and let the kernel function $K(u)$ be of order s. If,*

$$h^* = an^{-\frac{2}{2s+3}},$$

then

$$E(\hat{\lambda}_j - \lambda_j^*)^2 = O(n^{-1}), \quad j = 1, 2, \dots, p.$$

Note that the parametric $O(n^{-1})$ rate is also obtained for $h = an^{-\alpha}$, $1/2s \leq \alpha \leq 1/3$.

For all symmetric kernels we have $s = 2$, and if $f_U(u)$ and $m^*(w)$ are three-times differentiable, then the asymptotically optimal choice of the smoothing parameter is $h^* = an^{-2/7}$. Note that the asymptotically optimal choice for estimating the derivative of $f_U(u)$ or $m^*(w)$ is $h \approx n^{-1/7}$. This is a larger value than that needed for estimating the average derivative which corresponds to the choice in Theorem 14.9.

The parametric $O(n^{-1})$ rate in Theorem 14.9 holds for a large class of nonparametric methods used to form an estimate $\hat{\ell}(u)$ as well as for various resampling schemes. There is an interesting issue which estimator is the best one. The theory presented above does not distinguish between these estimators and the second-order asymptotics is needed. Hence, we wish to obtain the asymptotic representation for $E(\hat{\lambda}_j - \lambda_j^*)^2$ of the form $O(n^{-1}) + O(n^{-\gamma})$, $\gamma > 1$, where $O(n^{-\gamma})$ represents the second-order asymptotic term of the estimation error. An estimator with the largest γ could be called the most efficient one.

14.7.3   The average derivative estimate for the additive Wiener model

In Section 14.4.4, we considered the multiple-input Wiener system of the additive form which is described by (14.87) and depicted in Figure 14.2. Once again, we use the notation λ_1^*, λ_2^*, $m_1^*(\bullet)$, and $m_2^*(\bullet)$ to indicate the true characteristics of the system.

It has been demonstrated in Section 14.4.4 that if the inputs between channels are independent, then the combination of the least squares semiparametric method with the marginal integration strategy yields the consistent estimates of the system characteristics. We wish to extend this result to the case of dependent inputs between individual channels, employing the flexibility of the average derivative approach.

Let us rewrite the description in (14.87) in the following equivalent form:

$$Y_n = M(\mathbf{U}_{1,n}, \mathbf{U}_{2,n}) + Z_n, \tag{14.141}$$

where $M(\mathbf{U}_{1,n}, \mathbf{U}_{2,n}) = m_1^*(W_{1,n}) + m_2^*(W_{2,n})$, with $W_{1,n} = \sum_{j=0}^{p} \lambda_{1,j}^* U_{1,n-j}$ and $W_{2,n} = \sum_{j=0}^{p} \lambda_{2,j}^* U_{2,n-j}$. The vectors $\mathbf{U}_{1,n}$ and $\mathbf{U}_{2,n}$ are defined as $\mathbf{U}_{1,n} = (U_{1,n}, U_{1,n-1}, \ldots, U_{1,n-p})^T$ and $\mathbf{U}_{2,n} = (U_{2,n}, U_{2,n-1}, \ldots, U_{2,n-p})^T$, respectively. Let $f_{U_1 U_2}(u_1, u_2)$ be the joint density function of the input process $\{(U_{1,n}, U_{2,n})\}$. It is then easy to observe that,

$$E\left\{\mathcal{D}_{\mathbf{U}_{1,n}} M(\mathbf{U}_{1,n}, \mathbf{U}_{2,n})\right\} = \lambda_1^* E\left\{m_1^{*(1)}(W_{1,n})\right\}, \tag{14.142}$$

and

$$E\left\{\mathcal{D}_{\mathbf{U}_{2,n}} M(\mathbf{U}_{1,n}, \mathbf{U}_{2,n})\right\} = \lambda_2^* E\left\{m_2^{*(1)}(W_{2,n})\right\}, \tag{14.143}$$

where λ_1^* and λ_2^* are the vectors of the impulse response sequences of the linear subsystems. In formulas (14.142) and (14.143), $\mathcal{D}_{\mathbf{U}_{1,n}} M(\mathbf{U}_{1,n}, \mathbf{U}_{2,n})$ and $\mathcal{D}_{\mathbf{U}_{2,n}} M(\mathbf{U}_{1,n}, \mathbf{U}_{2,n})$ denote the partial derivatives of $M(\mathbf{U}_{1,n}, \mathbf{U}_{2,n})$ with respect to $\mathbf{U}_{1,n}$ and $\mathbf{U}_{2,n}$, respectively.

Denoting by $f(\mathbf{u}_1, \mathbf{u}_2)$ the joint density function of the random vector $(\mathbf{U}_{1,n}, \mathbf{U}_{2,n})$ and recalling Theorem 14.7 we note that the left-hand sides of (14.142) and (14.143) are equal to,

$$-E\left\{Y_n \frac{\mathcal{D}_{\mathbf{U}_{1,n}} f(\mathbf{U}_{1,n}, \mathbf{U}_{2,n})}{f(\mathbf{U}_{1,n}, \mathbf{U}_{2,n})}\right\}, \quad \text{and} \quad -E\left\{Y_n \frac{\mathcal{D}_{\mathbf{U}_{2,n}} f(\mathbf{U}_{1,n}, \mathbf{U}_{2,n})}{f(\mathbf{U}_{1,n}, \mathbf{U}_{2,n})}\right\},$$

respectively.

Then, by this, the normalization $\lambda_{1,0}^* = \lambda_{2,0}^* = 1$, and the decomposition

$$f(\mathbf{U}_{1,n}, \mathbf{U}_{2,n}) = \prod_{i=0}^{p} f_{U_1 U_2}(U_{1,n-i}, U_{2,n-i})$$

we obtain the following formulas for λ_1^* and λ_2^*:

$$\lambda_{1,j}^* = \frac{E\left\{Y_n \ell_1\left(U_{1,n-j}, U_{2,n-j}\right)\right\}}{E\left\{Y_n \ell_1\left(U_{1,n}, U_{2,n}\right)\right\}}, \tag{14.144}$$

$$\lambda_{2,j}^* = \frac{E\left\{Y_n \ell_2\left(U_{1,n-j}, U_{2,n-j}\right)\right\}}{E\left\{Y_n \ell_2\left(U_{1,n}, U_{2,n}\right)\right\}} \tag{14.145}$$

$j = 1, 2, \ldots, p$, where,

$$\ell_1(u_1, u_2) = \frac{\frac{\partial}{\partial u_1} f_{U_1 U_2}(u_1, u_2)}{f_{U_1 U_2}(u_1, u_2)}, \quad \ell_2(u_1, u_2) = \frac{\frac{\partial}{\partial u_2} f_{U_1 U_2}(u_1, u_2)}{f_{U_1 U_2}(u_1, u_2)}. \quad (14.146)$$

The assumption required for the existence of formulas (14.144) and (14.145) is that

$$E\left\{m_1^{*(1)}(W_{1,n})\right\} \neq 0 \quad \text{and} \quad E\left\{m_2^{*(1)}(W_{2,n})\right\} \neq 0. \quad (14.147)$$

This is equivalent to the result in Theorem 14.8, generalized to the two-channel Wiener system with dependent channels. Note that the assumption in (14.147) is always met if the nonlinearities are odd functions and the density functions of the interconnecting signals $W_{1,n}$ and $W_{2,n}$ are symmetric.

The following example gives some further insight into the formulas established in (14.144), (14.145).

Example 14.2 Let the input signal (U_1, U_2) be the bivariate, zero-mean Gaussian random vector with the parameters $\text{var}(U_1) = \sigma_1^2$, $\text{var}(U_2) = \sigma_2^2$ and $\text{corr}(U_1, U_2) = \rho$, $|\rho| < 1$. Then, simple algebra shows that

$$\ell_1(u_1, u_2) = -(1 - \rho^2)^{-1}\left(\frac{u_1}{\sigma_1^2} - \rho\frac{u_2}{\sigma_1 \sigma_2}\right),$$

$$\ell_2(u_1, u_2) = -(1 - \rho^2)^{-1}\left(\frac{u_2}{\sigma_2^2} - \rho\frac{u_1}{\sigma_1 \sigma_2}\right).$$

This leads to the following formulas for $\lambda_{1,j}^*$ and $\lambda_{2,j}^*$:

$$\lambda_{1,j}^* = \frac{E\{Y_n \omega_1(U_{1,n-j}, U_{2,n-j})\}}{E\{Y_n \omega_1(U_{1,n}, U_{2,n})\}}, \quad \lambda_{2,j}^* = \frac{E\{Y_n \omega_2(U_{1,n-j}, U_{2,n-j})\}}{E\{Y_n \omega_2(U_{1,n}, U_{2,n})\}},$$

where $\omega_1(u_1, u_2) = \frac{u_1}{\sigma_1} - \rho\frac{u_2}{\sigma_2}$ and $\omega_2(u_1, u_2) = \frac{u_2}{\sigma_2} - \rho\frac{u_1}{\sigma_1}$. Hence, $\lambda_{1,j}^*$ depends on the linear combination of the correlation between the output $\{Y_n\}$ and the inputs $\{U_{1,n}\}$ and $\{U_{2,n}\}$. This linear combination is given by,

$$\frac{1}{\sigma_1}E\{Y_n U_{1,n-j}\} - \frac{\rho}{\sigma_2}E\{Y_n U_{2,n-j}\}.$$

For small values of the correlation coefficient ρ, the second term is negligible.

In the case of independent inputs, that is, if $\rho = 0$, $\lambda_{1,j}^*$ is determined merely by the correlation between $\{Y_n\}$ and $\{U_{1,n}\}$. Since an analogous discussion can be carried out for the second channel, we have

$$\lambda_{1,j}^* = \frac{E\{Y_n U_{1,n-j}\}}{E\{Y_n U_{1,n}\}}, \quad \lambda_{2,j}^* = \frac{E\{Y_n U_{2,n-j}\}}{E\{Y_n U_{2,n}\}}.$$

This is a generalization of the result obtained in Chapter 9 for the one-channel Wiener system with a Gaussian input.

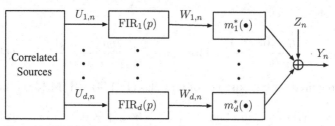

Figure 14.16 Multiple-input additive Wiener model.

The aforementioned theory can easily be generalized to the d-input additive Wiener system. Figure 14.16 depicts such a system with the input process $\{\mathbf{U}_n = (U_{1,n}, \ldots, U_{d,n})^T\}$ that possesses the density function $f_{\mathbf{U}}(\mathbf{u})$, $\mathbf{u} = (u_1, \ldots, u_d)^T$. The dynamical subsystems are of a finite impulse response type and denoted by $FIR_i(p)$, $i = 1, \ldots, d$. Then we can write the analogue of formula (14.144) that relates the impulse response sequence of the rth channel with the nonlinear correlation of the input and output processes, that is,

$$\lambda_{r,j}^* = \frac{E\{Y_n \ell_r(\mathbf{U}_{n-j})\}}{E\{Y_n \ell_r(\mathbf{U}_n)\}}, \qquad j = 1, \ldots, p, \tag{14.148}$$

where

$$\ell_r(\mathbf{u}) = \frac{\frac{\partial}{\partial u_r} f_{\mathbf{U}}(\mathbf{u})}{f_{\mathbf{U}}(\mathbf{u})}, \tag{14.149}$$

for $r = 1, \ldots, d$. Owing to formulas (14.148) and (14.149) it is straightforward to propose various estimators of the linear subsystem $\{\lambda_{r,j}^*, j = 1, \ldots, p\}$ characterizing the rth channel.

Indeed, let $\{(\mathbf{U}_1, Y_1), \ldots, (\mathbf{U}_n, Y_n)\}$ be the training set of the input–output signals. Then, the counterpart of the estimate $\hat{\lambda}_j$ in (14.138) and (14.139) takes the following form (for the d-input additive Wiener system):

$$\hat{\lambda}_{r,j} = \frac{n^{-1} \sum_{t=j+1}^{n} Y_t \hat{\ell}_r(U_{t-j})}{n^{-1} \sum_{t=1}^{n} Y_t \hat{\ell}_r(U_t)}, \qquad j = 1, \ldots, p \tag{14.150}$$

where

$$\hat{\ell}_r(\mathbf{u}) = \frac{(nh^{d+1})^{-1} \sum_{t=1}^{n} K_r\left(\frac{\mathbf{u}-\mathbf{U}_t}{h}\right)}{(nh^d)^{-1} \sum_{t=1}^{n} K\left(\frac{\mathbf{u}-\mathbf{U}_t}{h}\right)}.$$

The kernel function $K_r(\mathbf{u})$ is defined as,

$$K_r(\mathbf{u}) = \frac{\partial}{\partial u_r} K(\mathbf{u}). \tag{14.151}$$

For the product type kernel $K(\mathbf{u}) = \prod_{j=1}^{d} k(u_j)$, where $k(u)$ is some univariate admissible kernel, the derivative can be easily evaluated as,

$$K_r(\mathbf{u}) = k^{(1)}(u_r) \prod_{j=1, j \neq r}^{d} k(u_j).$$

The asymptotic theory of the average derivative kernel estimates presented in Section 14.7.1 can be carried over to the present case, with further refinements due to the dependent nature of the output process. Nevertheless, the result in (14.119) (see also the discussion leading to formula (14.121)) can be fully applied here. This yields the following theorem.

THEOREM 14.10 *Suppose that* $E\{m_r^{*(1)}(W_{r,n})\} \neq 0$, $r = 1, \ldots, d$. *Let* $Df_U(\mathbf{u})$ *and* $m_r^{*(1)}(w)$, $r = 1, \ldots, d$, *be s-times differentiable*, $s > 1$ *and let the kernel function* $K(\mathbf{u})$ *be of order s. If*

$$h^* = an^{-\frac{2}{2s+d+2}},$$

then,

$$E\left(\hat{\lambda}_{r,j} - \lambda_{r,j}^*\right)^2 = O\left(n^{-\min\left(1, \frac{4s}{2s+d+2}\right)}\right), j = 1, \ldots, p; r = 1, \ldots, d. \quad (14.152)$$

This theorem says that in order to estimate the characteristic of the linear subsystem of the rth channel we need to apply smoothing conditions to the input density as well as the nonlinear characteristics in all of the channels. Thus there is a strong dependence between the accuracy in recovering individual channels. Hence, the smoothness degree, s, required in Theorem 14.10 could be a reflection of the smoothness of the roughest nonlinearity present in the system. The rate in Theorem 14.10 reveals (see (14.122)) that s must be no smaller than $1 + d/2$ in order to achieve the parametric rate $O(n^{-1})$.

Having established the estimates of the linear part of the additive Wiener system, we can proceed to the problem of recovering the nonlinearities in the additive structure. This can be done in the manner outlined in Section 14.4. Hence, first we have to form the d-dimensional pilot-kernel-regression estimate,

$$\hat{M}(\mathbf{w}; \lambda_1, \ldots, \lambda_d) = \frac{(nh^d)^{-1} \sum_{j=1}^n Y_j K\left(\frac{\mathbf{w} - \mathbf{W}_j(\lambda_1, \ldots, \lambda_d)}{h}\right)}{(nh^d)^{-1} \sum_{j=1}^n K\left(\frac{\mathbf{w} - \mathbf{W}_j(\lambda_1, \ldots, \lambda_d)}{h}\right)},$$

where $\mathbf{W}_j(\lambda_1, \ldots, \lambda_d) = (W_{1,j}(\lambda_1), \ldots, W_{d,j}(\lambda_d))^T$ is the vector of the interconnecting signals when the impulse-response functions are $\lambda_1, \ldots, \lambda_d$ (see (14.89)).

Under standard assumptions the estimate $\hat{M}(\mathbf{w}; \lambda_1, \ldots, \lambda_d)$ can converge to the regression function $M(\mathbf{w}; \lambda_1, \ldots, \lambda_d) = E\{Y_n | \mathbf{W}_n(\lambda_1, \ldots, \lambda_d) = \mathbf{w}\}$. It is also clear that,

$$M(\mathbf{w}; \lambda_1^*, \ldots, \lambda_d^*) = \sum_{j=1}^d m_j^*(w_j).$$

Since $\lambda_1^*, \ldots, \lambda_d^*$ can be consistently estimated by $\hat{\lambda}_1, \ldots \hat{\lambda}_d$ given in (14.150), we can expect (see the arguments given in Section 14.4.2 B) that $\hat{M}(\mathbf{w}; \hat{\lambda}_1, \ldots, \hat{\lambda}_d)$ tends to $M(\mathbf{w}; \lambda_1^*, \ldots, \lambda_d^*)$. Hence, the estimate $\hat{M}(\mathbf{w}; \hat{\lambda}_1, \ldots, \hat{\lambda}_d)$ can recover the additive function $\sum_{j=1}^d m_j^*(w_j)$. In order to extract the individual components of the system we can apply the marginal integration method discussed in Chapter 13. Thus, we can define the following estimate of $m_1^*(w_1)$ (other components are estimated in an analogous way):

$$\hat{m}_1(w_1) = \int_{R^{d-1}} \hat{M}(\mathbf{w}; \hat{\lambda}_1, \ldots, \hat{\lambda}_d) \psi(w_2, \ldots, w_d) dw_2 \ldots dw_d$$

for some given $(d-1)$-dimensional weight function $\psi(w_2, \ldots, w_d)$.

Under the assumptions of Theorem 14.10, we can establish the asymptotic rate for $\hat{m}_1(w_1)$. This is an involved process requiring the techniques developed in Chapter 13 concerning the marginal integration estimate and the linearization method used in Section 14.4.2 B. This combined with the result of Theorem 14.10 gives the following formula for the mean-squared error of $\hat{m}_1(w_1)$:

$$E(\hat{m}_1(w_1) - m_1^*(w_1))^2 = c_1 n^{-\min(1, \frac{4s}{2s+d+2})} + c_2 n^{-\frac{2s}{2s+1}} + o_2. \tag{14.153}$$

The first expression in (14.153) comes from the linearization term and therefore is equal to the mean-squared error of the estimates $\hat{\lambda}_1, \ldots, \hat{\lambda}_d$ established in Theorem 14.10. The second term in (14.153), on the other hand, describes the usual rate for recovering the one dimensional function having s derivatives. The comparison of these two terms implies that if $d \leq 2s$, then we have the optimal rate,

$$E(\hat{m}_1(w_1) - m_1^*(w_1))^2 = O\left(n^{-\frac{2s}{2s+1}}\right).$$

On the other hand, if $d > 2s$, we have,

$$E(\hat{m}_1(w_1) - m_1^*(w_1))^2 = O\left(n^{-\frac{4s}{2s+d+2}}\right).$$

Thus, we lose the dimension-independent rate for the characteristics (and input density) possessing the smoothness $s < d/2$.

In Chapter 13, we analyzed the additive Hammerstein system and we found that there is an estimation method exhibiting the separation property, that is, we can estimate a nonlinearity in a given channel with the optimal rate $O(n^{-\frac{2s}{2s+1}})$ regardless of d and independent of the roughness of the nonlinearities in the remaining channels. This is in stark contrast to the estimation problem examined in this section, demonstrating again that Wiener type systems are much more complicated to identify than Hammerstein cascades.

14.7.4   The average derivative estimate for semiparametric multivariate Hammerstein models

In this section, we wish to consider the application of the average derivative estimation method to a multiple-input semiparametric Hammerstein system (depicted in Figure 14.1 in the case of two inputs). Thus, let the system have the input–output relationship given in (14.93).

In Section 14.5, the semiparametric least squares method for estimating the system parameters (λ^*, γ^*) was examined. It has been noted that the parameter γ^* is more difficult to identify. In fact, the impulse response sequence λ^* of the linear subsystem can be estimated via the correlation method, independently of γ^* and the system nonlinearity

$g^*(\bullet)$. Hence, let us focus on a direct method for estimating γ^* using the average derivative method proposed in Section 14.7.1.

To this end, let us write (14.93) in the following equivalent form:

$$Y_n = \sum_{t=0}^{p} \lambda_t^* M(\mathbf{U}_{n-t}) + Z_n, \tag{14.154}$$

where $M(\mathbf{u}) = g^*(\gamma^{*T}\mathbf{u})$, $\mathbf{u} \in R^d$. Let us assume that the input process $\{\mathbf{U}_n\}$ is an iid sequence with the density $f_\mathbf{U}(\mathbf{u})$. Then, since $\lambda_0^* = 1$, we can readily obtain that $M(\mathbf{u}) = E\{Y_n|\mathbf{U}_n = \mathbf{u}\}$. Next we observe that,

$$E\{\mathcal{D}M(\mathbf{U}_n)\} = \gamma^* E\{g^{*(1)}(W_n)\},$$

where $W_n = \gamma^{*T}\mathbf{U}_n$. Using Theorem 14.7, we obtain,

$$\gamma^* E\{g^{*(1)}(W_n)\} = -E\left\{Y_n \frac{\mathcal{D}f_\mathbf{U}(\mathbf{U}_n)}{f_\mathbf{U}(\mathbf{U}_n)}\right\}. \tag{14.155}$$

The normalization $\gamma_1^* = 1$ and the assumption,

$$E\left\{g^{*(1)}(W_n)\right\} \neq 0, \tag{14.156}$$

yield the following formula for γ_j^*:

$$\gamma_j^* = \frac{E\{Y_n \ell_j(\mathbf{U}_n)\}}{E\{Y_n \ell_1(\mathbf{U}_n)\}}, \tag{14.157}$$

where

$$\ell_j(\mathbf{u}) = \frac{\frac{\partial}{\partial u_j} f_\mathbf{U}(\mathbf{u})}{f_\mathbf{U}(\mathbf{u})}, \tag{14.158}$$

for $j = 2, \ldots, d$.

Example 14.3 Let us consider the input process $\{\mathbf{U}_n\}$ having the d-dimensional Gaussian distribution $N_d(0, \Sigma)$. Simple algebra reveals that,

$$\frac{\mathcal{D}f_\mathbf{U}(\mathbf{u})}{f_\mathbf{U}(\mathbf{u})} = -\Sigma^{-1}\mathbf{u}.$$

This gives the following formula for γ_j^*:

$$\gamma_j^* = \frac{E\{Y_n(\Sigma^{-1}\mathbf{U}_n)_j\}}{E\{Y_n(\Sigma^{-1}\mathbf{U}_n)_1\}}, \tag{14.159}$$

where $(\Sigma^{-1}\mathbf{U}_n)_j$ denotes the jth coordinate of the vector $\Sigma^{-1}\mathbf{U}_n$.

It is worth noting that for independent inputs, that is, when $\Sigma = \text{diag}(\sigma_1^2, \ldots, \sigma_d^2)$, formula (14.158) takes the form,

$$\gamma_j^* = \frac{\sigma_1^2}{\sigma_j^2} \frac{E\{Y_n U_{j,n}\}}{E\{Y_n U_{1,n}\}}.$$

The above formulas for γ_j^* hold subject to the condition in (14.156). Let us note that the random variable $W_n = \gamma^{*T} U_n$ is Gaussian $N(0, \tau^2)$ with $\tau^2 = \gamma^{*T} \Sigma \gamma^*$. Integration by parts yields,

$$E\{g^{*(1)}(W_n)\} = \frac{1}{\tau^2} E\{W_n g^*(W_n)\}.$$

Therefore, (14.156) is equivalent to the requirement that $E\{W_n g^*(W_n)\} \neq 0$. This always holds if $g^*(w)$ is an odd function.

The estimation theory can now be easily developed starting with the formulas in (14.157) and (14.158). Similar to what was done in Section 14.7.3, we can estimate γ_j^* by,

$$\hat{\gamma}_j = \frac{n^{-1} \sum_{t=1}^n Y_t \hat{\ell}_j(U_t)}{n^{-1} \sum_{t=1}^n Y_t \hat{\ell}_1(U_t)}, \tag{14.160}$$

where

$$\hat{\ell}_j(\mathbf{u}) = \frac{(nh^{d+1})^{-1} \sum_{t=1}^n K_j\left(\frac{\mathbf{u}-\mathbf{U}_t}{h}\right)}{(nh^d)^{-1} \sum_{t=1}^n K\left(\frac{\mathbf{u}-\mathbf{U}_t}{h}\right)}.$$

The d-dimensional kernel $K_j(\mathbf{u})$ is the derivative of the kernel $K(\mathbf{u})$ with respect to the variable u_j; see (14.151) for further comments on such kernels.

Following the discussion in Section 14.7.1 (see (14.119)) and that leading to Theorem 14.10, we can easily arrive at the following rate of convergence theorem.

THEOREM 14.11 *Suppose that* $E\{g^{*(1)}(W_n)\} \neq 0$. *Let* $\mathcal{D} f_U(\mathbf{u})$ *and* $g^{*(1)}(w)$ *be s-times differentiable,* $s > 1$, *and let the kernel function* $K(\mathbf{u})$ *be of order s. If,*

$$h^* = an^{-\frac{2}{2s+d+2}},$$

then,

$$E(\hat{\gamma}_j - \gamma_j^*)^2 = O\left(n^{-\min\left(1, \frac{4s}{2s+d+2}\right)}\right), \tag{14.161}$$

$j = 1, 2, \ldots, d$.

The conclusions that can be drawn from this theorem are identical to those obtained in Theorem 14.10. Hence, we can estimate the parameter γ^*, which characterizes the nonlinear subsystem of the d-dimensional input of the semiparametric Hammerstein model, with a precision that depends on the smoothness of the nonlinearity $g^*(w)$ and input density $f_U(\mathbf{u})$. The dimensionality of the input has a critical impact on the accuracy of the estimate $\hat{\gamma}$. In fact, for $d > 2(s - 1)$, the rate $O(n^{-\frac{4s}{2s+d+2}})$ deteriorates with d. For symmetric kernels the rate is $O(n^{-\frac{8}{d+5}})$ and it becomes half, that is, $O(n^{-\frac{1}{2}})$, of the optimal parametric rate $O(n^{-1})$ for $d = 11$.

The remaining problem of estimation of the nonlinearity $g^*(w)$ can easily be tackled by our basic two-stage strategy which, in the context of the semiprametric Hammerstein system, has already been discussed in Section 14.5. The first step is to form the pilot kernel estimate of the regression function $g(w; \gamma) = E\{Y_n | \gamma^T \mathbf{U}_n = w\}$ (see (14.95)), that is, we define,

$$\hat{g}(w; \gamma) = \frac{\sum_{t=1}^{n} Y_t K \left(\frac{w - \gamma^T \mathbf{U}_t}{h} \right)}{\sum_{t=1}^{n} K \left(\frac{w - \gamma^T \mathbf{U}_t}{h} \right)}.$$

In the second step, we plug the estimate $\hat{\gamma}$ into $\hat{g}(w; \gamma)$, that is, we estimate $g^*(w)$ by $\hat{g}(w) = \hat{g}(w; \hat{\gamma})$. Using the arguments leading to (14.153) we can obtain the following theorem concerning the rate of convergence for $\hat{g}(w)$.

THEOREM 14.12 *Let all of the assumptions of Theorem 14.11 be satisfied. Then, for $d \leq 2s$ and if,*

$$h^* = an^{-\frac{1}{2s+1}}$$

we have

$$E(\hat{g}(w) - g^*(w))^2 = O\left(n^{-\frac{2s}{2s+1}} \right). \tag{14.162}$$

If in turn $d > 2s$ and if,

$$h^* = an^{-\frac{2}{2s+d+2}}$$

then we have

$$E(\hat{g}(w) - g^*(w))^2 = O\left(n^{-\frac{4s}{2s+d+2}} \right). \tag{14.163}$$

Once again we observe that the rate of convergence for estimating the univariate function $g^*(w)$ depends on the dimensionality of the input signal. In fact, it is known that for symmetric kernels, and $g^*(w)$ and $f_U(\mathbf{u})$ possessing three derivatives, the optimal rate for recovering $g^*(w)$ is $O(n^{-\frac{4}{5}})$. If, however, $d > 4$, the rate obtained in Theorem 14.12 is $O(n^{-\frac{8}{d+6}})$. This is clearly a much slower rate, and it is reduced to $O(n^{-\frac{2}{5}})$–half of the optimal rate $O(n^{-\frac{4}{5}})$ for $d = 14$.

14.7.5 The average derivative estimate for semiparametric multivariate parallel models

In this final section on the average derivative estimates, we briefly examine the parallel multivariate block-oriented system, which was examined in Chapter 13. Figure 14.17 depicts the system being governed by the following input–output formula:

$$Y_n = m^*(\mathbf{U}_n) + \sum_{j=0}^{p} \lambda_j^* V_{n-j} + Z_n, \tag{14.164}$$

where $\mathbf{U}_n$ is the d-dimensional input random signal and $\lambda_0^* = 1$. Using the usual notation $\mathbf{V}_n = (V_n, \ldots, V_{n-p})^T$ and $\lambda^* = (\lambda_0^*, \ldots, \lambda_p^*)^T$, let us note that the first two terms on the right-hand side of (14.164) can be written as $M(\mathbf{U}_n, \mathbf{V}_n) = m^*(\mathbf{U}_n) + \lambda^{*T} \mathbf{V}_n$, that is, $M(\mathbf{U}_n, \mathbf{V}_n)$ is the regression function of Y_n on $(\mathbf{U}_n, \mathbf{V}_n)$.

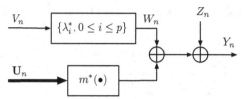

Figure 14.17 Semiparametric multiple-input parallel model.

Another important observation is that the average derivative of $M(\mathbf{U}_n, \mathbf{V}_n)$ with respect to $\mathbf{V}_n$ is equal to λ^*, that is,

$$E\left\{\mathcal{D}_{\mathbf{V}_n} M\left(\mathbf{U}_n, \mathbf{V}_n\right)\right\} = \lambda^*. \tag{14.165}$$

Recalling the result of Theorem 14.7 we can readily obtain our basic formula for λ^* resulting from (14.165),

$$\lambda^* = -E\left\{Y_n \frac{\mathcal{D}_{\mathbf{V}_n} f\left(\mathbf{U}_n, \mathbf{V}_n\right)}{f(\mathbf{U}_n, \mathbf{V}_n)}\right\}, \tag{14.166}$$

where $f(\mathbf{u}, \mathbf{v})$ is the joint density function of $(\mathbf{U}_n, \mathbf{V}_n) \in R^{d+p+1}$.

Let us now assume that the input process $\{(\mathbf{U}_n, V_n)\}$ is *iid* with the joint density function $f_{UV}(\mathbf{u}, v)$. Let also $f_V(v)$ be the marginal density of the input $\{V_n\}$. Then we can obtain the following decomposition for the density of $(\mathbf{U}_n, \mathbf{V}_n)$

$$f(\mathbf{U}_n, \mathbf{V}_n) = f_{U,V}(\mathbf{U}_n, V_n) \prod_{i=1}^{p} f_V(V_{n-i}).$$

Also for $j \geq 1$, we have

$$\frac{\partial}{\partial V_{n-j}} f(\mathbf{U}_n, \mathbf{V}_n) = f_{U,V}(\mathbf{U}_n, V_n) f_V^{(1)}(V_{n-j}) \prod_{\substack{i=1 \\ i \neq j}}^{p} f_V(V_{n-i}).$$

This and (14.166) yield,

$$\lambda_j^* = -E\left\{Y_n \ell\left(V_{n-j}\right)\right\}, \tag{14.167}$$

where $\ell(v) = \frac{f_V^{(1)}(v)}{f_V(v)}$. It is interesting to note that the solution for λ_j^* does not depend on the input signal $\{\mathbf{U}_n\}$ though $\{V_n\}$ and $\{\mathbf{U}_n\}$ are statistically dependent.

It can easily be calculated that if V_n is Gaussian $N(0, \tau^2)$, then we have $\ell(v) = -v/\tau^2$ and the formula in (14.167) reads as,

$$\lambda_j^* = \frac{E\{Y_n V_{n-j}\}}{\tau^2}. \tag{14.168}$$

Recall, this formula was obtained in Chapter 12 in the direct way without using the concept of the average derivative of the regression function.

The aforementioned developments readily yield corresponding estimates of λ_j^*. Replacing the expected value in (14.167) by its empirical counterpart and then plugging into (14.167) an estimate of $\ell(v)$, we can define

$$\hat{\lambda}_j = -\frac{1}{n} \sum_{t=j+1}^{n} Y_t \hat{\ell}(V_{t-j}), \qquad (14.169)$$

where $\hat{\ell}(v)$ is a nonparametric estimate of $\ell(v)$; see Section 14.7.2 for various estimates of $\ell(v)$. The asymptotic theory of the estimate $\hat{\lambda}_j$ in (14.169) can be established by following the discussion leading to the result in Theorem 14.9. The estimate can reach the parametric $O(n^{-1})$ rate subject to the conditions similar to those in Theorem 14.9. The main difference is that we do not need the differentiability of $m^*(\mathbf{u})$.

Regarding the nonlinear part of the system this can be directly obtained via the following identity (see Chapter 13):

$$m^*(\mathbf{u}) = E\{Y_n|\mathbf{U}_n = \mathbf{u}\} - E\{V_n|\mathbf{U}_n = \mathbf{u}\}. \qquad (14.170)$$

Hence, two regression functions are to be estimated in order to recover $m^*(\mathbf{u})$. This leads to a nonparametric estimate of $m^*(\mathbf{u})$. Note that (14.170) holds independently of λ^* and estimating the projection $m(\mathbf{u}; \lambda) = E\{Y_n - \lambda^T \mathbf{V}_n|\mathbf{U}_n = \mathbf{u}\}$, see (14.100), is not required in this case. This decoupling property, between the problem of estimating linear and nonlinear parts, holds if the input process $\{(\mathbf{U}_n, V_n)\}$ is a sequence of independent random vectors. If this is not the case, our two-stage strategy may be found to be useful.

14.8 Concluding remarks

In this chapter, we have examined semiparametric block-oriented models, which are intermediate cases between fully parametric and fully nonparametric models. The semiparametric model is usually obtained by restricting a class of linear dynamic subsystems and nonlinear characteristics to the representation $(\theta^*, \mathbf{g}^*(\bullet))$, where θ^* is a finite dimensional parameter and $\mathbf{g}^*(\bullet)$ are nonparametric functions, normally being a set of univariate nonlinearities.

Throughout this chapter, we have confined linear dynamics to finite impulse response filters, that is, we have dealt with block-oriented systems with finite memory. The general state-space parameterization of an infinite memory linear system,

$$\begin{cases} \mathbf{X}_{n+1} = A_{\theta^*}\mathbf{X}_n + \mathbf{b}_{\theta^*}\mathbf{U}_n \\ W_n = \mathbf{c}_{\theta^*}^T\mathbf{X}_n, \end{cases}$$

would also be possible to use in our considerations. In fact, there is not any difficulty in implementing a version of our two-stage least squares based algorithm for estimating θ^* and system nonlinearities. The asymptotic theory of such estimates would require extensions of our technical arguments to the case of general dependent processes with fast decaying correlation functions. This seems to be feasible, because there exists generalization of the least squares method to dependent data [192]. In addition, nonparametric estimates have been proven to be consistent for mixing, Markov, and long-range dependent stochastic processes [26], [89], and [139].

The least-squares method commonly used in our approach to identification of semi-parametric models can also be formally applied in the case of dependent input signals. Again, the asymptotic theory of the resulting estimates would need to be augmented. It is worth noting that the direct identification method based on the average derivative estimation has explicitly exploited the independence of the input signal. Nevertheless, the average derivative method has been found to be very useful since it avoids any optimization procedures and its accuracy depends mostly on the smoothness of the input density. In the case of multivariate systems, the average derivative estimates suffer somewhat from the curse of dimensionality. The dependence of the accuracy of the estimators on the dimensionality of the input signal is seen at a certain critical dimension, the value of which is determined by the smoothness of the input density and the smoothness of system nonlinearities.

Let us finally comment that the semiparametric method can be applied to a larger class of block-oriented systems than those examined in this chapter. One such challenging case would be the sandwich system, introduced in Chapter 12.

14.9   Auxiliary results, lemmas, and proofs

Before proving Theorems 14.3 and 14.5 we need some auxiliary results obtained from various sources in the literature. They can be found useful in various problems involving nonparametric and parametric components.

14.9.1   Auxiliary Results

LEMMA 14.1 *(Scheffes theorem). Let $\{f_n(x)\}$ be a sequence of probability density functions defined on the set $A \subseteq R^d$. Suppose that $f(x)$ is a density function on A, such that,*

$$f_n(x) \rightarrow f(x), \quad almost\ all\ x \in A\ as\ n \rightarrow \infty.$$

Then,

$$\int_A |f_n(x) - f(x)| dx \rightarrow 0 \quad as\ n \rightarrow \infty.$$

This is a classical result due to Scheffe [272], see also Devroye and Lugosi [70].

The lemma below involves the concept of lower semicontinuous functions. We say that the function $f : R \rightarrow R$ is lower semicontinuous if

$$\liminf_{z \rightarrow x} f(z) \geq f(x), \quad x \in R.$$

Note that the class of lower semicontinuous functions may include some discontinuous functions.

LEMMA 14.2 *Consider for any bounded function $h : R \to R$ the following mapping:*

$$g(x) = \int_{-\infty}^{\infty} h(y)f(y|x)dy,$$

where $\{f(y|x) : x \in R\}$ is a class of conditional density functions parametrized by $x \in R$. Then, $g(x)$ is continuous on R if and only if the function,

$$x \mapsto \int_{O} f(y|x)dy,$$

is lower semicontinuous for every measurable set O.

This lemma is a special case of the general result in [209] (Proposition 6.1.1) where the conditional density $f(y|x)$ can be replaced by a general conditional probability measure. It is worth noting that the continuity of $g(x)$ is implied by the continuity of $x \mapsto \int_{O} f(y|x)dy$.

In our studies of the convergence rate of parametric estimates of semiparametric models, we need the following lemma concerning the existence of derivatives of integrals depending on a vector of parameters. Hence, let $g : R \times \Theta \to R$ be a measurable function defined on $R \times \Theta$, where Θ is a compact subset of R^d.

LEMMA 14.3 *Let $g(\bullet, \bullet)$ be a function defined on $R \times \Theta$ that satisfies the following conditions:*

(a1) For each fixed $\theta \in \Theta$, the mapping $z \mapsto g(z, \theta)$ is measurable and $\int_{-\infty}^{\infty} |g(z, \theta_0)|dz < \infty$ for some $\theta_0 \in \Theta$.

(a2) The partial derivative $\frac{\partial g(z, \theta)}{\partial \theta}$ exists for every $(z, \theta) \in R \times \Theta$.

(a3) There is a nonnegative function $w(z)$ with $\int_{-\infty}^{\infty} w(z)dz < \infty$ and such that,

$$\left\| \frac{\partial g(z, \theta)}{\partial \theta} \right\| \le w(z) \quad \text{for all } (z, \theta) \in R \times \Theta.$$

Then, the function,

$$G(\theta) = \int_{-\infty}^{\infty} g(z, \theta)dz$$

is differentiable at every point $\theta \in \Theta$.

It is worth noting that this result holds also for higher derivatives with appropriate changes. The proof of this lemma can be found in [7].

The next lemma gives a general result regarding the estimation of the derivatives of a function with the derivatives of an estimate. Let us denote $\|f\|_q = \{\int_a^b |f(x)|^q dx\}^{1/q}$, $1 \le q \le \infty$, $-\infty \le a < b \le \infty$.

LEMMA 14.4 *(Yatracos theorem). Let $m(x)$ be a real valued function defined on an interval $[a, b]$. Let $\hat{m}_n(x)$ be a consistent estimate of $m(x)$ determined from a training set of size n. Let both $m(x)$ and $\hat{m}_n(x)$ have p continuous derivatives on $[a, b]$. Then for $1 \le q \le \infty$ and $1 \le s \le p$ we have,*

$$\|\hat{m}_n^{(s)} - m^{(s)}\|_q \le c_1 \beta_n^{p-s+\varepsilon} + c_2 \beta_n^{-s} \|\hat{m}_n - m\|_q,$$

where c_1, c_2 are constants, $\lim_{n\to\infty} \beta_n = 0$ and $\varepsilon > 0$ is an arbitrary small number.

The result given in Lemma 14.4 is a version of a slightly more general statement proved by Yatracos [330]. An important consequence of this result is that if one chooses,

$$\beta_n = c_2 \left\{ \|\hat{m}_n - m\|_q \right\}^{\frac{1}{p+\varepsilon}},$$

then we have,

$$\|\hat{m}_n^{(s)} - m^{(s)}\|_q = O\left(\left\{ \|\hat{m}_n - m\|_q \right\}^{\frac{p-s+\varepsilon}{p+\varepsilon}} \right).$$

This fact allows one to determine the convergence and rate of convergence of the derivatives of the estimate $\hat{m}_n(x)$ from the corresponding results of the estimate $\hat{m}_n(x)$. Indeed, note that this result holds for $s = p$.

14.9.2 Lemmas

In this section, we give some specific results related directly to problems examined in this chapter. The following lemma concerns the smoothness of the regression function $m(w; \lambda)$ defined in (14.42). Note that since $W_n(\lambda) = \lambda^T U_n$, $U_n = (U_n, \dots, U_{n-p})^T$, $m(w; \lambda)$ is a function of a single variable $w = \lambda^T u$.

LEMMA 14.5 *Let Assumptions A1 and A2 hold. Then $m(w; \lambda)$ is a continuous function in w.*

Proof. Let $f(w; \lambda)$ be the probability density function of $W_n(\lambda)$, whereas $f(z|w; \lambda)$ be the conditional probability density function of W_n on $W_n(\lambda)$. Due to Lemma 14.2 and Assumption A2, we may conclude that

$$m(w; \lambda) = \int_{-\infty}^{\infty} m^*(z) f(z|w; \lambda) dz$$

is continuous if

$$w \mapsto \int_O f(z|w; \lambda) dz \tag{14.171}$$

is lower continuous for every measurable set O. The continuity of the mapping in (14.171) results from Assumption A1 and Lemma 14.1. ∎

The next result has already been proved in Section 12.2.6, see Lemma 12.5.

LEMMA 14.6 *Let $\xi_1, \xi_2, \dots, \xi_n$ a sequence of random variables such that $E\{\xi_i\} = 0$ and $E\xi_i^2 < \infty$ for all i. Assume that for a nonnegative integer p the random variables ξ_i and ξ_j are independent whenever $|i - j| > p$. Then for $n > p$*

$$E\left(\sum_{j=1}^{n} \xi_j \right)^2 \leq (p+1) \sum_{j=1}^{n} E\xi_j^2. \tag{14.172}$$

We also need the following elementary lemma.

LEMMA 14.7 *Let X and η be independent random variables with density functions $f_X(\bullet)$ and $f_\eta(\bullet)$, respectively. Then,*

$$E\{X|X + \eta = v\} = \frac{\int_{-\infty}^{\infty} x f_\eta(v - x) f_X(x) dx}{\int_{-\infty}^{\infty} f_\eta(v - x) f_X(x) dx}.$$

14.9.3 Proofs

Proof of Theorem 14.3: Owing to the general result of Theorem 14.1, we first need to verify that the limit criterion $Q(\lambda)$ is a continuous function. Assumptions **A1** and **A2** and Lemma 14.5 yield the continuity of $m(w; \lambda)$. This readily implies the continuity of $Q(\lambda)$.

As we have already pointed out (see Theorem 14.1), the consistency $\hat{\lambda} \xrightarrow{n} \lambda^*(P)$ results from the following uniform convergence:

$$\sup_{\lambda \in \Lambda} |\hat{Q}_n(\lambda) - Q(\lambda)| \xrightarrow{n} 0, (P).$$

This, in turn, is implied by,

$$\sup_{\lambda \in \Lambda} |\hat{Q}_n(\lambda) - \bar{Q}(\lambda)| \xrightarrow{n} 0, (P),$$ (14.173)

and

$$\sup_{\lambda \in \Lambda} |\bar{Q}(\lambda) - Q(\lambda)| \xrightarrow{n} 0, (P),$$ (14.174)

where $\bar{Q}(\lambda)$ is an average of $\hat{Q}_n(\lambda)$; see (14.62).

The uniform convergence in (14.173) can be analyzed by the general theory of empirical processes and this entails verification of the Glivienko–Cantelli condition for the following class of functions:

$$\{\Psi(\mathbf{U}, Y; \lambda) : \lambda \in \Lambda\},$$ (14.175)

where $\Psi(\mathbf{U}, Y; \lambda) = (Y - \hat{m}(W(\lambda); \lambda))^2$, $W(\lambda) = \lambda^T \mathbf{U}$. The sufficient conditions for this requirement are given in (14.31). First of all, we need to verify that $\lambda \mapsto \Psi(\mathbf{U}, Y; \lambda)$ is continuous. This is guaranteed by the continuity of the kernel function (Assumption **A4**) provided that the denominator in formula (14.47) is not zero, which can easily be achieved by using some truncation arguments in (14.47), for example, by adding the factor $1/n$ to the denominator. Furthermore, due to Assumption **A4** and Lemma 14.6, we can show that

$$P\left\{\sum_{j \in I_1} K\left(\frac{w - W_j(\lambda)}{h}\right) = 0\right\} \le e^{\frac{-c(w; \lambda) n_1}{p + 1}},$$

where $c(w; \lambda) = \int_{w-rh}^{w+rh} f(v; \lambda) dv = 2rh f(w; \lambda) + o(1)$ as $h \to 0$. Hence, the probability that the denominator in (14.47) is equal to zero decays exponentially fast.

To check the integrability condition of (14.31), let us observe that for every $(\mathbf{U}, Y)$ from T_2 we have,

$$\Psi(\mathbf{U}, Y; \lambda) \leq 2Z^2 + 2(m^*(W) - \hat{m}(W(\lambda); \lambda))^2$$
$$\leq 2Z^2 + 4m^{*2}(W) + 4\hat{m}^2(W(\lambda); \lambda). \qquad (14.176)$$

Recalling the definition of $\hat{m}(w; \lambda)$ (see (14.47)), we readily obtain, by Jensen's inequality, that the last term on the right-hand side of (14.176) is bounded by,

$$4 \sum_{j \in I_1} Y_j^2 K\left(\frac{W(\lambda) - W_j(\lambda)}{h}\right) / \sum_{j \in I_1} K\left(\frac{W(\lambda) - W_j(\lambda)}{h}\right). \qquad (14.177)$$

Since the kernel weights are smaller than one, we can bound (14.177) by $4 \sum_{j \in I_1} Y_j^2$. Noting next that $Y_j = m^*(W_j) + Z_j$ and by virtue of (14.176), we can conclude that $\Psi(\mathbf{U}, Y; \lambda)$ has the following uniform bound:

$$\Psi(\mathbf{U}, Y; \lambda) \leq 2Z^2 + 4m^{*2}(W) + 8 \sum_{j \in I_1} m^{*2}(W_j) + 8 \sum_{j \in I_1} Z_j^2.$$

Since $E Z_j^2 < \infty$ and $E|m^*(W_j)|^2 < \infty$, the integrability condition is thus met.

Concerning the convergence in (14.174) let us observe that by virtue of (14.62), we wish to verify whether

$$E\left\{|\hat{m}(W(\lambda); \lambda) - m(W(\lambda); \lambda)|^2 | T_1\right\} \xrightarrow{n} 0, (P)$$

converges uniformly over $\lambda \in \Lambda$. This convergence can be deduced from the aforementioned results, that is, the continuity of the mappings $\lambda \mapsto \hat{m}(W(\lambda); \lambda)$, $\lambda \mapsto m(W(\lambda); \lambda)$ and the fact that $E|\hat{m}(W(\lambda); \lambda)|^2 \leq c$ for some constant c depending on $E Z^2$ and $E|m^*(W)|^2$.

The proof of Theorem 14.3 has been completed. ∎

Proof of Theorem 14.5: Owing to the discussion in Section 14.4.2 it suffices to prove the validity of (14.69) and (14.73). Let us begin by proving (14.69). First we write $\hat{m}(w; \lambda^*)$ as $\hat{m}(w; \lambda^*) = \hat{r}(w; \lambda^*)/\hat{f}(w; \lambda^*)$, see (14.70). It is clear that we need only to examine the term:

$$\hat{r}(w; \lambda^*) = (n_1^{-1} h^{-1}) \sum_{j \in I_1} Y_j K\left(\frac{w - W_j}{h}\right).$$

Let us first observe that

$$E\hat{r}(w; \lambda^*) = h^{-1} E\left\{m^*(W_n) K\left(\frac{w - W_n}{h}\right)\right\}.$$

Then due to Lemma A.1 we obtain that

$$\lim_{h \to 0} E\hat{r}(w; \lambda^*) = m^*(w) f_W(w),$$

at every point w where both $m^*(w)$ and $f_W(w)$ are continuous. The continuity is guaranteed by Assumptions **A1** and **A2**. It remains now to evaluate $\mathrm{var}\{\hat{r}(w; \lambda^*)\}$. By virtue of Lemma 14.6 we get,

$$\text{var}\{\hat{r}(w; \lambda^*)\} \leq (p+1)(n_1 h^2)^{-1} \text{var}\left\{Y_n K\left(\frac{w - W_n}{h}\right)\right\}.$$

Then by the fact that $E Y_n^2 < \infty$, we obtain

$$\text{var}\{\hat{r}(w; \lambda^*)\} = O\left((n_1 h)^{-1}\right)$$

at every point w where both $m^*(w)$ and $f_W(w)$ are continuous. This proves the claim in (14.69).

In order to prove (14.73), we first note that

$$E \bar{J}_2(w) = h^{-1} E\left\{m^*(W_n)U_{n-t} K\left(\frac{w - W_n}{h}\right)\right\}$$

$$= h^{-1} E\left\{m^*(W_n)a_0(W_n) K\left(\frac{w - W_n}{h}\right)\right\},$$

where

$$a_0(w) = E\{U_{n-t}|W_n = w\}. \tag{14.178}$$

Since $E|U_n| < \infty$ and $E|Y_n| < \infty$, we obtain

$$\lim_{h \to 0} E \bar{J}_2(w) = m^*(w)a_0(w)f_W(w) \tag{14.179}$$

at every point w where $m^*(w)$, $f_W(w)$, and $a_0(w)$ are continuous. The continuity is implied by Assumptions **A1** and **A2**. To see that, let us write $a_0(w)$ as follows:

$$a_0(w) = \frac{1}{\lambda_t^*} E\{\bar{U}_n|\bar{U}_n + \eta_n = w\},$$

where $\bar{U}_n = \lambda_t^* U_{n-t}$ and $\eta_n = U_n + \sum_{\substack{s=1 \\ s \neq t}}^{p} \lambda_s^* U_{n-s}$. Since $\bar{U}_n$ and η_n are independent we can apply Lemma 14.7, that is, we obtain after some simple algebraic manipulations that

$$a_0(w) = \frac{\int_{-\infty}^{\infty} x f_{\eta_n}(w - \lambda_t^* x) f_U(x) dx}{f_W(w)},$$

where $f_{\eta_n}(w)$ is a density function of the random variable η_n. Due to Assumption **A1** we see that $a_0(w)$ is a continuous function at every $w \in \mathbf{R}$ where $f_W(w) > 0$. To complete our proof it remains to consider $\text{var}\{\bar{J}_2(w)\}$. Since $Y_j = m^*(W_j) + Z_j$ it suffices to consider the variance of the following term:

$$B(w) = n_1^{-1} h^{-1} \sum_{j \in I_1} K\left(\frac{w - W_j}{h}\right) m^*(W_j)U_{j-t}. \tag{14.180}$$

By virtue of Lemma 14.6, we have

$$\text{var}\{B(w)\} \leq (p+1)n_1^{-1}h^{-2} \text{ var} \left\{ K\left(\frac{w-W_n}{h}\right) m^*(W_n)U_{n-t} \right\}.$$

Then, reasoning as in the first part of the proof we can easily show that

$$\text{var}\{B(w)\} = O\left((n_1 h)^{-1}\right),$$

at every point w where both $m^*(w)$ and $f_W(w)$ are continuous.

This completes the proof of Theorem 14.5. ∎

14.10   Bibliographical notes

Semiparametric models have been extensively examined in the econometric literature, see, for example, Härdle, Müller, Sperlich, and Werwatz [152], Ruppert, Wand, and Carroll [259], and Yatchev [329]. There, they have been introduced as more flexible extension of the standard linear regression model and popular models include partial linear and multiple-index models. These are static models and this chapter can be viewed as the generalization of these types of models to dynamic nonlinear block-oriented systems. In fact, the partially linear models fall into the category of parallel models, whereas multiple-index models correspond to Hammerstein–Wiener connections. Semiparametric models have recently been introduced in the nonlinear time series literature, see Fan and Yao [89], and Gao and Tong [93]. Some empirical results on the identification of the semiparametric partially linear model have been reported by Espinozo, Suyken, and De Moor [83]; see also Bruls, Chou, Haverkamp, and Verhaegen [34] for the fully parametric version of such models. Comprehensive studies of semiparametric Hammerstein-Wiener models have been given in Hasiewicz and Mzyk [157], Pawlak [230], and Pawlak, Hasiewicz, and Wachel [236].

The basics of M-estimators and their extension to the semiparametric inference are described in van der Vaart [295]. The theory of the uniform convergence of empirical processes and its importance in statistical learning and nonparametric inference can be found in Devroye, Györfi, and Lugosi [69], Györfi, Kohler, Krzyżak, and Walk [140], and Vapnik [299]. Various resampling schemes and their applications are examined in Devroye [67], Devroye and Lugosi [70], and Zoubir and Iskander [334].

The generalization of the parametric nonlinear least squares method, see Jennrich [171], Lai [192], to semiparametric models is examined in the seminal paper of Ichimura [169]. In Powell, Stock, and Stoker [240] the average derivative approach for estimating semiparametric static models was introduced; see also Härdle and Stoker [153] and Hristache, Juditsky, and Spokoiny [165] for further advances of this method. The problem of nonparametric estimation of derivatives is discussed in Mack and Müller [202], Schuster and Yakowitz [274], and Yatracos [330].

The use of the average derivative method for nonlinear system identification is studied in Pawlak [230].

The approximation property of multichannel Wiener models as a faithful representation for fading memory nonlinear systems has been elaborated on in Boyd and Chua [29], Sanberg [268], and Sandilya and Kulkarni [269]. The importance of multichannel and additive Wiener models in modeling of biological, physiological, and mechanical systems has been pointed out in Bendat [16], Marmarelis and Marmarelis [207], Verhaegen and Westwick [301], Westwick and Verhaegen [315], and Westwick and Kearney [316].

Appendix A Convolution and kernel functions

A.1   Introduction

In this appendix, both $\varphi(\bullet)$ and $K(\bullet)$ are Borel measurable functions defined over the whole real line R. Our first purpose is to examine the following convolution:

$$\varphi_h(x) = \frac{1}{h} \int \varphi(\xi) K \left(\frac{x - \xi}{h} \right) d\xi.$$

The function $K(\bullet)$ is called a kernel. We are interested in kernels for which

$$\lim_{h \to 0} \varphi_h(x) = \varphi(x) \int K(\xi) d\xi. \qquad (A.1)$$

One may expect that also the integrated error vanishes, that is, that

$$\lim_{h \to 0} \int (\varphi_h(x) - \varphi(x))^2 dx = 0, \qquad (A.2)$$

provided that $\int K(\xi) d\xi = 1$.

Since

$$\varphi_h(x) = \frac{1}{h} \int \varphi(x - \xi) K \left(\frac{\xi}{h} \right) d\xi,$$

and

$$\varphi(x) \int K(\xi) d\xi = \frac{1}{h} \int \varphi(x) K \left(\frac{\xi}{h} \right) d\xi,$$

we can write

$$\varphi_h(x) - \varphi(x) \int K(\xi) d\xi = \frac{1}{h} \int (\varphi(x - \xi) - \varphi(x)) K \left(\frac{\xi}{h} \right) d\xi. \qquad (A.3)$$

Therefore, to prove (A.1), it suffices to show that the quantity in (A.3) converges to zero as $h \to 0$.

A.2   Convergence

A.2.1   Pointwise convergence

LEMMA A.1 *Let $\int |\varphi(x)| dx < \infty$. Let the kernel $K(\bullet)$ satisfy the following restrictions:*

$$\sup_x |K(x)| < \infty, \tag{A.4}$$

$$\int |K(x)| dx < \infty, \tag{A.5}$$

$$|x| K(x) \to 0 \text{ as } |x| \to \infty. \tag{A.6}$$

Then (A.1) holds at every point x of continuity of $\varphi(\bullet)$.

Proof. With no loss of generality, let $h > 0$. Let $\varphi(\bullet)$ be continuous at x. Thus, for any $\eta > 0$, there exists $\delta > 0$, such that $|\varphi(x - \xi) - \varphi(x)| < \eta$ if $|\xi| < \delta$. Since the absolute value of the quantity on the right-hand side in (A.3) is bounded by

$$\frac{1}{h} \int_{|\xi| \leq \delta} |\varphi(x - \xi) - \varphi(x)| \left| K\left(\frac{\xi}{h}\right) \right| d\xi + \frac{1}{h} \int_{|\xi| > \delta} |\varphi(x - \xi) - \varphi(x)| \left| K\left(\frac{\xi}{h}\right) \right| d\xi, \tag{A.7}$$

we thus find it is not greater than

$$\eta \int |K(\xi)| d\xi + \frac{1}{h} \int_{|\xi| > \delta} |\varphi(x - \xi)| \left| K\left(\frac{\xi}{h}\right) \right| d\xi + |\varphi(x)| \frac{1}{h} \int_{|\xi| > \delta} \left| K\left(\frac{\xi}{h}\right) \right| d\xi.$$

Since η can be arbitrarily small, so can the first term. The second one is not greater than

$$\frac{1}{h} \sup_{|\xi| > \delta} \left| K\left(\frac{\xi}{h}\right) \right| \int_{|\xi| > \delta} |\varphi(x - \xi)| d\xi \leq \frac{1}{\delta} \left[\frac{\delta}{h} \sup_{|\xi| > \delta/h} |K(\xi)| \right] \int |\varphi(\xi)| d\xi$$

$$\leq \frac{1}{\delta} \left[\sup_{|\xi| > \delta/h} |\xi K(\xi)| \right] \int |\varphi(\xi)| d\xi$$

and, due to (A.6), converges to zero as $h \to 0$. The last term is bounded by $|\varphi(x)| \int_{|\xi| > \delta/h} |K(\xi)| d\xi$ and, by (A.5), also converges to zero as $h \to 0$. The lemma follows. ∎

For a bounded $\varphi(\bullet)$, the class of possible kernels can be enlarged by dropping (A.6), which is shown in the following lemma:

LEMMA A.2 *Let $\sup_x |\varphi(x)| < \infty$. Let the kernel $K(\bullet)$ satisfy (A.4) and (A.5). Then (A.1) takes place at every point x of continuity of $\varphi(\bullet)$.*

Proof. The quantity in (A.7) is not greater than

$$\eta \frac{1}{h} \int_{|\xi| \leq \delta} \left| K\left(\frac{\xi}{h}\right) \right| d\xi + 2 \sup_{\xi} |\varphi(\xi)| \frac{1}{h} \int_{|\xi| > \delta} \left| K\left(\frac{\xi}{h}\right) \right| d\xi$$

$$\leq \eta \int |K(\xi)| d\xi + 2 \sup_{\xi} |\varphi(\xi)| \int_{|\xi| \geq \delta/h} |K(\xi)| d\xi.$$

Since η can be arbitrarily small and, due to (A.5), $\int_{|\xi| \geq \delta/h} |K(\xi)| d\xi \to 0$ as $h \to 0$, the proof has been completed. ∎

The next lemma says that the examined convergence can take place not only at continuity points but also at every Lebesgue point of $\varphi(\bullet)$, that is, at almost every (with respect to the Lebesgue measure) point x. We recall that x is said to be a Lebesgue point of $\varphi(\bullet)$ if

$$\lim_{h \to 0} \frac{1}{2h} \int_{-h}^{h} |\varphi(x - \xi) - \varphi(x)| d\xi = 0. \tag{A.8}$$

If $\varphi(\bullet)$ is locally integrable, almost every point is a Lebesgue point of $\varphi(\bullet)$. Every continuity point of $\varphi(\bullet)$ is also its Lebesgue point.

LEMMA A.3 *Let $\int |\varphi(x)| dx < \infty$. Let the kernel $K(\bullet)$ satisfy (A.4) and let*

$$|x|^{1+\varepsilon} K(x) \to 0 \text{ as } |x| \to \infty \tag{A.9}$$

for some $\varepsilon > 0$. Then (A.1) takes place at every Lebesgue point of $\varphi(\bullet)$, and, a fortiori, at almost every (with respect to the Lebesgue measure) point x and at every continuity point of $\varphi(\bullet)$, as well.

Proof. For simplicity, $h > 0$. From (A.4) and (A.9), it follows that

$$|K(x)| \leq \frac{\kappa}{(1 + |x|)^{1+\varepsilon}}$$

for some κ. Let x be a Lebesgue point of $\varphi(\bullet)$, that is, a point at which (A.8) holds. With no loss of generality we assume that $x = 0$ and $\varphi(0) = 0$, that is, that

$$\lim_{h \to 0} \frac{1}{2h} \int_{-h}^{h} |\varphi(\xi)| d\xi = 0.$$

Denote $F(\xi) = \int_0^\xi |\varphi(\zeta)| d\zeta$ and observe that, since $x = 0$ is a Lebesgue point of $\varphi(\bullet)$,

$$\lim_{\xi \to 0} \frac{F(\xi) - F(-\xi)}{2\xi} = 0.$$

It means that, for any $\eta > 0$, there exists $\delta > 0$ such that $|F(\xi) - F(-\xi)| \leq 2\xi\eta$ if $|\xi| \leq \delta$.

Since $x = 0$ and $\varphi(0) = 0$, the quantity in (A.3), that is,

$$\frac{1}{h} \int \varphi(\xi) K\left(\frac{\xi}{h}\right) d\xi,$$

is bounded in absolute value by

$$\kappa \int |\varphi(\xi)| \frac{h^\varepsilon}{(h + |\xi|)^{1+\varepsilon}} d\xi$$

$$\leq \int_{-\delta}^{\delta} |\varphi(\xi)| \frac{h^\varepsilon}{(h + |\xi|)^{1+\varepsilon}} d\xi + \int_{|\xi| > \delta} |\varphi(\xi)| \frac{h^\varepsilon}{(h + |\xi|)^{1+\varepsilon}} d\xi. \tag{A.10}$$

Integrating the first term by parts, we find it equal to

$$\frac{h^\varepsilon}{(h+\delta)^{1+\varepsilon}}\left[F(\delta)-F(-\delta)\right]+(1+\varepsilon)\int_{-\delta}^{\delta}F(\xi)\frac{\xi}{|\xi|}\frac{h^\varepsilon}{(h+|\xi|)^{2+\varepsilon}}d\xi.$$

The first term in the obtained expression converges to zero as $h \to 0$, while the other is bounded by

$$2\eta\,(1+\varepsilon)\int_0^\delta \xi\frac{h^\varepsilon}{(h+\xi)^{2+\varepsilon}}d\xi = 2\eta\,(1+\varepsilon)\int_0^{\delta/h}\frac{\xi}{(1+\xi)^{2+\varepsilon}}d\xi \le 2\eta\,(1+\varepsilon)c,$$

where

$$c=\int_0^\infty\frac{\xi}{(1+\xi)^{2+\varepsilon}}d\xi < \infty,$$

and can be made arbitrarily small by selecting η small enough. The other integral in (A.10) is bounded by

$$\frac{h^\varepsilon}{\delta^{1+\varepsilon}}\int|\varphi(\xi)|d\xi$$

and converges to zero as $h \to 0$. Thus the lemma follows. ∎

In Lemmas A.1 and A.2, convergence at the continuity points of $\varphi(\bullet)$ is shown while Lemma A.3 shows convergence at almost every point. To shed some light on the relation between the sets where the lemmas hold we refer to the following theorem, see Wheeden and Zygmund [318, Theorem (5.54)].

THEOREM A.1 *A bounded function $\varphi(\bullet)$ is almost everywhere continuous in a finite interval (a, b) if and only if it is Riemann integrable on the interval.*

In particular, from the theorem, it follows that a Lebesgue integrable function may not be continuous almost everywhere.

Denoting the Lebesgue measure by λ, we can write

$$\varphi_h(x)=\frac{\int\varphi(\xi)K\left(\frac{x-\xi}{h}\right)d\xi}{\int K\left(\frac{x-\xi}{h}\right)d\xi}=\frac{\int\varphi(\xi)K\left(\frac{x-\xi}{h}\right)\lambda(d\xi)}{\int K\left(\frac{x-\xi}{h}\right)\lambda(d\xi)}.$$

The next lemma, which can be found in Greblicki, Krzyżak, and Pawlak [122], shows that $\varphi_h(x)$ converges to $\varphi(x)$ even if the Lebesgue measure is replaced by any general probability measure. This proves useful when examining distribution-free properties of our estimates. The probability measure μ is arbitrary, that is, it may or may not have a density.

LEMMA A.4 *Let $H(\bullet)$ be a nonnegative nonincreasing Borel function defined on $[0, \infty)$, continuous and positive at $t = 0$ and such that*

$$tH(t) \to 0 \text{ as } t \to \infty. \tag{A.11}$$

Let, for some c_1 and c_2,

$$c_1 H(|u|) \leq K(u) \leq c_2 H(|u|).$$

Let μ be any probability measure. Then,

$$\lim_{h \to 0} \frac{\int \varphi(\xi) K\left(\frac{x - \xi}{h}\right) \mu(d\xi)}{\int K\left(\frac{x - \xi}{h}\right) \mu(d\xi)} = \varphi(x) \qquad (A.12)$$

for almost every (μ) $x \in R$.

Proof. It will be convenient to denote $S_x(h) = \{\xi : |x - \xi| \leq h\}$. Thus, in particular, $\lambda(S_x(h)) = 2h$.

For the sake of simplicity let $h > 0$. Clearly,

$$\left| \frac{\int \varphi(\xi) K\left(\frac{x - \xi}{h}\right) \mu(d\xi)}{\int K\left(\frac{x - \xi}{h}\right) \mu(d\xi)} - \varphi(x) \right| \leq \frac{c_2}{c_1} \frac{\int H\left(\frac{|x - \xi|}{h}\right) |\varphi(x) - \varphi(\xi)| \mu(d\xi)}{\int H\left(\frac{|x - \xi|}{h}\right) \mu(d\xi)}.$$

$$(A.13)$$

Since $H(t) = \int_0^\infty I_{\{H(t) > s\}}(s)ds$, we can write

$$\int H\left(\frac{|x - \xi|}{h}\right) \mu(d\xi) = \int_0^\infty \mu(A_{t,h})dt$$

and

$$\int H\left(\frac{|x - \xi|}{h}\right) |\varphi(x) - \varphi(\xi)| \mu(d\xi) = \int_0^\infty \left[\int_{A_{t,h}} |\varphi(x) - \varphi(\xi)| \mu(d\xi) \right] dt,$$

where

$$A_{t,h} = \left\{ \xi : H\left(\frac{|x - \xi|}{h}\right) > t \right\}$$

is an interval centered at x. Because of this, we can rewrite the second quotient on the right-hand side in (A.13) in the following form:

$$\frac{\int_0^\infty \left[\int_{A_{t,h}} |\varphi(x) - \varphi(\xi)| \mu(d\xi) \right] dt}{\int_0^\infty \mu(A_{t,h})dt} = V_1(x) + V_2(x),$$

where

$$V_1(x) = \frac{\int_0^\delta \left[\int_{A_{t,h}} |\varphi(x) - \varphi(\xi)| \mu(d\xi) \right] dt}{\int_0^\infty \mu(A_{t,h})dt}$$

and

$$V_2(x) = \frac{\int_\delta^\infty \left[\int_{A_{t,h}} |\varphi(x) - \varphi(\xi)| \mu(d\xi) \right] dt}{\int_0^\infty \mu(A_{t,h}) dt}$$

for any $\delta > 0$. In the remaining part of the proof $\delta = \varepsilon h$ with $\varepsilon > 0$.

Observe that

$$\int_0^\delta \left[\int_{A_{t,h}} |\varphi(x) - \varphi(\xi)| \mu(d\xi) \right] dt \leq (c_3 + |\varphi(x)|)\delta$$

with $c_3 = \int |\varphi(\xi)| \mu(d\xi)$. Since $H(\bullet)$ is positive and continuous at the origin, there exist $r > 0$ and $c > 0$, such that

$$H(|x|) \geq \begin{cases} c, & \text{for } |x| \leq r \\ 0, & \text{otherwise,} \end{cases}$$

which implies

$$\int_0^\infty \mu(A_{t,h}) dt = \int H\left(\frac{|x - \xi|}{h} \right) \mu(d\xi) \geq c\mu \left(S_x (rh) \right).$$

Thus,

$$V_1(x) \leq \delta \frac{c_3 + |\varphi(x)|}{\lambda \left(S_x (rh) \right)} \frac{\lambda \left(S_x (rh) \right)}{\mu \left(S_x (rh) \right)} = \frac{\delta}{h} \frac{c_3 + |\varphi(x)|}{2r} \frac{\lambda \left(S_x (rh) \right)}{\mu \left(S_x (rh) \right)}.$$

Finally, recalling the definition of δ, we find the quantity bounded by

$$\varepsilon \frac{c_3 + |\varphi(x)|}{2r} \frac{\lambda \left(S_x (rh) \right)}{\mu \left(S_x (rh) \right)}.$$

Applying Lemma A.13 in Section A.4 we conclude that the above quantity, and consequently $V_1(x)$, can be made arbitrarily small for almost all (μ) x by selecting ε small enough.

In turn,

$$V_2(x) \leq \sup_{t \geq \delta} \frac{\int_{A_{t,h}} |\varphi(x) - \varphi(\xi)| \mu(d\xi)}{\int_{A_{t,h}} \mu(d\xi)}. \tag{A.14}$$

It is clear that $H(\bullet)$ is majorized by a function $G(\bullet)$ invertible in the product $[0, \infty) \times (0, \infty)$ such that $tG(t) \to 0$ as $t \to \infty$. The last property is equivalent to the following: $vG^{-1}(v) \to 0$ as $v \to 0$. Thus the radii of intervals $A_{t,h}$, $t \geq \delta$, are not greater than $hG^{-1}(\delta)$, which, by the definition of δ, is equal to $hG^{-1}(\varepsilon h)$ and converges to zero as $h \to 0$. Therefore, by Lemma A.12, $V_2(x) \to 0$ as $h \to 0$ for almost every (μ) x. The lemma follows. ■

Lemma A.4 leads to next result.

LEMMA A.5 *Let the kernel $K(\bullet)$ satisfy the restrictions of Lemma A.4. Then, for any probability measure μ,*

$$\lim_{h \to 0} \frac{h}{\int K\left(\frac{x-\xi}{h}\right)\mu(d\xi)} \quad \text{is finite}$$

for almost every (μ) $x \in R$.

Proof. From the fact that $K(\bullet)$ is positive and continuous at the origin it follows that there exist $r > 0$ and $c > 0$ such that

$$K(x) \geq \begin{cases} c, & \text{for } |x| \leq r \\ 0, & \text{otherwise,} \end{cases}$$

which implies

$$\int K\left(\frac{x-\xi}{h}\right)\mu(d\xi) \geq c\mu\left(S_x\left(rh\right)\right),$$

where $S_x\left(h\right) = \{\xi : |x - \xi| < h\}$. Thus

$$\frac{h}{\int K\left(\frac{x-\xi}{h}\right)\mu(d\xi)} \geq \frac{1}{2cr}\frac{\lambda(S_x(rh))}{\mu(S_x(rh))}.$$

Application of Lemma A.13 completes the proof. ■

A.2.2   Convergence rate

By imposing smoothness restrictions on $\varphi(\bullet)$ and selecting an appropriate kernel we examine the speed of convergence in (A.1). In this section $\varphi(\bullet)$ has q derivatives and its qth derivative is square integrable, that is, $\int (\varphi^{(q)}(x))^2 dx < \infty$. In addition to those in Lemma A.1, the kernel satisfies the following additional restrictions:

$$\int x^i K(x)dx = 0, \quad \text{for } i = 1, 2, \ldots, q - 1, \tag{A.15}$$

and

$$\int |x^{q-1/2}K(x)|dx < \infty. \tag{A.16}$$

It follows from (A.3) that

$$\varphi_h(x) - \varphi(x)\int K(\xi)d\xi = \frac{1}{h}\int (\varphi(x + \xi h) - \varphi(x))K(-\xi)d\xi.$$

As $\varphi(\bullet)$ has q derivatives, expanding the function in a Taylor series, we get

$$\varphi(x + h\xi) - \varphi(x) = \sum_{j=1}^{q-1} \frac{(h\xi)^j}{j!}\varphi^{(j)}(x) + \frac{1}{(q-1)!}\int_x^{x+h\xi}(x + h\xi - \eta)^{q-1}\varphi^{(q)}(\eta)d\eta$$

and using (A.16) obtain

$$\varphi_h(x) - \varphi(x) \int K(\xi) d\xi$$

$$= \frac{1}{(q-1)!} \int \left[\int_x^{x+h\xi} (x + h\xi - \eta)^{q-1} \varphi^{(q)}(\eta) d\eta \right] K(-\xi) d\xi.$$

Since, by the Schwartz inequality, the inner integral is not greater than

$$M \left(\int_x^{x+h\xi} (x + h_n\xi - \eta)^{2q-2} d\eta \right)^{1/2} = M \left(\int_0^{h\xi} \eta^{2q-2} d\eta \right)^{1/2}$$

$$= \frac{M}{(2q-1)^{1/2}} (h\xi)^{q-1/2},$$

where $M = (\int (\varphi^{(q)}(x))^2 dx)^{1/2}$, bearing (A.15) in mind we get

$$\left| \varphi_h(x) - \varphi(x) \int K(\xi) d\xi \right| \le \frac{M h^{q-1/2}}{(2q-1)^{1/2}(q-1)!} \int |\xi^{q-1/2} K(\xi)| d\xi.$$

Finally,

$$\left| \varphi_h(x) - \varphi(x) \int K(\xi) d\xi \right| = O(h^{q-1/2}) \tag{A.17}$$

as $h \to 0$, where the bound is independent of x.

If $\varphi^{(q)}(\bullet)$ is bounded convergence is faster. In such a case, we write

$$\varphi(x + h\xi) - \varphi(x) = \sum_{j=1}^{q-1} \frac{(h\xi)^j}{j!} \varphi^{(j)}(x) + \frac{(h\xi)^q}{(q-1)!} \varphi^{(q)}(x + \theta h\xi)$$

with $0 < \theta < 1$. Thus if $\sup_x |\varphi^{(q)}(x)| < \infty$,

$$\left| \varphi_h(x) - \varphi(x) \int K(\xi) d\xi \right| \le \frac{h^q}{(q-1)!} \sup_x |\varphi^{(q)}(x)| \int |\xi^p K(\xi)| d\xi.$$

Hence,

$$\left| \varphi_h(x) - \varphi(x) \int K(\xi) d\xi \right| = O(h^q), \tag{A.18}$$

provided that $\int |\xi^p K(\xi)| d\xi < \infty$.

The improved rate of convergence of the convolution integral $\varphi_h(x)$ requires higher order kernels. In Section 3.4, we have discussed direct methods for generating higher order kernels. Yet another simple way of obtaining such kernels is based on the concept of kernel twicing. Hence, it can be easily shown that if $K(\bullet)$ is an admissible kernel of order q then $2K(\bullet) - (K * K)(\bullet)$ is the admissible kernel of order $2q$. Thus, starting with a basic kernel of order two, we can generated kernels of higher order by convolving the kernel with itself.

A.2.3   Integrated error

Before showing that the integrated error converges to zero, we verify the following lemma:

LEMMA A.6 *If* $\int \varphi^2(x)dx < \infty$, *then*

$$\lim_{h \to 0} \int (\varphi(x + h) - \varphi(x))^2 dx = 0.$$

Proof. Every simple function can be represented as a linear combination of characteristic functions of a finite number of some intervals. Thus the hypothesis holds for any such function. Moreover, for any square integrable $\varphi(\bullet)$ there exists a sequence of simple functions $\varphi_n(\bullet)$ such that $\int (\varphi_n(x) - \varphi(x))^2 \to 0$ as $n \to \infty$. Since each $\varphi_n(\bullet)$ is simple, $\int (\varphi_n(x + h) - \varphi_n(x))^2 dx \to 0$ as $h \to 0$. From this and the inequality

$$\int (\varphi(x + h) - \varphi(x))^2 dx$$

$$\leq 2 \int (\varphi(x + h) - \varphi_n(x + h))^2 dx + 4 \int (\varphi_n(x + h) - \varphi_n(x))^2 dx$$

$$+ 4 \int (\varphi_n(x) - \varphi(x))^2 dx$$

$$= 4 \int (\varphi_n(x + h) - \varphi_n(x))^2 dx + 6 \int (\varphi_n(x) - \varphi(x))^2 dx$$

it follows that

$$\limsup_{h \to 0} \int (\varphi(x + h) - \varphi(x))^2 dx \leq 6 \int (\varphi_n(x) - \varphi(x))^2 dx$$

for every n. Letting n tend to infinity completes the proof. ∎

LEMMA A.7 *Let* $\int \varphi^2(x)dx < \infty$. *If the kernel satisfy (A.4) and, moreover,* $\int K(x)dx = 1$, *then (A.2) holds.*

Proof. It follows from (A.3) that

$$|\varphi_h(x) - \varphi(x)| \leq \int (\varphi(x - \xi) - \varphi(x)) \left[\frac{1}{h} K \left(\frac{\xi}{h} \right) \right]^{1/2} \left[\frac{1}{h} K \left(\frac{\xi}{h} \right) \right]^{1/2} d\xi.$$

Squaring both sides and using the Schwartz inequality, we get

$$(\varphi_h(x) - \varphi(x))^2 \leq \int (\varphi(x - \xi) - \varphi(x))^2 \left| \frac{1}{h} K \left(\frac{\xi}{h} \right) \right| d\xi \int \left| \frac{1}{h} K \left(\frac{\eta}{h} \right) \right| d\eta,$$

which leads to

$$\int (\varphi_h(x) - \varphi(x))^2 dx \leq \int |K(\eta)| d\eta \iint (\varphi(x - \xi) - \varphi(x))^2 \frac{1}{h} K \left(\frac{\xi}{h} \right) d\xi dx.$$

Changing the order of integration, we find the double integral in the expression equal to

$$\int \phi(\xi) \left| \frac{1}{h} K \left(\frac{\xi}{h} \right) \right| d\xi = \int_{|\xi| < \delta} \phi(\xi) \left| \frac{1}{h} K \left(\frac{\xi}{h} \right) \right| d\xi + \int_{|\xi| \geq \delta} \phi(\xi) \left| \frac{1}{h} K \left(\frac{\xi}{h} \right) \right| d\xi$$

with $\phi(\xi) = \int(\varphi(x - \xi) - \varphi(x))^2 dx$ and some $\delta > 0$. Since, from Lemma A.6, it follows that $\phi(\xi) \to 0$ as $\xi \to 0$ for any $\varepsilon > 0$, there exists $\delta > 0$ such that $|\phi(\xi)| \leq \varepsilon$ if $|\xi| < \delta$. For such δ, the first integral is bounded by

$$\varepsilon \int_{|\xi|<\delta} \left| \frac{1}{h} K \left(\frac{\xi}{h} \right) \right| d\xi \leq \varepsilon.$$

Since $\phi(\bullet)$ is bounded, the other is not greater than $\sup_t \phi(t) \int_{|\xi|\geq\delta/h} |K(\xi)| d\xi$, which converges to zero as $h \to 0$. The lemma follows. ∎

A.3   Applications to probability

In this section, we wish to show how the results of Section A.2 can be applied directly to some quantities in the probability theory involving average operators.

Suppose that X is a random variable and has a probability density $f(\bullet)$. We now rewrite some results presented in Section A.2 in a new context. From Lemma A.1, we obtain

LEMMA A.8 *Let $E|\varphi(X)| < \infty$. If $K(\bullet)$ satisfies the appropriate assumptions of Lemma A.1, then*

$$\lim_{h \to 0} \frac{1}{h} E \left\{ \varphi(X) K \left(\frac{x - X}{h} \right) \right\} = \varphi(x) f(x) \int K(\xi) d\xi$$

and

$$\sup_{h>0} \frac{1}{h} E \left| \varphi(X) K \left(\frac{x - X}{h} \right) \right| \quad \text{is finite}$$

at every x where both $\varphi(\bullet)$ and $f(\bullet)$ are continuous. In particular,

$$\lim_{h \to 0} \frac{1}{h} E K \left(\frac{x - X}{h} \right) = f(x) \int K(\xi) d\xi$$

and

$$\sup_{h>0} \frac{1}{h} E \left| K \left(\frac{x - X}{h} \right) \right| \quad \text{is finite}$$

at every x where $f(\bullet)$ is continuous.

LEMMA A.9 *Let $E|\varphi(X)| < \infty$. If $K(\bullet)$ satisfies the appropriate assumptions of Lemma A.3, then*

$$\lim_{h \to 0} \frac{1}{h} E \left\{ \varphi(X) K \left(\frac{x - X}{h} \right) \right\} = \varphi(x) f(x) \int K(\xi) d\xi$$

and

$$\sup_{h>0} \frac{1}{h} E \left| \varphi(X) K \left(\frac{x - X}{h} \right) \right| \quad \text{is finite.}$$

at every Lebesgue point x of both $f(\bullet)$ and $\varphi(\bullet)$, and, a fortiori, at almost every (with respect to the Lebesgue measure) x, and at every continuity point x of both $f(\bullet)$ and $\varphi(\bullet)$, as well. In particular,

$$\lim_{h\to 0}\frac{1}{h}E K\left(\frac{x-X}{h}\right) = f(x)\int K(\xi)d\xi$$

and

$$\sup_{h>0}\frac{1}{h}E\left|K\left(\frac{x-X}{h}\right)\right| \text{ is finite}$$

at every Lebesgue point of $f(\bullet)$, and, a fortiori, at almost every (with respect to the Lebesgue measure) point x and at every continuity point of $f(\bullet)$, as well.

In the next two lemmas X has an arbitrary distribution and, in particular, may not have a density. The probability measure of X is denoted by μ.

LEMMA A.10 *Let $E|\varphi(X)| < \infty$ and let $K(\bullet)$ satisfy the assumptions of Lemma A.4. Then, at almost every (μ) point x,*

$$\lim_{h\to 0}\frac{E\left\{\varphi(X)K\left(\frac{x-X}{h}\right)\right\}}{E K\left(\frac{x-X}{h}\right)} = \varphi(x)$$

and

$$\limsup_{h\to 0}\left|\frac{E\left\{\varphi(X)K\left(\frac{x-X}{h}\right)\right\}}{E K\left(\frac{x-X}{h}\right)}\right| \text{ is finite.}$$

From Lemma A.5, we obtain

LEMMA A.11 *If $K(\bullet)$ satisfies assumptions of Lemma A.4, then*

$$\lim_{h\to 0}\frac{h}{E K\left(\frac{x-X}{h}\right)} \text{ is finite}$$

at almost (μ) point x.

A.4   Lemmas

In the next two lemmas, which can be found in Wheeden and Zygmund [318, Theorem (10.49) and Corollary (10.50)], μ is any probability measure.

LEMMA A.12 *If $\int |\varphi(x)| \mu(dx) < \infty$, then*

$$\lim_{h \to 0} \frac{\displaystyle\int_{S_x(h)} \varphi(\xi) \mu(d\xi)}{\mu(S_x(h))} = \varphi(x) \qquad (A.19)$$

for almost every (μ) x.

LEMMA A.13 *For almost every (μ) x,*

$$\lim_{h \to 0} \frac{\lambda(S_x(h))}{\mu(S_x(h))}$$

is finite.

Observe that (A.19) can be rewritten in the following form:

$$\lim_{h \to 0} \frac{\displaystyle\int_{S_x(h)} \varphi(\xi) \mu(d\xi)}{\displaystyle\int_{S_x(h)} \mu(d\xi)} = \varphi(x)$$

equivalent to

$$\lim_{h \to 0} \frac{\displaystyle\int \varphi(\xi) W\left(\frac{x - \xi}{h}\right) \mu(d\xi)}{\displaystyle\int \varphi(\xi) W\left(\frac{x - \xi}{h}\right) \mu(d\xi)} = \varphi(x),$$

with

$$W(x) = \begin{cases} 1, & \text{for } |x| < 1 \\ 0, & \text{otherwise.} \end{cases}$$

Since

$$\int \varphi(\xi) W\left(\frac{x - \xi}{h}\right) \mu(d\xi) = E\left\{\varphi(X) W\left(\frac{x - X}{h}\right)\right\}$$

and

$$\int W\left(\frac{x - \xi}{h}\right) \mu(d\xi) = E\left\{W\left(\frac{x - X}{h}\right)\right\},$$

we finally conclude that (A.12) takes place almost everywhere (μ). Therefore, compared to Lemma A.12, Lemma A.4 significantly enlarges the class of applicable kernels.

Appendix B Orthogonal functions

B.1   Introduction

In this chapter, we present Fourier (that is, trigonometric), Legendre, Laguerre, and Hermite orthogonal functions and discuss some properties of orthogonal expansions. In Section B.6 we show fundamental properties of Haar wavelets.

Legendre, Laguerre, and Hermite polynomials, particular cases of Jacobi polynomials, denoted temporarily as $W_k(x)$, $k = 0, 1, \ldots$, are orthogonal, each with respect to some nonnegative weight function $\lambda(x)$ on some interval, say (a, b). It means that

$$\int_a^b W_m(x)W_n(x)\lambda(x)dx = \begin{cases} 0, & \text{for } n \neq m, \\ c_n, & \text{for } n = m, \end{cases}$$

with $c_n \neq 0$. For

- $a = -1$, $b = 1$, and $\lambda(x) = 1$, $W_n(x)$ is the Legendre polynomial denoted by $P_n(x)$,
- $a = 0$, $b = \infty$, and $\lambda(x) = e^{-x}$, $W_n(x)$ is the Laguerre polynomial $L_n(x)$,
- $a = -\infty$, $b = \infty$, and $\lambda(x) = e^{-x^2}$, $W_n(x)$ is the Hermite polynomial $H_n(x)$.

Clearly, functions $w_n(x) = c_n^{-1/2}\lambda^{1/2}(x)W_n(x)$ are orthonormal, that is,

$$\int_a^b w_m(x)w_n(x)dx = \begin{cases} 0, & \text{for } n \neq m, \\ 1, & \text{for } n = m. \end{cases}$$

The set $\{w_k(x); k = 0, 1, \ldots\}$ of functions is also called an orthonormal system.

For any integrable $\varphi(\bullet)$ such that $\int_a^b |\varphi(x)|dx < \infty$, writing

$$\varphi(x) \sim \sum_{k=0}^{\infty} a_k w_k(x),$$

we mean that $a_k = \int_a^b \varphi(y)w_k(y)dy$ and call $\sum_{k=0}^{\infty} a_k w_k(x)$ an orthogonal expansion of the function $\varphi(x)$. The sum

$$\sigma_n(x) = \sum_{k=0}^{n} a_k w_k(x)$$

is a partial sum of the expansion. Each orthogonal system we deal with is complete, which means that

$$\lim_{n\to\infty} \int_a^b (\sigma_n(x) - \varphi(x))^2 dx = 0$$

for any $\varphi(\bullet)$ such that $\int_a^b \varphi^2(x)dx < \infty$.

The partial sum can be rewritten in the following kernel form:

$$\sigma_n(x) = \int_a^b K_n(x, y)\varphi(y)dy,$$

where

$$K_n(x, y) = \sum_{k=0}^{n} w_k(x)w_k(y)$$

is the corresponding kernel of the orthogonal system. The kernel $K_n(x, y)$ is often referred to as the reproducing kernel of the orthonormal basis $\{w_k(x)\}$.

An important problem is the pointwise convergence of the expansion, that is, convergence

$$\lim_{n\to\infty} \sigma_n(x) = \varphi(x)$$

at a fixed point $x \in (a, b)$. The convergence of the expansion in the Hermite system at a point is strictly related to the expansion of the function in the trigonometric system in an arbitrarily small neighborhood of the point. Though not exactly the same, a similar property holds for the other orthogonal systems. These facts are presented in consecutive sections in so called equiconvergence theorems.

By imposing some smoothness restrictions on $\varphi(\bullet)$, we present upper bounds for the Fourier coefficient a_k and also for pointwise

$$\sigma_n(x) - \varphi(x) = \sum_{k=n+1}^{\infty} a_k w_k(x)$$

and global

$$\int_a^b (\sigma_n(x) - \varphi(x))^2 dx = \sum_{k=n+1}^{\infty} a_k^2$$

errors. Consecutive sections are devoted to expansions in the trigonometric, Legendre, Laguerre, and Hermite orthogonal systems. For $\varphi(\bullet)$ having q derivatives and satisfying some additional restrictions, the results are summarized in Table B.1.

Table B.1. Pointwise and global errors for various orthogonal systems

	$\sigma_n(x) - \varphi(x)$	$\int(\sigma_n(x) - \varphi(x))^2 dx$
Fourier	$O(n^{-q+1/2})$	$O(n^{-2q})$
Legendre	$O(n^{-q+1/2})$	$O(n^{-2q})$
Laguerre	$O(n^{-q/2+1/4})$	$O(n^{-q})$
Hermite	$O(n^{-q/2+1/4})$	$O(n^{-q})$

B.2 Fourier series

The system of trigonometric functions

$$\frac{1}{\sqrt{2\pi}}, \frac{1}{\sqrt{\pi}} \cos x, \frac{1}{\sqrt{\pi}} \sin x, \frac{1}{\sqrt{\pi}} \cos 2x, \frac{1}{\sqrt{\pi}} \sin 2x, \ldots$$

is complete and orthonormal over the interval $[-\pi, \pi]$. For an integrable function $\varphi(\bullet)$,

$$S_n(x) = \frac{1}{2}a_0 + \sum_{k=1}^{n}(a_k \cos kx + b_k \sin kx)$$

with

$$a_0 = \frac{1}{\pi} \int_{-\pi}^{\pi} \varphi(x) dx,$$

and

$$a_k = \frac{1}{\pi} \int_{-\pi}^{\pi} \varphi(x) \cos(kx) dx, \quad b_k = \frac{1}{\pi} \int_{-\pi}^{\pi} \varphi(x) \sin(kx) dx,$$

$k = 1, 2, \ldots$, is the partial sum of the Fourier expansion in the trigonometric series

$$\varphi(x) \sim \frac{1}{2}a_0 + \sum_{k=1}^{\infty}(a_k \cos kx + b_k \sin kx). \tag{B.1}$$

Since the system is complete,

$$\lim_{n \to \infty} \int_{-\pi}^{\pi} (S_n(x) - \varphi(x))^2 dx = 0$$

for any $\varphi(\bullet)$ such that $\int_{-\pi}^{\pi} \varphi^2(x) dx < \infty$.

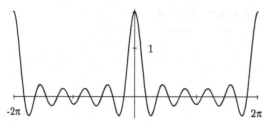

Figure B.1 Dirichlet kernel $D_5(x)$.

For the partial sum, we get

$$S_n(x) = \frac{1}{2\pi} \int_{-\pi}^{\pi} \varphi(y)dy + \frac{1}{\pi} \sum_{k=1}^{n} \int_{-\pi}^{\pi} \varphi(y)(\cos ky \cos kx + \sin ky \sin kx)dy$$

$$= \frac{1}{\pi} \int_{-\pi}^{\pi} \varphi(y) \left(\frac{1}{2} + \sum_{k=1}^{n} \cos(k(x-y)) \right) dy$$

$$= \frac{1}{\pi} \sum_{k=1}^{n} \int_{-\pi}^{\pi} \varphi(x+y) \left(\frac{1}{2} + \sum_{k=1}^{n} \cos ky \right) dy.$$

Since

$$\frac{1}{\pi} \left(\frac{1}{2} + \sum_{k=1}^{n} \cos kx \right) = D_n(x),$$

where

$$D_n(x) = \frac{\sin\left(n + \frac{1}{2}\right)x}{2\pi \sin \frac{1}{2}x} \tag{B.2}$$

is the Dirichlet kernel (see Figure B.1), we finally obtain

$$S_n(x) = \int_{-\pi}^{\pi} D_n(y)\varphi(x+y)dy = \int_{-\pi}^{\pi} D_n(y-x)\varphi(y)dy. \tag{B.3}$$

In passing, we notice $\int_{-\pi}^{\pi} D_n(x)dx = 1$.

According to Riemann's principle of localization, see Sansone [270, Chapter II, Theorem 14], Davis [64, Theorem 12.1.13], or [14], the convergence of $S_n(x_0)$ to $\varphi(x_0)$ depends only upon the behavior of $\varphi(\bullet)$ in an arbitrarily small neighborhood of x_0. Continuity of $\varphi(\bullet)$ at x_0 doesn't guarantee, however, the convergence. An example presented by du Bois-Reymond, see Davis [64, Theorem 14.4.15], shows that there exist continuous functions whose Fourier series are divergent at a point.

There are a number of various sufficient conditions yielding the convergence of $S_n(x_0)$ to $\varphi(x_0)$. We present Dini's theorem, see Sansone [270, Chapter II, Theorem 16].

THEOREM B.1 (DINI) *Let $\int_{-\pi}^{\pi} |\varphi(x)|dx < \infty$. If there exists $\varepsilon > 0$, such that*

$$\int_{0}^{\varepsilon} \left| \frac{\varphi(x+t) + \varphi(x-t) - 2\varphi(x)}{t} \right| dt < \infty,$$

then

$$\lim_{n \to \infty} S_n(x) = \varphi(x). \tag{B.4}$$

As a corollary from Dini's theorem, we get the following:

COROLLARY B.1 *If $\int_{-\pi}^{\pi} |\varphi(x)| dx < \infty$, then (B.4) holds at every $x \in (-\pi, \pi)$, where $\varphi(\bullet)$ is differentiable.*

To present another corollary, we recall that $\varphi(\bullet)$ is said to satisfy a Lipschitz condition at x if there exists a constant c such that

$$|\varphi(x) - \varphi(y)| \le c|x - y|^r,$$

$0 < r \le 1$, in some neighborhood of x.

COROLLARY B.2 *If $\int_{-\pi}^{\pi} |\varphi(x)| dx < \infty$, then (B.4) holds at every $x \in (-\pi, \pi)$, where $\varphi(\bullet)$ satisfies a Lipschitz condition.*

The result given below is due to Carleson [39] and Hunt [167], for $s = 2$ and $s > 1$, respectively.

THEOREM B.2 *If $\int_{-\pi}^{\pi} |\varphi(x)|^s dx < \infty$ for some $s > 1$, then (B.4) holds at almost every $x \in (-\pi, \pi)$.*

Restriction $s > 1$ cannot be weakened to $s = 1$, since there exist integrable functions whose trigonometric series diverge at every point, see Kolmogoroff [176] or [14].

The expression

$$s_n(x) = \frac{S_0(x) + S_1(x) + \cdots + S_{n-1}(x)}{n}$$

is called the Cesàro mean of partial sums. Since

$$\frac{1}{n} \sum_{k=0}^{n-1} D_k(x) = \frac{1}{n} \sum_{k=0}^{n-1} \frac{\sin\left(k + \frac{1}{2}\right) x}{2\pi \sin \frac{1}{2} x} = F_n(x),$$

where

$$F_n(x) = \frac{\sin^2 \frac{n}{2} x}{2\pi n \sin^2 \frac{1}{2} x} \tag{B.5}$$

is the Fejér kernel (see Figure B.2) we get

$$s_n(x) = \int_{-\pi}^{\pi} F_n(y) \varphi(x + y) dy = \int_{-\pi}^{\pi} F_n(y - x) \varphi(y) dy. \tag{B.6}$$

Observe that $\int_{-\pi}^{\pi} F_n(x) dx = 1$ and

$$D_n^2(x) = \left(n + \frac{1}{2}\right) \frac{1}{\pi} F_{2n+1}(x). \tag{B.7}$$

The next theorem on Cesàro summability of the Fourier expansion is rather well-known.

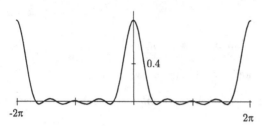

Figure B.2 The Fejér kernel.

THEOREM B.3 (LEBESGUE) *If $\int_{-\pi}^{\pi} |\varphi(x)| dx < \infty$, then*

$$\lim_{n \to \infty} s_n(x) = \varphi(x)$$

at every point where (B.4) holds, at every continuity point x of $\varphi(\bullet)$, and at almost every $x \in (-\pi, \pi)$.

The next lemma is a simple consequence of the aforementioned results.

LEMMA B.1 *If $\int_{-\pi}^{\pi} |\varphi(x)| dx < \infty$, then*

$$\lim_{n \to \infty} \frac{1}{n} \int_{-\pi}^{\pi} D_n^2(y - x)\varphi(y) dy = \frac{1}{\pi}\varphi(x)$$

at every point where (B.4) holds, at every continuity point x of $\varphi(\bullet)$, and at almost every $x \in (-\pi, \pi)$.

Let us define the following kernels:

$$\mathfrak{D}_n(x) = \frac{\sin\left(n + \frac{1}{2}\right) x}{\pi x} \tag{B.8}$$

and

$$\mathfrak{F}_n(x) = \frac{2 \sin^2 \frac{n}{2} x}{\pi n x^2}. \tag{B.9}$$

Clearly,

$$\mathfrak{D}_n^2(x) = \left(n + \frac{1}{2}\right) \frac{1}{\pi} \mathfrak{F}_n(x). \tag{B.10}$$

Moreover, let

$$\mathfrak{S}_n(x) = \int_{-\pi}^{\pi} \mathfrak{D}_n(y)\varphi(x + y) dy \tag{B.11}$$

and

$$\mathfrak{s}_n(x) = \int_{-\pi}^{\pi} \mathfrak{F}_n(y)\varphi(x + y) dy. \tag{B.12}$$

Compare (B.8), (B.9), (B.10), (B.11), and (B.12) with (B.2), (B.5), (B.7), (B.3), and (B.6), respectively.

LEMMA B.2 *Let $\int_{-\pi}^{\pi} |\varphi(x)| dx < \infty$. Then,*

$$\lim_{n \to \infty} (\mathfrak{S}_n(x) - S_n(x)) = 0,$$

and

$$\lim_{n \to \infty} (\mathfrak{s}_n(x) - s_n(x)) = 0.$$

Proof. Since

$$D_n(x) - \mathfrak{D}_n(x) = \frac{1}{\pi} g(x) \sin\left(n + \frac{1}{2}\right) x,$$

where

$$g(x) = \frac{1}{2 \sin \frac{1}{2}x} - \frac{1}{x}$$

with $g(0) = \lim_{x \to 0} g(x) = 0$, to verify the first part of the lemma it suffices to show that

$$\lim_{N \to \infty} \int_{-\pi}^{\pi} \varphi(x + y) g(y) \sin(Ny) dy = 0.$$

Since $g(x)$ is continuous in the interval $[-\pi, \pi]$, to do that we apply the Riemann-Lebesgue theorem, see Sansone [270, Chapter II, Section 4, Theorem 12] or Davis [64, Theorem 12.1.10], according to which

$$\lim_{n \to \infty} \int_a^b \varphi(x) \sin(nx) dx = 0$$

if $\int_a^b |\varphi(x)| dx < \infty$.

The verification of the other part of the lemma is similar since

$$F_n(x) - \mathfrak{F}_n(x) = \frac{1}{\pi n} h(x) \sin^2 \frac{n}{2} x,$$

where

$$h(x) = \frac{1}{2 \sin^2 \frac{1}{2}x} - \frac{2}{x^2}.$$

Thus, the lemma follows. ∎

Owing to Lemma B.2 we can make the following two remarks.

REMARK B.1 *Dini's Theorem B.1, as well as Corollaries B.1, B.2, and B.2, all remain true if $S_n(x)$ is replaced by $\mathfrak{S}_n(x)$.*

REMARK B.2 *The Lebesgue Theorem B.3 on Cesàro summability of the Fourier expansion and Lemma B.1 are still true if $s_n(x)$ is replaced by $\mathfrak{s}_n(x)$. In particular, if $\int_{-\pi}^{\pi} |\varphi(x)| dx < \infty$, then*

$$\lim_{n \to \infty} \frac{1}{n} \int_{-\pi}^{\pi} \mathfrak{D}_n^2(y - x) \varphi(y) dy = \frac{1}{\pi} \varphi(x)$$

at every x where (B.4) holds, at every continuity point x of φ(•), and at almost every $x \in (-\pi, \pi)$.

At every x where (B.4) holds, we have

$$S_n(x) - \varphi(x) = \sum_{k=n+1}^{\infty} (a_k \cos kx + b_k \sin kx).$$

Thus, the quality of the pointwise approximation of $\varphi(x)$ with $S_n(x)$ depends upon the speed where coefficients a_k and b_k diminish with k increasing to infinity. Assuming that $\varphi(-\pi) = \varphi(\pi) = 0$ and $\varphi(•)$ is differentiable and then integrating by parts, we get

$$a_k = -\frac{1}{\pi k} \int_{-\pi}^{\pi} \varphi'(x) \sin(kx) dx = -\frac{1}{k} \gamma_{1k}$$

and

$$b_k = \frac{1}{\pi k} \int_{-\pi}^{\pi} \varphi'(x) \cos(kx) dx = \frac{1}{k} \delta_{1k},$$

where

$$\gamma_{1k} = \frac{1}{\pi} \int_{-\pi}^{\pi} \varphi'(x) \sin(kx) dx,$$

$$\delta_{1k} = \frac{1}{\pi} \int_{-\pi}^{\pi} \varphi'(x) \cos(kx) dx.$$

In general, if $\varphi^{(m)}(-\pi) = \varphi^{(m)}(\pi) = 0$ for $m = 1, \ldots, q - 1$, and $\int_{-\pi}^{\pi} (\varphi^{(q)}(x))^2 dx < \infty$,

$$|a_k| = \frac{1}{k^q} |\gamma_{qk}| \text{ and } |b_k| = \frac{1}{k^q} |\gamma_{qk}|,$$

where γ_{qk} and δ_{qk} are Fourier coefficients of $\varphi^{(q)}(x)$ such that $\sum_{k=0}^{\infty} \gamma_{qk}^2 < \infty$ and $\sum_{k=0}^{\infty} \delta_{qk}^2 < \infty$. Therefore,

$$\sum_{k=n+1}^{\infty} (a_k^2 + b_k^2) = \sum_{k=n+1}^{\infty} k^{-2q} (\gamma_{qk}^2 + \delta_{qk}^2) = O(n^{-2q}),$$

which means that

$$\int_{-\pi}^{\pi} (S_n(x) - \varphi(x))^2 dx = \sum_{k=n+1}^{\infty} (a_k^2 + b_k^2) = O(n^{-2q}). \tag{B.13}$$

Moreover,

$$\left| \sum_{k=n+1}^{\infty} (a_k \cos kx + b_k \sin kx) \right| \leq \sum_{k=n+1}^{\infty} (|a_k| + |b_k|) \leq \sum_{k=n+1}^{\infty} k^{-q} (|\gamma_{qk}| + |\delta_{qk}|)$$

$$\leq \left(\sum_{k=n+1}^{\infty} k^{-2q} \right)^{1/2} \left(\sum_{k=n+1}^{\infty} (\gamma_{qk}^2 + \delta_{qk}^2) \right)^{1/2}.$$

Thus, at every x where (B.4) holds,

$$|S_n(x) - \varphi(x)| = O(n^{-q+1/2}).\tag{B.14}$$

For any $\varphi(\bullet)$ such that $\int_{-\pi}^{\pi} |\varphi(x)| dx < \infty$, applying the Schwartz inequality and (B.7), we get

$$\left(\int_{-\pi}^{\pi} D_n(y-x) \varphi(y) dy \right)^2 \leq \left(\int_{-\pi}^{\pi} |D_n(y-x)| \sqrt{|\varphi(y)|} \sqrt{|\varphi(y)|} dy \right)^2$$

$$\leq \left(n + \frac{1}{2} \right) \frac{1}{\pi} \int_{-\pi}^{\pi} F_n(y-x) |\varphi(y)| dy \int_{-\pi}^{\pi} |\varphi(y)| dy,$$

which gives

$$\int_{-\pi}^{\pi} D_n(y-x) \varphi(y) dy = O(n^{1/2}).$$

For similar reasons,

$$\int_{-\pi}^{\pi} \mathfrak{D}_n(y-x) \varphi(y) dy = O(n^{1/2}).\tag{B.15}$$

Sometimes, it is convenient to apply the complex form of the trigonometric series. The system of functions

$$\frac{1}{\sqrt{2\pi}} e^{ikx} = \frac{1}{\sqrt{2\pi}} \cos kx + i \frac{1}{\sqrt{2\pi}} \sin kx,$$

$k = 0, \pm 1, \pm 2, \ldots$, is orthonormal over the interval $[-\pi, \pi]$. Hence,

$$\varphi(x) \sim \sum_{k=-\infty}^{\infty} c_k e^{ikx}$$

with

$$c_k = \frac{1}{2\pi} \int_{-\pi}^{\pi} e^{\overline{ikx}} \varphi(x) dx = \frac{1}{2\pi} \int_{-\pi}^{\pi} e^{-ikx} \varphi(x) dx.$$

The relation among coefficients a_k, b_k and c_k is of the following form:

$$\frac{1}{2} a_0 = c_0, \quad a_k = c_k + c_{-k}, \quad b_k = i(c_k - c_{-k}).$$

Thus,

$$\frac{1}{2} a_0 + \sum_{k=1}^{n} (a_k \cos kx + b_k \sin kx) = c_0 + \sum_{k=1}^{n} c_k e^{ikx} + \sum_{k=1}^{n} c_{-k} e^{-ikx},$$

which implies

$$S_n(x) = \sum_{k=-n}^{n} c_k e^{ikx}.\tag{B.16}$$

Further relations between the decay of Fourier coefficients and the smoothness of a given function are examined in [14], [200], and [335].

B.3   Legendre series

Legendre polynomials $P_0(x)$, $P_1(x)$, ... are complete and orthogonal over the interval $[-1, 1]$, see, for example, Sansone [270] or Szegö [288]. They satisfy Rodrigues' formula

$$P_k(x) = \frac{1}{2^k k!} \frac{d^k}{dx^k} (x^2 - 1)^k.$$

First, a few are

$$P_0(x) = 1,$$
$$P_1(x) = x,$$
$$P_2(x) = \frac{1}{2}(3x^2 - 1),$$
$$P_3(x) = \frac{1}{2}(5x^3 - 3x),$$
$$P_4(x) = \frac{1}{8}(35x^4 - 30x^2 + 3).$$

Moreover, for $|x| \leq 1$,

$$|P_k(x)| \leq 1, \tag{B.17}$$

$$P_k(1) = (-1)^k P_k(-1) = 1, \tag{B.18}$$

and

$$|P_k'(x)| \leq k, \tag{B.19}$$

see Sansone [270, Chapter III, Sections 11 and 17]. The polynomials have the following property:

$$(2k + 1)P_k(x) = P_{k+1}'(x) - P_{k-1}'(x). \tag{B.20}$$

Since $\int_{-1}^{1} P_k^2(x)dx = 1/(2k + 1)$, the system

$$p_0(x), \; p_1(x), \; p_2(x), \ldots,$$

where

$$p_k(x) = \sqrt{\frac{2k + 1}{2}} P_k(x),$$

is orthonormal over the interval $[-1, 1]$.

Denoting

$$k_n(x, y) = \sum_{k=0}^{n} p_k(x)p_k(y),$$

we write the Christoffel–Darboux formula:

$$k_n(x, y) = \frac{n + 1}{2} \frac{P_n(x)P_{n+1}(y) - P_{n+1}(x)P_n(y)}{y - x}. \tag{B.21}$$

Clearly,

$$\sigma_n(x) = \sum_{k=0}^{n} a_k p_k(x)$$

with

$$a_k = \int_{-1}^{1} \varphi(x) p_k(x) dx$$

is the partial expansion of $\varphi(\bullet)$ in the Legendre system.

The next theorem due to Hobson, see Szegö [288, Theorem 9.1.2] or Sansone [270, Chapter III, Section 14], is on pointwise convergence of $\sigma_n(x)$ to $\varphi(x)$. The convergence is strictly related to the convergence of the trigonometric expansion of $\varphi(\cos\xi)\sqrt{\sin\xi}$ at a point $\xi = \arccos x$, $0 < \xi < \pi$. In the theorem, $S_n(\cos\xi)$ is the nth partial sum of the trigonometric series of $\varphi(\cos\xi)\sqrt{\sin\xi}$ in any open interval containing $\xi = \arccos x$, that is,

$$S_n(\cos\xi) = \int_{\xi_0 - \delta}^{\xi_0 + \delta} D_n(\eta - \xi)\varphi(\cos\eta)\sqrt{\sin\eta}\, d\eta$$

with any $\delta > 0$.

THEOREM B.4 (EQUICONVERGENCE) *Let $\int_{-1}^{1} |\varphi(x)| dx < \infty$ and*

$$\int_{-1}^{1} \frac{1}{\sqrt[4]{1-x^2}} |\varphi(x)| dx < \infty. \tag{B.22}$$

Then,

$$\lim_{n \to \infty} \left(\sigma_n(x) - \frac{1}{\sqrt[4]{1-x^2}} S_n(x) \right) = 0 \tag{B.23}$$

at every $x \in (-1, 1)$, such that $S_n(x)$ converges.

Since

$$\left(\int_{-1}^{1} \frac{1}{\sqrt[4]{1-x^2}} |\varphi(x)| dx \right)^2 \leq \int_{-1}^{1} \frac{1}{\sqrt{1-x^2}} dx \int_{-1}^{1} \varphi^2(x) dx = \pi \int_{-1}^{1} \varphi^2(x) dx,$$

restriction (B.22) is satisfied if $\int_{-1}^{1} \varphi^2(x) dx < \infty$. On the other hand, (B.22) implies $\int_{-1}^{1} |\varphi(x)| dx < \infty$. Notice, moreover, that $\int_{-1}^{1} (1-x^2)^{-1/4} |\varphi(x)| dx = \int_{0}^{\pi} |\varphi(\cos\xi)|\sqrt{\sin\xi}\, d\xi$.

From the theorem, Corollaries B.1, B.2, and Theorem B.2, we present the following three corollaries:

COROLLARY B.3 *If $\varphi(\bullet)$ satisfies the restrictions of Theorem B.4, then*

$$\lim_{n \to \infty} \sigma_n(x) = \varphi(x) \tag{B.24}$$

at every $x \in (-1, 1)$, where $\varphi(\bullet)$ is differentiable.

COROLLARY B.4 *If $\varphi(\bullet)$ satisfies the restrictions of Theorem B.4, then (B.24) holds at every $x \in (-1, 1)$, where $\varphi(\bullet)$ satisfies a Lipschitz condition.*

COROLLARY B.5 *If $\varphi(\bullet)$ satisfies the restrictions of Theorem B.4 and, in addition, $\int_{-1}^{1} |\varphi(x)|^s dx < \infty$ with some $s > 1$, then (B.24) holds at almost every $x \in (-1, 1)$.*

Applying (B.20), bearing (B.18) in mind, assuming $\int_{-1}^{1} (\varphi'(x))^2 dx < \infty$ and $\varphi(-1) = \varphi(1) = 0$, we integrate by parts to get

$$\int_{-1}^{1} \varphi(x) P_k(x) dx = -\frac{1}{2k+1} \int_{-1}^{1} \varphi'(x)(P_{k+1}(x) - P_{k-1}(x)) dx.$$

Since $P_k(x) = \sqrt{2/(2k+1)} p_k(x)$, therefore we have

$$a_k = -\frac{1}{2k+1}(\alpha_k b_{k+1} - \beta_k b_{k-1}) = -\frac{1}{2k+1} b_{1k},$$

where $b_k = \int_{-1}^{1} \varphi'(x) p_k(x) dx$ and $b_{1k} = \alpha_k b_{k+1} - \beta_k b_{k-1}$, α_k and β_k are positive sequences such that $\alpha_k \leq 1$ and $\beta_k \leq 3$. This leads to $a_k = O(k^{-1}) b_{1k}$, where $\sum_{k=0}^{\infty} b_{1k}^2 < \infty$. Assuming that further derivatives exists we can repeat the reasoning. In general, if $\varphi(\bullet)$ has q derivatives, $\int_{-1}^{1} (\varphi^{(q)}(x))^2 dx < \infty$, and $\varphi^{(i)}(-1) = \varphi^{(i)}(1) = 0$, $i = 0, 1, \ldots, q-1$, we obtain $a_k = O(k^{-q}) b_{qk}$ with some b_{qk}'s, such that $\sum_{k=0}^{\infty} b_{qk}^2 < \infty$. Consequently,

$$\int_{-1}^{1} (\sigma_n(x) - \varphi(x))^2 dx = \sum_{k=n+1}^{\infty} a_k^2 = O(n^{-2q}).$$

At every $x \in (-1, 1)$ where (B.24) holds, we have

$$|\sigma_n(x) - \varphi(x)| \leq \sum_{k=n+1}^{\infty} |a_k p_k(x)| = \sum_{k=n+1}^{\infty} O\left(k^{-q+1/2}\right) |b_{qk}||P_k(x)|$$

$$= \sum_{k=n+1}^{\infty} O(k^{-q}) |b_{qk}|,$$

since

$$|P_n(x)| \leq 4 \sqrt{\frac{2}{\pi}} \frac{1}{\sqrt{n}} \frac{1}{\sqrt[4]{1-x^2}} \tag{B.25}$$

for $-1 \leq x \leq 1$, see Sansone [270, Chapter III, Section 10, Stieltjes formula (14)]. As the obtained quantity is of order $(\sum_{k=n+1}^{\infty} k^{-2q})^{1/2} (\sum_{k=n+1}^{\infty} b_{qk}^2)^{1/2}$, we finally get

$$|\sigma_n(x) - \varphi(x)| = O(n^{-q+1/2}). \tag{B.26}$$

LEMMA B.3 *Let $\varphi(\bullet)$ satisfy the restrictions of Theorem B.4. Then,*

$$\lim_{n \to \infty} \frac{1}{n} \int_{-1}^{1} k_n^2(x, y) \varphi(y) dy = \frac{1}{\pi \sqrt{1-x^2}} \varphi(x),$$

for almost every $x \in (-1, 1)$, and at every continuity point x of $\varphi(\bullet)$.

Proof. Defining $x = \cos \xi$ and $y = \cos \eta$, we begin with the following equality:

$$
k_n(\cos \xi, \cos \eta)
$$

$$
= \frac{1}{\sqrt{\sin \xi \sin \eta}} \left[\frac{\sin [(n+1)(\eta - \xi)]}{2\pi \sin \frac{1}{2}(\eta - \xi)} + \frac{\sin [(n+1)(\xi + \eta) - \frac{\pi}{2}]}{2\pi \sin \frac{1}{2}(\eta + \xi)} + R_n(\xi, \eta) \right]
$$

holding for $0 < \xi < \pi$, $0 < \eta < \pi$, where $R_n(\xi, \eta) = O(1)$, see Szegö [288, (9.3.5)].
The bound for $R_n(\xi, \eta)$ is uniform for $\varepsilon < \xi < \pi - \varepsilon$ and $\varepsilon < \eta < \pi - \varepsilon$, any $\varepsilon > 0$.
Therefore

$$
k_n(\cos \xi, \cos \eta) = \frac{1}{\sqrt{\sin \xi \sin \eta}} \left[D_{n+1/2}(\xi - \eta) + r_n(\xi, \eta) \right],
$$

where $r_n(\xi, \eta) = O(1)$, and, as a consequence, see (B.7),

$$
k_n^2(\cos \xi, \cos \eta)
$$

$$
= \frac{1}{\sin \xi \sin \eta}(n+1) \left[\frac{1}{\pi} F_{2n+2}(\xi - \eta) + 2r_n(\xi, \eta) D_{n+1/2}(\xi - \eta) + r_n^2(\xi, \eta) \right]
$$

which gives

$$
\int_{-|x|-\varepsilon}^{|x|+\varepsilon} k_n^2(x, y)\varphi(y)dy = \int_{\delta}^{\pi-\delta} k_n^2(\cos \xi, \cos \eta)\varphi(\cos \eta) \sin \eta d\eta
$$

$$
= (n+1)\frac{1}{\pi \sin \xi} \int_{\delta}^{\pi-\delta} F_{2n+2}(\xi - \eta)\varphi(\cos \eta)d\eta
$$

$$
+ \frac{2}{\sin \xi} \int_{\delta}^{\pi-\delta} r_n(\xi, \eta) D_{n+1/2}(\xi - \eta)\varphi(\cos \eta)d\eta
$$

$$
+ \frac{1}{\sin \xi} O(1) \int_{0}^{\pi} |\varphi(\cos \eta)|d\eta,
$$

where $\delta = \arccos(|x| + \varepsilon)$. By Theorem B.3 on Cesàro summability of the Fourier
expansion, for almost every ξ, and for every continuity point of $\varphi(\cos \xi)$,

$$
\lim_{n \to \infty} \int_{\delta}^{\pi-\delta} F_{2n+2}(\xi - \eta)\varphi(\cos \eta)d\eta = \varphi(\cos \xi). \tag{B.27}
$$

In turn,

$$
\left| \int_{\delta}^{\pi-\delta} r_n(\xi, \eta) D_{n+1/2}(\xi - \eta)\varphi(\cos \eta)d\eta \right|
$$

$$
= O(1) \int_{\delta}^{\pi-\delta} |D_{n+1/2}(\xi - \eta)||\varphi(\cos \eta)|d\eta.
$$

Moreover,

$$\int_0^\pi |D_{n+1/2}(\xi - \eta)||\varphi(\cos \eta)|d\eta$$

$$\leq \left(\int_0^\pi D_{n+1/2}^2(\xi - \eta)|\varphi(\cos \eta)|d\eta \right)^{1/2} \left(\int_0^\pi |\varphi(\cos \eta)|d\eta \right)^{1/2}$$

$$= \sqrt{\frac{n+1}{\pi}} \left(\int_0^\pi F_{2n+2}(\xi - \eta)|\varphi(\cos \eta)|d\eta \right)^{1/2} \left(\int_0^\pi |\varphi(\cos \eta)|d\eta \right)^{1/2}$$

$$= O(n^{1/2})$$

at every ξ where (B.27) holds. Since

$$\frac{1}{\sin \xi}\varphi(\cos \xi) = \frac{1}{\sqrt{1-x^2}}\varphi(x),$$

we have shown that

$$\lim_{n \to \infty} \frac{1}{n} \int_{-|x|-\varepsilon}^{|x|+\varepsilon} k_n^2(x, y)\varphi(y)dy = \frac{1}{\pi \sqrt{1-x^2}}\varphi(x)$$

for almost every x.

To complete the proof, we need to examine

$$\int_{|x|+\varepsilon}^1 k_n^2(x, y)\varphi(y)dy \text{ and } \int_{-1}^{-|x|-\varepsilon} k_n^2(x, y)\varphi(y)dy. \tag{B.28}$$

Recalling Christoffel–Darboux formula (B.21) and using (B.25), we get

$$k_n(x, y) = \frac{n+1}{2} \left(P_n(x)\frac{P_{n+1}(y)}{y-x} - P_{n+1}(x)\frac{P_n(y)}{y-x} \right)$$

$$= O(n^{1/2}) \left(\left| \frac{P_{n+1}(y)}{y-x} \right| + \left| \frac{P_n(y)}{y-x} \right| \right) = O(n^{1/2})(|P_{n+1}(y)| + |P_n(y)|),$$

since $|x - y| > 0$. Thus, using (B.25) again, we find

$$\int_{1-\varepsilon}^1 k_n^2(x, y)\varphi(y)dy = O(n) \int_{1-\varepsilon}^1 (P_{n+1}^2(y) + P_n^2(y))|\varphi(y)|dy$$

$$= O(1) \int_{1-\varepsilon}^1 \frac{1}{\sqrt[4]{1-y^2}}|\varphi(y)|dy.$$

Since the other integral in (B.28) is of the same order, the proof is completed. ∎

The Legendre polynomials are members of a larger class of orthogonal polynomials defined on the interval $[-1, 1]$. The ultraspherical (Gegenbauer) polynomials consists of polynomials orthogonal with respect to the weight function $\lambda(x, \alpha) = (1 - x^2)^{\alpha-1/2}$, where $\alpha > -1/2$. The case $\alpha = 1/2$ gives the Legendre polynomials, whereas $\alpha = 0$ corresponds to the Chebyshev polynomials of the first kind. It is an interesting fact that among of all orthogonal polynomials with the weight $\lambda(x, \alpha)$, the Chebyshev polynomials of the first kind are optimal in the sense of the uniform metric on $[-1, 1]$, that is, they give the lowest uniform error for any continuous function on $[-1, 1]$. This result is due to Rivlin and Wilson [253].

B.4   Laguerre series

Laguerre polynomials $L_0, L_1, \ldots$ satisfy the following Rodrigues' formula:

$$e^{-x} L_k(x) = \frac{1}{k!} \frac{d^k}{dx^k} (x^k e^{-x}).$$

It is not difficult to check that

$$L_0(x) = 1,$$
$$L_1(x) = -x + 1,$$
$$L_2(x) = \frac{1}{2}x^2 - 2x + 1,$$
$$L_3(x) = -\frac{1}{6}x^3 + \frac{3}{2}x^2 - 3x + 1,$$
$$L_4(x) = \frac{1}{24}x^4 - \frac{2}{3}x^3 + 3x^2 - 4x + 1,$$

and so on. In general,

$$L_k(x) = \sum_{i=0}^{k} (-1)^i \frac{1}{i!} \binom{k}{i} x^i. \tag{B.29}$$

The system of functions

$$l_k(x) = L_k(x) e^{-x/2},$$

$k = 0, 1, \ldots$, is complete and orthonormal on the real half line $[0, \infty)$.

The Christoffel–Darboux summation formula is of the following form:

$$k_n(x, y) = (n + 1) e^{-x/2} e^{-y/2} \frac{L_{n+1}(x) L_n(y) - L_n(x) L_{n+1}(y)}{y - x}$$
$$= (n + 1) \frac{l_{n+1}(x) l_n(y) - l_n(x) l_{n+1}(y)}{y - x},$$

where

$$k_n(x, y) = e^{-x/2} e^{-y/2} \sum_{k=0}^{n} L_k(x) L_k(y) = \sum_{k=0}^{n} l_k(x) l_k(y)$$

is the kernel of the system.

Let

$$\sigma_n(x) = \sum_{k=0}^{n} a_k l_k(x),$$

where

$$a_k = \int_0^\infty \varphi(x) l_k(x) dx.$$

In Szegö [288, Theorem 9.1.5] or Sansone [270, Chapter IV, Section 11], see also Uspensky [294], we find a result relating pointwise convergence of the Fourier and Laguerre expansions. In the next theorem

$$\mathfrak{S}_n(\sqrt{x}) = \int_{\sqrt{x}-\delta}^{\sqrt{x}+\delta} \varphi(y^2) \mathfrak{D}_N(\sqrt{x} - y) dy$$

where $N = 2\sqrt{n} - 1/2$, and where $\mathfrak{D}_N(x)$ is as in (B.8). Clearly,

$$S_n(\sqrt{x}) = \int_{\sqrt{x}-\delta}^{\sqrt{x}+\delta} \varphi(y^2) D_N(\sqrt{x} - y) dy$$

is the Nth partial sum of the trigonometric expansion of $\varphi(y^2)$ in the trigonometric series in the interval $(\sqrt{x} - \delta, \sqrt{x} + \delta)$ with any $\delta > 0$. Relations between asymptotic properties of $\mathfrak{S}_n(\sqrt{x})$ and $S_n(\sqrt{x})$ are explained in Lemma B.2.

THEOREM B.5 (EQUICONVERGENCE) *Let $\int |\varphi(x)| e^{-x/2} dx < \infty$. Then, at any $x > 0$,*

$$\lim_{n \to \infty} (\sigma_n(x) - \mathfrak{S}_n(\sqrt{x})) = 0.$$

Applying Lemma B.2, we obtain

THEOREM B.6 (EQUICONVERGENCE) *Let $\int |\varphi(x)| e^{-x/2} dx < \infty$. Then, at any $x > 0$,*

$$\lim_{n \to \infty} (\sigma_n(x) - S_n(\sqrt{x})) = 0.$$

According to the theorem, at any point x_0, $x_0 > 0$, the Laguerre expansion of $\varphi(\bullet)$ behaves like the Fourier expansion of $\varphi(x^2)$ in an arbitrarily small neighborhood of $\sqrt{x_0}$.

Recalling Theorem B.1, we obtain the following two theorems.

THEOREM B.7 *Let $\int_0^\infty |\varphi(x)| e^{-x/2} dx < \infty$. At every $x \in (0, \infty)$, where $\varphi(\bullet)$ is differentiable,*

$$\lim_{n \to \infty} \sigma_n(x) = \varphi(x). \tag{B.30}$$

THEOREM B.8 *Let $\int_0^\infty |\varphi(x)| e^{-x/2} dx < \infty$. At every $x \in (0, \infty)$, where $\varphi(\bullet)$ satisfies a Lipschitz condition, (B.30) holds.*

Applying Theorem B.2, we get

THEOREM B.9 *If $\int_0^\infty |\varphi(x)| e^{-x/2} dx < \infty$, then (B.30) holds for almost every $x \in (0, \infty)$.*

Now, assuming that $\varphi(\bullet)$ is smooth we determine the expansion coefficients. To achieve this goal, we apply associated Laguerre polynomials $L_k^{(\alpha)}(x)$, where for $\alpha > -1$, $L_k^{(\alpha)}(x)$ is defined as follows:

$$L_0^{(\alpha)}(x) = 1$$

and

$$e^{-x} x^\alpha L_k^{(\alpha)}(x) = \frac{1}{k!} \frac{d^k}{dx^k} (x^{k+\alpha} e^{-x}), \text{ for } k = 1, 2, \dots.$$

In the sequel, we use the integer values of α, that is, let $m = 0, 1, 2, \dots$. Functions

$$l_k^{(m)}(x) = \frac{1}{\sqrt{(k+1)\cdots(k+m)}} x^{m/2} e^{-x/2} L_k^{(m)}(x),$$

$k = 0, 1, \dots$, constitute a complete orthonormal system over the half real line $[0, \infty)$. Clearly, $L_k^{(0)}(x) = L_k(x)$. Notice that the symbol $^{(m)}$ does not denote a derivative. To avoid misunderstandings not $^{(q)}$ but d^q/dx^q stands for a derivative of the order q.

The polynomials $\{L_k^{(m)}(x)\}$ satisfy the following differential equation:

$$x \frac{d^2}{dx^2} L_k^{(m)}(x) + (m - x + 1) \frac{d}{dx} L_k^{(m)}(x) + k L_k^{(m)}(x) = 0,$$

(see Sansone [270, Chapter IV, Section 1]), which can be written as

$$-\frac{1}{k} \frac{d}{dx} \left(x^{m+1} e^{-x} \frac{d}{dx} L_k^{(m)}(x) \right) = -e^{-x} x^m L_k^{(m)}(x). \tag{B.31}$$

We assume that, for some $\delta > 0$,

$$\lim_{x \to \infty} \varphi(x) e^{-x/2+\delta} = 0 \tag{B.32}$$

and $\lim_{x \to 0} x\varphi(x) = 0$. The latter, together with the fact that

$$L_k(0) = \frac{(k+m)!}{k! m!} = O(k^m)$$

as $k \to \infty$, implies

$$\lim_{x \to 0} x\varphi(x) L_{k-1}^{(m+1)}(x) = 0.$$

We begin writing

$$\int_0^\infty \varphi(x) l_k^{(m)}(x) dx$$

$$= \frac{1}{\sqrt{(k+1)\cdots(k+m)}} \int_0^\infty \varphi(x) x^{m/2} e^{-x/2} L_k^{(m)}(x) dx$$

$$= \frac{1}{\sqrt{(k+1)\cdots(k+m)}} \int_0^\infty \left(\varphi(x) x^{-m/2} e^{x/2} \right) \left(x^m e^{-x} L_k^{(m)}(x) \right) dx.$$

Integrating by parts, using (B.31) and the fact that

$$\frac{d}{dx} L_k^{(m)}(x) = -L_{k-1}^{(m+1)}(x),$$

we find the integral is equal to

$$\frac{1}{k}\int_0^\infty \left[\frac{d}{dx}\left(\varphi(x)x^{-m/2}e^{x/2}\right)\right]\left(x^{m+1}e^{-x}\frac{d}{dx}L_k^{(m)}(x)\right)dx$$

$$= \frac{1}{k}\int_0^\infty \left(x\varphi'(x)+\frac{1}{2}x\varphi(x)-m\varphi(x)\right)x^{m/2}e^{-x/2}\frac{d}{dx}L_k^{(m)}(x)dx$$

$$= \frac{1}{k}\int_0^\infty \left(x\varphi'(x)+\frac{1}{2}x\varphi(x)-m\varphi(x)\right)x^{m/2}e^{-x/2}L_{k-1}^{(m+1)}(x)dx$$

$$= \frac{\sqrt{k(k+1)\cdots(k+m)}}{k}\int_0^\infty \left(x\varphi'(x)+\frac{1}{2}x\varphi(x)-m\varphi(x)\right)l_{k-1}^{(m+1)}(x)dx.$$

Finally,

$$\int_0^\infty \varphi(x)l_k^{(m)}(x)dx = \frac{1}{\sqrt{k}}\int_0^\infty \left(x\varphi'(x)+\frac{1}{2}x\varphi(x)-m\varphi(x)\right)l_{k-1}^{(m+1)}(x)dx.$$

Thus, denoting

$$\Phi_m\varphi(x) = x^{1+m/2}e^{-x/2}\frac{d}{dx}\left(\varphi(x)x^{-m/2}e^{x/2}\right),$$

we find

$$\Phi_m\varphi(x) = x\varphi'(x)+\frac{1}{2}x\varphi(x)-m\varphi(x) = \frac{1}{2}(2xD+x-2m)\varphi(x),$$

where D stands for d/dx, and write

$$\int_0^\infty \varphi(x)l_k^{(m)}(x)dx = \frac{1}{\sqrt{k}}\int_0^\infty \left(\Phi_m\varphi(x)\right)l_{k-1}^{(m+1)}(x)dx$$

$$= \frac{1}{2\sqrt{k}}\int_0^\infty \left[(2xD+x-2m)\,\varphi(x)\right]l_{k-1}^{(m+1)}(x)dx.$$

Therefore,

$$a_k = \int_0^\infty \varphi(x)l_k(x)dx$$

$$= \frac{1}{\sqrt{k(k-1)\cdots(k-q+1)}}\int_0^\infty \left(\Phi_{q-1}\cdots\Phi_1\Phi_0\varphi(x)\right)l_{k-q}^{(q)}(x)dx,$$

provided that (B.32) holds and $\lim_{x\to 0}x^s\varphi^{(i)}(x) = 0$, for $i = 0, 1, 2, \ldots, q-1$. Hence,

$$a_k = \frac{1}{\sqrt{k(k-1)\cdots(k-q+1)}}b_{k-q}^{(q)} = O(1)\frac{1}{k^{q/2}}b_{k-q}^{(q)},$$

where

$$b_{qk} = \int_0^\infty \left(\Phi_{q-1}\cdots\Phi_0 g(x)\right)l_k^{(q)}(x)dx$$

$$= \frac{1}{2^q\sqrt{k}}\int_0^\infty \left[\left(\prod_{s=0}^{q-1}(2xD+x-2s)\right)\varphi(x)\right]l_k^{(q)}(x)dx$$

is the appropriate coefficient of the expansion of

$$\Phi_{q-1} \cdots \Phi_0 \varphi(x) = \left(\prod_{s=0}^{q-1} (2xD + x - 2s) \right) \varphi(x)$$

in the series $l_k^{(q)}(x)$.

If $\int_0^\infty (\Phi_{q-1} \cdots \Phi_0 g(x))^2 dx < \infty$, $\sum_{k=0}^\infty (b_{qk})^2 < \infty$, then

$$\sum_{k=n+1}^\infty a_k^2 = O(1) \sum_{k=n+1}^\infty \frac{1}{k^q}(b_{q,k-q})^2 = O(1)\frac{1}{n^q} \sum_{k=0}^\infty (b_{qk})^2.$$

Consequently, we have the following bound for the integrated squared error

$$\int_0^\infty (\sigma_n(x) - \varphi(x))^2 dx = O(n^{-q}).$$

Concerning the pointwise error, we have (see (B.34))

$$\sum_{k=n+1}^\infty |a_k||l_k(x)| = O(1) \sum_{k=n+1}^\infty \frac{1}{k^{q/2+1/4}}|b_{q,k-q}|$$

$$\leq O(1) \left(\sum_{k=n+1}^\infty \frac{1}{k^{q+1/2}} \right)^{1/2} \left(\sum_{k=n+1}^\infty (b_{q,k-q})^2 \right)^{1/2}$$

$$= O(n^{-q/2+1/4}).$$

This implies that at every point $x \in (0, \infty)$ where the expansion converges to $\varphi(x)$,

$$|\sigma_n(x) - \varphi(x)| = O(n^{-q/2+1/4}). \tag{B.33}$$

For our next developments we need some inequalities for Laguerre polynomials. On the interval $0 < x < a$, any a, we have

$$L_n(x) = x^{-1/4} O(n^{-1/4}), \tag{B.34}$$

$$L_n^{(-1)}(x) = O(n^{-3/4}), \tag{B.35}$$

see Szegö [288, formulas (7.6.9) and (7.6.10)]. On the interval $a \leq x < \infty$, any $a > 0$,

$$e^{-x/2} L_n(x) = x^{-1/4} O(n^{-1/12}), \tag{B.36}$$

$$e^{-x/2} L_n^{(-1)}(x) = x^{1/4} O(n^{-7/12}), \tag{B.37}$$

see Szegö [288, formula (8.91.2)]. Moreover,

$$L_n(x) = L_{n+1}(x) - L_{n+1}^{(-1)}(x), \tag{B.38}$$

see also Szegö [288, formula (5.1.13)].

Passing to the next lemma we notice that if $\int_0^\infty |\varphi(x)|dx < \infty$ and $\lim_{x \to 0} \varphi(x)$ is finite, then $\int_0^1 x^{-1/2}|\varphi(x)|dx < \infty$.

LEMMA B.4 *If $\int_0^\infty |\varphi(x)|dx < \infty$ and $\int_0^1 x^{-1/2}|\varphi(x)|dx < \infty$, then*

$$\lim_{n\to\infty} \frac{1}{\sqrt{n}} \int_0^\infty k^2(x,y)\varphi(y)dy = \frac{1}{\pi\sqrt{x}}\varphi(x)$$

at every $x > 0$ where $\varphi(\bullet)$ is continuous and at almost every $x \in (0, \infty)$.

Proof. We recall the Christoffel–Darboux formula and write $k_n(x,y) = e^{-x/2}$ $e^{-y/2}K_n(x,y)$, where

$$K_n(x,y) = (n+1)\frac{L_{n+1}(x)L_n(y) - L_n(x)L_{n+1}(y)}{y-x}$$

$$= (n+1)\frac{L_{n+1}(x)L_{n+1}^{(-1)}(y) - L_{n+1}^{(-1)}(x)L_{n+1}(y)}{x-y}.$$

We substituted $L_{n+1}(x) - L_{n+1}^{(-1)}(x)$ for $L_n(x)$ and did similarly for $L_n(y)$, see (B.38).

Further considerations hold for every $x > 0$, where $\varphi(\bullet)$ is continuous and for almost every $x > 0$. Let $0 < \varepsilon < x$.

On the interval $|x - y| \le \varepsilon$

$$K_n(x,y) = e^x \frac{1}{2\sqrt{y}}\mathfrak{D}_N\left(\sqrt{y} - \sqrt{x}\right) + O(1)$$

with $N = 2\sqrt{n} - 1/2$, where $\mathfrak{D}_N(x)$ is as in (B.8), see Szegö [288, formula (9.5.6)]. Hence

$$K_n^2(x,y) = e^{2x}\frac{1}{4\pi}\left(N + \frac{1}{2}\right)\frac{1}{y}\mathfrak{F}_N\left(\sqrt{y} - \sqrt{x}\right)$$

$$+ O(1)e^x \frac{1}{\sqrt{y}}\mathfrak{D}_N\left(\sqrt{y} - \sqrt{x}\right) + O(1),$$

see (B.10). Since

$$\int_{|x-y|\le\varepsilon} \varphi(y)e^{-y}\frac{1}{y}\mathfrak{F}_N\left(\sqrt{y} - \sqrt{x}\right) dy = 2\int_{\sqrt{x-\varepsilon}}^{\sqrt{x+\varepsilon}} \frac{1}{v}\varphi(v^2)e^{-v^2}\mathfrak{F}_N\left(v - \sqrt{x}\right) dv$$

converges to $2x^{-1/2}e^{-x}\varphi(x)$ as $N \to \infty$, applying (B.15), we obtain

$$\lim_{n\to\infty} \frac{1}{\sqrt{n}} \int_{|x-y|\le\varepsilon} k^2(x,y)\varphi(y)dy = \frac{1}{\pi\sqrt{x}}\varphi(x).$$

Let $0 < y \le x - \varepsilon$. Using (B.34) and (B.35), we obtain $K_n(x,y) = O(1)\left(1 + y^{-1/4}\right)$, which leads to $\int_0^\varepsilon k_n^2(x,y)\varphi(y)dy = O(1)\int_0^\varepsilon y^{-1/2}|\varphi(y)|dy$. Hence

$$\lim_{n\to\infty} \frac{1}{\sqrt{n}} \int_0^{x-\varepsilon} \varphi(y)k^2(x,y)dy = 0.$$

For $x + \varepsilon \le y$, we apply (B.36) and (B.37), to obtain

$$e^{-x/2}e^{-y/2}\frac{L_{n+1}^{(-1)}(x)L_{n+1}(y)}{x-y} = y^{-1}O(n^{-7/12})y^{1/4}O(n^{-1/12}) = O(n^{-2/3})$$

and

$$e^{-x/2}e^{-y/2}\frac{L_{n+1}^{(-1)}(x)L_{n+1}(y)}{x-y} = y^{-1}O(n^{-7/12})y^{-1/4}O(n^{-1/12}) = O(n^{-2/3}).$$

Hence,

$$\lim_{n\to\infty}\frac{1}{\sqrt{n}}\int_{x+\varepsilon}^{\infty}k_n^2(x,y)\varphi(y)dy = 0,$$

which completes the proof. ∎

It is worth mentioning that the orthonormal Laguerre functions $\{l_k(x)\}$ satisfy (see Szegö [288]), the following often used convolution property:

$$l_n(x) * l_m(x) = l_{n+m}(x) + l_{n+m+1}(x).$$

Furthemore, it is proved in Budke [36] that the Laguerre functions define the only complete and orthonormal system of functions on $[0, \infty)$ satisfying the convolution property.

B.5   Hermite series

A function $H_k(x)$ satisfying Rodrigues' formula

$$e^{-x^2}H_k(x) = (-1)^k\frac{d^k}{dx^k}e^{-x^2},$$

$n = 0, 1, \ldots$, is the kth Hermite polynomial, see Szegö [288, Chapter V], or Sansone [270, Chapter IV]. Notice, however, that definitions in these references differ by the factor $(-1)^n$. One can easily calculate

$$H_0(x) = 1,$$
$$H_1(x) = 2x,$$
$$H_2(x) = 4x^2 - 2,$$
$$H_3(x) = 8x^3 - 12x,$$
$$H_4(x) = 16x^4 - 48x^2 + 12,$$
$$H_5(x) = 32x^5 - 160x^3 + 120x,$$

and so on. Functions

$$h_k(x) = \frac{1}{\sqrt{2^k k!\sqrt{\pi}}}H_k(x)e^{-x^2/2},$$

$k = 0, 1, 2, \ldots$, constitute a Hermite system which is complete and orthonormal on the entire real line R. For Hermite functions the following inequalities hold (see Szegö [288] Theorem 8.91.3),

$$\max_{-\infty < x < \infty} |h_k(x)| \leq \frac{c}{(k+1)^{1/12}}, \tag{B.39}$$

$$\max_{-a \leq x \leq a} |h_k(x)| \leq \frac{d(a)}{(k+1)^{1/4}}, \tag{B.40}$$

for any $a \geq 0$. Moreover,

$$H_k'(x) = 2k H_{k-1}(x), \tag{B.41}$$

(see Szegö [288], formula (5.5.10)) which leads to

$$h_k'(x) = \sqrt{\frac{k}{2}} h_{k-1}(x) - \sqrt{\frac{k+1}{2}} h_{k+1}(x).$$

Thus,

$$\max_{-\infty < x < \infty} |h_k'(x)| = O(k^{5/12}) \tag{B.42}$$

and

$$\max_{-a \leq x \leq a} |h_k'(x)| = O(k^{1/4}) \tag{B.43}$$

for any $a \geq 0$.

According to the Christoffel–Darboux formula,

$$\sum_{k=0}^{n} \frac{1}{2^k k!} H_k(x) H_k(y) = \frac{H_{n+1}(x) H_n(y) - H_n(x) H_{n+1}(y)}{2^{n+1} n! (y - x)},$$

which is equivalent to

$$k_n(x, y) = \sqrt{\frac{n+1}{2}} \frac{h_{n+1}(x) h_n(y) - h_n(x) h_{n+1}(y)}{y - x}, \tag{B.44}$$

where

$$k_n(x, y) = \sum_{k=0}^{n} h_k(x) h_k(y)$$

is the kernel of the Hermite system.

Let

$$\sigma_n(x) = \sum_{k=0}^{n} a_k h_k(x),$$

where

$$a_k = \int_{-\infty}^{\infty} \varphi(x) h_k(x) dx.$$

The next result is due to Uspensky [294], see also see Szegö [288, Theorem 9.1.6] or Sansone [270, Chapter IV, Section 10]. In the next theorem

$$\mathfrak{S}_n(x) = \int_{x-\delta}^{x+\delta} \varphi(y)\mathfrak{D}_N(x-y)dy,$$

where $N = 2\sqrt{n} - 1/2$, and where $\mathfrak{D}_N(x)$ is as in (B.8). Clearly,

$$S_n(x) = \int_{x-\delta}^{x+\delta} \varphi(y)D_N(x-y)dy$$

is the Nth partial sum of the trigonometric expansion of $\varphi(y)$ in the trigonometric series in the interval $(x - \delta, x + \delta)$ for any $\delta > 0$. Relations between asymptotic properties of $\mathfrak{S}_n(x)$ and $S_n(x)$ are established in Lemma B.2.

THEOREM B.10 (EQUICONVERGENCE) *Let $\int_{-\infty}^{\infty} |\varphi(x)|e^{-x^2/2}dx < \infty$. Then, at every $x \in (-\infty, \infty)$,*

$$\lim_{n\to\infty}(\sigma_n(x) - \mathfrak{S}_n(x)) = 0.$$

Application of Lemma B.2 yields

THEOREM B.11 (EQUICONVERGENCE) *Let $\int_{-\infty}^{\infty} |\varphi(x)|e^{-x^2/2}dx < \infty$. Then, at every $x \in (-\infty, \infty)$,*

$$\lim_{n\to\infty}(\sigma_n(x) - S_n(x)) = 0.$$

Thus, at any point the Hermite expansion of $\varphi(\bullet)$ behaves like the Fourier expansion of $\varphi(\bullet)$ in an arbitrarily small neighborhood of the point.

Recalling Theorem B.1, we obtain the next two theorem.

THEOREM B.12 *Let $\int_{-\infty}^{\infty} |\varphi(x)|e^{-x^2/2}dx < \infty$. At every point $x \in R$ where $\varphi(\bullet)$ is differentiable,*

$$\lim_{n\to\infty}\sigma_n(x) = \varphi(x). \tag{B.45}$$

THEOREM B.13 *Let $\int_{-\infty}^{\infty} |\varphi(x)|e^{-x^2/2}dx < \infty$. At every point x where $\varphi(\bullet)$ satisfies a Lipschitz condition, (B.45) holds.*

Revoking Theorem B.2, we get

THEOREM B.14 *If $\int_{-\infty}^{\infty} |\varphi(x)|e^{-x^2/2}dx < \infty$, then (B.45) holds for almost every $x \in R$.*

To evaluate the expansion coefficients for smooth functions, we use (B.41), and we obtain

$$\int_{-\infty}^{\infty} \varphi(x)H_k(x)e^{-x^2/2}dx = \frac{1}{2(k+1)}\int_{-\infty}^{\infty} \varphi(x)e^{-x^2/2}H'_{k+1}(x)dx.$$

Assuming that $\lim_{|x|\to\infty} \varphi(x)e^{-x^2/2} H_{k+1}(x) = 0$ and integrating by parts, we find that the quantity is equal to

$$\frac{1}{2(k+1)} \int_{-\infty}^{\infty} \left(\varphi(x)e^{-x^2/2}\right)' H_{k+1}(x)dx$$

$$= -\frac{1}{2(k+1)} \int_{-\infty}^{\infty} (\varphi'(x) - x\varphi(x))e^{-x^2/2} H_{k+1}(x)dx$$

Thus,

$$a_k = -\frac{1}{\sqrt{2(k+1)}} b_{1,k+1},$$

where $b_{1,k} = -\int_{-\infty}^{\infty}[(D-x)\,\varphi(x)]h_k(x)dx$ with D standing for d/dx.

In general, if, for $m = 0, 1, \ldots, q-1$, $\lim_{|x|\to\infty} \varphi^{(m)}(x)e^{-x^2/2+\delta} = 0$ with some $\delta > 0$,

$$a_k = (-1)^q \frac{1}{2^{q/2}\sqrt{(k+1)(k+2)(k+q)}} b_{q,k+q} = O\left(\frac{1}{k^{q/2}}\right) b_{q,k+q},$$

where $b_{q,k} = \int_{-\infty}^{\infty}[(D-x)^q\,\varphi(x)]h_k(x)dx$. If, moreover,

$$\int_{-\infty}^{\infty}[(D-x)^q\,\varphi(x)]^2 dx < \infty,$$

we have $\sum_{k=0}^{\infty} b_{q,k}^2 < \infty$, and consequently we obtain

$$\sum_{k=n+1}^{\infty} a_k^2 = \sum_{k=n+1}^{\infty} O(k^{-q}) b_{q,k+q}^2 = O(1)\frac{1}{n^q} \sum_{k=n+1}^{\infty} b_{q,k+q}^2.$$

Thus,

$$\int_{-\infty}^{\infty} (\sigma_n(x) - \varphi(x))^2 dx = O(n^{-q}).$$

Moreover,

$$\sum_{k=n+1}^{\infty} |a_k h_k(x)| = \sum_{k=n+1}^{\infty} O\left(\frac{1}{k^{q/2+1/4}}\right) |b_{q,k+q}|.$$

Since the sum is bounded by

$$O(1)\left(\sum_{k=n+1}^{\infty} \frac{1}{k^{q+1/2}}\right)^{1/2} \left(\sum_{k=n+1}^{\infty} b_{q,k+q}^2\right)^{1/2} = O(1)\left(\sum_{k=n+1}^{\infty} \frac{1}{k^{q+1/2}}\right)^{1/2},$$

at every x where (B.45) holds, we get

$$|\sigma_n(x) - \varphi(x)| = O(n^{-q/2+1/4}). \tag{B.46}$$

LEMMA B.5 *Let $\int_{-\infty}^{\infty} |\varphi(x)|dx < \infty$. Then,*

$$\lim_{n\to\infty} \frac{1}{\sqrt{n}} \int_{-\infty}^{\infty} k_n^2(x, y)\varphi(y)dy = \frac{\sqrt{2}}{\pi}\varphi(x)$$

at every continuity point x of $\varphi(\bullet)$ and at almost every $x \in (-\infty, \infty)$.

Proof. The considerations hold for every continuity point x of $\varphi(\bullet)$ and for almost every point x. For any fixed x and any $\varepsilon > 0$, in the interval $x - \varepsilon \le y \le x + \varepsilon$

$$k_n(x, y) = \mathfrak{D}_{N-1/2}(y - x) + \frac{1}{N}O(1)$$

with $2N = \sqrt{2n+3} + \sqrt{2n+1}$, where $\mathfrak{D}_N(x)$ is as in (B.8), see Sansone [270, Chapter IV, Section 10]. Consequently,

$$k_n^2(x, y) = \frac{N}{\pi}\mathfrak{F}_{2N}(y - x) + \frac{1}{N}O(1)\mathfrak{D}_{N-1/2}(y - x) + \frac{1}{N^2}O(1),$$

see (B.7). Thus, recalling Remark B.2 and noticing that $N/n^{1/2} \to \sqrt{2}$ as $n \to \infty$, we get

$$\lim_{n\to\infty} \frac{1}{\sqrt{n}} \frac{N}{\pi} \int_{|x-y|\le\varepsilon} \mathfrak{F}_{2N}(y - x)\varphi(y)dy = \frac{\sqrt{2}}{\pi}\varphi(x).$$

Applying (B.15), we come to the conclusion that

$$\lim_{n\to\infty} \frac{1}{\sqrt{n}} \int_{|x-y|\le\varepsilon} k_n^2(x, y)\varphi(y)dy = \frac{\sqrt{2}}{\pi}\varphi(x).$$

Moreover, by (B.39) and (B.40),

$$\max_{|x-y|>\varepsilon} |k_n(x, y)| \le \frac{1}{\varepsilon\sqrt{2}}(n + 1)^{1/2} \max_{|x-y|>\varepsilon} |h_{n+1}(y)h_n(x) - h_n(y)h_{n+1}(x)|$$

$$\le \frac{1}{\varepsilon\sqrt{2}}(n + 1)^{1/2}\frac{2d(x)c}{(n + 1)^{1/4+1/12}} = \frac{1}{\varepsilon}O(n^{1/6}),$$

which yields

$$\lim_{n\to\infty} \frac{1}{\sqrt{n}} \int_{|x-y|>\varepsilon} k_n^2(x, y)\varphi(y)dy = 0$$

and completes the proof. ■

B.6   Wavelets

In this section, we discuss orthogonal wavelets. From a certain viewpoint there is some basic difference between wavelets and orthogonal Fourier, Legendre, Laguerre, and Hermite orthogonal polynomials. All polynomial expansions behaves very similarly. Convergence of Legendre, Laguerre, and Hermite expansions to expanded functions at a point is strictly related to those of the Fourier one in a neighborhood of the point, see equiconvergence theorems B.4, B.6, and B.11. This is not the case with wavelets.

In our considerations we deal with Haar wavelets. Let $\phi(x) = I_{[0,1)}(x)$, where I is the indicator function, that is, where

$$I_A(x) = \begin{cases} 1, & \text{for } x \in A, \\ 0, & \text{otherwise.} \end{cases}$$

For a fixed m being a nonnegative integer and $n = 0, 1, \ldots, 2^{m-1}$, we define

$$\phi_{mn}(x) = 2^{m/2} \phi \left(2^m x - n \right).$$

Clearly,

$$\phi_{mn}(x) = I_{A_{mn}}(x),$$

where

$$A_{mn} = \left[\frac{n}{2^m}, \frac{n+1}{2^m} \right)$$

The set A_{mn} is the support of $\phi_{mn}(\bullet)$. For $n \neq k$, $A_{mn} \cap A_{mk} = \emptyset$, which means that $\phi_{mn}(\bullet)$ and $\phi_{mk}(\bullet)$ do not overlap. Thus, for $n \neq k$,

$$\int_0^1 \phi_{mn}(x) \phi_{mk}(x) = 0.$$

Hence, functions $\phi_{mn}(\bullet)$ and $\phi_{mk}(\bullet)$ are, for $n \neq k$, orthogonal. Since $\cup_{n=0}^{2^m-1} A_{mn} = [0, 1)$, then $\sum_{n=0}^{2^m-1} \phi_{mn}(x) = \phi(x) = I_{[0,1)}(x)$. Observe that, for each m, we have 2^m functions

$$\phi_{m0}, \phi_{m1}, \ldots, \phi_{m,2^m-1}.$$

The number of functions increases with m as shown below:

$$\phi_{00},$$
$$\phi_{10}, \quad \phi_{11},$$
$$\phi_{20}, \quad \phi_{21}, \quad \phi_{22} \quad \phi_{23},$$
$$\phi_{30}, \quad \phi_{31}, \quad \phi_{32}, \quad \phi_{33}, \quad \phi_{34}, \quad \phi_{35}, \quad \phi_{36}, \quad \phi_{37},$$
$$\vdots \qquad \vdots \qquad \vdots \qquad \vdots \qquad \vdots \qquad \vdots \qquad \vdots \qquad \vdots$$
$$\phi_{m0}, \quad \phi_{m1}, \quad \phi_{m2}, \quad \phi_{m3}, \quad \phi_{m4}, \quad \phi_{m5}, \quad \phi_{m6}, \quad \phi_{m7}, \quad \cdots \quad \phi_{m,2^m-1}.$$

Since support of $\phi_{mn}(\bullet)$, that is, A_{mn} has the length $1/2^m$, we have

$$\int_{-1}^1 \phi_{mn}^2(x) dx = \frac{1}{2^m}.$$

Therefore, the system

$$\varphi_{m0}, \varphi_{m1}, \ldots, \varphi_{m,2^m-1}$$

with $\varphi_{mn}(x) = 2^{m/2} \phi_{mn}(x)$ is orthonormal.

For obvious reasons,

$$\sigma_m(x) = \sum_{n=0}^{2^m-1} c_{mn}\varphi_{mn}(x)$$

with

$$c_{mn} = \int_0^1 \varphi(x)\varphi_{mn}(x)dx$$

is the mth partial expansion of $\varphi(\bullet)$ into the wavelet series. Clearly,

$$\sigma_m(x) = 2^m \sum_{n=0}^{2^m-1} a_{mn}\phi_{mn}(x)$$

with

$$a_{mn} = \int_0^1 \varphi(x)\phi_{mn}(x)dx.$$

THEOREM B.15 *If $\int_0^1 |\varphi(x)|dx < \infty$, then*

$$\sigma_m(x) \to f(x) \text{ as } m \to \infty$$

at every point $x \in (0, 1)$ where $\varphi(\bullet)$ is continuous.

Proof. Fix $x \in (0, 1)$ and assume that $\varphi(\bullet)$ is continuous at the point. The points belongs to exactly one of the intervals A_{mn}, $n = 0, 1, \ldots, 2^m - 1$. Denoting the interval by A_{mn_x}, we can write

$$\sigma_m(x) = 2^m a_{mn_x} = \frac{1}{2^{-m}} \int_{A_{mn_x}} \varphi(y)dy.$$

Since the length of A_{nm_x} is 2^{-m}, by the mean-value theorem,

$$\sigma_m(x) \to \varphi(x) \text{ as } m \to \infty$$

which completes the proof. ∎

If, $\varphi(\bullet)$ satisfies the Lipschitz condition, that is, if

$$|\varphi(x) - \varphi(y)| \le c|x - y|$$

for all $x, y \in [0, 1]$, then

$$|\sigma_m(x) - \varphi(x)| \le \frac{1}{2^{-m}} \int_{A_{mn_x}} |\varphi(y) - \varphi(x)|dy \le \frac{c}{2^{-m}} \int_{A_{mn_x}} |x - y|dy \le \frac{c}{2^m}.$$

Hence,

$$|\sigma_m(x) - \varphi(x)| = O\left(\frac{1}{2^m}\right) \text{ as } m \to \infty. \tag{B.47}$$

Let

$$k_m(x, y) = \sum_{n=0}^{2^m-1} \varphi_n(x)\varphi_n(y) = 2^m \sum_{n=0}^{2^m-1} \phi_n(x)\phi_n(y)$$

denote the wavelet kernel function.

LEMMA B.6 *If $\int_0^1 |\varphi(x)|dx < \infty$, then*

$$\lim_{m\to\infty} \frac{1}{2^m} \int_0^1 k_m^2(x, y)\varphi(y)dy = \varphi(x)$$

at every point $x \in (0, 1)$ where $\varphi(\bullet)$ is continuous.

Proof. Fix $x \in (0, 1)$. Bearing in mind that $\phi_{mn}(x) = I_{A_{mn}}(x)$, we can write

$$k_m(x, y) = 2^m \sum_{n=0}^{2^m-1} \phi_n(x)\phi_n(y) = 2^m \sum_{n=0}^{2^m-1} I_{A_{mn}}(x)I_{A_{mn}}(y).$$

Since x belongs to exactly one of the intervals A_{mn}, $n = 0, 1, \ldots, 2^m - 1$, denoting the interval by A_{mn_x} we find the kernel equal to $2^m I_{A_{mn_x}}(y)$. Thus

$$\frac{1}{2^m} \int_0^1 k_m^2(x, y)\varphi(y)dy = 2^m \int_0^1 I_{A_{mn_x}}(y)\varphi(y)dy = 2^m \int_{A_{mn_x}} \varphi(y)dy$$

which converges to $\varphi(x)$ as $m \to \infty$ at every continuity point of $\varphi(\bullet)$, see the proof of Theorem B.15. The proof is completed. ∎

For other wavelets and their mathematical properties, see Daubechies [60], Chui [52], Walter and Shen [308], Ogden [223], Mallat [204] or Wojtaszczyk [324].

Appendix C Probability and statistics

C.1　White noise

C.1.1　Discrete time

LEMMA C.1 *If* $\{U_n\}$ *is a stationary white random process, then, for any functions* $f(\bullet)$, $g(\bullet)$, $\varphi(\bullet)$, *and* $\psi(\bullet)$,

$$\text{cov}\left[f(U_q)\varphi(U_i), g(U_r)\psi(U_j)\right]$$

$$=\begin{cases}
\text{cov}\left[f(U)\varphi(U), g(U)\psi(U)\right], & \text{for } q=i=r=j, \\
E\varphi(U)\,\text{cov}\left[f(U), g(U)\psi(U)\right], & \text{for } q=r=j, q\neq i, \\
E\psi(U)\,\text{cov}\left[f(U)\varphi(U), g(U))\right], & \text{for } q=i=r, q\neq j, \\
Eg(U)\,\text{cov}\left[f(U)\varphi(U), \psi(U)\right], & \text{for } q=i=j, q\neq r, \\
Ef(U)\,\text{cov}\left[\varphi(U), g(U)\psi(U)\right], & \text{for } i=r=j, q\neq i, \\
E\left\{f(U)g(U)\right\}E\left\{\varphi(U)\psi(U)\right\} \\
\quad -Ef(U)Eg(U)E\varphi(U)E\psi(U), & \text{for} q=r, i=j, q\neq i, \\
E\left\{f(U)\psi(U)\right\}E\left\{\varphi(U)g(U)\right\} \\
\quad -Ef(U)Eg(U)E\varphi(U)E\psi(U), & \text{for } q=j, i=r, q\neq i, \\
0, & \text{otherwise.}
\end{cases}$$

Proof. The proof is straightforward. ∎

The next lemma is a simple consequence of Lemma C.1.

LEMMA C.2 *Let* $\{U_n\}$ *be a stationary white random process, such that* $Em(U)=0$. *For any functions* $\varphi(\bullet)$, *and* $\psi(\bullet)$,

$$\text{cov}\left[m(U_q)\varphi(U_i), m(U_r)\psi(U_j)\right]$$

$$=\begin{cases}
\text{cov}\left[m(U)\varphi(U), m(U)\psi(U)\right], & \text{for } q=i=r=j, \\
E\varphi(U)E\left\{m^2(U)\psi(U)\right\}, & \text{for } q=r=j, q\neq i, \\
E\psi(U)E\left\{m^2(U)\varphi(U)\right\}, & \text{for } q=i=r, q\neq j, \\
\text{var}\left[m(U)\right]E\left\{\varphi(U)\psi(U)\right\}, & \text{for } q=r, i=j, q\neq i, \\
E\left\{m(U)\psi(U)\right\}E\left\{m(U)\varphi(U)\right\}, & \text{for } q=j, i=r, q\neq i, \\
0, & \text{otherwise.}
\end{cases}$$

Using Lemma C.2, we can verify Lemma C.3.

LEMMA C.3 *Let $\{U_n\}$ be a stationary white random process such that $EU = 0$ and $Em(U) = 0$. Then,*

$$\operatorname{cov}\left[m(U_q)U_i, m(U_r)U_j\right] = \begin{cases} \operatorname{var}\left[Um(U)\right], & \text{for } q = i = r = j, \\ \sigma_U^2 \operatorname{var}\left[m(U)\right], & \text{for } q = r, i = j, q \neq i, \\ E^2\left\{Um(U)\right\}, & \text{for } q = j, i = r, q \neq i, \\ 0, & \text{otherwise.} \end{cases}$$

C.1.2   Continuous time

LEMMA C.4 *If $\{U(t)\}$ is a stationary white random process, then, for any functions $f(\bullet)$, $g(\bullet)$, $\varphi(\bullet)$, and $\psi(\bullet)$,*

$$\begin{aligned}
\operatorname{cov}&\left[f(U(t))\varphi(U(\xi)), g(U(\tau))\psi(U(\eta))\right] \\
&= \delta(t - \xi)\delta(\xi - \tau)\delta(\tau - \eta)\operatorname{cov}\left[f(U)\varphi(U), g(U)\psi(U)\right] \\
&\quad + \delta(t - \tau)\delta(\tau - \eta)E\varphi(U)\operatorname{cov}\left[f(U), g(U)\psi(U)\right] \\
&\quad + \delta(t - \xi)\delta(\xi - \tau)E\psi(U)\operatorname{cov}\left[f(U)\varphi(U), g(U)\right] \\
&\quad + \delta(t - \xi)\delta(\xi - \eta)Eg(U)\operatorname{cov}\left[f(U)\varphi(U), \psi(U)\right] \\
&\quad + \delta(\xi - \tau)\delta(\tau - \eta)Ef(U)\operatorname{cov}\left[\varphi(U), g(U)\psi(U)\right] \\
&\quad + \delta(t - \tau)\delta(\xi - \eta)\left[E\left\{f(U)g(U)\right\}E\left\{\varphi(U)\psi(U)\right\} \right. \\
&\quad \left. - Ef(U)Eg(U)E\varphi(U)E\psi(U)\right] \\
&\quad + \delta(t - \eta)\delta(\xi - \tau)\left[E\left\{f(U)\psi(U)\right\}E\left\{\varphi(U)g(U)\right\} \right. \\
&\quad \left. - Ef(U)Eg(U)E\varphi(U)E\psi(U)\right].
\end{aligned}$$

LEMMA C.5 *Let $\{U(t)\}$ be a stationary white random process such that $Em(U) = 0$. For any functions $\varphi(\bullet)$, and $\psi(\bullet)$,*

$$\begin{aligned}
\operatorname{cov}&\left[m(U(t))\varphi(U(\xi)), m(U(\tau))\psi(U(\eta))\right] \\
&= \delta(t - \xi)\delta(\xi - \tau)\delta(\tau - \eta)\operatorname{cov}\left[m(U)\varphi(U), m(U)\psi(U)\right] \\
&\quad + \delta(t - \tau)\delta(\tau - \eta)E\varphi(U)E\left\{m^2(U)\psi(U)\right\} \\
&\quad + \delta(t - \xi)\delta(\xi - \tau)E\psi(U)E\left\{m^2(U)\varphi(U)\right\} \\
&\quad + \delta(t - \tau)\delta(\xi - \eta)\operatorname{var}\left[m(U)\right]E\left\{\varphi(U)\psi(U)\right\} \\
&\quad + \delta(t - \eta)\delta(\xi - \tau)E\left\{m(U)\psi(U)\right\}E\left\{\varphi(U)m(U)\right\}.
\end{aligned}$$

LEMMA C.6 *Let $\{U(t)\}$ be a stationary white random process such that $Em(U) = 0$. Then,*

$$\begin{aligned}
\operatorname{cov}&\left[m(U(t))U(\xi), m(U(\tau))U(\eta)\right] \\
&= \delta(t - \xi)\delta(\xi - \tau)\delta(\tau - \eta)\operatorname{var}\left[Um(U)\right] \\
&\quad + \delta(t - \tau)\delta(\xi - \eta)\sigma_U^2 \operatorname{var}\left[m(U)\right] \\
&\quad + \delta(t - \eta)\delta(\xi - \tau)E^2\left\{Um(U)\right\}.
\end{aligned}$$

C.2   Convergence of random variables

DEFINITION C.1 *A sequence $\{X_n\}$ of random variables is said to converge to zero in probability if*

$$\lim_{n \to \infty} P\{|X_n| > \varepsilon\} = 0,$$

for every $\varepsilon > 0$.

We write

$$X_n = O_P(\alpha_n) \quad \text{as } n \to \infty$$

and say that X_n converges to zero in probability as fast as $O_P(\alpha_n)$ if

$$\lambda_n \frac{1}{\alpha_n} |X_n| \to 0 \text{ in probability as } n \to \infty$$

for every number sequence λ_n converging to zero.

LEMMA C.7 *If $E(X_n - a)^2 = O(\alpha_n)$, then, for any $\varepsilon > 0$,*

$$P\{|X_n - a| > \varepsilon\} = O(\alpha_n) \text{ as } n \to \infty$$

and

$$|X_n - a| = O_P(\sqrt{\alpha_n}) \text{ as } n \to \infty.$$

Proof. Fix $\varepsilon > 0$. Since $EX_n \to a$ as $n \to \infty$, $|EX_n - a| < \varepsilon/2$ for n large enough. Thus,

$$P\{|X_n - a| > \varepsilon\} \le P\{|X_n - EX_n| + |EX_n - a| > \varepsilon\}$$
$$\le P\{|X_n - EX_n| > \varepsilon/2\}$$

which, by Chebyshev's inequality, is not greater than

$$\frac{4 \operatorname{var}[X_n]}{\varepsilon^2} \le \frac{4}{\varepsilon^2} E(X_n - a)^2 = O(\alpha_n).$$

For the same reasons, for n large enough,

$$P\left\{ \frac{\lambda_n}{\sqrt{\alpha_n}} |X_n - a| > \varepsilon \right\} = P\left\{ |X_n - a| > \varepsilon \frac{\sqrt{\alpha_n}}{\lambda_n} \right\} \le \frac{4}{\varepsilon^2} \frac{\lambda_n^2}{\alpha_n} E(X_n - a)^2$$
$$= \lambda_n^2 O(1),$$

which converges to zero as $n \to \infty$. ∎

We shall now examine a quotient Y_n/X_n of random variables. The next lemma, in part, can be found in Greblicki and Krzyżak [121] or Greblicki and Pawlak [124].

LEMMA C.8 *If, for any $\varepsilon > 0$,*

$$P\{|Y_n - a| > \varepsilon\} = O(\alpha_n),$$

$$P\{|X_n - b| > \varepsilon\} = O(\beta_n)$$

with $b \neq 0$, then

$$P\left\{\left|\frac{Y_n}{X_n} - \frac{a}{b}\right| > \varepsilon\left|\frac{a}{b}\right|\right\} = O(\gamma_n)$$

and

$$\left|\frac{Y_n}{X_n} - \frac{a}{b}\right| = O_P(\sqrt{\gamma_n}) \text{ as } n \to \infty,$$

where $\gamma_n = \max(\alpha_n, \beta_n)$.

Proof. Fix $\varepsilon > 0$. For simplicity, let $a \geq 0$ and $b > 0$. We begin with the following inequality:

$$\left|\frac{Y_n}{X_n} - \frac{a}{b}\right| \leq \left|\frac{Y_n}{X_n}\right|\left|\frac{X_n - b}{b}\right| + \left|\frac{Y_n - a}{b}\right|. \tag{C.1}$$

Hence,

$$|Y_n - a| \leq \frac{\varepsilon}{2 + \varepsilon}a$$

and

$$|X_n - b| \leq \frac{\varepsilon}{2 + \varepsilon}b$$

imply

$$\left|\frac{Y_n}{X_n} - \frac{a}{b}\right| \leq \varepsilon\frac{a}{b}.$$

Thus,

$$P\left\{\left|\frac{Y_n}{X_n} - \frac{a}{b}\right| > \varepsilon\frac{a}{b}\right\}$$

$$\leq P\left\{|Y_n - a| > \frac{\varepsilon}{2 + \varepsilon}a\right\} + P\left\{|X_n - b| > \frac{\varepsilon}{2 + \varepsilon}b\right\}.$$

Applying Lemma C.7, we complete the proof. ∎

Combining Lemmas C.7 and C.8, we get

LEMMA C.9 *If $E(Y_n - a)^2 = O(\alpha_n)$, and $E(X_n - b)^2 = O(\beta_n)$ with $b \neq 0$, then*

$$P\left\{\left|\frac{Y_n}{X_n} - \frac{a}{b}\right| > \varepsilon\left|\frac{a}{b}\right|\right\} = O(\gamma_n)$$

and

$$\left| \frac{Y_n}{X_n} - \frac{a}{b} \right| = O_P(\sqrt{\gamma_n}) \text{ as } n \to \infty$$

with $\gamma_n = \max(\alpha_n, \beta_n)$.

From (C.1), it follows that

$$\left(\frac{Y_n}{X_n} - \frac{a}{b} \right)^2 \leq 2 \left(\frac{Y_n}{X_n} \right)^2 \frac{1}{b^2} (X_n - b)^2 + 2 \frac{1}{b^2}(Y_n - b)^2.$$

Thus, if $|Y_n/X_n| \leq c$,

$$E \left(\frac{Y_n}{X_n} - \frac{a}{b} \right)^2 \leq 2 \frac{1}{b^2} E(Y_n - b)^2 + 2c^2 \frac{1}{b^2} E (X_n - b)^2, \qquad (\text{C.2})$$

which leads to the following lemma:

LEMMA C.10 *If* $E(Y_n - a)^2 = O(\alpha_n)$, $E(X_n - b)^2 = O(\beta_n)$ *with* $b \neq 0$, *and*

$$\left| \frac{Y_n}{X_n} \right| \leq c$$

for some c, then

$$E \left(\frac{Y_n}{X_n} - \frac{a}{b} \right)^2 = O(\gamma_n),$$

with $\gamma_n = \max(\alpha_n, \beta_n)$.

In this book, we have often encountered the problem of dealing with an estimate $\hat{m}$ being of the ratio form, that is, $\hat{m} = \frac{\hat{g}}{\hat{f}}$. For such estimates we can show that $\hat{m}$ tends (in probability) to a limit value m. This is done by proving that $\hat{g}$ and $\hat{f}$ converge to g and f, respectively, such that $m = \frac{g}{f}$. In order to evaluate the variance and bias of $\hat{m}$, we have often used the following decomposition called the ratio trick

$$\hat{m} = m + \frac{\hat{g} - m\hat{f}}{f} + \frac{(\hat{m} - m)(f - \hat{f})}{f}.$$

Since the second term is the product of two tending to zero expressions, the leading terms in the asymptotic variance and bias of $\hat{m}$ are controlled by the first expression in the above decomposition, that is, we can write

$$\text{var}(\hat{m}) \approx \text{var} \left(\frac{\hat{g} - m\hat{f}}{f} \right)$$

and

$$E(\hat{m}) \approx m + E \left(\frac{\hat{g} - m\hat{f}}{f} \right).$$

If one wishes to find the higher order contributions to the variance and bias of $\hat{m} = \frac{\hat{g}}{\hat{f}}$, it may apply the following identity established by Pawlak [228]:

$$\frac{1}{u} = \sum_{i=0}^{p}(-1)^i \frac{(u - u_0)^i}{u_0^{i+1}} + (-1)^{p+1}\frac{(u - u_0)^{p+1}}{uu_0^{p+1}},$$

for $p \geq$ and any $u_0 \neq 0$. A simple consequence of this identity is the following useful inequality

$$\text{var}\left(\frac{Y}{X}\right) \leq \frac{\text{var}(Y)}{E^2(X)},$$

for any random variables X and Y, such that the ratio is well-defined.

C.3 Stochastic approximation

Let $Y(\xi)$ be a random variable whose distribution depends on a nonrandom parameter ξ. To find ξ^*, which solves a regression equation

$$EY(\xi) = a$$

with given a, Robbins and Monro [254] have proposed the following stochastic approximation procedure:

$$\xi_n = \xi_{n-1} - \gamma_n(Y(\xi_{n-1}) - a)$$

for some ξ_0, where $\{\gamma_n\}$ is a number sequence and γ_n is called gain. Denoting $Y(\xi) = m(\xi) + Z(\xi)$ with $Z(\xi) = Y(\xi) - EY(\xi)$ having zero mean and interpreted as disturbance, we can say that the equation $m(\xi) = a$ is solved with a procedure

$$\xi_n = \xi_{n-1} - \gamma_n[m(\xi_{n-1}) + Z(\xi_{n-1}) - a]$$

in which observations of the regression $m(\bullet)$ are corrupted by Z.

In particular, to find EY one can apply the procedure

$$\xi_n = \xi_{n-1} - \gamma_n(\xi_{n-1} - Y_n). \tag{C.3}$$

For $\gamma_n = n^{-1}$, $\xi_n = n^{-1}\sum_{i=1}^{n} Y_i$ is just the mean of observations.

In turn, to estimate EY/EX the following modified procedure can be employed:

$$\xi_n = \xi_{n-1} - \gamma_n(X_n\xi_{n-1} - Y_n). \tag{C.4}$$

One can expect that ξ_n converges to ξ for which $E\{X_n\xi - Y_n\} = 0$. The ξ solving the equation is certainly equal to EY/EX. Observe that the procedure finds a quotient but performs operations of adding, subtracting, and multiplying only, but not dividing.

Rewriting (C.4) as

$$\xi_n = \xi_{n-1} - \gamma_n X_n\left(\xi_{n-1} - \frac{Y_n}{X_n}\right),$$

we can say that, whereas the gain is deterministic in (C.3), in (C.4) it is random.

C.4 Order statistics

In the process of ordering random variables,

$$X_1, X_2, \ldots, X_n$$

are arranged in increasing order to yield a new sequence

$$X_{(1)}, X_{(2)}, \ldots, X_{(n)},$$

where $X_{(1)} \leq X_{(2)} \leq \cdots \leq X_{(n)}$. These new random variables are called order statistics, $X_{(i)}$ is called the ith order statistics. The intervals

$$(-\infty, X_{(1)}), (X_{(1)}, X_{(2)}), \ldots, (X_{(n)}, \infty)$$

are called sample blocks, whereas the length $X_{(i+1)} - X_{(i)}$ is named spacing. Numbers

$$F(X_{(1)}), F(X_{(2)}) - F(X_{(1)}), \ldots, F(X_{(n-1)}) - F(X_{(n-2)}), F(X_{(n)}),$$

where F is the distribution function of X_is, are called coverages.

In this section, $X_1, X_2, \ldots, X_n$ are independent random variables distributed uniformly over the interval $(0, 1)$. In such a case, $P\{X_{(i-1)} = X_{(i)}\} = 0$ and $X_{(1)} < X_{(2)} < \cdots < X_{(n)}$ almost surely. We give some facts concerning distributions and moments of $X_{(i)}$'s and spacings $(X_{(i+1)} - X_{(i)})$'s.

The theory of order statistics can be found in Arnold, Balakrishnan, and Nagaraja [8], Balakrishnan and Cohen [13], David [61], or Wilks [322]. An analysis of spacings is in Darling [59] or Pyke [243].

C.4.1 Distributions and moments

Denoting by $F_i(x)$ the distribution function of $X_{(i)}$, for $x \in [0, 1]$, we get

$$
\begin{aligned}
F_i(x) &= P\{X_{(i)} < x\} \\
&= P\{\text{at least } i \text{ of } X_1, X_2, \ldots, X_n \text{ are in the interval } (0, x)\} \\
&= \sum_{j=i}^{n} P\{\text{exactly } j \text{ of } X_1, X_2, \ldots, X_n \text{ are in the interval } (0, x)\} \\
&= \sum_{j=i}^{n} \binom{n}{j} x^j (1 - x)^{n-j} \\
&= \int_0^x \frac{n!}{(i - 1)!(n - i)!} t^{i-1} (1 - t)^{n-i} dt.
\end{aligned}
$$

Hence, denoting the density of $X_{(i)}$ as $f_i(x)$, we can write

$$f_i(x) = \frac{n!}{(i - 1)!(n - i)!} x^{i-1} (1 - x)^{n-i}$$

for $x \in [0, 1]$. After some calculations, we can obtain

$$E X_{(i)}^m = \frac{n!}{(n+m)!} \frac{(i+m-1)!}{(i-1)!}$$

and, a fortiori,

$$E X_{(i)} = \frac{i}{n+1},$$

$$E X_{(i)}^2 = \frac{i(i+1)}{(n+1)(n+2)},$$

and

$$\text{var}[X_{(i)}] = i \frac{n+1-i}{(n+2)(n+1)^2}.$$

Denoting by $F_{ij}(x, y)$ the joint distribution of $(X_{(i)}, X_{(j)})$, with $i < j$, we obtain, for $x \geq y$,

$$F_{ij}(x, y) = F_j(x)$$

and, for $x < y$,

$$\begin{aligned}
F_{ij}(x, y) &= P\left\{X_{(i)} \leq x, X_{(j)} \leq y\right\} \\
&= P\{\text{at least } i \text{ of } X_1, \ldots, X_n \text{ are in } [0, x] \\
&\quad \text{and at least } j \text{ of } X_1, \ldots, X_n \text{ are in } [0, y]\} \\
&= \sum_{p=j}^{n} \sum_{q=i}^{p} P\{\text{exactly } i \text{ of } X_1, \ldots, X_n \text{ are in } [0, x] \\
&\quad \text{and exactly } j \text{ of } X_1, \ldots, X_n \text{ are in } [0, y]\} \\
&= \sum_{p=j}^{n} \sum_{q=i}^{p} \frac{n!}{q!(p-q)!(n-p)!} x (y-x)^{p-q} (1-y)^{n-p},
\end{aligned}$$

which, for $0 \leq x < y \leq 1$, is equal to

$$\int_0^x \int_t^y \frac{n!}{(i-1)!(j-i-1)!(n-j)!} t^{i-1} (\tau - t)^{j-i-1} (1-\tau) d\tau \, dt.$$

Hence, the density denoted as $f_{ij}(x, y)$ has the following form:

$$f_{ij}(x, y) = \frac{n!}{(i-1)!(j-i-1)!(n-j)!} x^{i-1} (y-x)^{j-i-1} (1-y)^{n-j}.$$

After rather arduous calculations, we obtain

$$E\left\{X_{(i)}^p X_{(j)}^q\right\} = \frac{n!}{(n+p+q)!} \frac{(i+p-1)!}{(i-1)!} \frac{(j+p+q-1)!}{(j+q-1)!}$$

for $i < j$. Hence, in particular,

$$E\left\{X_{(i)} X_{(j)}\right\} = \frac{i(j+1)}{(n+1)(n+2)}, \quad \text{for } i < j$$

and

$$\text{cov}\left[X_{(i)}, X_{(j)}\right] = i\frac{n+1-j}{(n+2)(n+1)^2}, \quad \text{for } i < j.$$

C.4.2   Spacings

Uniform spacings

Setting $X_{(0)} = 0$ and $X_{(n+1)} = 1$, we define $n + 1$ intervals $(X_{(i-1)}, X_{(i)})$ called sample blocks. Their lengths $d_i = X_{(i)} - X_{(i-1)}$ are called spacings. Certainly $d_1 + d_2 + \cdots + d_{n+1} = 1$. We can say that the interval $[0, 1]$ is split into $n + 1$ subintervals of random length.

It is obvious that the distribution function of d_i is independent of i and that the distribution of (d_i, d_j) with $i \neq j$ is the same as of (d_1, d_2). In general, the uniform spacings are interchangeable random variables. Therefore, denoting by $F_d(x)$ the distribution function of d_i we find, for $0 \leq x \leq 1$,

$$\begin{aligned}
F_d(x) &= P\{d_i < x\} = P\{d_1 < x\} = P\{X_{(1)} < x\} \\
&= P\{\text{at least one of } X_1, X_2, \ldots, X_n \text{ is in the interval } (0, x)\} \\
&= 1 - P\{\text{all } X_1, X_2, \ldots, X_n \text{ are outside the interval } (0, x)\} \\
&= 1 - P^n\{X_1 \text{ is outside the interval } (1 - x, 1)\} \\
&= 1 - (1 - x)^n.
\end{aligned}$$

Thus,

$$F_d(x) = \begin{cases} 1 - (1 - x)^n, & \text{for } 0 \leq x \leq 1 \\ 0, & \text{otherwise} \end{cases}$$

and

$$f_d(x) = \begin{cases} n(1 - x)^{n-1}, & \text{for } 0 \leq x \leq 1 \\ 0, & \text{otherwise,} \end{cases}$$

where $f_d(x)$ is the density of d_i. Therefore, the random variable d_i has a beta distribution $Be(1, n)$.

Observe that if $X_1, \ldots, X_{n-1}$ are distributed uniformly over the interval $(0, 1 - y)$, the distribution function of d_i is

$$1 - \left(1 - \frac{x}{1 - y}\right)^{n-1} = 1 - \frac{(1 - x - y)^{n-1}}{(1 - y)^{n-1}}$$

for $x \in (0, 1 - y)$, and the density equals

$$\begin{cases} \dfrac{n-1}{(1 - y)^{n-1}}(1 - x - y)^{n-2}, & \text{for } 0 \leq x \leq 1 - y \\ 0, & \text{otherwise.} \end{cases}$$

From this and the fact that the conditional density of d_i conditioned on $d_j = y$ is the same as that of spacings from $n - 1$ random variables distributed uniformly on $(0, 1 - y)$, it follows that the conditional density equals

$$f_{d_i|d_j}(x|y) = \begin{cases} \dfrac{n-1}{(1-y)^{n-1}}(1-x-y)^{n-2}, & \text{for } 0 \le x \le 1-y \\ 0, & \text{otherwise.} \end{cases}$$

Thus, denoting the joint density of (d_i, d_j) by $f_{d_i d_j}(x, y)$, we write

$$f_{d_i d_j}(x, y) = \begin{cases} n(n-1)(1-x-y)^{n-2}, & \text{for } 0 \le x, 0 \le y, 0 \le x+y \le 1 \\ 0, & \text{otherwise.} \end{cases}$$

It is not difficult to compute

$$E d_i^p = \frac{p!}{(n+1)\cdots(n+p)}$$

and, a fortiori,

$$E d_i = \frac{1}{n+1},$$

$$E d_i^2 = \frac{2}{(n+1)(n+2)},$$

and

$$E d_i^3 = \frac{6}{(n+1)(n+2)(n+3)}.$$

Moreover, for $i \ne j$,

$$E\{d_i d_j\} = \frac{1}{(n+1)(n+2)}.$$

Clearly,

$$E\left\{(X_{(i)} - X_{(i-1)})|X_{(i)}\right\} = E\left\{d_i|X_{(i)}\right\} = \frac{1}{i}X_{(i)}$$

and

$$E\left\{(X_{(i)} - X_{(i-1)})^2|X_{(i)}\right\} = E\left\{d_i^2|X_{(i)}\right\} = \frac{2}{i(i+1)}X_{(i)}^2 \tag{C.5}$$

Denote by

$$M_n = \max_{1 \le i \le n+1}(X_{(i)} - X_{(i-1)}) = \max_{1 \le i \le n+1} d_i$$

the longest of the subintervals the interval $(0, 1)$ is split in. In probability and almost sure convergence of M_n to zero has been shown by Darling [59] and Slud [279], respectively.

LEMMA C.11 *For uniform spacings,*

$$\lim_{n \to \infty} \frac{n}{\log n} M_n = 1 \text{ almost surely,}$$

and, a fortiori,

$$\lim_{n \to \infty} M_n = 0 \text{ almost surely.}$$

Proof. The lemma follows immediately from Slud [279] who showed that $n M_n - \log n = O(\log \log n)$ almost surely. ∎

General spacings
In this subsection, $X_1, X_2, \ldots, X_n$ are independent random variables with a distribution function $F(x)$ and a probability density $f(x)$.

Since $F(x)$ is monotonic, $F(X_{(1)}) \le F(X_{(2)}) \le \cdots \le F(X_{(n)})$. Moreover, since $f(x)$ exists, ties, that is, events that $X_{(i-1)} = X_{(i)}$ have zero probability. Hence, $F(X_{(1)}) < F(X_{(2)}) < \cdots < F(X_{(n)})$. Denoting $X_{(0)} = -\infty$ and $X_{(n+1)} = +\infty$, we can say that the real line is split into intervals $(X_{(i)} - X_{(i-1)})$ while the interval $[0, 1]$ into corresponding subintervals $(F(X_{(i)}) - F(X_{(i-1)}))$.

Let $b_i = F(X_{(i)}) - F(X_{(i-1)})$ be the length of the subinterval. Since $F(X)$ is distributed uniformly over the interval $(0, 1)$, b_is are just uniform spacings, that is, distributed like d_is in the previous section. Thus, for any density $f(x)$,

$$b_i = \int_{X_{(i-1)}}^{X_{(i)}} f(x) dx$$

has a beta distribution $Be(1, n)$. Hence,

$$E b_i^p = \frac{p!}{(n+1) \cdots (n+p)}$$

and, a fortiori,

$$E b_i = \frac{1}{n+1},$$

$$E b_i^2 = \frac{2}{(n+1)(n+2)},$$

$$E b_i^3 = \frac{6}{(n+1)(n+2)(n+3)}.$$

Moreover, for $i \ne j$,

$$E\{b_i b_j\} = \frac{1}{(n+1)(n+2)}.$$

C.4.3   Integration and random spacings

Let $\varphi(x)$ be a Riemann integrable function over the interval $[0, 1]$, let $0 < x_1 < x_2 < \cdots < x_n$ and also let $x_0 = 0$, $x_{n+1} = 1$. It is well-known that if $\max_{1 \leq i \leq n+1}(x_{i+1} - x_i) \to 0$ as $n \to \infty$,

$$\lim_{n \to \infty} \sum_{i=0}^{n}(x_{i+1} - x_i)\varphi(x_i) = \int_0^1 \varphi(x)dx.$$

Therefore, for a uniform order statistics $X_{(1)}, X_{(2)}, \ldots, X_{(n)}$, defining $S_n = \sum_{i=0}^{n}(X_{(i+1)} - X_{(i)})\varphi(X_{(i)})$ and applying Lemma C.11, we obtain

$$\lim_{n \to \infty} S_n = \int_0^1 \varphi(x)dx \text{ almost surely.}$$

References

[1] I. A. Ahmad and P. E. Lin, Nonparametric sequential estimation of multiple regression function, *Bulletin of Mathematical Statistics*, **17**:63–75, 1976.

[2] H. Al-Duwaish, M. N. Karim, and V. Chandrasekar, Use of multilayer feed forward neural networks in identification and control of Wiener model, *IEEE Proceedings–Control Theory and Applications*, **143**:255–258, 1996.

[3] G. L. Anderson and R. J. P. de Figueiredo, An adaptive orthogonal-series estimator for probability density functions, *Annals of Statistics*, **8**:347–376, 1980.

[4] D. W. K. Andrews, Non-strong mixing autoregressive processes, *Journal of Applied Probability*, **21**:930–934, 1984.

[5] D. W. K. Andrews, A nearly independent, but non-strong mixing, triangular array, *Journal of Applied Probability*, **22**:729–731, 1985.

[6] A. Antoniadis and G. Oppenheim, Wavelets and statistics, in *Lecture Notes in Statistics*, Volume 103, New York: Springer-Verlag, 1995.

[7] M. Apostol, *Mathematical Analysis*, Addison Wesley, 1974.

[8] B. C. Arnold, N. Balakrishnan, and H. N. Nagaraja, *A First Course in Order Statistics*, New York: Wiley, 1992.

[9] M. Bagnoli and T. Bergstrom, Log-concave probability and its applications, *Economic Theory*, **26**:445–469, 2005.

[10] E. W. Bai, An optimal two-stage identification algorithm for Hammerstein–Wiener nonlinear systems, *Automatica*, **34**:333–338, 1998.

[11] E. W. Bai, Identification of linear systems with hard input nonlinearities of known structure, *Automatica*, **38**:853–860, 2002.

[12] E. W. Bai, Frequency domain identification of Hammerstein models, *IEEE Trans. on Automatic Control*, **48**:530–542, 2003.

[13] N. Balakrishnan and A. C. Cohen, *Order Statistics and Inference*, Boston: Academic Press, 1991.

[14] N. K. Bari, *A Treatise on Trigonometric Series*, New York: Pergamon Press, 1964.

[15] J. F. Barrett and D. G. Lampard, An expansion for some second-order probability distributions and its application to noise problems, *IRE Trans. Information Theory*, **1**:10–15, 1955.

[16] J. S. Bendat, *Nonlinear System Analysis and Identification from Random Data*, New York: Wiley, 1990.

[17] J. Benedetti, On the nonparametric estimation of regression function, *Journal of Royal Statistical Society B*, **39**:248–253, 1977.

[18] N. J. Bershad, P. Celka, and S. McLaughlin, Analysis of stochastic gradient identification of Wiener–Hammerstein systems for nonlinearities with Hermite polynomial expansions, *IEEE Trans. on Signal Processing*, **49**:1060–1072, 2001.

[19] S. A. Billings, Identification of nonlinear systems–A survey, *IEEE Proceedings*, **127**:272–285, 1980.

[20] S. A. Billings and S. Y. Fakhouri, Identification of nonlinear systems using the Wiener model, *Electronics Letters*, **13**:502–504, 1977.

[21] S. A. Billings and S. Y. Fakhouri, Identification of a class of nonlinear systems using correlation analysis, *IEEE Proceedings*, **125**:691–697, 1978.

[22] S. A. Billings and S. Y. Fakhouri, Theory of separable processes with applications to the identification of nonlinear systems, *IEEE Proceedings*, **125**:1051–1058, 1978.

[23] S. A. Billings and S. Y. Fakhouri, Identification of non-linear systems using correlation analysis and pseudorandom inputs, *International Journal of Systems Science*, **11**:261–279, 1980.

[24] S. A. Billings and S. Y. Fakhouri, Identification of systems containing linear dynamic and static nonlinear elements, *Automatica*, **18**:15–26, 1982.

[25] J. Bleuez and D. Bosq, Conditons nécasaires et suffisantes de convergence de l'estimateur de la densité par la méthode des fonctions orthogonales, *Comptes Rendus de l'Académie Science de Paris*, **279**:157–159, 1976.

[26] D. Bosq, *Nonparametric Statistics for Stochastic Processes: Estimation and Prediction*, New York: Springer-Verlag, 1998.

[27] M. Boutayeb and M. Darouach, Recursive identification method for MISO Wiener–Hammerstein model, *IEEE Trans. on Automatic Control*, **40**:287–291, 1995.

[28] S. Boyd and L. O. Chua, Uniqueness of a basic nonlinear structure, *IEEE Trans. on Circuits and Systems*, **30**:648–651, 1983.

[29] S. Boyd and L. O. Chua, Fading memory and the problem of approximating nonlinear operators with Volterra series, *IEEE Trans. on Circuits and Systems*, **32**:1150–1171, 1985.

[30] L. Breiman and J. H. Friedman, Estimating optimal transformations for multiple regression and correlation, *Journal of the American Statistical Association*, **80**:580–619, 1985.

[31] D. R. Brillinger, The identification of a particular nonlinear time series system, *Biometrica*, **64**:509–515, 1977.

[32] D. R. Brillinger, *Time Series: Data Analysis and Theory*, San Francisco: Holden-Day, 1981.

[33] P. J. Brockwell and R. A. Davies, *Time Series: Theory and Methods*, New York: Springer, 1987.

[34] J. Bruls, C. T. Chou, B. R. J. Haverkamp, and M. Verhaegen, Linear and nonlinear system identification using separable least-squares, *European Journal of Control*, **5**:116–128, 1999.

[35] H. Buchner, J. Benesty, and W. Kellermann, Multichannel frequency-domain adaptive filtering with application to multichannel acoustic echo cancellation, in J. Benesty and Y. Huang, editors, *Adaptive Signal Processing: Applications to Real-World Problems*, pages 95–128, New York: Springer-Verlag, 2002.

[36] G. Budke, On a convolution property characterizing the Laguerre functions, *Monatshefte für Mathematik*, **107**:281–285, 1989.

[37] T. Cacoullos, Estimation of a multivariate density, *Annals of the Institute of Statistical Mathematics*, **18**:179–190, 1965.

[38] E. Capobianco, Hammerstein system representation of financial volatility processes, *The European Physical Journal B*, **27**:201–211, 2002.

[39] L. Carleson, On convergence and growth of partial sums of Fourier series, *Acta Mathematica*, **116**:135–157, 1966.

[40] D. A. Castanon and D. Teneketzis, Distributed estimation algorithms for nonlinear systems, *IEEE Trans. on Automatic Control*, **30**:418–425, 1985.

[41] P. Celka and P. Colditz, Nonlinear nonstationary Wiener model of infant EEG seizures, *IEEE Trans. on Biomedical Engineering*, **49**:556–564, 2002.

[42] N. N. Cencov, Evaluation of an unknown distribution density from observations, *Soviet Mathematics*, **3**:1559–1562, 1962, translated from Dokl. Akad. Nauk SSSR, 147, pp. 45–48.

[43] K. H. Chan, J. Bao, and W. J. Whiten, Identification of MIMO Hammerstein systems using cardinal spline functions, *Journal of Process Control*, **16**:659–670, 2006.

[44] H. F. Chen, Recursive identification for Wiener model with discontinuous piece-wise linear function, *IEEE Trans. on Automatic Control*, **51**:390–400, 2006.

[45] H. W. Chen, Modeling and identification of parallel nonlinear systems: structural classification and parameter estimation methods, *Proceedings of the IEEE*, **83**:39–66, 1995.

[46] R. Chen and R. Tsay, Nonlinear additive ARX models, *Journal of the American Statistical Association*, **88**:955–967, 1993.

[47] S. Chen and S. A. Billings, Representation of non-linear systems: the NARMAX model, *International Journal of Control*, **49**:1012–1032, 1989.

[48] S. Chen, S. A. Billings, C. Cowan, and P. Grant, Practical identification of NARMAX models using radial basis functions, *International Journal of Control*, **52**:1327–1350, 1999.

[49] T. W. Chen and Y. B. Tin, Laguerre series estimate of a regression function, *Tamkang Journal of Management Sciences*, **8**:15–19, 1987.

[50] K. F. Cheng and P. E. Lin, Nonparametric estimation of a regression function, *Zeitschrift für Wahrscheinlichtstheorie und vervandte Gebiete*, **57**:223–233, 1981.

[51] C. K. Chu and J. S. Marron, Choosing a kernel regression estimator, *Statistical Science*, **6**:404–436, 1991.

[52] C. K. Chui, *An Introduction to Wavelets*, Boston: Academic Press, 1992.

[53] K. L. Chung, On a stochastic approximation method, *Annals of Mathematical Statistics*, **25**:463–483, 1954.

[54] R. M. Clark, Non-parametric estimation of a smooth regression function, *Journal of Royal Statistical Society B*, **99**:107–113, 1977.

[55] M. G. Collomb, Quelques proprietés de la méthode du noyau pour l'estimation non paramétrique de la régresion en un point fixé, *Comptes Rendus de l'Académie Science de Paris, Series A*, **285**:289–292, 1977.

[56] R. G. Corlis and R. Luus, Use of residuals in the identification and control of two-input, signal-output systems, *Ind. Eng. Chem. Fundamentals*, **8**:246–253, 1969.

[57] B. R. Crain, A note on density estimation using orthogonal expansions, *Journal of American Statistical Association*, **68**:964–965, 1968.

[58] S. Csörgö and J. Mielniczuk, Density estimation in the simple proportional hazards model, *Statistics and Probability Letters*, **6**:419–426, 1988.

[59] D. A. Darling, On a class of problems related to the random division of an interval, *Annals of Mathematical Statistics*, **24**:239–253, 1953.

[60] I. Daubechies, *Ten Lectures on Wavelets*, Philadelphia: SIAM, 1992.

[61] H. A. David, *Order Statistics*, New York: Wiley, 1981.

[62] H. I. Davies, Strong consistency of a sequential estimator of a probability density function, *Bulletin of Mathematical Statistics*, **16**:49–54, 1973.

[63] H. I. Davies and E. J. Wegman, Sequential nonparametric density estimation, *IEEE Trans. on Information Theory*, **21**:619–628, 1975.

[64] P. J. Davis, *Interpolation and Approximation*, New York: Blaisdell Publishing Company, 1963.

[65] P. Deheuvels, Sur une famille d'estimateurs de la densité d'une variable al eatoire, *Comptes Rendus de l'Académie Science de Paris, Series A*, **276**:1013–1015, 1973.

[66] A. C. den Brinker, A comparison of results from parameter estimations of impulse responses of the transient visual system, *Biological Cybernetics*, **61**:139–151, 1989.

[67] L. Devroye, Automatic pattern recognition: a study of the probability of error, *IEEE Trans. on Pattern Analysis and Machine Intelligence*, **10**:530–543, 1988.

[68] L. Devroye and L. Györfi, *Nonparametric Density Estimation: The L_1 View*, New York: Wiley, 1985.

[69] L. Devroye, L. Györfi, and G. Lugosi, *A Probabilistic Theory of Pattern Recognition*, New York: Springer-Verlag, 1996.

[70] L. Devroye and G. Lugosi, *Combinatorial Methods in Density Estimation*, New York: Springer-Verlag, 2001.

[71] L. P. Devroye, On the almost everywhere convergence of nonparametric regression function estimates, *Annals of Statistics*, **9**:1310–1319, 1981.

[72] L. P. Devroye and T. J. Wagner, The L_1 convergence of kernel density estimates, *Annals of Statistics*, **7**:1136–1139, 1979.

[73] L. P. Devroye and T. J. Wagner, Distribution-free consistency results in nonparametric discrimination and regression function estimation, *Annals of Statistics*, **9**:231–239, 1980.

[74] L. P. Devroye and T. J. Wagner, On the L_1 convergence of kernel estimators of regression functions with application in discrimination, *Zeitschrift für Wahrscheinlichtstheorie und Vervandte Gebiete*, **51**:15–25, 1980.

[75] P. Diaconis and M. Shahshahani, On nonlinear functions of linear combinations, *SIAM Journal on Scientific Computing*, **5(1)**:175–191, 1984.

[76] C. Dunkl and Y. Xu, *Orthogonal Polynomials of Several Variables*, Cambridge, UK: Cambridge University Press, 2001.

[77] R. Dykstra, An algorithm for restricted least squares regression, *Journal of the American Statistical Association*, **77**:621–628, 1983.

[78] S. Efromovich, *Nonparametric Curve Estimation: Methods, Theory, and Applications*, New York: Springer-Verlag, 1999.

[79] H. E. Emara-Shabaik, K. A. F. Moustafa, and J. H. S. Talaq, On identification of parallel block-cascade models, *International Journal of Systems Science*, **26**:1429–438, 1995.

[80] R. C. Emerson, M. J. Korenberg, and M. C. Citron, Identification of complex-cell intensive nonlinearities in a cascade model of cat visual cortex, *Biological Cybernetics*, **66**:291–300, 1992.

[81] M. Enqvist and L. Ljung, Linear approximations of nonlinear FIR systems for separable input processes, *Automatica*, **41**:459–473, 2005.

[82] E. Eskinat, S. H. Johnson, and W. L. Luyben, Use of Hammerstein models in identification of nonlinear systems, *American Institute of Chemical Engineers Journal*, **37**:255–268, 1991.

[83] M. Espinozo, J. A. Suyken, and B. D. Moor, Kernel-based partially linear models and nonlinear identification, *IEEE Trans. on Automatic Control*, **50**:1602–1606, 2005.

[84] R. L. Eubank, J. D. Hart, and P. Speckman, Trigonometric series regression estimators with an application to partially linear models, *Journal of Multivariate Analysis*, **32**:70–83, 1990.

[85] V. Fabian, Stochastic approximation of minima with improved asymptotic speed, *Annals of Mathematical Statistics*, **38**:191–200, 1967.

[86] S. Y. Fakhouri, S. A. Billings, and C. N. Wormald, Analysis of estimation errors in the identification of non-linear systems, *International Journal of Systems Science*, **12**:205–225, 1981.

[87] J. Fan, Local linear regression smoothers and their minimax efficiencies, *Annals of Statistics*, **21**:196–216, 1993.

[88] J. Fan and I. Gijbels, *Local Polynomial Modeling and Its Application*, London: Chapman and Hall, 1996.

[89] J. Fan and Q. Yao, *Nonlinear Time Series: Nonparametric and Parametric Methods*, New York: Springer-Verlag, 2003.

[90] T. Gałkowski and L. Rutkowski, Nonparametric recovery of multivariate functions with applications to system identification, *Proceedings of the IEEE*, **73**:942–943, 1985.

[91] P. G. Gallman, An iterative method for the identification of nonlinear systems using a Uryson model, *IEEE Trans. on Automatic Control*, **20**:771–775, 1975.

[92] P. G. Gallman, A comparison of two Hammerstein model identification algorithms, *IEEE Trans. on Automatic Control*, **21**:125–127, 1976.

[93] J. Gao and H. Tong, Semiparametric non-linear time series model selection, *Journal of Royal Statistical Society B*, **66**:321–336, 2004.

[94] A. B. Gardiner, Identification of processes containing single-valued non-linearities, *International Journal of Control*, **18**:1029–1039, 1973.

[95] A. Georgiev and W. Greblicki, Nonparametric function recovering from noisy observations, *Journal of Statistical Planning and Inference*, **13**:1–14, 1986.

[96] A. A. Georgiev, Kernel estimates of functions and their derivatives with applications, *Statistics and Probability Letters*, **2**:45–50, 1984.

[97] A. A. Georgiev, On the recovery of functions and their derivatives from imperfect measurements, *IEEE Trans. on Systems, Man, and Cybernetics*, **14**:900–904, 1984.

[98] A. A. Georgiev, Speed of convergence in nonparametric kernel estimation of a regression function and its derivatives, *Annals of Institute of Statistical Mathematics*, **36**:455–462, 1984.

[99] A. A. Georgiev, Nonparametric kernel algorithm for recovering of functions from noisy measurements with applications, *IEEE Trans. on Automatic Control*, **30**:782–784, 1985.

[100] A. A. Georgiev, Proprietés asymptotiques d'un estimateur fonctionnel nonparamétrique, *Comptes Rendus de l'Académie Science de Paris*, **300**:407–410, 1985.

[101] A. A. Georgiev, Asymptotic properties of the multivariate Nadaraya–Watson regression function estimate: the fixed design case, *Statistics and Probability Letters*, **7**:35–40, 1989.

[102] G. B. Giannakis and E. Serpendin, A bibliography on nonlinear system identification, *Signal Processing*, **81**:533–580, 2001.

[103] F. Giri, F. Z. Chaoui, and Y. Rochidi, Parameter identification of a class of Hammerstein plants, *Automatica*, **37**:749–756, 2001.

[104] J. C. Gomez and E. Baeyens, Subspace identification of multivariable Hammerstein and Wiener models, *European Journal of Control*, **11**:127–136, 2005.

[105] W. Greblicki, Asymptotycznie optymalne algorytmy rozpoznawania i identyfikacji w warunkach probabilistycznych, *Prace Naukowe Instytutu Cybernetyki Technicznej, 18, Seria: Monografie Nr 3(18)*:90, 1974, (in Polish).

[106] W. Greblicki, Nonparametric system identification by orthogonal series, *Problems of Control and Information Theory*, **8**:67–73, 1979.

[107] W. Greblicki, Non-parametric orthogonal series identification of Hammerstein systems, *International Journal of Systems Science*, **20**:2355–2367, 1989.

[108] W. Greblicki, Nonparametric identification of Wiener systems, *IEEE Trans. on Information Theory*, **38**:1487–1493, 1992.

[109] W. Greblicki, Nonparametric identification of Wiener systems by orthogonal series, *IEEE Trans. on Automatic Control*, **39**:2077–2086, 1994.

[110] W. Greblicki, Nonlinearity estimation in Hammerstein systems based on ordered observations, *IEEE Trans. on Signal Processing*, **44**:1224–1233, 1996.

[111] W. Greblicki, Nonparametric approach to Wiener system identification, *IEEE Trans. on Circuits and Systems I: Fundamental Theory and Applications*, **44**:538–545, 1997.

[112] W. Greblicki, Continuous-time Wiener system identification, *IEEE Trans. on Automatic Control*, **43**:1488–1493, 1998.

[113] W. Greblicki, Recursive identification of continuous-time Wiener systems, *International Journal of Control*, **72**:981–989, 1999.

[114] W. Greblicki, Continuous-time Hammerstein system identification, *IEEE Trans. on Automatic Control*, **45**:1232–1236, 2000.

[115] W. Greblicki, Recursive identification of Wiener systems, *International Journal of Applied Mathematics and Computer Science*, **11**:977–991, 2001.

[116] W. Greblicki, Recursive identification of continuous-time Hammerstein systems, *International Journal of Systems Science*, **33**:969–977, 2002.

[117] W. Greblicki, Stochastic approximation in nonparametric identification of Hammerstein systems, *IEEE Trans. on Automatic Control*, **47**:1800–1810, 2002.

[118] W. Greblicki, Hammerstien system identification with stochastic approximation, *International Journal of Modelling and Simulation*, **24**:131–138, 2004.

[119] W. Greblicki, Nonlinearity recovering in Wiener system driven with correlated signal, *IEEE Trans. on Automatic Control*, **49**:1805–1810, 2004.

[120] W. Greblicki, Continuous-time Hammerstein system identification from sampled data, *IEEE Trans. on Automatic Control*, **51**:1195–1200, 2006.

[121] W. Greblicki and A. Krzyżak, Asymptotic properties of kernel estimates of a regression function, *Journal of Statistical Planning and Inference*, **4**:81–90, 1980.

[122] W. Greblicki, A. Krzyżak, and M. Pawlak, Distribution-free pointwise consistency of kernel regression estimate, *Annals of Statistics*, **12**:1570–1575, 1985.

[123] W. Greblicki and M. Pawlak, Hermite series estimate of a probability density and its derivatives, *Journal of Multivariate Analysis*, **15**:174–182, 1984.

[124] W. Greblicki and M. Pawlak, Fourier and Hermite series estimates of regression functions, *Annals of the Institute of Statistical Mathematics*, **37**:443–454, 1985.

[125] W. Greblicki and M. Pawlak, Pointwise consistency of the Hermite series density estimates, *Statistics and Probability Letters*, **3**:65–69, 1985.

[126] W. Greblicki and M. Pawlak, Identification of discrete Hammerstein systems using kernel regression estimates, *IEEE Trans. on Automatic Control*, **31**:74–77, 1986.

[127] W. Greblicki and M. Pawlak, Hammerstein system identification by non-parametric regression estimation, *International Journal of Control*, **45**:343–354, 1987.

[128] W. Greblicki and M. Pawlak, Necessary and sufficient consistency conditions for a recursive kernel regression estimate, *Journal of Multivariate Analysis*, **23**:67–76, 1987.

[129] W. Greblicki and M. Pawlak, Nonparametric identification of Hammerstein systems, *IEEE Trans. on Information Theory*, **35**:409–418, 1989.

[130] W. Greblicki and M. Pawlak, Recursive identification of Hammerstein systems, *Journal of the Franklin Institute*, **326**:461–481, 1989.

[131] W. Greblicki and M. Pawlak, Nonparametric identification of a cascade nonlinear time series system, *Signal Processing*, **22**:61–75, 1991.

[132] W. Greblicki and M. Pawlak, Nonparametric identification of a particular nonlinear time series system, *IEEE Trans. on Signal Processing*, **40**:985–989, 1992.

[133] W. Greblicki and M. Pawlak, Cascade non-linear system identification by a non-parametric method, *International Journal of Systems Science*, **25**:129–153, 1994.

[134] W. Greblicki and M. Pawlak, Dynamic system identification with order statistics, *IEEE Trans. on Information Theory*, **40**:1474–1489, 1994.

[135] W. Greblicki and M. Pawlak, Nonparametric recovering nonlinearities in block-oriented systems with the help of Laguerre polynomials, *Control-Theory and Advanced Technology*, **10**:771–791, 1994.

[136] W. Greblicki, D. Rutkowska, and L. Rutkowski, An orthogonal series estimate of time-varying regression, *Annals of the Institute of Statistical Mathematics*, **35**:215–228, 1983.

[137] W. Greblicki and L. Rutkowski, Density-free Bayes risk consistency of nonparametric pattern recognition procedures, *Proceedings of the IEEE*, **69**:482–483, 1981.

[138] L. Györfi, Recent results on nonparametric regression estimate and multiple classification, *Problems of Control and Information Theory*, **10**:43–52, 1977.

[139] L. Györfi, W. Härdle, P. Sarda, and P. Vieu, *Nonparametric Curve Estimation From Time Series*, New York: Springer-Verlag, 1989.

[140] L. Györfi, M. Kohler, A. Krzyżak, and H. Walk, *A Distribution-Free Theory of Nonparametric Regression*, New York: Springer-Verlag, 2002.

[141] L. Györfi and H. Walk, On the strong universal consistency of a recursive regression estimate by Pál Révész, *Statistics and Probability Letters*, **31**:177–183, 1997.

[142] N. D. Haist, F. H. I. Chang, and R. Luus, Nonlinear identification in the presence of correlated noise using a Hammerstein model, *IEEE Trans. on Automatic Control*, **18**:553–555, 1973.

[143] P. Hall, Estimating a density on the positive half line by the method of orthogonal series, *Annals of Institute of Statistical Mathematics*, **32**:351–362, 1980.

[144] P. Hall, Comparison of two orthogonal series methods of estimating a density and its derivatives on an interval, *Journal of Multivariate Analysis*, **12**:432–449, 1982.

[145] P. Hall, Measuring the efficiency of trigonometric series estimates of a density, *Journal of Multivariate Analysis*, **13**:234–256, 1983.

[146] P. Hall, On the rate of convergence of orthogonal series density estimators, *Journal of Royal Statistical Society B*, **48**:115–122, 1986.

[147] P. Hall and I. V. Keilegom, Using difference-based methods for inference in nonparametric regression with time-series errors, *Journal of Royal Statistical Society B*, **65**:443–456, 2003.

[148] A. Hammerstein, Nichtlinear Integralgleichungen nebst Anwendungen, *Acta Mathematica*, **54**:117–176, 1930.

[149] T. Han and S. Amari, Statistical inference under multiterminal data compression, *IEEE Trans. on Information Theory*, **44**:2300–2324, 1998.

[150] W. Härdle, *Applied Nonparametric Regression*, Cambridge, UK: Cambridge University Press, 1990.

[151] W. Härdle, G. Kerkyacharian, D. Picard, and A. Tsybakov, *Wavelets, Approximation, and Statistical Applications*, New York: Springer-Verlag, 1998.

[152] W. Härdle, M. Müller, S. Sperlich, and A. Werwatz, *Nonparametric and Semiparametric Models*, New York: Springer-Verlag, 2004.

[153] W. Härdle and T. M. Stoker, Investigating smooth multiple regression by the method of average derivatives, *Journal of the American Statistical Association*, **84**:986–995, 1989.

[154] Z. Hasiewicz, Hammerstein system identification by the Haar multiresolution approximation, *International Journal of Adaptive Control and Signal Processing*, **13**:691–717, 1999.

[155] Z. Hasiewicz, Modular neural networks for non-linearity recovering by the Haar approximation, *Neural Networks*, **13**:1107–1133, 2000.

[156] Z. Hasiewicz, Non-parametric estimation of non-linearity in a cascade time-series system by multiscale approximation, *Signal Processing*, **81**:791–807, 2001.

[157] Z. Hasiewicz and G. Mzyk, Combined parametric-nonparametric identification of Hammerstein systems, *IEEE Trans. on Automatic Control*, **49**:1370–1375, 2004.

[158] Z. Hasiewicz, M. Pawlak, and P. Śliwiński, Nonparametric identification of nonlinearities in block-oriented systems by orthogonal wavelets with compact support, *IEEE Trans. on Circuits and Systems I: Fundamental Theory and Applications*, **52**:427–442, 2005.

[159] Z. Hasiewicz and P. Śliwiński, Identification of non-linear characteristics of a class of block-oriented non-linear systems via Daubechies wavelet-based models, *International Journal of Systems Science*, **33**:1121–1141, 2002.

[160] Z. Hasiewicz and P. Śliwiński, *Wavelets with Compact Support: Application to Nonparametric System Identification* (in Polish), Warsaw: EXIT, 2005.

[161] T. Hastie and R. Tibshirani, *Generalized Additive Models*, London: Chapman and Hall, 1990.

[162] N. W. Hengartner and S. Sperlich, Rate optimal estimation with the integration method in the presence of many variables, *Journal of Multivariate Analysis*, **95**:246–272, 2005.

[163] P. S. C. Heuberger, P. M. J. V. den Hof, and B. Wahlberg, editors, *Modelling and Identification with Rational Orthogonal Basis*, New York: Springer-Verlag, 2005.

[164] A. K. Hosni and G. A. Gado, Mean integrated square error of a recursive estimator of a probability density, *Tamkang Journal of Mathematics*, **18**:83–89, 1987.

[165] M. Hristache, A. Juditsky, and Van Spokoiny, Direct estimation of the index coefficient in a single-index model, *Annals of Statistics*, **29**:595–623, 2001.

[166] J. T. Hsu and K. D. T. Ngo, A Hammerstein-based dynamic model for hysteresis phenomenon, *IEEE Trans. on Power Electronics*, **12**:406–413, 1997.

[167] R. A. Hunt, On the convergence of Fourier series, in *Orthogonal Expansions and their Continuous Analogues*, pages 235–255, Carbondale: Southern Illinois University Press, 1968.

[168] I. W. Hunter and M. J. Korenberg, The identification of nonlinear biological systems: Wiener and Hammersten cascade models, *Biological Cybernetics*, **55**:135–144, 1986.

[169] H. Ichimura, Semiparametric least squares (SLS) and weighted SLS estimation of single-index models, *Journal of Econometrics*, **58**:71–120, 1993.

[170] E. Isogai, Nonparametric estimation of a regression function by delta sequences, *Annals of the Institute of Statistical Mathematics*, **42**:699–708, 1990.

[171] R. Jennrich, Asymptotic properties of nonlinear least squares estimators, *Annals of Mathematical Statistics*, **40**:633–643, 1969.

[172] M. C. Jones, S. J. Davies, and B. U. Park, Versions of kernel-type regression estimators, *Journal of the American Statistical Association*, **89**:815–832, 1994.

[173] A. Kalafatis, N. Arifin, L. Wang, and W. R. Cluett, A new approach to the identification of ph processes based on the Wiener model, *Chemical Engineering Science*, **50**:3693–3701, 1995.

[174] A. D. Kalafatis, L. Wang, and W. R. Cluett, Identification of Wiener-type nonlinear systems in a noisy environment, *International Journal of Control*, **66**:923–941, 1997.

[175] V. A. Kaminskas, Parameter estimation of discrete systems of the Hammerstein class, *Automation and Remote Control*, **36**:63–69, 1975.

[176] A. Kolmogoroff, Une série de Fourier-Lebesgue divergente presque partout, *Fundamenta Matematicae*, **4**:324–328, 1923.

[177] M. J. Korenberg and I. W. Hunter, The identification of nonlinear biological systems: LNL cascade models, *Biological Cybernetics*, **55**:125–134, 1986.

[178] M. J. Korenberg and I. W. Hunter, Two methods for identifying Wiener cascades having noninvertible static nonlinearities, *Annals of Biomedical Engineering*, **27**:793–804, 1999.

[179] A. S. Kozek and M. Pawlak, Distribution-free consistency of kernel non-parametric M-estimators, *Statistics and Probability Letters*, **58**:343–353, 2002.

[180] R. Kronmal and M. Tarter, The estimation of probability densities and cumulatives by Fourier series methods, *Journal of the American Statistical Association*, **63**:925–952, 1968.

[181] A. Krzyżak, Identification of discrete Hammerstein systems by the Fourier series regression estimate, *International Journal of Systems Science*, **20**:1729–1744, 1989.

[182] A. Krzyżak, On estimation of a class of nonlinear systems by the kernel regression estimate, *IEEE Trans. on Information Theory*, **36**:141–152, 1990.

[183] A. Krzyżak, Global convergence of the recursive kernel regression estimates with applications in classification and nonlinear system estimation, *IEEE Trans. on Information Theory*, **38**:1323–1338, 1992.

[184] A. Krzyżak, On nonparametric estimation of nonlinear dynamic systems by the Fourier series estimate, *Signal Processing*, **52**:299–321, 1996.

[185] A. Krzyżak and M. A. Partyka, On identification of block oriented systems by nonparametric techniques, *International Journal of Systems Science*, **24**:1049–1066, 1993.

[186] A. Krzyżak and M. Pawlak, Universal consistency results for Wolverton–Wagner regression function estimate with application in discrimination, *Problems of Control and Information Theory*, **12**:33–42, 1983.

[187] A. Krzyżak and M. Pawlak, Almost everywhere convergence of recursive kernel regression function estimates, *IEEE Trans. on Information Theory*, **30**:91–93, 1984.

[188] A. Krzyżak and M. Pawlak, The pointwise rate of convergence of the kernel regression estimate, *Journal of Statistical Planning and Inference*, **16**:159–166, 1987.

[189] A. Krzyżak, J. Z. Sąsiadek, and B. Kégl, Non-parametric identification of dynamic nonlinear systems by a Hermite series approach, *International Journal of Systems Science*, **32**:1261–1285, 2001.

[190] F. C. Kung and D. H. Shih, Analysis and identification of Hammerstein model non-linear delay systems using block-pulse function expansions, *International Journal of Control*, **43**:139–147, 1986.

[191] S. L. Lacy, R. S. Erwin, and D. S. Bernstein, Identification of Wiener systems with known noninvertible nonlinearities, *Trans. of the ASME*, **123**:566–571, 2001.

[192] T. L. Lai, Asymptotic properties of nonlinear least squares estimates in stochastic regression models, *Annals of Statistics*, **22**:1917–1930, 1994.

[193] Z. Q. Lang, A nonparametric polynomial identification algorithm for the Hammerstein system, *IEEE Trans. on Automatic Control*, **42**:1435–1441, 1997.

[194] R. Leipnik, The effect of instantaneous nonlinear devices on cross-correlation, *IRE Trans. Information Theory*, **4**:73–76, 1958.

[195] H. E. Liao and W. A. Sethares, Suboptimal identification of nonlinear ARMA models using an orthogonality approach, *IEEE Trans. on Circuits and Systems – I: Fundamental Theory and Applications*, **42**:14–22, 1995.

[196] E. Liebscher, Hermite series estimators for probability densities, *Metrika*, **37**:321–343, 1990.

[197] O. Linton and J. P. Nielsen, A kernel method of estimating structured nonparametric regression based on marginal integration, *Biometrika*, **82**:93–100, 1995.

[198] L. Ljung, *System Identification: Theory for the User*, Englewood Cliffs: Prentice-Hall, 1987.

[199] L. Ljung and A. Vicino, editors, Special issue on system identification, *IEEE Trans. on Automatic Control*, **50**, October 2005.

[200] G. G. Lorentz, Fourier-Koeffizienten und Funktionenklassen, *Mathematische Zeitschrift*, **51**:135–149, 1948.

[201] Y. P. Mack and H. G. Müller, Convolution type estimator for nonparametric regression, *Statistics and Probability Letters*, **7**:229–239, 1989.

[202] Y. P. Mack and H. G. Müller, Derivative estimation in nonparametric regression with random predictor variable, *Sankhā Series A*, **51**:59–71, 1989.

[203] P. M. Mäkilä, LTI approximation of nonlinear systems via signal distribution theory, *Automatica*, **42**:917–928, 2006.

[204] S. Mallat, *A Wavelet Tour of Signal Processing*, San Diego: Academic Press, 1998.

[205] E. Mammen, O. Linton, and J. Nielsen, The existence and asymptotic properties of a backfitting projection algorithm under weak conditions, *Annals of Mathematical Statistics*, **27**:1443–1490, 1999.

[206] E. Mammen, J. Marron, B. Turlach, and M. Wand, A general projection framework for constrained smoothing, *Statistical Science*, **16**:232–248, 2001.

[207] P. Z. Marmarelis and V. Z. Marmarelis, *Analysis of Physiological Systems: The White Noise Approach*, New York: Plenum Press, 1978.

[208] V. J. Mathews and G. L. Sicuranza, *Polynomial Signal Processing*, New York: Wiley, 2000.

[209] S. P. Meyn and R. L. Tweedie, *Markov Chains and Stochastic Stability*, New York: Springer-Verlag, 1993.

[210] M. A. Mirzahmedov and S. A. Hasimov, On some properties of density estimation, in *European Meeting of Statisticians*, Colloqia Mathematica Societatis János Bolay, pages 535–546, Budapest (Hungary), 1972.

[211] H. G. Müller, Density adjusted smoothers for random design nonparametric regression, *Statistics and Probability Letters*, **36**:161–172, 1997.

[212] H. G. Müller and K. S. Song, Identity reproducing multivariate nonparametric regression, *Journal of Multivariate Analysis*, **46**:237–253, 1993.

[213] H. G. Müller and U. Stadtmüller, Variable bandwidth kernel estimators of regression curves, *Annals of Statistics*, **15**:182–201, 1987.

[214] H. G. Müller and U. Stadtmüller, On variance function estimation with quadratic forms, *Journal of Statistical, Planning, and Inference*, **35**:213–231, 1993.

[215] E. Nadaraya, On regression estimators, *Theory of Probability and Its Applications*, **9**:157–159, 1964.

[216] K. S. Narendra and P. G. Gallman, An iterative method for identification of nonlinear systems using a Hammerstein model, *IEEE Trans. on Automatic Control*, **11**:546–550, 1966.

[217] A. E. Ndorsjö, Cramér-Rao bounds for a class of systems described by Wiener and Hammerstein models, *International Journal of Control*, **68**:1067–1083, 1997.

[218] O. Nells, *Nonlinear System Identification*, New York: Springer-Verlag, 2001.

[219] C. L. Nikias and A. P. Petropulu, *Higher-Order Spectra Analysis: A Nonlinear Signal Processing Framework*, Oppenheim Series in Signal Processing, New York: Prentice-Hall, 1993.

[220] B. Ninness and S. Gibson, Quantifying the accuracy of Hammerstein model estimation, *Automatica*, **38**:2037–2051, 2002.

[221] J. P. Norton, *An Introduction to Identification*, London: Academic Press, 1986.

[222] A. H. Nuttall, Theory and application of the separable class of random processes, Technical Report 343, MIT, 1958.

[223] R. T. Ogden, *Essential Wavelets for Statistical Applications and Data Analysis*, Boston: Birkhäuser, 1997.

[224] J. Opsomer, Y. Wang, and Y. Yang, Nonparametric regression with correlated errors, *Statistical Science*, **16**:134–153, 2001.

[225] A. Papoulis and S. U. Pillai, *Probability, Random Variables and Stochastic Processes*, Boston: McGraw-Hill, 4th edition, 2002.

[226] E. Parzen, On estimation of a probability density and mode, *Annals of Mathematical Statistics*, **33**:1065–1076, 1962.

[227] R. S. Patwardhan, S. Lakshminarayanan, and S. L. Shah, Constrained nonlinear MPC using Hammerstein and Wiener models: PLS framework, *American Institute of Chemical Engineers Journal*, **44**:1611–1622, 1998.

[228] M. Pawlak, On the almost everywhere properties of the kernel regression estimate, *Annals of the Institute of Statistical Mathematics*, **43**:311–326, 1991.

[229] M. Pawlak, On the series expansion approach to the identification of Hammerstein system, *IEEE Trans. on Automatic Control*, **36**:763–767, 1991.

[230] M. Pawlak, Direct identification algorithms of semiparametric Wiener systems, Technical Report, University of Manitoba, 2007.

[231] M. Pawlak, Nonparametric estimation of a nonlinearity in the sandwich system, Technical Report, University of Manitoba, 2007.

[232] M. Pawlak, Nonparametric identification of generalized Hammerstein systems, Technical Report, University of Manitoba, 2007.

[233] M. Pawlak and W. Greblicki, Nonparametric estimation of a class of nonlinear time series models, in G. Roussas, editor, *Nonparametric Functional Estimation and Related Topics*, pages 541–552, Dordrecht: Kluwer, 1991.

[234] M. Pawlak and Z. Hasiewicz, Nonlinear system identification by the Haar multiresolution analysis, *IEEE Trans. on Circuits and Systems I: Fundamental Theory and Applications*, **45**:945–961, 1998.

[235] M. Pawlak and Z. Hasiewicz, Nonparametric identification of non-linear one and two-channel systems by multiscale expansions, Technical Report, University of Manitoba, 2006.

[236] M. Pawlak, Z. Hasiewicz, and P. Wachel, On nonparametric identification of Wiener systems, *IEEE Trans. on Signal Processing*, **55**:482–492, 2007.

[237] M. Pawlak, R. Pearson, B. Ogunnaike, and F. Doyle, Nonparametric identification of generalized Hammerstein models, in *IFAC Symposium on System Identification*, Santa Barbara: IFAC, 2000.

[238] D. B. Percival and A. T. Walden, *Wavelet Methods for Time Series*, Cambridge, UK: Cambridge University Press, 2000.

[239] B. Portier and A. Oulidi, Nonparametric estimation and adaptive control of functional autoregressive models, *SIAM Journal on Control and Optimization*, **39**:411–432, 2000.

[240] J. L. Powell, J. H. Stock, and T. M. Stoker, Semiparametric estimation of index coefficients, *Econometrica*, **57**:1403–1430, 1989.

[241] B. L. S. Prakasa Rao, *Nonparametric Functional Estimation*, Orlando: Academic Press, 1983.

[242] M. B. Priestley and M. T. Chao, Nonparametric function fitting, *Journal of Royal Statistical Society B*, **34**:385–392, 1972.

[243] R. Pyke, Spacings, *Journal of the Royal Statistical Society B*, **27**:395–436, 1965.

[244] T. Quatieri, D. A. Reynold, and G. C. O'Leary, Estimation of handset nonlinearity with application to speaker recognition, *IEEE Trans. on Speech and Audio Processing*, **8**:567–583, 2000.

[245] E. Rafajłowicz, Nonparametric orthogonal series estimators of regression: A class attaining the optimal convergence rate in L_2, *Statistics and Probability Letters*, **5**:219–224, 1987.

[246] E. Rafajłowicz, Nonparametric least squares estimation of a regression function, *Statistics*, **19**:349–358, 1988.

[247] E. Rafajłowicz and R. Schwabe, Equidistributed designs in nonparametric regression, *Statistica Sinica*, **13**:129–142, 2003.

[248] E. Rafajłowicz and E. Skubalska-Rafajłowicz, FTT in calculating nonparametric regression estimate based on trigonometric series, *Applied Mathematics and Computer Science*, **3**:713–720, 1993.

[249] J. C. Ralston, A. M. Zoubir, and B. Boashash, Identification of a class of conlinear systems under stationary non-Gaussian excitation, *IEEE Trans. on Signal Processing*, **45**:719–735, 1997.

[250] P. Révész, How to apply the method of stochastic approximation in the non-parametric estimation of a regression function, *Mathematische Operationforschung, Series Statistics*, **8**:119–126, 1977.

[251] L. Reytö and P. Révész, Density estimation and pattern classification, *Problems of Control and Information Theory*, **2**:67–80, 1973.

[252] J. Rice, Approximation with convex constraints, *SIAM Journal on Applied Mathematics*, **11**:15–32, 1963.

[253] T. J. Rivlin and M. Wilson, An optimal property of Chebyshev expansions, *Journal of Approximation Theory*, **2**:312–317, 1969.

[254] H. Robbins and S. Monro, A stochastic approximation method, *Annals of Mathematical Statistics*, **22**:400–407, 1951.

[255] P. R. Robinson, Nonparametric estimation for time series models, *Journal of Time Series*, **4**:185–208, 1983.

[256] M. Rosenblatt, Remarks on some nonparametric estimates of a density function, *Annals of Mathematical Statistics*, **27**:832–837, 1956.

[257] M. Rosenblatt, Curve estimates, *Annals of Mathematical Statistics*, **42**:1815–1842, 1971.

[258] N. Rozario and A. Papoulis, Some results in the application of polyspectra to certain nonlinear communication systems, in *Workshop on Higher-Order Spectral Analysis*, pages 37–41, Vail, CO, 1989.

[259] D. Ruppert, M. Wand, and R. Carroll, *Semiparametric Regression*, Cambridge, UK: Cambridge University Press, 2003.

[260] L. Rutkowski, On system identification by nonparametric function fitting, *IEEE Trans. on Automatic Control*, **27**:225–227, 1982.

[261] L. Rutkowski, On nonparametric identification with prediction of time-varying systems, *IEEE Trans. on Automatic Control*, **29**:58–60, 1984.

[262] L. Rutkowski, Nonparametric identification of quasi-stationary systems, *Systems and Control Letters*, **6**:33–35, 1985.

[263] L. Rutkowski, Application of multiple Fourier series to identification of multivariate nonstationary systems, *International Journal of Systems Science*, **20**:1993–2002, 1989.

[264] L. Rutkowski, Non-parametric learning algorithms in time-varying environments, *Signal Processing*, **18**:129–137, 1989.

[265] L. Rutkowski, Identification of MISO nonlinear regressions in the presence of a wide class of disturbances, *IEEE Trans. on Information Theory*, **37**:214–216, 1991.

[266] L. Rutkowski, Multiple Fourier series procedures for extraction of nonlinear regressions from noisy data, *IEEE Trans. on Signal Processing*, **41**:3062–3065, 1993.

[267] L. Rutkowski and E. Rafajłowicz, On optimal global rate of convergence of some nonparametric identification procedures, *IEEE Trans. on Automatic Control*, **34**:1089–1091, 1989.

[268] I. W. Sanberg, Approximation theorems for discrete-time systems, *IEEE Trans. Circuits and Systems*, **38**:564–566, 1991.

[269] S. Sandilya and S. Kulkarni, Nonparametric control algorithms for nonlinear fading memory systems, *IEEE Trans. on Automatic Control*, **46**:1117–1121, 2001.

[270] G. Sansone, *Orthogonal Functions*, New York: Dover, 1959.

[271] G. Scarano, D. Caggiati, and G. Jacovitti, Cumulant series expansion of hybrid nonlinear moments of *n* variates, *IEEE Trans. on Signal Processing*, **41**:486–489, 1993.

[272] H. Scheffe, A useful convergence theorem for probability distributions, *Annals of Mathematical Statistics*, **18**:434–458, 1947.

[273] J. Schoukens, R. Pintelon, T. Dobrowiecki, and Y. Rolain, Identification of linear systems with nonlinear distortions, *Automatica*, **41**:491–504, 2005.

[274] E. Schuster and S. Yakowitz, Contributions to the theory of nonparametric regression with application to system identification, *Annals of Statistics*, **7**:139–149, 1979.

[275] S. C. Schwartz, Estimation of probability density by an orthogonal series, *Annals of Mathematical Statistics*, **38**:1261–1265, 1967.

[276] D. H. Shih and F. C. Kung, The shifted Legendre approach to non-linear system analysis and identification, *International Journal of Control*, **42**:1399–1410, 1985.

[277] B. W. Silverman, *Density Estimation for Statistics and Data Analysis*, London: Chapman and Hall, 1985.

[278] J. S. Simonoff, *Smoothing Methods in Statistics*, New York: Springer-Verlag, 1996.

[279] E. Slud, Entropy and maximal spacings for random partitions, *Zeitschrift für Wahrscheinlichtstheorie und vervandte Gebiete*, **41**:341–352, 1978.

[280] T. Söderström and P. Stoica, *System Identification*, New York: Prentice Hall, 1989.

[281] D. F. Specht, Series estimation of a probability density function, *Technometrics*, **13**:409–424, 1971.

[282] C. Spiegelman and J. Sacks, Consistent window estimation in nonparametric regression, *Annals of Statistics*, **8**:240–246, 1980.

[283] W. J. Staszewski, Identification of non-linear systems using multi-scale ridges and skeletons of the wavelet transform, *Journal of Sound and Vibration*, **214**:639–658, 1998.

[284] C. J. Stone, Consistent nonparametric regression, *Annals of Statistics*, **47**:595–645, 1977.

[285] C. J. Stone, Optimal rates of convergence for nonparametric estimators, *Annals of Statistics*, **8**:1348–1360, 1980.

[286] C. J. Stone, Optimal global rates of convergence for nonparametric regression, *Annals of Statistics*, **10**:1040–1053, 1982.

[287] A. Sugiyama, Y. Joncour, and A. Hirano, A stereo echo canceler with correct echo-path identification based on input-sliding techniques, *IEEE Trans. on Signal Processing*, **49**:2577–2587, 2001.

[288] G. Szegö, *Orthogonal Polynomials*, Providence, RI: American Mathematical Society, 3rd ed. 1974.

[289] M. A. L. Thathachar and S. Ramaswamy, Identification of a class of non-linear systems, *International Journal of Control*, **18**:741–752, 1973.

[290] D. Tjøstheim, Non-linear time series: A selective review, *Scandinavian Journal of Statistics*, **21**:97–130, 1994.

[291] H. Tong, A personal overview of non-linear time series analysis from a chaos perspective, *Scandinavian Journal of Statistics*, **22**:399–445, 1995.

[292] F. Tricomi, *Integral Equations*, New York: Dover, 1957.

[293] S. Trybuła, Identification theory of electric power systems, *Systems Science*, **10**:1–33, 1984.

[294] J. V. Uspensky, On the development of arbitrary functions in series of Hermite's and Laguerre's polynomials, *Annals of Mathematics*, **28**:593–619, 1927.

[295] A. van der Vaart, *Asymptotic Statistics*, Cambridge, UK: Cambridge University Press, 1998.

[296] G. Vandersteen, Y. Rolain, and J. Schoukens, Non-parametric estimation of the frequency-response functions of the linear blocks of a Wiener-Hammerstein model, *Automatica*, **33**:1351–1355, 1997.

[297] J. Van Ryzin, Non-parametric Bayesian decision procedure for pattern classification with stochastic learning, in *Transactions of the Fourth Prague Conference on Information Theory, Statistical Decision Functions and Random Processes*, pages 479–494, Prague, 1965.

[298] J. Van Ryzin, Bayes risk consistency of classification procedure using density estimation, *Sankhyā*, **28**:261–270, 1966.

[299] V. Vapnik, *Statistical Learning Theory*, New York: Wiley, 1998.

[300] S. Verdu, *Multiuser Detection*, Cambridge, UK: Cambridge University Press, 1998.

[301] M. Verhaegen and D. Westwick, Identifying MIMO Hammerstein systems in the context of subspace model identification methods, *International Journal of Control*, **63**:331–349, 1996.

[302] B. Vidakovic, *Statistical Modeling by Wavelets*, New York: Wiley, 1999.

[303] A. J. Viollaz, Asymptotic distribution of L_2-norms of the deviation of density function estimates, *Annals of Statistics*, **8**:322–346, 1980.

[304] R. Viswanathan and P. S. Varshney, Distributed detection with multiple sensors – Part I: Fundamentals, *Proceedings of the IEEE*, **85**:54–63, 1997.

[305] J. Vörös, Recursive identification of Hammerstein systems with discontinuous nonlinearities containing dead-zone, *IEEE Trans. on Automatic Control*, **48**:2203–2206, 2003.

[306] G. Wahba, Optimal convergence properties of variable knots, kernel and orthogonal series methods for density estimation, *Annals of Statistics*, **3**:15–29, 1975.

[307] G. Wahba, Data-based optimal smoothing of orthogonal series density estimates, *Annals of Statistics*, **9**:146–156, 1981.

[308] G. Walter and X. Shen, *Wavelets and Other Orthogonal Systems with Applications*, Boca Raton: Chapman & Hall/CRC Press, 2nd edition, 2001.

[309] G. G. Walter, Properties of Hermite series density estimation of probability density, *Annals of Statistics*, **5**:1258–1264, 1977.

[310] M. P. Wand and M. C. Jones, *Kernel Smoothing*, London: Chapman and Hall, 1995.

[311] M. T. Wasan, *Stochastic Approximation*, Cambridge, UK: Cambridge University Press, 1969.

[312] G. S. Watson, Smooth regression analysis, *Sankhyā*, **26**:359–372, 1964.

[313] G. S. Watson, Density estimation by orthogonal series, *Annals of Mathematical Statistics*, **40**:1496–1498, 1969.

[314] E. J. Wegman and H. I. Davies, Remarks on some recursive estimators of a probability density, *Annals of Statistics*, **7**:316–327, 1979.

[315] D. Westwick and M. Verhaegen, Identifying MIMO Wiener systems using subspace model identification methods, *Signal Processing*, **52**:235–258, 1996.

[316] D. T. Westwick and R. Kearney, *Identification of Nonlinear Physiological Systems*, Piscataway: IEEE Press, 2003.

[317] D. T. Westwick and R. E. Kearney, A new algorithm for the identification of multiple input Wiener systems, *Biological Cybernetics*, **68**:75–85, 1992.

[318] R. L. Wheeden and A. Zygmund, *Measure and Integral*, New York: Dekker, 1977.

[319] T. Wigren, Recursive prediction error identification using the nonlinear Wiener model, *Automatica*, **29**:1011–1025, 1993.

[320] T. Wigren, Convergence analysis of recursive identification algorithms based on nonlinear Wiener model, *IEEE Trans. on Automatic Control*, **39**:2191–2206, 1994.

[321] T. Wigren, Circle criteria in recursive identification, *IEEE Trans. on Automatic Control*, **42**:975–979, 1997.

[322] S. S. Wilks, *Mathematical Statistics*, New York: Wiley, 1962.

[323] A. S. Willsky, M. G. Bello, D. A. Castanon, B. C. Levy, and G. C. Verghese, Combining and updating of local estimates and regional maps along sets of one-dimensional tracks, *IEEE Trans. on Automatic Control*, **27**:799–813, 1982.

[324] P. Wojtaszczyk, *A Mathematical Introduction to Wavelets*, Cambridge, UK: Cambridge University Press, 1997.

[325] C. T. Wolverton and T. J. Wagner, Asymptotically optimal discriminant functions for pattern recognition, *IEEE Trans. on Information Theory*, **15**:258–265, 1969.

[326] C. T. Wolverton and T. J. Wagner, Recursive estimates of probability density, *IEEE Trans. on Systems Science and Cybernetics*, **5**:307, 1969.

[327] S. Yakowitz, J. E. Krimmel, and F. Szidarovszky, Weighted Monte Carlo integration, *SIAM Journal on Numerical Analysis*, **15**:1289–1300, 1978.

[328] H. Yamato, Sequential estimation of a continuous probability density function and mode, *Bulletin of Mathematical Statistics*, **14**:1–12, 1971.

[329] A. Yatchev, *Semiparametric Regression for the Applied Econometrician*, Cambridge, UK: Cambridge University Press, 2003.

[330] Y. Yatracos, On the estimation of the derivative of a function with the derivative of an estimate, *Journal of Multivariate Analysis*, **28**:172–175, 1989.

[331] X. Zhu and D. Seborg, Nonlinear predictive control based on Hammerstein models, *Control Theory and Application*, **11**:564–575, 1994.

[332] Y. Zhu, *Multivariable System Identification*, Amsterdam: Pergamon Press, 2001.

[333] Y. Zhu, Estimation of N-L-N Hammerstein–Wiener model, *Automatica*, **38**:1607–1614, 2002.

[334] A. Zoubir and D. Iskander, *Bootstrap Techniques for Signal Processing*, Cambridge, UK: Cambridge University Press, 2004.

[335] A. Zygmund, *Trigonometric Series*, Cambridge, UK: Cambridge University Press, 1959.

Index

additive approximation(s), 219, 229, 231–235, 239, 243, 244, 248, 284, 285
additive Hammerstein model (system), 175, 244, 304
additive model(s), 181, 222, 228, 229, 237–240, 242, 245, 249, 285, 286
additive modeling, 227, 231, 233, 248, 249
additive regression, 233
additive system(s), 182, 219, 220, 222, 285
additive Wiener model (system), 255, 286, 300, 302, 303, 317
Afirin, N., 122
Ahmad, I. A., 43
Al-Duwaish, H., 122
Anderson, G. L., 78
Arnold, B. C., 365
automatic choice (method), 165, 167
average derivative estimate (estimation), 290, 291, 293, 300, 304, 307, 310
average operation, 183–185
averaging operation, 209

backfitting algorithm, 237, 238, 249
Bai, E. W., 10
Balakrishnan, N., 365
bandwidth(s), 12, 20, 144, 145, 165–167, 210, 223, 224, 228, 238, 241, 242, 245–247, 271, 273, 279
Bendat, J. S., 10, 122
Benedetti, J., 99
Bernstein, D. S., 122
Billings, S. A., 10, 122
Bleuez, J., 78
block-oriented systems (structures), xii, 149, 157, 173, 220, 222, 242, 248, 250, 251, 255, 256, 259, 309, 310, 316
Bosq, D., 78, 112
Boutayeb, M., 10
Brillinger, D. R., 122, 191, 193, 220
Brockwell, P. J., 2
Bussgang's theorem, 193, 194

Cacoullos, T., 29
Capobianco, E., 10
Carleson, L., 335

Celka, P., 122
Cencov, N. N., 78
Cesáro summation, 156, 164
Chandrasekar, V., 122
Chang, F. H. I., 9
Chao, M. T., 99
Chaoui, F. Z., 10
Chebyshev polynomials, 344
Chen, H. F., 122
Chen, T. W., 78
Cheng, K. F., 99
Christoffel–Darboux formula, 340, 344, 350, 352
Chui, C. K., 358
Chung, K. L., 51
Citron, M. C., 10, 122
Clark, R. M., 99
Cluett, W. R., 122
Cohen, A. C., 365
Colditz, P., 122
Collomb, M. G., 29
consistency, 1, 5, 6, 11, 13, 29–31, 46, 82, 92, 102, 114, 119, 124, 141, 212, 219, 226, 237, 246, 254, 259–262, 273, 275–277, 285, 287, 313
continuous-time Hammerstein system, 101, 102, 112
continuous-time Wiener system, xii, 143, 144
convergence analysis, 219, 272, 283, 289
convergence rates, 1, 237, 250, 252
convolution property, 242, 351
correlated input, 79
correlation method, 253, 254, 286, 295, 304
coverages, 365
Crain, B. R., 78
cross-validation, 167
Csörgo, S., 29
cumulant(s), 160, 161, 173, 186, 187, 189–194, 200, 201, 203, 205, 210, 220
curse of dimensionality, xii, 178, 188, 218, 219, 222, 228, 257, 284, 310

Darling, D. A., 81, 365, 368
Darouach, M., 10
data splitting, 251, 279
data-driven choice, 155
Daubechies scaling functions, 153

Daubechies, I., 358
David, H. A., 365
Davies, H. I., 43
Davis, P. J., 334, 337
Davis, R. A., 2
de Figueiredo, R. J. P., 78
decoupling property, 309
Deheuvels, P., 29
den Brinker, A. C., 122
Devroye, L. P., 29, 43
Dini's theorem, 334, 335, 337
direct estimates, 286, 290
Dirichlet kernel, 63, 86, 107, 334
Discrete-time Wiener system, 113
distribution-free, 11, 29, 226, 322
dynamic subsystem identification, 8, 102, 119

Emara-ShaBaik, H. E., 10
Emerson, R. C., 10, 122
equiconvergence theorems, 332, 355
Erwin, R. S., 122
Eskinat, E., 10
Eubank, R. L., 99

Fabian, V., 51
fading memory, 255, 282, 284, 317
Fakhouri, S. Y., 10, 122
Fan, J., 29
Fejér kernel, 15, 335, 336
Fourier series, 126, 333, 334
Fourier series estimate, 61, 66, 86
Fourier transform, 9, 155, 199, 206

Gado, G. A., 43
Gallman, P. G., 9, 10
Gardiner, A. B., 10
Gauss–Weierstrass kernel, 14, 15, 18
Gałkowski, T., 99
generalized kernel estimate, 152, 156, 168
generalized Wiener model, 218
Georgiev, A. A., 99
Giannakis, G. B., 2, 10
Gibson, S., 10
Giri, F., 10
Glivienko–Cantelli condition, 313
Greblicki, W., 29, 43, 58, 78, 99, 100, 112, 142, 148, 322, 362
Györfi, L., 1, 29, 58

Härdle, W., 1, 29
Hájek projection, 236
Haar wavelets, 69, 331, 356
Haist, N. D., 9
Hall, P., 78
Hammerstein system(s) (models), xi, xii, 3–5, 7, 9, 10, 29, 43, 78, 100–102, 112, 117, 124, 130, 151, 157–162, 173, 175–177, 179–181, 184, 186, 191,

197–199, 208, 218, 243–248, 252, 253, 286, 304, 306
Hammerstein, A., 10
Hart, J. D., 99
Hasiewicz, Z., 79
Hasimov, S. A., 78
Hermite polynomials (functions), 10, 61, 68, 194, 201, 202, 331, 351, 352
Hermite series, xi, 73, 78, 128, 351
Hermite series estimate, 68–70
Hosni, A. K., 43
Hsu, T. J., 10
Hunt, R. A., 335
Hunter, I. W., 10, 122

identifiability, 228–232, 236, 253, 259, 265, 286, 288
integrated mean-squared error, 166
internally corrected (modified, normalized) kernel estimate, 240, 241, 247, 249
inverse regression, 185, 186, 252, 264, 282
Isogai, E., 99

Johnson, S. H., 10
Jones, M. C., 1

Kégl, B., 78, 142
Kalafatis, A. D., 122
Kaminskas, V. A., 10
Karim, M. N., 122
Kearney, R. E., 122
kernel algorithm, xi, xii, 11, 30, 44, 84, 103, 123
kernel density estimate, 154, 240, 291
kernel estimate(s), 10–12, 29, 30, 43, 44, 60, 61, 70, 81, 87, 100, 103, 123, 124, 141, 142, 148, 152, 156, 165, 168, 182, 210, 212, 223, 226, 238–242, 245–247, 249, 268–271, 273, 276, 279, 282, 283, 285, 289, 291, 292, 298, 303, 307, 388
kernel functions, 11, 14, 19, 152, 153, 223, 225, 226, 242, 272, 319
kernel methods, 12, 29, 223, 226
kernel regression estimate, 29, 210, 223, 244, 246, 252, 264, 287, 288, 303
kernel twicing, 326
Kohler, M., 1
Kolmogoroff, A., 335
Korenberg, M. J., 10, 122
Kozek, A. S., 29
Kronmal, R., 78
Krzyżak, A., 1, 29, 43, 78, 142, 322, 362
Kung, D. H., 10

Lacy, S. L., 122
Laguerre polynomial(s) (functions), 66, 331, 345, 346, 349, 351
Laguerre series, 66, 78, 79, 156, 345
Laguerre series estimate, 66

Lakshminarayanan, S., 10, 122
Lang, Z. Q., 29
Lebesgue kernel, 15
Lebesgue point, 14, 32, 47, 105, 321, 329
Legendre polynomial(s), 19, 331, 340, 344
Legendre series, 20, 60, 64, 68, 88, 128, 340
Legendre series estimate, 64, 88
Legendre series estimates, 68
Liao, H. E., 10
Liebscher, E., 78
Lin, P. E., 43, 99
linear approximation, 178, 198, 199, 221
Ljung, L., 2
log-concave densities, 296
Luus, R., 9
Luyben, W. L., 10

M-estimators, 256, 260, 316
Müller, H. G., 29, 99, 100
Mack, Y. P., 100
Mallat, S., 358
Mammen, E., 122
marginal integration, 181, 182, 189, 231, 239, 240,
    244, 246, 248, 249, 285, 286, 300, 303
Marron, J. S., 122
maximum likelihood, 256, 296
mean integrated squared error, 71, 247
mean square error, 61
Mielniczuk, J., 29
Mirzahmedov, M. A., 78
monotonicity preserving algorithms, 117
Monro, S., 364
Monte Carlo integration (method), 239
Moustafa, K. A. F., 10
multichannel Hammerstein system, 245, 248
multichannel Wiener system, 255
multidimensional orthogonal polynomials, 226
multiple-regression function, 191
multiple-input Hammerstein system, 243
multiplicative model, 178, 182
multiresolution, 152, 153, 168
multiresolution kernel(s), 153, 168
multivariate kernels, 212
multivariate nonlinear models, 222
multivariate nonparametric regression, 219, 222
multivariate orthonormal bases, 225
multivariate systems, 222, 228, 242, 310

Nadaraya, E., 29
Nadaraya–Watson estimator, 12
Nagaraya, H. N., 365
Narendra, K. S., 9
Ngo, K. D. T., 10
Ninness, B., 10
noninvertible nonlinearities, 251, 255
nonlinear dynamics, 149, 173, 175, 177, 179, 191,
    198, 199, 217, 218, 220

nonparametric models, 2, 309
nonparametric regression, xi, 1, 29, 149, 151, 160,
    167, 219, 220, 222, 223, 226, 228, 240, 245, 251,
    254, 264
Nordsjö, A. E., 122
Norton, J. P., 2
nuisance characteristics, 157, 220

Ogden, R. T., 358
oracle property, 259
ordered (order) statistics, xii, 1, 80, 201, 268–270,
    365, 370
ordered observations, xi, 80
orthogonal polynomials (functions), 19, 20, 64, 202,
    208, 226, 331, 344, 355
orthogonal series, xi, xii, 1, 10, 59, 60, 85, 99,
    100, 106, 123, 126, 142, 152, 203, 209,
    220
orthogonal series estimate, 85, 99, 100, 126, 142,
    203, 209

parabolic kernel, 14
parallel model (system), xii, 150, 151, 154–158,
    161, 162, 168, 170–172, 220, 250, 251, 263, 287,
    288, 307, 308, 316
parallel-series model(s), 160, 220
parametric estimation (methods), 2, 6, 122, 221,
    260, 272
Partyka, M. A., 29
Parzen, E., 29
Patwardhan, R. S., 10, 122
Pawlak, M., 29, 43, 78, 79, 100, 322, 362
pilot estimate, 165, 190, 259, 267, 269
plug-in estimate (technique), 156, 165, 167, 209,
    292
pointwise consistency (convergence), 11, 61, 320,
    332, 341, 346
Poisson kernel, 15
Prakasa Rao, B. L. S., 1, 29
predictive error, 190, 191, 267, 271
Priestley, M. B., 99
projection pursuit, 284
Pyke, R., 365

Révész, P., 43, 58
Rafajłowicz, E., 99
Ramaswamy, S., 10
ratio trick, 165, 363
rectangular kernel, 12, 14, 21, 34, 124
recursive kernel estimate, 43, 268
regression function, xi, 11, 60, 78, 99, 151, 158,
    159, 163, 167, 173, 179–181, 183, 185, 189–191,
    202, 209, 210, 212, 217, 219, 222, 223, 226–228,
    233, 236–239, 242–244, 250, 251, 257, 258, 260,
    265–268, 270, 273, 283, 286, 288–291, 293, 294,
    303, 307–309, 312
Rejtö, L., 43

resampling, 167, 251, 259, 267, 274, 289, 292, 298, 299, 316
Riemann–Lebesgue theorem, 337
Robbins, H., 364
Rochidi, Y., 10
Rodrigues' formula, 64, 66, 68, 340, 345, 351
Rolain, Y., 10, 122
Rosenblatt, M., 29
Rutkowska, D., 78
Rutkowski, L., 78, 99

Söderström, T., 2
Sąsiadek, J. Z., 78, 142
Sacks, J., 29
Sansone, G., 334, 337, 340–342, 346, 347, 351, 353, 355
Schoukens, J., 10, 122
Schuster, E., 99
Schwabe, R., 99
Schwartz, S. C., 78
semiparametric estimation, 255
semiparametric least squares, 191, 251, 304
semirecursive estimates, 125, 126
semirecursive kernel estimate, 30, 148, 268
Serpendin, E., 2, 10
Sethares, W. A., 10
Shah, S. L., 10, 122
Shih, D. H., 10
Silverman, B. W., 29
Simonoff, J. S., 1
Skubalska-Rafajłowicz, E., 99
Sliwiński, P., 79
Slud, E., 81, 368, 369
Song, K. S., 29
spacing(s), 82, 365, 367, 369, 370
Specht, D. F., 78
Speckman, P., 99
Spiegelman, C., 29
Stadtmüller, U., 99
steepest descent, 159, 258, 269
stochastic approximation, xi, 44, 46, 364
Stoica, P., 2
Stone, C. J., 29
strong mixing, 7, 112
Szegö, G., 142, 340, 341, 343, 346, 349, 351–353

Talaq, J. H. S., 10
Tarter, M., 78
Thathachar, M. A. L., 10
Tin, Y. B., 78
Toeplitz lemma, 39–41, 43, 51

transfer function, 9, 155, 159, 177, 208, 246
triangle kernel, 14
Tricomi, F. G., 10
trigonometric orthonormal system, 61, 106
trigonometric series, 62, 86, 106, 107, 333, 335, 341, 346, 353
trigonometric series estimate, 62, 106, 107
truncation parameter, 152, 155, 156, 168, 190, 203, 204, 224, 225
Turlach, B. A., 122

uniform consistency, 261
Uryson model, 161, 220
Uspensky, J. V., 346, 353

Van Ryzin, J., 29, 78
Vandersteen, G., 10, 122
Vapnik–Chervonenkis class, 274
Verhaegen, M., 10, 122
Viollaz, A. J., 78
Vörös, J., 10

Wagner, T. J., 29, 43
Wahba, G., 29, 78
Walk, H., 1, 58
Walter, G. G., 78
Wand, M. P., 1, 122
Wang, L., 122
Wasan, M. T., 44
Watson, G. S., 29, 78
wavelet estimate(s), 1, 69
wavelet expansion, 69, 79
wavelets, xi, 10, 61, 69, 79, 331, 355, 356, 358
Wegman, E. J., 43
Westwick, D., 10, 122
Wheeden, R. L., 322, 329
Wiener system(s) (model), xii, 1, 113, 114, 117, 118, 122, 143, 144, 149, 185, 186, 208, 218, 251, 252, 254, 255, 264–267, 269, 278, 279, 282–286, 294, 295, 300–303, 316, 317, 387
Wigren, T., 122
Wilks, S. S., 365
Wolverton, C. T., 43
Wormald, C. N., 10, 122

Yakowitz, S., 99
Yamato, H., 43
Yatracos theorem, 311

Zhu, Y., 10
Zygmund, A., 322, 329

Printed in the United States
by Baker & Taylor Publisher Services